Experimental, Theoretical, Numerical and Big-Data-Based Investigations on Characterizations for Geomaterials

Experimental, Theoretical, Numerical and Big-Data-Based Investigations on Characterizations for Geomaterials

Editors

Shaofeng Wang
Linqi Huang
Tianshou Ma
Jie Zhou
Changjie Zheng

MDPI

Basel • Beijing • Wuhan • Barcelona • Belgrade • Novi Sad • Cluj • Manchester

Editors

Shaofeng Wang
School of Resources and
Safety Engineering
Central South University
Changsha
China

Linqi Huang
School of Resources and
Safety Engineering
Central South University
Changsha
China

Tianshou Ma
School of Petroleum
Engineering
Southwest Petroleum
University
Chengdu
China

Jie Zhou
Civil Engineering
Tongji University
Shanghai
China

Changjie Zheng
College of Civil Engineering
Fujian University of
Technology
Fuzhou
China

Editorial Office
MDPI
St. Alban-Anlage 66
4052 Basel, Switzerland

This is a reprint of articles from the Special Issue published online in the open access journal *Materials* (ISSN 1996-1944) (available at: www.mdpi.com/journal/materials/special_issues/BXR92M8N1Q).

For citation purposes, cite each article independently as indicated on the article page online and as indicated below:

Lastname, A.A.; Lastname, B.B. Article Title. *Journal Name* **Year**, *Volume Number*, Page Range.

ISBN 978-3-0365-9124-7 (Hbk)
ISBN 978-3-0365-9125-4 (PDF)
doi.org/10.3390/books978-3-0365-9125-4

Contents

Preface

Rock and rock-like materials such as concrete, soil, and underground backfilling materials are considered to be geomaterials. Geomaterials are essential for life due to human construct extraction, mining, storage, and transport areas in the Earth's crust for raw material. Drilling and excavations of underground openings in the Earth's crust are requirements for the exploitation and utilization of mineral resources, energy resources, and underground spaces. The deepest drilling depth has exceeded 12 km, and the deepest underground excavation now operates mines with depths exceeding 4 km to 5 km. Drilling, excavation, and rock support processes largely rely on the physical and mechanical properties of geomaterials. Rock excavations are faced with some instability phenomena, such as caving, rock bursts, slabbing, large deformation, and zonal disintegration, posing a serious threat to the safety of mining and tunneling operations. Rock drilling also encounters many challenges deep underground. Deformation, fracture, failure, and fragmentation are the different stages of geomaterials, the monitoring and control of which are essential for ensuring drilling and excavation safety. Therefore, understanding the response processes of geomaterials during drilling and excavation activities depends on the precise characterizations of geomaterials. However, it remains a significant challenge to fully understand the deformation/fracture/failure/fragmentation behaviors and mechanisms of geomaterials in complex underground environments.

This Special Issue aimed to collect research papers and review articles encompassing in situ observations, laboratory experiments, theoretical analyses, numerical simulations, and big-data-based analyses concerning the characterization of geomaterials.

Shaofeng Wang, Linqi Huang, Tianshou Ma, Jie Zhou, and Changjie Zheng
Editors

materials

MDPI

Editorial

Experimental, Theoretical, Numerical and Big-Data-Based Investigations on Characterizations for Geomaterials

Shaofeng Wang

School of Resources and Safety Engineering, Central South University, Changsha 410083, China; sf.wang@csu.edu.cn

Rock and rock-like materials such as concrete, soil, and underground backfilling materials are considered to be geomaterials. Geomaterials are essential for life and engineering because they are important for human construct extraction, mining, storage, and transport areas in the Earth's crust. Essential material that has to be faced in mining processes is rock or geomaterials. Drilling, blasting, cutting, crushing and rock support processes in geotechnical, mining, petroleum, natural gas and geothermal engineering rely largely on the physical and mechanical properties of rock. Additionally, the quarrying and machining processes of natural stone depends in a greater extent on the physical and mechanical properties of natural rock materials. In mineral processing operations, physical and mechanical properties of ore rock are important parameters at various stages such as grinding, milling, and gathering. Soil and concrete are also important geomaterials. Soil behaviour can determine the long-term stabilities of civil and geotechnical engineering. Concrete, when used together with the rock structures in excavation and backfilling operations, as support or lining construct for surround rock, can significantly affect the behaviour of the rock structure. Drilling and excavations of underground openings in the Earth's crust are requirements for the exploitation and utilization of mineral resources, energy resources, and underground spaces. The deepest drilling depth has exceeded 12 km, and the deepest underground excavation now operates mines with depths exceeding 4 km to 5 km. Drilling, excavation, and rock support processes are essential operations for forming openings in soil or rock, which largely rely on the associated properties of geomaterials and can significantly influence the safety and stability of underground engineering. Rock excavations are faced with some instability phenomena, such as caving, rockburst, slabbing, large deformation, and zonal disintegration, posing a serious threat to the safety of mining and tunneling operations. Rock drilling also encounters many challenges deep underground. Deformation, fracture, failure, and fragmentation are the different stages of geomaterial behaviour, the monitoring and control of which are essential for ensuring drilling and excavation safety and support stability. Therefore, understanding the response processes of geomaterials during drilling and excavation activities for exploitations of georesources and geoenergy depends on the precise characterisations for geomaterials. In addition, stress conditions in geomaterials are typically characterised by combined static and dynamic loads, consisting of the excavation-induced high stress concentration around openings and strong dynamic disturbances from drilling and blasting, excavation unloading, caving, and fault slippage. It is well-recognized that geomaterial has multi-scale structures, from minerals, particles, fractures, fissures, joints, and stratification, to fault, and involves multi-scale fracture processes. Meanwhile, there are many multi-physics coupling processes in geomaterial, such as the coupled thermo–hydromechanical–chemical interaction in porous geomaterials. However, it remains a significant challenge to fully understand the deformation/fracture/failure/fragmentation behaviours and mechanisms of geomaterials in complex underground environments with complex stress conditions, multi-physics processes, and multi-scale changes.

Citation: Wang, S. Experimental, Theoretical, Numerical and Big-Data-Based Investigations on Characterizations for Geomaterials. *Materials* **2023**, *16*, 1727. https://doi.org/10.3390/ma16041727

Received: 17 January 2023
Revised: 6 February 2023
Accepted: 12 February 2023
Published: 20 February 2023

This Special Issue aims to call for research papers and review articles encompassing in situ observations, laboratory experiments, theoretical analyses, numerical simulations, and big-data-based analyses concerning characterisations for geomaterials.

Potential topics include, but are not limited to, the following:

(1) Characterisations of geomaterials;
(2) Drilling and blasting performances of rock and rock-like materials;
(3) Cuttabilities of geomaterials;
(4) Fracabilities of geomaterials;
(5) Permeability properties of geomaterials;
(6) Dynamic response of geomaterial;
(7) Geomaterial behaviours;
(8) Geomaterial strength;
(9) Geomaterial deformation;
(10) Geomaterial fracture;
(11) Stress wave propagation in geomaterials;
(12) Thermodynamic properties of geomaterials;
(13) Beneficial utilisations of geomaterials.

Potential methodologies for studying geomaterials include, but are not limited to, the following:

(1) Experimental investigation;
(2) Field observation and monitoring;
(3) Theoretical analyses;
(4) Geostatistics;
(5) Numerical simulation;
(6) Big-data-based methods;
(7) Data mining and deep learning.

The guest editors hope that the selected papers for this Special Issue will help scholars and researchers in pushing forward and progressing in experimental, theoretical, numerical, and big-data-based investigations on characterisations for geomaterials, and provide some valuable information and recommendations for geotechnical, mining, tunnelling, and drilling engineers.

Acknowledgments: We express our thanks to the authors for their contributions, as well as to the journal Mathematics and MDPI for their support during this work.

Conflicts of Interest: The authors declare no conflict of interest.

materials

MDPI

Article

Predicting the Temperature-Dependent Long-Term Creep Mechanical Response of Silica Sand-Textured Geomembrane Interfaces Based on Physical Tests and Machine Learning Techniques

Zhiming Chao [1,2,3,†], Haoyu Wang [3], Hanwen Hu [4,†], Tianchen Ding [4] and Ye Zhang [4,*]

1 Shanghai Estuarine and Coastal Science Research Center, Shanghai 201201, China
2 Institute of Water Sciences and Technology, Hohai University, Nanjing 211106, China
3 College of Ocean Science and Engineering, Shanghai Maritime University, Shanghai 200135, China
4 Mentverse Ltd., 25 Cabot Square, Canary Wharf, London E14 4QZ, UK
* Correspondence: ye.zhang@mentverse.co.uk; Tel.: +44-7557782175
† These authors contributed equally to this work.

Abstract: Preciously assessing the creep mechanical response of sand–geomembrane interfaces is vital for the design of relevant engineering applications, which is inevitable to be influenced by temperature and stress statuses. In this paper, based on the self-developed temperature-controlled large interface shear apparatus, a series of long-term creep shear tests on textured geomembrane–silica sand interfaces in different temperatures, normal pressure, and creep shear pressure were conducted, and a database compiled from the physical creep shear test results is constructed. By adopting the database, three disparate machine learning algorithms of the Back Propagation Artificial Neural Network (BPANN), the Support Vector Machine (SVM) and the Extreme Learning Machine (ELM) were adopted to assess the long-term creep mechanical properties of sand–geomembrane interfaces while also considering the influence of temperature. Then, the forecasting results of the different algorithms was compared and analyzed. Furthermore, by using the optimal machine learning model, sensitivity analysis was carried out. The research indicated that the BPANN model has the best forecasting performance according to the statistics criteria of the Root-Mean-Square Error, the Correlation Coefficient, Wilmot's Index of Agreement, and the Mean Absolute Percentage Error among the developed models. Temperature is the most important influence factor on the creep interface mechanical properties, followed with time. The research findings can support the operating safety of the related engineering facilities installed with the geomembrane.

Keywords: sand; creep shear mechanical response; SVM; BPANN

Citation: Chao, Z.; Wang, H.; Hu, H.; Ding, T.; Zhang, Y. Predicting the Temperature-Dependent Long-Term Creep Mechanical Response of Silica Sand-Textured Geomembrane Interfaces Based on Physical Tests and Machine Learning Techniques. *Materials* **2023**, *16*, 6144. https://doi.org/10.3390/ma16186144

Academic Editor: Krzysztof Schabowicz

Received: 14 August 2023
Revised: 5 September 2023
Accepted: 7 September 2023
Published: 10 September 2023

1. Introduction

As an excellent water-proofing material, geomembrane is extensively used as construction material in various engineering applications [1–5]. With the development of a geomembrane manufacturing technology, different types of this material are available. Among them, textured geomembrane is the most popular in practical engineering applications because textured geomembrane can form a strong interface mechanical performance when in contact with other construction material, such as soil, due to the existence of texture on the geomembrane surfaces [6,7]. Although compared to other types of geomembrane, the interfaces between textured geomembrane and soil have better stability, but the textured geomembrane–soil interface is still the weakest component in engineering facilities [8–13]. Thus, the correct estimation of the interface mechanical properties between textured geomembrane and soil is critical for the operating safety of relative engineering facilities.

In general, the service life of geomembrane in engineering infrastructure is over decades [14,15]. During the operation of engineering applications, the textured geomembrane–soil interface is often subjected to constant creep shear stress resulting from the overlaying construction material [16,17]. The long-term creep interface deformation has an obvious difference with the short-term deformation caused by the rapid loading of shear stress. The occurrence of creep shear deformation is more hidden and the deformation can rise rapidly in a short period, which has a more significant hazard on the safety of engineering facilities [18,19]. It highlights the necessity of preciously predicting the long-term creep mechanical response of textured geomembrane–soil interfaces.

In reality, the engineering environment is complex, and the buildings installed with geomembrane are often subjected to temperature loadings [20–25]. For example, due to the exothermal reaction of buried waste biodegradation, the inside temperature of landfills can reach 60~80 °C [26,27]. As a common water-proofing material, it is inevitable that geomembrane installed in landfills will experience a high temperature environment. Since the main raw material of producing geomembrane is thermal-softening plastic materials such as nylon, polyethylene, etc., in elevated temperatures, the softening of geomembrane may occur [28]. This has an obvious influence on the interface mechanical properties between geomembrane and soil [16]. Especially for textured geomembrane, the softening of texture can remarkably change the interaction between geomembrane and soil to result in a considerable impact on the interface's mechanical response [29,30]. However, due to the limitation of the temperature-controlled interface shear apparatus, at present, the research involving the assessment of the temperature-dependent mechanical properties for geomembrane–soil interfaces is rare, let alone the investigation about the long-term creep mechanical behavior of textured temperature–soil interfaces by considering the temperature effects.

The interaction between textured geomembrane and soil is complicated, which leads to the multiple influence factors on the mechanical performance of textured geomembrane–soil interfaces [31–33]. Especially for the temperature-dependent creep shear mechanical response, it requires to consider the factors of time and temperature, which further increases the complexity of the interface action mechanism. The complex interaction and multiple impact factors cause the difficulty of predictive modelling for the long-term creep mechanical behavior of textured geomembrane–soil interfaces by considering the temperature effects and adopting the traditional estimation methods, such as the mathematical statistical approach, etc. [15,34]. Due to the development in computer technology, in recent years, the machine learning techniques are extensively adopted to replicate the complex action mechanism by considering multiple influence factors [30,35–41]. In the field of civil engineering, machine learning modeling techniques have found widespread use, including but not limited to the following:

(1) estimating rock permeability [42,43];
(2) predicting interface shear strength [44–46];
(3) assessing cement mortar permeability [47,48].

The existing research manifesting the machine learning methods can describe the complex relationships between plentiful factors with high precision and efficiency. However, in the published research, the investigation involving the modelling of the temperature-dependent long-term creep mechanical response of the textured geomembrane–soil interface by adopting machine learning techniques is not reported.

In this paper, the effectiveness of the three disparate machine learning algorithms in assessing the temperature–dependent long-term creep interface mechanical response between textured geomembrane and silica sand are compared and analyzed, including the Backpropagation Artificial Neural Network (BPANN), the Support Vector Machine (SVM), and the Extreme Learning Machine (ELM). Moreover, by using the optimal machine learning predictive model, the sensitivity analysis of different influence factors on the creep shear mechanical response of textured geomembrane–silica sand interfaces is conducted to determine the relative importance. The research outcomes can provide an effective tool for estimating the temperature-dependent long-term creep mechanical properties of textured geomembrane–sand interfaces, which can provide a reference for the construction of related engineering services.

2. Physical Experiment

To obtain the data for machine learning modeling, a self-developed large interface direct shear apparatus was utilized, and the tested geomembrane is shown in Figure 1. The temperature–controlled creep shear tests on silica sand–textured geomembrane interfaces were conducted at 25 kPa, 50 kPa, and 100 kPa under normal stress, and the creep shear stress level was 50%, 70%, and 90% of the monotonic peak shear strength for the interface under the corresponding normal stress. For each test condition, three different temperatures of 30 °C, 60 °C, and 200 °C were adopted. The specific test procedure is as follows:

(1) Extured geomembrane and silica sand sample was installed;
(2) The interface temperature was adjusted to the predetermined value and kept stable during the whole test;
(3) The interface was consolidated under the corresponding normal stress for 12 h;
(4) A certain shear stress was imposed on the interfaces and kept stable during the whole test;
(5) The test was terminated until the interface failed or the test duration (6 days) ran out.

The properties of the test material are listed in Table 1, and the test scheme is shown in Table 2.

Figure 1. The tested textured geomembrane.

Table 1. The properties of the test sample.

Textured geomembrane (Flat die extruded)	Thickness (mm)	2.0
	Textured height (mm)	0.26
	Fracturing strength (N/mm)	16.2
	Yield strength (N/mm)	22.3
	Yield elongation rate (%)	12.2
	Fracturing elongation rate (%)	120
	Puncture strength (N)	195
Silica sand	Particle size range (mm)	0.075~2
	Density (g/cm^3)	1.50
	Optimum water content (%)	10
	Uniformity coefficient	3.327
	Curvature coefficient	0.3
	Median particle size (mm)	0.785

Table 2. Test scheme.

Test Sample	Normal Pressure (kPa)	Temperature (°C)	Creep Shear Stress
Silica sand–textured geomembrane interface	25, 50, 100	30, 60, 200	50%, 70%, and 90% of the monotonic interface peak shear strength

3. Machine Learning Algorithm

By adopting the afore physical test results, three different kinds of machine learning algorithms, BPANN, SVM and ELM, were selected to establish the predictive models on the temperature-dependent creep shear mechanical response of textured geomembrane–sand interfaces by adopting the program software Matlab_R2022b_Windows. The introduction of the adopted machine learning algorithms is presented below.

3.1. BPANN

Artificial Neural Network is the most prevalent machine learning algorithm, which can be divided as different kinds [49–51]. Among them, the BPANN is the most extensively adopted in civil engineering areas [10,52]. The BPANN model is composed of the input, hidden, and output layers. In this research, the input parameters are normal pressure, creep shear stress, temperature, and time, which corresponds to four joints in the input layer. The output parameter is the creep shear displacement, which corresponds to one joint in the output layer. The joint number of the hidden layer is ascertained by adopting the exhaustive approach, which is 5 in this research. The schematic diagram of the adopted BPANN model structure is shown in Figure 2. The activation function and network training algorithm in the BPANN model adopts the Hyperbolic Tangent Sigmoid Transfer Function and the Levenberg-Marquardt Backpropagation Algorithm, respectively.

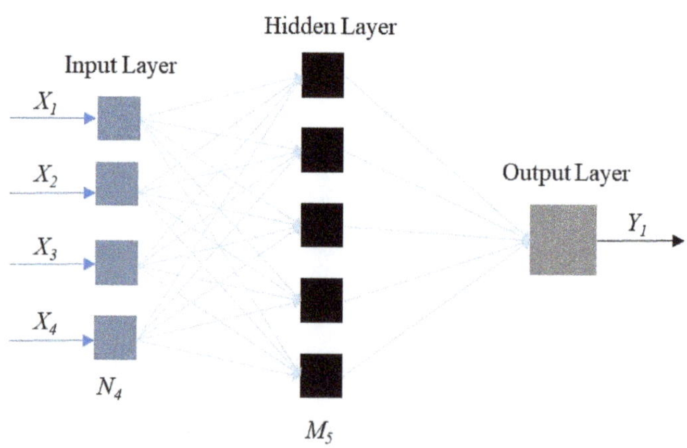

Figure 2. The schematic diagram of BPANN model.

3.2. SVM

SVM is another widely adopted machine learning algorithm, which can describe the complex relationship between multiple variables based on the limited number of sample data [53,54]. Also, SVM can project low-dimensional data to high-dimensional data by adopting the kernel function so that the non-linear modelling can be transformed into linear modelling [55,56]. Additionally, SVM can adopt the k-fold cross-validation method (k-CV) in conducting modelling. The theory of k-CV is to divide the original data as equal groups. Among the groups, k-1 groups are adopted as the training data, and the left group is adopted for validation. By selecting different data groups as the testing dataset, the train and validation procedure is repetitive with k times, and the average predictive outcomes of the k times are used as the final value. In this research, the value of k is taken as 10.

3.3. ELM

ELM originates from the ANN model, with the structure being similar with the feedforward ANN [57]. However, in some cases, compared to the ANN model, ELM can achieve complex modelling in a short period, with low computation cost. Like ANN, the joint number of the ELM hidden layer also impacts the forecasting behavior significantly. Therefore, the same exhaustive method is adopted to determine the hidden layer joint number in ELM as 5. For the input and output layer joint number, they are 4 and 1, corresponding to the four input parameters and one output parameter, respectively. The activation function adopts the Logarithmic Sigmoid Function.

4. Machine Learning Modelling

4.1. Establishment of the Database

Based on the aforementioned test results, the database was established. The database includes 27,000 datasets. Each dataset consists of four input parameters of normal pressure, creep shear pressure, time, and temperature, and one output parameter of creep shear displacement. The range of normal pressure is from 25 kPa to 100 kPa, and the time is from 0 min to 8640 min, with the temperature ranging from 30 °C to 200 °C. The parameters were chosen because they have large influence on the creep mechanical response of textured geomembrane–soil interfaces, as indicated by the existing research [1,58]. The statistics for the input and output parameters is listed in Table 3. The data distribution for the database is shown in Figure 3, with the x-axis and the y-axis representing the value of the input parameters and the corresponding data group number, respectively.

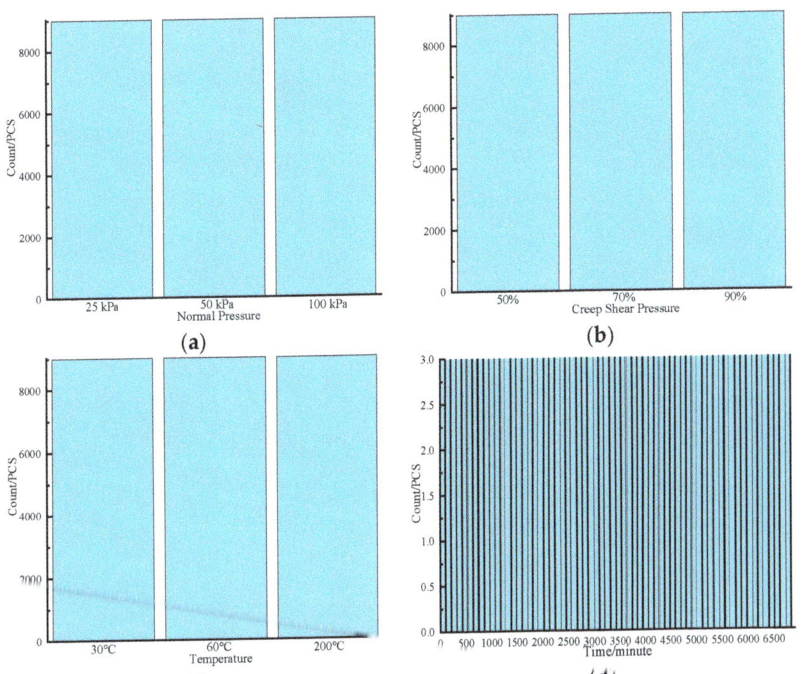

Figure 3. Data distribution of the complied database. (**a**) Normal pressure; (**b**) creep shear pressure; (**c**) temperature; and (**d**) time.

Table 3. The statistics.

	Parameters	Minimum Value	Maximum Value	Mean Value	Standard Deviation
Input	Normal pressure (kPa)	25	100	59	30.6
	Creep shear pressure	50% of the peak shear strength	90% of the peak shear strength	70% of the peak shear strength	20.39
	Time (minute)	0	8640	4320	650
	Temperature (°C)	30	200	95	20.72
Output	Creep shear displacement (mm)	0	47.20	24.50	6.40

4.2. Data Processing

Before machine learning modelling, it was required to divide the database into the training data, to train the models, and the testing data, to validate the predictive performance of the models. In this paper, the division ratio is set as 80% (21,600 groups) for the training data and 20% (5400 groups) for the testing data. Also, to improve the predictive precision and efficiency, the input data were normalized by adopting Equation (1).

$$x_{Normalised} = \frac{2(x - x_{\min})}{x_{\max} - x_{\min}} - 1 \tag{1}$$

where $x_{Normalised}$ and x indicates the normalized and original values, respectively, and $x_{\min}$ and $x_{\max}$ indicates the minimum and maximum values, respectively.

4.3. Precision Assessment Indexes

To quantitatively analyze the forecasting accuracy of the constructed machine learning models, the following four assessment indexes were adopted: Root-Mean-Square Error (RMSE), Coefficient of Determination (R^2), Mean Absolute Percentage Error (MAPE), and Wilmot's Index of Agreement (WI), as presented in Equations (2)–(5).

$$\text{RMSE} = \sqrt{\sum_{i=1}^{n} \frac{(y_i - f_i)^2}{n}} \tag{2}$$

where n indicates the data number, y_i indicates the measured data, and f_i indicates the assessing value. The less the RMSE value is, the more precise the model.

$$R^2 = 1 - \frac{\sum_{i=1}^{n} (y_i - f_i)^2}{\sum_{i=1}^{n} (y_i - \bar{y})^2} \tag{3}$$

where $\bar{y}$ indicates the average measured value.

$$\text{MAPE} = \frac{100\%}{n} \sum_{i=1}^{n} \frac{|y_i - f_i|}{y_i} \tag{4}$$

$$\text{WI} = 1 - \frac{\sum_{i=1}^{n} (y_i - f_i)^2}{\sum_{i=1}^{n} \left(\left|f_i - \bar{y}\right| + \left|y_i - \bar{y}\right|\right)^2} \tag{5}$$

5. Result and Analysis

5.1. Determining the Optimal Hidden Layer Joint Number

As aforementioned, the hidden layer joint number has a significant influence on the assessing results of the BPANN and ELM models. To determine the optimum number of hidden layer joints, the exhaustive approach was adopted with RMSE as the assessment index. The results are shown in Figure 4.

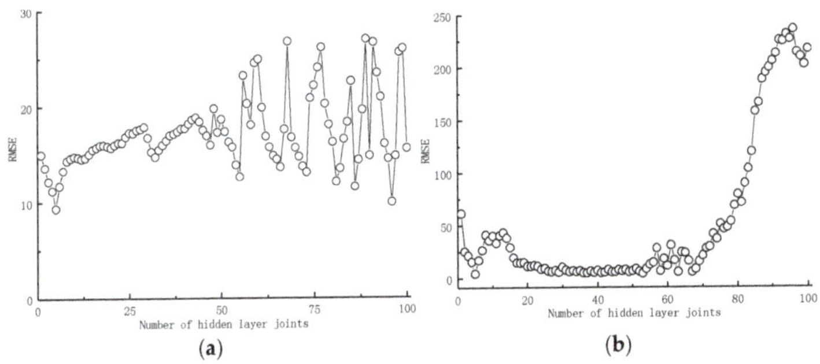

Figure 4. The optimization process. (**a**) BPANN; and (**b**) ELM.

Based on Figure 4, the hidden layer joint number has a considerable impact on the RMSE value of the BPANN and ELM models, ranging from 4.23 to 26.97 and from 236.24 to 4.51, respectively. By comparing the assessing results of the BPANN and ELM models containing various hidden layer joint numbers, it is found that when the hidden layer joint number is 5, both the BPANN and ELM models reach the high estimation precision of 4.23 and 4.51, respectively. Thus, in this research, both the BPANN and ELM models adopt five hidden layer joints.

5.2. Assessment Results of Various Machine Learning Models

The forecasting precision of different machine learning models on the train and test datasets, based on the indexes of RMSE, MAPE, WI, and R^2, is shown in Figures 5–8.

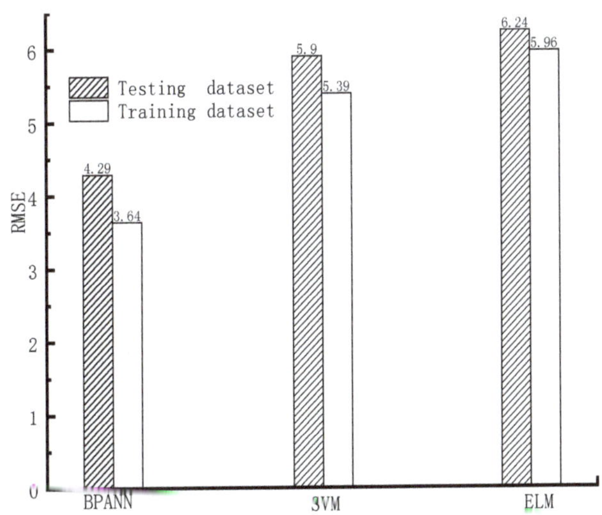

Figure 5. RMSE value.

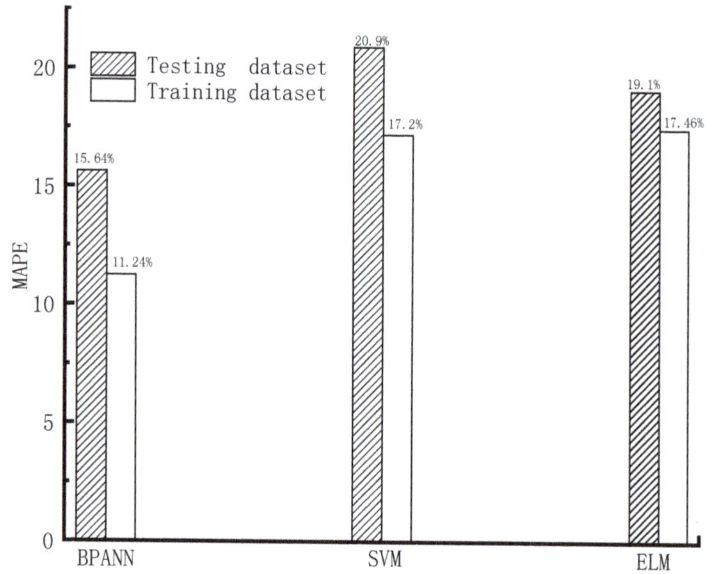

Figure 6. MAPE value.

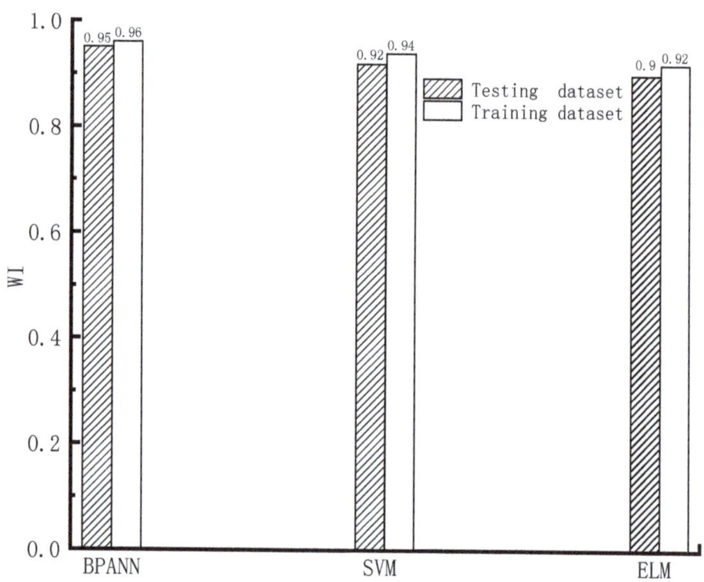

Figure 7. WI value.

As shown in Figures 5–8, based on the assessment indexes of RMSE, MAPE, WI, and R^2, the assessing precision of the BPANN model is the highest among the models in the aspect of assessing the training dataset. In the quantitative analysis, the BPANN model achieved the lowest RMSE (3.64) and MAPE (11.24%) and the largest R^2 (0.99) and WI (0.96). It is followed by the SVM model, and its forecasting performance is better than the ELM model. For example, the RMSE value and the MAPE value of the ELM model (5.96 and 17.46%, respectively) is higher than that of the SVM model (5.39 and 17.2%, respectively).

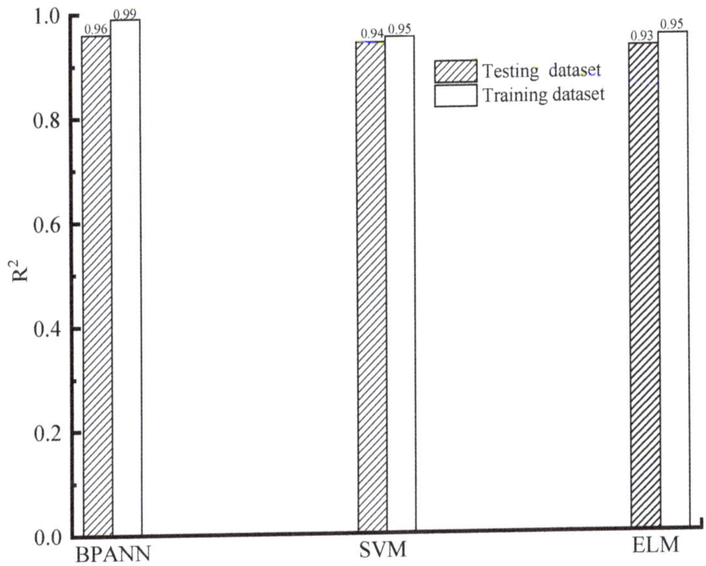

Figure 8. R^2 value.

In the aspect of the forecasting behavior on the testing datasets, as presented in Figures 5–8, like the outcomes on the training datasets, the assessing precision of the BPANN model is the best among the different algorithms. More specifically, among the models, the RMSE and MAPE values of the BPANN model is the most low, with 4.29 and 15.64%, respectively, and the R^2 and WI values of the BPANN model is the largest, with 0.96 and 0.95, respectively. For the SVM and ELM models, their forecasting performance is different based on the different assessment indexes. For example, the RMSE value of the ELM model (6.24) is higher than that of the SVM model (5.9), which indicates the superior performance of the SVM model than the ELM model. In comparison, the MAPE value of the ELM model (19.1%) is less than that of the SVM model (20.9%), which indicates the better precision of the ELM model than that of the SVM model.

5.3. Sensitivity Analysis

To investigate the relative significance of different factors on influencing the creep shear mechanical response of textured geomembrane–silica sand interfaces, the Garson's Algorithm was adopted to quantitatively assess the relative significance for the input parameters on the long-term creep shear mechanical properties of the interfaces. The formula for Garson's Algorithm [59,60] is presented in Equation (6), and the calculated relative significance for the input parameters is depicted in Figure 9.

$$R_{ik} = \frac{\sum\limits_{j=1}^{L} \left(\left| W_{ij} W_{jk} \right| / \sum\limits_{r=1}^{N} \left| W_{rj} \right| \right)}{\sum\limits_{i=1}^{N} \sum\limits_{j=1}^{L} \left(\left| W_{ij} W_{jk} \right| / \sum\limits_{r=1}^{N} \left| W_{rj} \right| \right)} \quad (6)$$

where R_{ik} indicates the parameter relative importance; W_{ij},W_{jk} indicates the connection weight between the input and hidden layers as well as the hidden and output layers; N indicates the input parameter number; and M indicates the output parameter number.

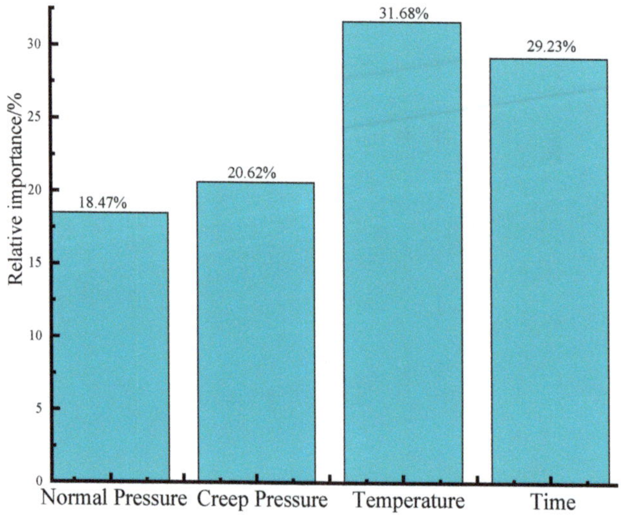

Figure 9. The sensitivity analysis results.

According to Figure 9, temperature has the highest influence on the long-term creep shear mechanical properties of textured geomembrane–sand interfaces, with the relative significance of 31.68%. It is followed by time, with the percentage of 29.23%. In comparison, normal pressure and creep shear pressure have a relatively small impact on the creep shear mechanical response of the interfaces, with the proportions of 18.47% and 20.62%, respectively.

6. Case Study

To verify the reliability of the established machine learning models, a case study is carried out to assess the creep shear displacement of a textured geomembrane–silica sand interface in specific normal pressure, creep shear pressure, temperature, and time by adopting the developed BPANN model. The predicting results were compared with the creep shear displacement measured in the laboratory tests to verify the precision of this model. The specific test conditions and the corresponding input parameters for the BPANN model are listed in Table 4. The predicted creep shear displacement of the interface by using the BPANN model and the measured value in the physical test is presented in Figure 10.

Table 4. The test conditions and the corresponding input values.

Test Condition	Value	Input Value
Temperature	50 °C, 90 °C	50, 90
Time	2000 min, 4000 min	2000, 4000
Normal stress	60 kPa, 100 kPa	60, 100
Creep shear pressure	70% of the peak shear strength	70

According to Figure 10, the assessed creep shear displacements of the textured geomembrane–silica sand interface in the specific normal pressure, creep shear pressure, temperature, and time obtained from the BPANN model are similar to the measured creep shear displacements in the physical test. It indicates the developed machine learning model can estimate the temperature-dependent long–term creep shear mechanical response of the textured geomembrane–silica sand interfaces preciously.

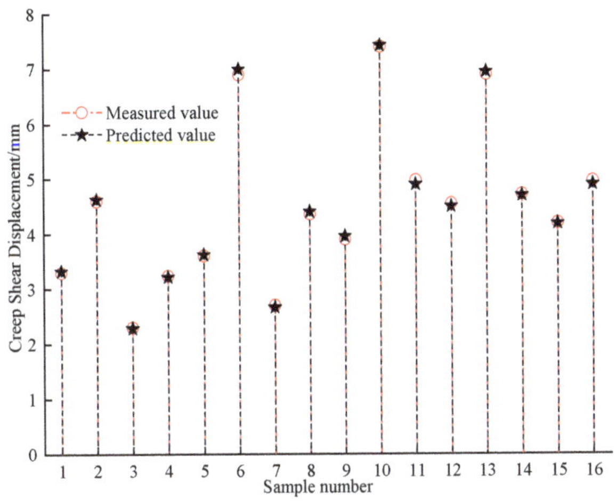

Figure 10. The measured and predicted creep shear displacement.

7. Limitation

Despite the research presenting precious findings, there are limitations for this research. (1) The adopted database for establishing the machine learning models consists of 27,000 datasets, which is not less. However, it still has the room to further enlarge, and the forecasting performance for the machine learning models can be continually enhanced if the bigger database is available; and (2) the influence factors on the creep shear mechanical properties of textured geomembrane–silica sand interfaces are many. Except for the selected input parameters in this research, the properties of textured geomembrane and sand also have impacts on the mechanical response. It should be considered in the future investigation.

8. Conclusions

In this paper, by adopting the self-developed large temperature-controlled interface shear apparatus, a series of long-term creep shear tests on textured geomembrane–silica sand interfaces in different temperatures, normal pressure, and creep shear pressure were conducted. Based on the physical test results, the machine learning models for assessing the temperature-dependent long-term creep shear mechanical response of textured geomembrane–silica sand interfaces were established by using three different machine learning algorithms of BPANN, SVM, and ELM. Then, the forecasting outcomes of the different machine learning models were compared and analyzed. After that, by using the optimal machine learning model combined with Garson's Algorithm, the sensitivity analysis was carried out to determine the relative significance of the input parameters on influencing the creep shear mechanical properties of the interfaces. In the end, the case study was conducted to verify the reliability for the results forecasted from the constructed machine learning model by comparing with the measured mechanical response of the interfaces in the physical tests.

The main research outcomes are as follows: The hidden layer joint number has a considerable impact on the assessing precision of the BPANN and ELM models. Compared to SVM and ELM model, the BPANN model has a superior performance in forecasting the temperature-dependent long-term creep shear mechanical response of textured geomembrane–silica sand interfaces based on the assessment indexes of RMSE, R, WI, and MAPE. The sensitivity analysis indicates the influence of temperature on the creep shear mechanical properties of textured geomembrane–silica sand interfaces is the largest, followed by time, and creep shear pressure and normal pressure have relative low influence.

For future research directions, it is advisable to consider the incorporation of additional factors into the modeling process, such as wet–dry cycling and the characterization of geomembranes (whether they are smooth or textured). Furthermore, the application of 3D printing technology can be explored to fabricate geomembranes with varying protrusion heights. Subsequent experiments can be conducted to gather extensive data, thus facilitating the expansion of the machine learning model's database. These steps are essential in the development of a machine learning model with improved accuracy and enhanced generalization capabilities for the prediction of long-term creep mechanical properties at geomembrane–soil interfaces.

Author Contributions: Conceptualization, Z.C.; Investigation, H.W. and H.H.; Resources, Z.C.; Data curation, H.H. and Y.Z.; Writing—original draft, Z.C.; Writing—review & editing, Z.C., H.W. and Y.Z.; Visualization, H.W., T.D. and Y.Z.; Supervision, T.D.; Funding acquisition, Z.C. All authors have read and agreed to the published version of the manuscript.

Funding: This research was funded by the consistent support of the National Natural Science Foundation of China (No. 52301327). The paper is also sponsored by the 2022 Open Project of Failure Mechanics and Engineering Disaster Prevention, Key Lab of Sichuan Province (No. FMEDP202209); the Shanghai Sailing Program (No. 22YF1415800, No. 23YF1416100); Finally, this project is also funded by the Shanghai Natural Science Foundation (No. 23ZR1426200), the China Postdoctoral Science Foundation (No. 2023M730929), The Shanghai Soft Science Key Project (No. 23692119700), the Key Laboratory of Ministry of Education for Coastal Disaster and Protection, Hohai University (No. 202302), and the Key Laboratory of Estuarine & Coastal Engineering, Ministry of Transport (No. KLECE202302).

Institutional Review Board Statement: Not applicable.

Informed Consent Statement: Not applicable.

Data Availability Statement: In this paper, all the data, models, and code used during the study appear in the submitted article.

Conflicts of Interest: The authors declare no conflict of interest.

References

1. Araújo, G.L.S.; Sánchez, N.P.; Palmeira, E.M.; Almeida, M.D.G.G.D. Influence of micro and macroroughness of geomembrane surfaces on soil-geomembrane and geotextile-geomembrane interface strength. *Geotext. Geomembr.* **2022**, *50*, 751–763. [CrossRef]
2. Abdelaal, F.; Solanki, R. Effect of geotextile ageing and geomembrane surface roughness on the geomembrane-geotextile interfaces for heap leaching applications. *Geotext. Geomembr.* **2022**, *50*, 55–68. [CrossRef]
3. Zainab, B.; Wireko, C.; Li, D.; Tian, K.; Abichou, T. Hydraulic conductivity of bentonite-polymer geosynthetic clay liners to coal combustion product leachates. *Geotext. Geomembr.* **2022**, *49*, 1129–1138. [CrossRef]
4. Zainab, B.; Tian, K. Effect of effective stress on hydraulic conductivity of bentonite–polymer geosynthetic clay liners to coal combustion product leachates. *Environ. Geotech.* **2021**, *40*, 1–12. [CrossRef]
5. Li, D.; Tian, K. Effects of prehydration on hydraulic conductivity of bentonite-polymer geosynthetic clay liner to coal com-bus-tion product leachate. *Geo-Congress* **2022**, *2022*, 568–577.
6. Rowe, R.K.; Fan, J. Effect of geomembrane hole geometry on leakage overlain by saturated tailings. *Geotext. Geomembr.* **2021**, *49*, 1506–1518. [CrossRef]
7. Rowe, R.K.; Jabin, F. Effect of prehydration, permeant, and desiccation on GCL/Geomembrane interface transmissivity. *Geotext. Geomembr.* **2021**, *49*, 1451–1469. [CrossRef]
8. Chao, Z.; Shi, D.; Fowmes, G. Mechanical behaviour of soil under drying–wetting cycles and vertical confining pressures. *Environ. Geotech.* **2023**, 1–9. [CrossRef]
9. Cui, C.-Y.; Zhang, S.-P.; Yang, G.; Li, X.-F. Vertical vibration of a floating pile in a saturated viscoelastic soil layer overlaying bedrock. *J. Central S. Univ.* **2016**, *23*, 220–232. [CrossRef]
10. Shu, Z.; Ning, B.; Chen, J.; Li, Z.; He, M.; Luo, J.; Dong, H. Reinforced moment-resisting glulam bolted connection with coupled long steel rod with screwheads for modern timber frame structures. *Earthq. Eng. Struct. Dyn.* **2023**, *52*, 845–864. [CrossRef]
11. Meng, K.; Cui, C.; Liang, Z.; Li, H.; Pei, H. A new approach for longitudinal vibration of a large-diameter floating pipe pile in visco-elastic soil considering the three-dimensional wave effects. *Comput. Geotech.* **2020**, *128*, 103840. [CrossRef]
12. Cui, C.; Xu, M.; Xu, C.; Zhang, P.; Zhao, J. An ontology-based probabilistic framework for comprehensive seismic risk evaluation of subway stations by combining Monte Carlo simulation. *Tunn. Undergr. Space Technol.* **2023**, *135*, 105055. [CrossRef]

13. Cui, C.; Meng, K.; Xu, C.; Liang, Z.; Li, H.; Pei, H. Analytical solution for longitudinal vibration of a floating pile in saturated porous media based on a fictitious saturated soil pile mode. *Comput. Geotech.* **2021**, *131*, 103942. [CrossRef]
14. Eldesouky, H.; Brachman, R. Calculating local geomembrane strains from gravel particle indentations with thin plate theory. *Geotext. Geomembr.* **2023**, *51*, 56–72. [CrossRef]
15. Francey, W.; Rowe, R.K. Importance of thickness reduction and squeeze-out Std-OIT loss for HDPE geomembrane fusion seams. *Geotext. Geomembr.* **2023**, *51*, 30–42. [CrossRef]
16. Abdelaal, F.B.; Rowe, R.K. Physical and mechanical performance of an HDPE geomembrane in 10 mining solutions with different pHs. *Can. Geotech. J.* **2023**, *60*, 978–993. [CrossRef]
17. Zhang, W.; Shi, D.; Shen, Z.; Wang, X.; Gan, L.; Shao, W.; Tang, P.; Zhang, H.; Yu, S. Effect of calcium leaching on the fracture properties of concrete. *Constr. Build. Mater.* **2023**, *365*, 130018. [CrossRef]
18. Ng, C.W.; Chen, R.; Coo, J.; Liu, J.; Ni, J.; Chen, Y.; Zhan, L.-T.; Guo, H.; Lu, B. A novel vegetated three-layer landfill cover system using recycled construction wastes without geomembrane. *Can. Geotech. J.* **2019**, *56*, 1863–1875. [CrossRef]
19. Cai, X.; Yuan, J.; Zhou, Z.; Pi, Z.; Tan, L.; Wang, P.; Wang, S.; Wang, S. Effects of hole shape on mechanical behavior and fracturing mechanism of rock: Implications for instability of underground openings. *Tunn. Undergr. Space Technol.* **2023**, *141*, 105361. [CrossRef]
20. Zhang, X.; Ma, Z.; Wu, Y.; Liu, J. Response of Mechanical Properties of Polyvinyl Chloride Geomembrane to Ambient Temperature in Axial Tension. *Appl. Sci.* **2021**, *11*, 10864. [CrossRef]
21. Zhou, Z.; Wang, P.; Cai, X.; Cao, W. Influence of Water Content on Energy Partition and Release in Rock Failure: Implications for Water-Weakening on Rock-burst Proneness. *Rock Mech. Rock Eng.* **2023**, *56*, 6189–6205. [CrossRef]
22. Zhou, Z.; Cai, X.; Li, X.; Cao, W.; Du, X. Dynamic Response and Energy Evolution of Sandstone Under Coupled Static–Dynamic Compression: Insights from Experimental Study into Deep Rock Engineering Applications. *Rock Mech. Rock Eng.* **2020**, *53*, 1305–1331. [CrossRef]
23. Cui, C.Y.; Meng, K.; Wu, Y.J.; Chapman, D.; Liang, Z. Dynamic response of pipe pile embedded in layered visco-elastic media with radial inho-mogeneity under vertical excitation. *Geomech. Eng.* **2018**, *16*, 609–618.
24. Cui, C.; Liang, Z.; Xu, C.; Xin, Y.; Wang, B. Analytical solution for horizontal vibration of end-bearing single pile in radially heterogeneous saturated soil. *Appl. Math. Model.* **2023**, *116*, 65–83. [CrossRef]
25. Cui, C.; Meng, K.; Xu, C.; Wang, B.; Xin, Y. Vertical vibration of a floating pile considering the incomplete bonding effect of the pile-soil interface. *Comput. Geotech.* **2022**, *150*, 104894. [CrossRef]
26. Samea, A.; Abdelaal, F.B. Effect of elevated temperatures on the degradation behaviour of elastomeric bituminous geomem-branes. *Geotext. Geomembr.* **2023**, *51*, 219–232. [CrossRef]
27. Cui, C.; Zhang, S.; Chapman, D.; Meng, K. Dynamic impedance of a floating pile embedded in poro-visco-elastic soils subjected to vertical harmonic loads. *Geomech. Eng.* **2018**, *15*, 793–803.
28. Abdelaal, F.; Morsy, M.; Rowe, R.K. Long-term performance of a HDPE geomembrane stabilized with HALS in chlorinated water. *Geotext. Geomembr.* **2019**, *47*, 815–830. [CrossRef]
29. Mhamed, M.; Sutcliffe, R.; Sun, X.; Feng, J.; Retta, E.A. Arabic sentiment analysis using GCL-based architectures and a customized regularization function. *Eng. Sci. Technol. Int. J.* **2023**, *43*, 101433.
30. Janiesch, C.; Zschech, P.; Heinrich, K. Machine learning and deep learning. *Electron. Mark.* **2021**, *31*, 685–695. [CrossRef]
31. Cazzuffi, D.; Gioffrè, D. Lifetime assessment of exposed PVC-P geomembranes installed on Italian dams. *Geotext. Geomembr.* **2020**, *48*, 130–136. [CrossRef]
32. Dong, Y.; Cui, L.; Zhang, X. Multiple-GPU parallelization of three-dimensional material point method based on single-root complex. *Int. J. Numer. Meth. Eng.* **2022**, *123*, 1481–1504. [CrossRef]
33. Zhao, G.; Wu, T.; Ren, G.; Zhu, Z.; Gao, Y.; Shi, M.; Ding, S.; Fan, H. Reusing waste coal gangue to improve the dispersivity and mechanical properties of dispersive soil. *J. Clean. Prod.* **2023**, *404*, 136993. [CrossRef]
34. Anjana, R.; Keerthana, S.; Arnepalli, D.N. Coupled effect of UV ageing and temperature on the diffusive transport of aqueous, vapour and gaseous phase organic contaminants through HDPE geomembrane. *Geotext. Geomembr.* **2023**, *51*, 316–329. [CrossRef]
35. Wang, F.; Zhang, D.; Huang, H.; Huang, Q. A phase-field-based multi-physics coupling numerical method and its application in soil–water inrush accident of shield tunnel. *Tunn. Undergr. Space Technol.* **2023**, *140*, 105233. [CrossRef]
36. Wang, F.; Huang, H.; Zhang, D.; Zhou, M. Cracking feature and mechanical behavior of shield tunnel lining simulated by a phase-field modeling method based on the spectral decomposition. *Tunn. Undergr. Space Technol.* **2021**, *119*, 104246. [CrossRef]
37. Wang, F.; Zhou, M.; Shen, W.; Huang, H.; He, J. Fluid-solid-phase multi-field coupling modeling method for hydraulic fracture of saturated brittle porous materials. *Eng. Fract. Mech.* **2023**, *286*, 109231. [CrossRef]
38. Zargar, O.; Pharr, M.; Muliana, A. Modeling and simulation of creep response of sorghum stems: Towards an understanding of stem geometrical and material variations. *Biosyst. Eng.* **2022**, *217*, 1–17. [CrossRef]
39. Li, D.; Jiang, Z.; Tian, K.; Ji, R. Prediction of hydraulic conductivity of sodium bentonite GCLs by machine learning approaches. *Environ. Geotech.* **2023**, 1–17. [CrossRef]
40. Zhang, W.; Shi, D.; Shen, Z.; Zhang, J.; Zhao, S.; Gan, L.; Li, Q.; Chen, Y.; Tang, P. Influence of chopped basalt fibers on the fracture performance of concrete subjected to calcium leaching. *Theor. Appl. Fract. Mech.* **2023**, *125*, 103934. [CrossRef]
41. Hart, G.L.W.; Mueller, T.; Toher, C.; Curtarolo, S. Machine learning for alloys. *Nat. Rev. Mater.* **2021**, *6*, 730–755. [CrossRef]

42. Chao, Z.; Dang, Y.; Pan, Y.; Wang, F.; Wang, M.; Zhang, J.; Yang, C. Prediction of the shale gas permeability: A data mining approach. *Geomech. Energy Environ.* **2023**, *33*, 100435. [CrossRef]
43. Zhang, W.; Shi, D.; Shen, Z.; Shao, W.; Gan, L.; Yuan, Y.; Tang, P.; Zhao, S.; Chen, Y. Reduction of the calcium leaching effect on the physical and mechanical properties of concrete by adding chopped basalt fibers. *Constr. Build. Mater.* **2023**, *365*, 130080. [CrossRef]
44. Chao, Z.; Shi, D.; Fowmes, G.; Xu, X.; Yue, W.; Cui, P.; Hu, T.; Yang, C. Artificial intelligence algorithms for predicting peak shear strength of clayey soil-geomembrane interfaces and experimental validation. *Geotext. Geomembr.* **2023**, *51*, 179–198. [CrossRef]
45. Dong, Y.; Wang, D.; Randolph, M.F. Investigation of impact forces on pipeline by submarine landslide using material point method. *Ocean Eng.* **2017**, *146*, 21–28. [CrossRef]
46. Sabitova, A.; Yarushina, V.M.; Stanchits, S.; Stukachev, V.; Khakimova, L.; Myasnikov, A. Experimental compaction and dilation of porous rocks during triaxial creep and stress relaxation. *Rock Mech. Rock Eng.* **2021**, *54*, 5781–5805. [CrossRef]
47. Pillai, A.G.; Gali, M.L. Digital image-based performance evaluation of GCL-sand interfaces under repeated shearing. *Geosynth. Int.* **2022**, 1–21. [CrossRef]
48. Fan, N.; Jiang, J.; Nian, T.; Dong, Y.; Guo, L.; Fu, C.; Tian, Z.; Guo, X. Impact action of submarine slides on pipelines: A review of the state-of-the-art since 2008. *Ocean Eng.* **2023**, *286*, 115532. [CrossRef]
49. Piegari, E.; De Donno, G.; Melegari, D.; Paoletti, V. A machine learning-based approach for mapping leachate contamination using geoelectrical methods. *Waste Manag.* **2023**, *157*, 121–129. [CrossRef]
50. Shao, W.; Xiong, Y.; Shi, D.; Xu, X.; Yue, W.; Soomro, M.A. Time dependent analysis of lateral bearing capacity of reinforced concrete piles combined with corrosion and scour. *Ocean Eng.* **2023**, *282*, 115065. [CrossRef]
51. Sanchez, D.; Gilchrist, D.; Yang, S.; Bandara, K.M.A.S.; Gamage, R.P.; Zheng, W. Experimental characterization of time-dependent mechanical behaviours of frac sand at high compressive stresses and implication on long-term proppant conductivity. *Geomech. Geophys. Geo-Energy Geo-Resour.* **2022**, *8*, 1–18. [CrossRef]
52. Camarena-Martinez, R.; Lizarraga-Morales, R.A.; Baeza-Serrato, R. Classification of Geomembranes as Raw Material for Defects Reduction in the Manufacture of Biodigesters Using an Artificial Neuronal Network. *Energies* **2021**, *14*, 7345. [CrossRef]
53. Fleury, M.P.; Kamakura, G.K.; Pitombo, C.S.; Cunha, A.L.B.; da Silva, J.L. Prediction of non-woven geotextiles' reduction factors for damage caused by the drop of backfill materials. *Geotext. Geomembr.* **2023**, *51*, 120–130. [CrossRef]
54. Wang, H.; Li, L.; Li, J.; Sun, D. Drained expansion responses of a cylindrical cavity under biaxial in situ stresses: Numerical in-vestigation with implementation of anisotropic S-CLAY1 model. *Can. Geotech. J.* **2023**, *60*, 198–212. [CrossRef]
55. Scholkopf, B.; Smola, A.J. *Learning with Kernels: Support Vector Machines, Regularization, Optimization, and Beyond*; MIT Press: Cambridge, MA, USA, 2001.
56. Zang, W.; Zheng, Y.; Zhang, Y.; Lin, X.; Li, Y.; Fernandez-Rodriguez, E. Numerical Investigation on a Diffuser-Augmented Horizontal Axis Tidal Stream Turbine with the Entropy Production Theory. *Mathematics* **2022**, *11*, 116. [CrossRef]
57. Witten, I.H.; Frank, E.; Hall, M.A. *Data Mining Fourth Edition: Practical Machine Learning Tools and Techniques*; Morgan Kaufmann Publishers Inc.: San Francisco, CA, USA, 2016.
58. Otchere, D.A.; Ganat, T.O.A.; Gholami, R.; Ridha, S. Application of supervised machine learning paradigms in the prediction of petroleum reservoir properties: Comparative analysis of ANN and SVM models. *J. Pet. Sci. Eng.* **2020**, *200*, 108182. [CrossRef]
59. Das, S.K.; Basudhar, P.K. Undrained lateral load capacity of piles in clay using artificial neural network. *Comput. Geotech.* **2006**, *33*, 454–459. [CrossRef]
60. Kanungo, D.P.; Sharma, S.; Pain, A. Artificial Neural Network (ANN) and Regression Tree (CART) applications for the indirect estimation of unsaturated soil shear strength parameters. *Front. Earth Sci.* **2014**, *8*, 439–456. [CrossRef]

materials

MDPI

Article

Investigation of Damage and Creep for Bedding's Carbonaceous Slate with Chemical Erosion Effect

Weihao Zeng [1], Zhenghong Chen [1,*], Yunpeng Xie [2] and Qiunan Chen [1,*]

1 School of Civil Engineering, Hunan University of Science and Technology, Xiangtan 411201, China
2 Research Center of Geotechnical and Structural Engineering, Shandong University, Jinan 250061, China
* Correspondence: zhchen@hnust.edu.cn (Z.C.); cqndoc@163.com (Q.C.)

Abstract: Tunnel projects in the southwestern mountainous area of China are in full swing. According to the tunnel burial depth, structural characteristics, and chemical erosion environments of the Lixiang railway tunnel, carbonaceous slate specimens obtained in the field were taken to experimentally investigate the physical, mechanical, and creep characteristics of the bedding's slate specimens after chemical erosion. The results of scanning electron microscopy (SEM) indicate that chemical erosion leads to internal damage in the carbonaceous slate specimens, and the internal damages are increasing with the increase of erosion days. Moreover, the specimens' ultrasonic test (UT) results prove that specimens with smaller bedding angles suffer a more serious erosion and induce more internal cracks. Under conventional triaxial compression conditions with 40 MPa of confining pressures, the conventional triaxial compressive strength (σ_s) decreases with the decrease of the bedding angle and the increase of erosion days, and the failure modes of the specimens are mainly controlled by oblique shear fractures and accompanied by the occurrence of slip dislocation fractures between the bedding inclination. Under creep conditions with 40 MPa of confining pressures, the final deformations of specimens are increasing with the increase of erosion days, which means the longer the erosion days, the greater the deformations. The failure modes of the specimens under creep conditions are controlled by shear fractures, and for the specimen with a 60° bedding angle and long-term erosion, there are block separations and many cavities along the shear planes. Therefore, more attention should be paid to prevent serious failure of the surrounding rock if the surrounding rock has a bedding angle of 60° or suffers long-term erosion.

Keywords: carbonaceous slate; bedding angle; chemical erosion; creep characteristic

Citation: Zeng, W.; Chen, Z.; Xie, Y.; Chen, Q. Investigation of Damage and Creep for Bedding's Carbonaceous Slate with Chemical Erosion Effect. *Materials* **2023**, *16*, 5163. https://doi.org/10.3390/ma16145163

Academic Editor: Rui Miranda Guedes

Received: 26 June 2023
Revised: 15 July 2023
Accepted: 19 July 2023
Published: 22 July 2023

1. Introduction

Under a high in situ stress environment, rock masses present complex mechanical behaviors [1,2]; therefore, in engineering practice, a deep understanding of the mechanical behaviors of rock masses is the key to ensuring the safety and operation of deep-buried tunnels [3,4]. During long-time crustal movement, most sedimentary rocks and some metamorphic rocks have obvious bedding structures that will have a great influence on mechanical behaviors and result in discontinuity, heterogeneity, and anisotropy of the rock [5–7]. According to the engineering geological survey, the bedding structure is obvious in the surrounding rock mass of the Lixiang railway [8]. Since the bedding angle is varied and the mechanical properties are different, the influence of the bedding structures on the stability and safety of rock engineering cannot be ignored. Until now, the influences of bedding structures on the physical and mechanical properties of rock specimens have been extensively investigated. For example, He et al. [9] studied the rockburst ejection velocity of unloading sandstones with different bedding angles, and they found that the specimen with 60° bedding angles is the easiest one to rockburst, but the corresponding rockburst intensity is the weakest. Horizontal bedding is the least prone to rockburst, while the rockburst is strong once it occurs. Zhang et al. [10] investigated the effect of bedding on the dynamic compressive properties of slate; the results indicate bedding plays an increasingly dominant role in the failure of samples as the dip angle increases, and there is a strain rate decrease when the bedding angle increases from 30° to 60°. Liu et al. [11] conducted both quasi-static and dynamic uniaxial compression tests on bedding coal specimens, and they found that the bedding effect on the coal behavior on the quasi-static tests was more prominent than that on the dynamic tests. Hao et al. [12] investigated the influences of confining pressure and bedding angles on the mechanical behavior of brittle slate; they concluded that the fracture angle of slate presents a U-shaped curve with the bedding angle increases from 0° to 90°, and when the bedding angle is 45°–60°, the slate specimens present the lowest strength, and when bedding angle is below 15°, it shows the greatest strength. Wasantha et al. [13] carried out uniaxial compression tests on bedded sandstone with layers of different bedding orientations; they obtained that the failures of rock specimens with shallow bedding angles could be more violent and devastative than the failures of rock with steeply oriented bedding. Through Brazilian disc splitting tests, Liu et al. [14] concluded that the tensile strength of the bedding's slate gradually decreases, with the bedding angle varying from 0° to 90°. Xia et al. [15] analyzed the influence of the geometrical properties of bedding planes on the direct shear strength characteristics, and they concluded two patterns of the shear fractures with a range of bedding plane geometries.

Except for the bedding structures, the chemical erosion environment is also encountered in the practical engineering of the Lixiang railway. Chemical erosion mainly affects the microstructure, particle adhesion, and mineral composition of the rock, which will change the rock's physical and mechanical properties, creep properties, and macroscopic cracks [16–18]. There has been considerable research on rock behavior under chemical erosion environments for rock specimens. For example, Zhang et al. [19] conducted creep tests on cracked granite under various confining pressures and different chemical solutions, and they found that an acid solution has a more obvious influence on the creep behavior than that of an alkaline solution. Hu et al. [20] proposed constitutive models to describe the mechanical responses of cement-based material subjected to a long-term chemo-mechanical coupling effect. Using double-torsion load–relaxation tests, Callahan et al. [21] evaluated the effect of the chemical environment on the fracture toughness and subcritical fracture growth index (SCI) in silicified fault rocks; reductions in the SCI in all aqueous environments, with a >60% reduction in the alkaline solutions, were observed. To evaluate the effect of chemical solutions on the frictional properties of quartz-rich sandstone, triaxial compression tests have been performed on sandstone by Feucht and Logan [22]; the results suggest that the frictional resistance to sliding of the sandstone seems to be primarily controlled by the ionic strength of the solutions, with secondary control by the pH value of the solutions. Li et al. [23] investigated the micro-damage evolution and macro-mechanical property degradation of limestone due to chemical effects during triaxial compression conditions; they concluded that the reason for the mechanical property degradation of the rock is that the chemical solutions change the porosity and micro-damage evolution process.

The above studies indicate that bedding structures and chemical erosion can both effectively influence the mechanical properties of rock. It can be found that most of the existing studies only considered the single effect of the bedding structures or the chemical erosion on the mechanical characteristics of rock under short-term stress conditions. However, the coupling effect of the bedding structures and chemical erosion will significantly affect the long-term stability of surrounding rock for deep-buried tunnels, whereas related research is rare. To fill this gap, in this study, the coupling effects of bedding structures and chemical erosion on the characteristics of slate specimens under long-term creep conditions were investigated. Accordingly, the physical, mechanical, and creep characteristics of the bedding's slate specimens after chemical erosion were obtained through the techniques of scanning electron microscopy (SEM), ultrasonic tests (UTs), conventional triaxial compression tests, and creep tests.

2. Experimental Setup

2.1. Test Apparatus

The conventional triaxial compression tests and the triaxial creeping tests were performed on a servo-controlled rock mechanics testing system, TYJ-2000 (as shown in Figure 1), which consists of axial loading platens, a triaxial pressure chamber, a confining pressure control system, and a software control platform. The TYJ-2000 rock mechanics testing system can conduct conventional rock triaxial compression tests and creep tests under high temperatures and high pressures. The maximum axial load of the testing system is 2000 KN, and the maximum confining pressure is 100 MPa.

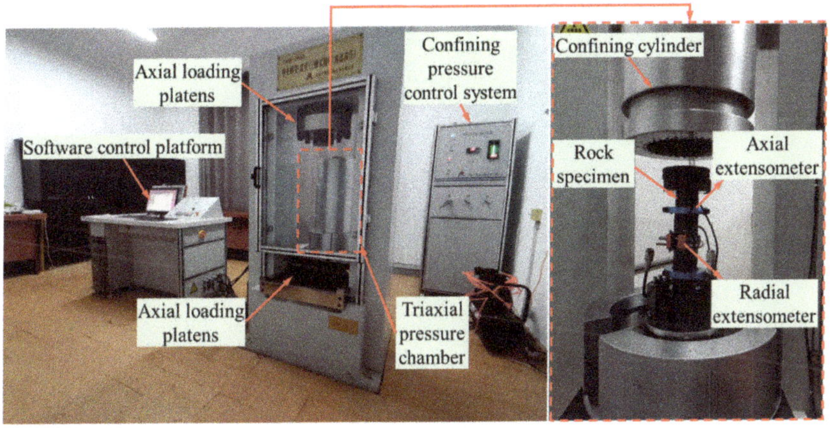

Figure 1. Photographs of the test apparatus.

2.2. Rock Specimen Preparation

Since the purpose of the research is to obtain the general damage and creep characteristics of the bedding's carbonaceous slate under the chemical erosion effect, all the investigated specimens should be alike, except for the bedding angle and erosion days. Therefore, rather than collect rock cores in the drill holes to obtain testing specimens, the specimens were cored from an uneroded carbonaceous slate block, which is collected in a certain tunnel along the Lixiang railway. The carbonaceous slate block is gray (black) with no visible cracks, its lithology is compact, and the bedding structure is obvious. According to our previous research results of an X-ray diffraction (XRD) experiment, the mineral compositions and their contents of this carbonaceous slate are listed in Table 1 [24]. The rock specimens were cored from the slate block using the water drill method (Figure 2a). According to the geological investigations for the surrounding rock mass in the tunnel, the angle between the carbonaceous slate strata and the vertical direction is mostly around 25° to 90°. Therefore, three typical bedding angles of 30°, 60°, and 90° were investigated to capture the general law of the bedding angle effect. The angle between the drill bit direction and the horizontal bedding structure of the slate block is 30°, 60°, and 90°, respectively. To do so, standard cylinder specimens (50 mm × 100 mm) with three different bedding angles of 30°, 60°, and 90° were prepared (as shown in Figure 2b). The prepared specimens contained no visible damage or defects. Moreover, the end surfaces of the specimens were carefully polished, and the testing processes were well-designed to comply with the International Society of Rock Mechanics (ISRM) suggested testing requirements [25].

Table 1. Contents and structure characteristics of mineral composition [24].

Mineral Composition	Quartz	White Mica	Albite	Graphite
Mineral Content	72%	16%	9%	<3%
Structure characteristics	Granular or dense aggregate, without cleavage	Scaly or flaky aggregate, perfectly cleaved	Glassy crystal, medium cleavage	Carbon crystal, perfectly cleaved

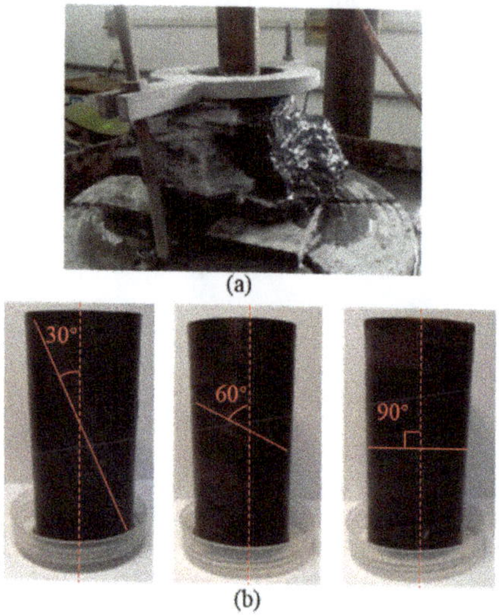

Figure 2. (**a**) Slate block; (**b**) specimens with different bedding angles.

According to the water quality investigation results of a tunnel along the Lixiang railway, the underground water solution in this area holds a pH value of 4.8 to 5.3 and contains a variety of ionic components, mainly including Mg^{2+}, Ca^{2+}, SO_4^{2-}, and Cl^-, indicating that the groundwater is a chemical erosion environment. To reproduce the chemical erosion effect of the SO_4^{2-}, Cl^-, and H^+ on the carbonaceous slate in this area, in this paper, dilute hydrochloric acid and $CaCl_2$ and Na_2SO_4 solutions were used to prepare the chemical solvent with a pH value of 4.8.

2.3. *Experimental Schemes*

The prepared specimens with three different bedding angles (30°, 60°, and 90°) were immersed in the prepared chemical solvent for 30 days' erosion, 60 days' erosion, and 90 days' erosion, respectively (Figure 3). After the immersion, the natural dry specimens (0 days' erosion) were taken as a reference, and conventional triaxial compression tests were carried out to obtain conventional mechanical properties. Moreover, triaxial creep tests were also conducted on specimens with three different bedding angles and four different erosion days to obtain creep characteristics. Accordingly, the specimens were divided into two groups: Group A for the conventional triaxial compression test and Group B for the triaxial creep test. The specific specimen grouping is shown in Table 2.

Figure 3. Photograph of the specimen's erosion process.

Table 2. Specimens for the tests.

	Specimen Number	Erosion Days	Bedding Angle (°)
Conventional triaxial compression tests (Group A)	A-30°-0D	0	30°
	A-30°-30D	30	
	A-30°-60D	60	
	A-30°-90D	90	
	A-60°-0D	0	60°
	A-60°-30D	30	
	A-60°-60D	60	
	A-60°-90D	90	
	A-90°-0D	0	90°
	A-90°-30D	30	
	A-90°-60D	60	
	A-90°-90D	90	
Triaxial creeping tests (Group B)	B-30°-90D	90	30°
	B-60°-30D	30	60°
	B-60°-60D	60	
	B-60°-90D	90	
	B-90°-90D	90	90°

Considering the measured in situ stress in the field of the deep-buried Lixiang railway tunnel is 39.8 MPa, the confining pressure was set as 40 MPa in the conventional triaxial compression test and triaxial creep test.

The specific experimental schemes in the conventional triaxial compression test and triaxial creep test were as follows:

(1) During the conventional triaxial compression tests, confining pressures were increased to 40 MPa at the rate of 0.10 MPa/s. The conventional triaxial compressive strength (σ_s) for specimens could be obtained by increasing the axial loads at a rate of 0.20 MPa/s until the specimens were damaged;

(2) During the triaxial creep test, confining pressures were increased to 40 MPa at the rate of 0.10 MPa/s. The axial pressure was step-loaded at a rate of 0.20 MPa/s, the first step was loaded to 0.4 times the conventional triaxial compressive strength (σ_s), and the subsequent load of each step increased by 0.1 times the σ_s until creep failure occurred in the specimen (as shown in Figure 4).

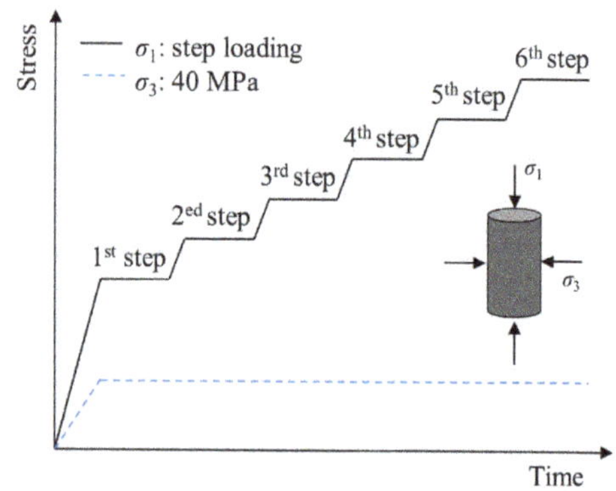

Figure 4. Stress paths for the triaxial creep test.

3. Physical Properties of the Bedding's Slate Specimens after Chemical Erosion

3.1. Microscopic Damage Properties of the Bedding's Slate Specimens after Chemical Erosion

Affected by the high in situ stress in the southwest mountainous area, the microstructure of the collected slate specimens is dense and compact. However, there are still many microscopic cracks that cannot be detected by the eyes, especially for specimens after chemical erosion. As shown in Figure 5, the chemical solvent will have a significant influence on the microscopic damage of the slate specimens. Therefore, after different erosion days by a chemical solvent, the microstructure of the slate specimens will be damaged to different degrees. The microstructure damages have an important effect on the physical and mechanical properties of rock. To figure out the microstructure damage properties of the slate specimens by a chemical solvent, they were analyzed by a Gemini Sigma 300 scanning electron microscope, which is a widely used technology to obtain the micromorphology properties of objects [26,27].

Figure 5. Photograph and schematic diagram of surface damage after chemical solvent erosion.

To conduct SEM, 20 thin slices (no more than 2 cm in length and width) of slate specimens with four different erosion days were taken along the bedding plane. Typical SEM images of the specimens under different erosion days are shown in Figure 6. According to the images, it can be found that the specimens without erosion (0 days' erosion) present a stepped structure, and the step surface is relatively flat and compact. For the specimens after 30 days' erosion, there are many small and loose blocks on the surface of the stepped structure, and some micro-cracks emerged in the specimen slice. For the specimens after 60 days' erosion, the edges and corners of the stepped structures are fuzzy and loose, showing a pit-like structure; moreover, the micro-cracks are the largest and most obvious among the four kinds of slices. For the specimens after 90 days' erosion, the stepped structure is mostly a deep pit structure, and a lot of micro-cracks are filled or buried by loose and small blocks. Therefore, it can be concluded that the internal cracks induced by chemical erosion are increasing with the increase of erosion days.

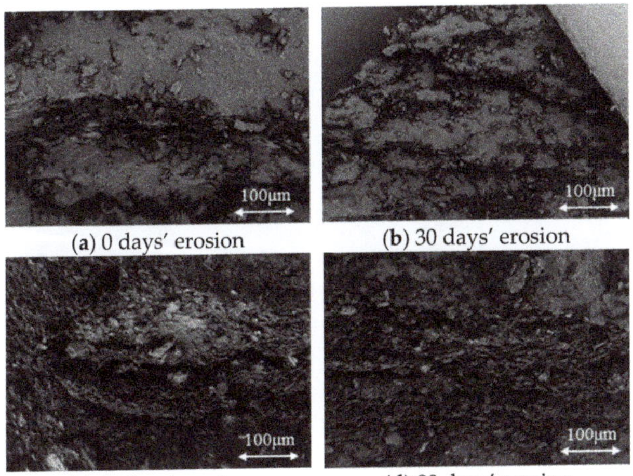

(**a**) 0 days' erosion (**b**) 30 days' erosion

(**c**) 60 days' erosion (**d**) 90 days' erosion

Figure 6. Typical SEM images of specimens under different erosion days.

3.2. Ultrasonic P-Wave Velocity of the Bedding's Slate Specimens after Chemical Erosion

Longitudinal wave velocity in the ultrasonic tests (UTs) is one of the important parameters that can reflect the difference in the rock's physical properties, such as the integrity, density, porosity, etc. [28,29] Therefore, by measuring the acoustic wave propagation velocity of the slate specimens, the integrity of the specimens can be analyzed (the internal cracks' enrichment degree), and the relationship between the wave velocity and the bedding angle and chemical erosion degree can be explored.

In this paper, a ZBL-U5200 nonmetallic ultrasonic detector was used to detect the P-wave velocity of slate specimens with different bedding angles and chemical erosion days. During the detection, petroleum jelly was used as the coupling material between the rock specimens and transducer. According to the working principle of the nonmetallic ultrasonic detector, if there are more internal cracks or serious degradation in the specimen, the detected wave velocity of the specimen is lower [30,31].

Figure 7 shows the wave velocity of specimens with different bedding angles and different erosion duration days. According to the descriptive statistics of three specimens under the same states (the same bedding angle and erosion days), it is found that the standard deviations of the wave velocity for the specimens are less than 0.1, which means the difference in the measured values is very small for specimens under the same states. Therefore, it is believed that the differences among the specimens under the same states are small.

Moreover, according to the analyses of the mean values for specimens under the same erosion days, the velocity of the ultrasonic longitudinal wave increases with the increase of the bedding angle. For example, for specimens under 30 days' erosion, the increasing amount of wave velocity is 0.121 (km/s) when the bedding angle increased from 30° to 90°. This means that specimens with smaller bedding angles suffered serious erosion and induced more internal cracks. On the other hand, for specimens with the same bedding angles, it shows that the velocity of the ultrasonic longitudinal waves increases with the increase of erosion days. For example, for specimens with a 30° bedding angle, the increasing amount of the wave velocity is 0.498 (km/s) when the erosion days increased from 0 days to 90 days. The principal reason for this phenomenon is that the greater the erosion days, the more liquid will fill the microscopic cracks in the rock specimens. The effect of liquid filling on acoustic waves is greater than that of internal cracks induced by chemical erosion.

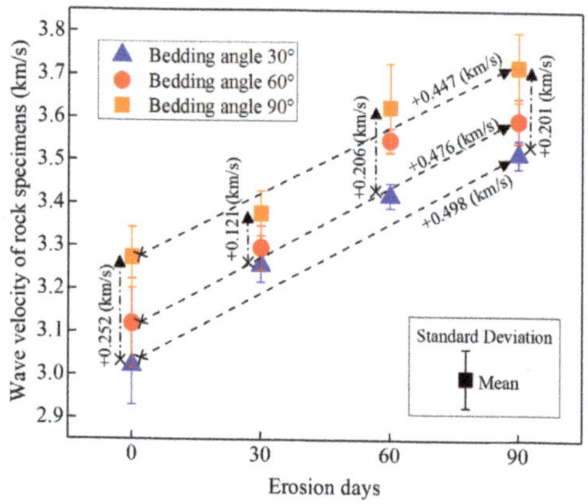

Figure 7. Wave velocity of specimens with different bedding angles with different erosion durations.

4. Conventional Triaxial Characteristics of the Bedding's Slate Specimens after Chemical Erosion

According to the above analysis of the physical properties of the bedding's slate specimens after chemical erosion, it is known that the chemical erosion changed the microstructures of the slate specimens; therefore, the macro-mechanical properties of the specimens will be changed as well. To investigate the compression characteristics of the bedding's slate specimens after chemical erosion under 40 MPa of in situ stress conditions, conventional triaxial compression tests with 40 MPa of confining pressure were conducted on specimens with different bedding angles and erosion days.

4.1. Conventional Deformation and Triaxial Strength Characteristics

The stress–strain curves of specimens under conventional triaxial compression tests are shown in Figure 8. Before the peak of deviation stress ($\sigma_1 - \sigma_3$), the stress–strain curves increased linearly, indicating elastic deformations of the specimens. After the peak of the deviation stress, the stress–strain curves of the specimens with different bedding angles and erosion days are diverse. For the specimens with 30° and 90° bedding angles, under short-term erosions (0 days' erosion and 30 days' erosion), the specimens' stress dropped sharply, with little strain increase, indicating a brittle failure, whereas, for the specimens with 30° and 90° bedding angles, under long-term erosions (60 days' erosion and 90 days' erosion), the specimens' stress drop after failure is small, and there is a great strain increase, indicating ductile failure. The variations of stress–strain for the specimens with a 60° bedding angle are discrepancies from that of the specimens with 30° and 90° bedding angles. Under four different erosion days, the specimens with a 60° bedding angle all showed brittle failure features in the stress–strain curve. However, the specimens with a 60° bedding angle showed greater axial and radial strain at the peak of the deviation stresses than the specimens with 30° and 90° bedding angles, and the values of the axial and radial strain at the peak of the deviation stresses increased with the increase of erosion days. The reasons for this discrepancy phenomenon for the specimens with a 60° bedding angle can be interpreted by the fracture modes analysis in the next section.

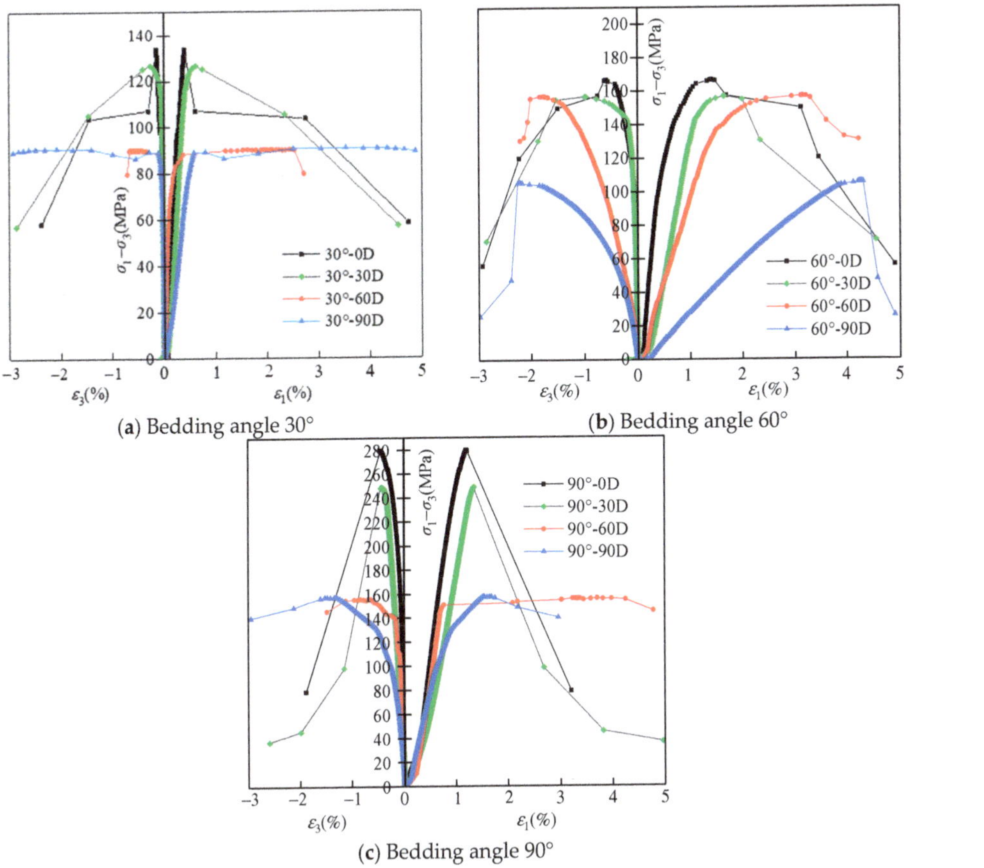

Figure 8. Stress–strain curves of specimens under triaxial compression tests.

Based on the stress–strain curves, the conventional triaxial compressive strength (σ_s) of specimens can be obtained by adding 40 MPa of confining pressure to the peak of deviation stresses. Figure 9 shows the relationship of the σ_s with the bedding angle and erosion days. For specimens under the same erosion days, the σ_s decreased with the decrease of the bedding angle. This result can be explained by the variation of the wave velocity for the specimens with different bedding angles; in Section 3, it proves that the specimens with a smaller bedding angle exhibited a lower wave velocity because they suffered serious erosion and induced more internal cracks; more internal cracks in specimens will induce a lower σ_s. On the other hand, for the specimens with the same bedding angle, the σ_s decreased with the increase of erosion days. This conclusion is consistent with the results of the SEM images for the specimens under different erosion days, as shown in Figure 6, which indicates long-term erosion will induce more internal cracks and loose structures, and loose structures will result in lower σ_s. Moreover, for the specimens with a 90° bedding angle under 0 and 30 days' erosion, their σ_s is greater than that of the other specimens; this is because they suffered slight erosion, and at the same time, the 90° bedding structure was vertical to the axial stress, which made them harder to break.

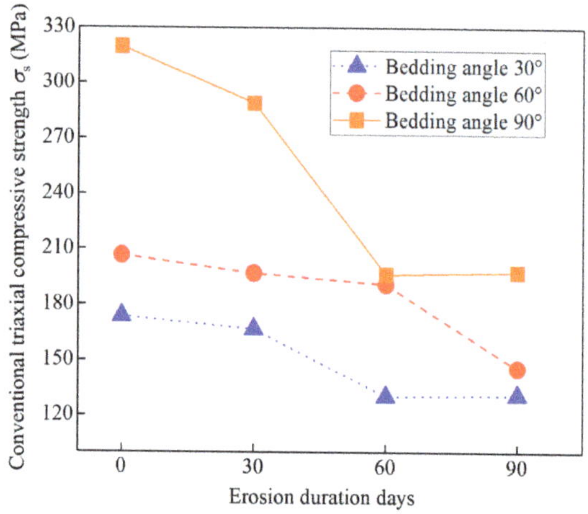

Figure 9. Conventional triaxial compressive strength with different bedding angles and erosion days.

4.2. Conventional Triaxial Failure Modes

A deep understanding of the failure behavior of rock is important for the prediction and protection of engineering disasters. In this paper, the failure modes of specimens with different bedding angles and erosion days under conventional triaxial conditions were analyzed and compared. Figure 10 shows the failure modes and failure sketches of specimens under triaxial compression tests. It can be found that under conventional triaxial conditions, for specimens with 30° bedding angles, the main fractures in the specimens are connected diagonally along the bedding inclination, and X-shaped small secondary cracks are visible; moreover, with the increase of erosion days, the fractures along the bedding inclination are more obvious, and secondary cracks are rarer. For specimens with 60° bedding angles, there are several cut-through fractures propagated along the bedding inclination, resulting in spalling failure, which explains why the specimens with 60° bedding angles show greater axial and radial strain than that of the specimens with 30° and 90° bedding angles. Furthermore, with the increase of erosion days, the cut-through fractures changed from scattered to concentrated. For specimens with 90° bedding angles, the specimens under short-term erosion (0 days' erosion and 30 days' erosion) are relatively broken and show a common X-type conjugate inclined to plane shear failure, whereas the specimens' failure under long-term erosion (60 days' erosion and 90 days' erosion) are controlled by shear failure, together with horizontal cracks.

In conclusion, the failure modes of specimens with different bedding angles and erosion days under conventional triaxial conditions are affected by the bedding's inclination angle. The failure modes of the specimens are mainly controlled by oblique shear fractures and accompanied by the occurrence of slip dislocation fractures between the bedding inclination. Especially with increasing erosion days, the fracture development along the bedding's inclination angle is gradually obvious, which indicates that chemical erosion accelerates the corrosion of the interlayer.

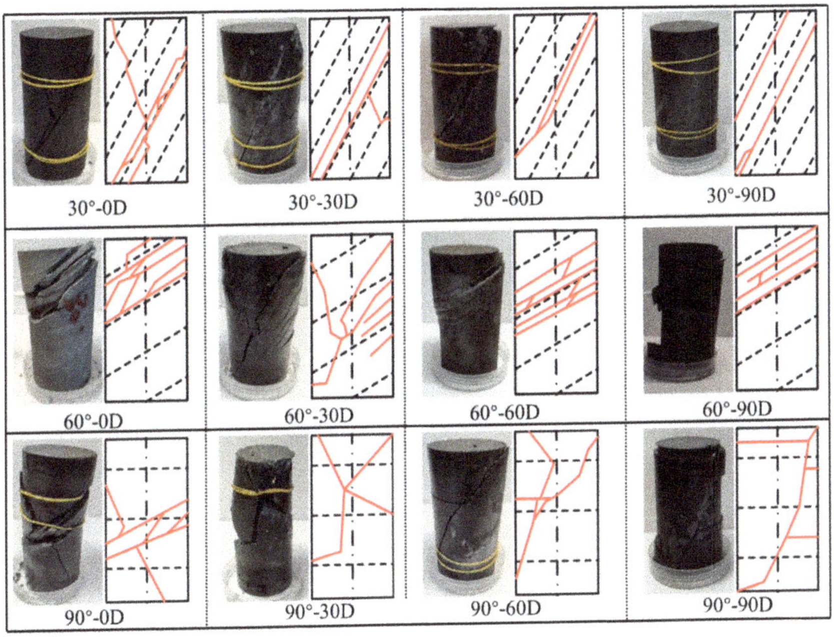

Figure 10. Failure modes and failure sketches of specimens under triaxial compression tests.

5. Creep Characteristics of the Bedding's Slate Specimens after Chemical Erosion

Through the above analyses of the physical properties and triaxial compression test results, the basic physical and mechanical characteristics of the bedding's slate specimens after chemical erosion are clear. In engineering practice, since the surrounding rock will deform greatly under high in situ stress and the deformation process will last for a long time, the long-term stability of the surrounding rock is the key to ensuring the safety and operation of deep-buried tunnels in high in situ stress environments, especially for the surrounding rock of the bedding's slate subjected to chemical erosion. Therefore, in this section, to obtain the long-term deformation and failure characteristics, the creep tests were carried out on specimens with different bedding angles and chemical erosion days.

5.1. Creep Deformation Characteristics

(1) Influence of bedding angles on creep deformation characteristics

According to the experimental scheme, the triaxial creep tests under a confining pressure of 40 MPa were carried out on the specimens with bedding angles of 30°, 60°, and 90° after 90 days of erosion. The strain–time curves of the creep tests for the specimens with different bedding angles after 90 days of erosion are plotted in Figure 11. It can be found that at the last step of loading, the creep of the specimen changed from a stable creep to an unstable creep, resulting in specimen failure. At the same time, an unstable accelerated creep occurred suddenly, indicating sudden brittle specimen failure.

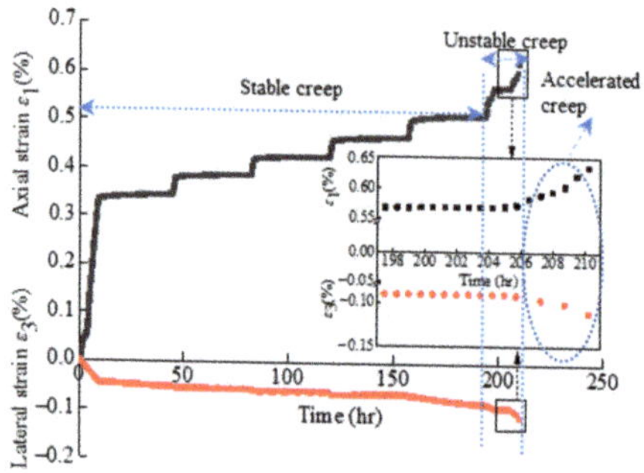

(**a**) Specimen 30°-90D

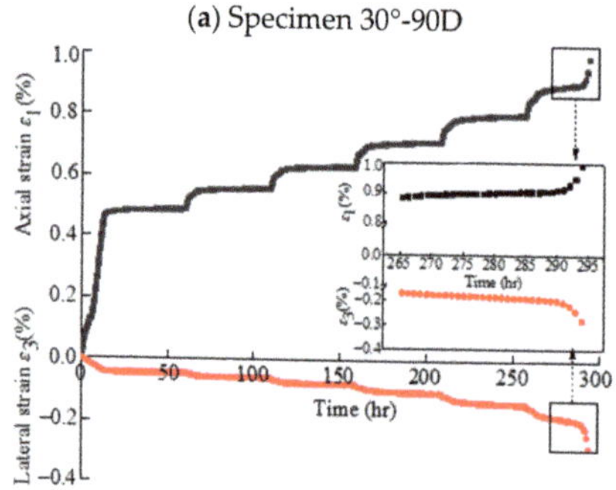

(**b**) Specimen 60°-90D

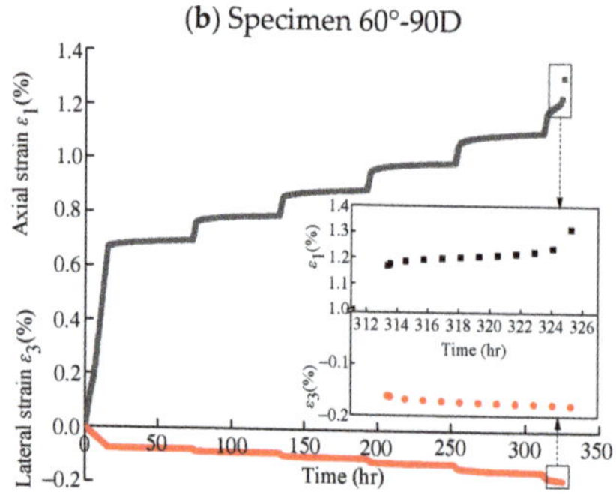

(**c**) Specimen 90°-90D

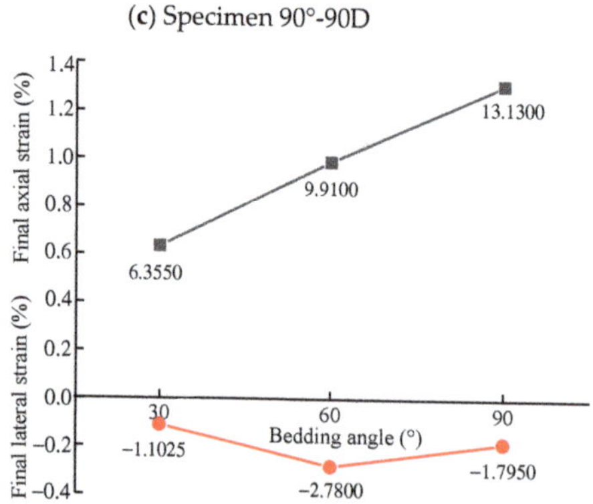

(**d**) Final strains of specimens with different bedding angles

Figure 11. Creep curves and final strains of specimens with different bedding angles after 90 days of erosion.

Moreover, comparing the strain–time curves of specimens with bedding angles of 30°, 60°, and 90° after 90 days' erosion, it is obvious that the axial deformations of accelerated creep stages increase with the increasing of the bedding angles; thus, the final axial strain increases (as shown in Figure 11d), whereas the lateral deformations of the accelerated creep stages are different. For the specimens of 30°-90D, the lateral deformations of the accelerated creep stage increase inconspicuously; for the specimens of 60°-90D, the lateral deformations of the accelerated creep stage increase obviously; for the specimens of 90°-90D, the lateral deformations show no accelerated creep stage. These variations of lateral deformations of the accelerated creep stage indicate the difference in the failure modes for specimens with different bedding angles under creep tests, which will be interpreted next in Section 5.2.

(2) Influence of erosion days on creep deformation characteristics

According to Sections 4.1 and 5.1, there are discrepancies in the deformation characteristics for specimens with a 60° bedding angle under conventional triaxial tests and creep tests compared with 30° and 90° bedding angles. Thus, triaxial creep tests under a confining pressure of 40 MPa were carried out on the specimens with 60° bedding angles after 0, 30, 60, and 90 days of erosion. The strain–time curves of the creep tests for specimens with different erosion days are plotted in Figure 12. It can be found that for the specimens under long-term erosion (60°-60D and 60°-90D), the specimens' unstable accelerated creep occurred more suddenly than that of the specimens under short-term erosion (60°-0D and 60°-30D), indicating a more sudden brittle specimen failure for the specimens under long-term erosion. Moreover, the final axial and final lateral deformations increase with the increase of erosion days (as shown in Figure 12e), which means the longer the erosion days, the greater the deformation.

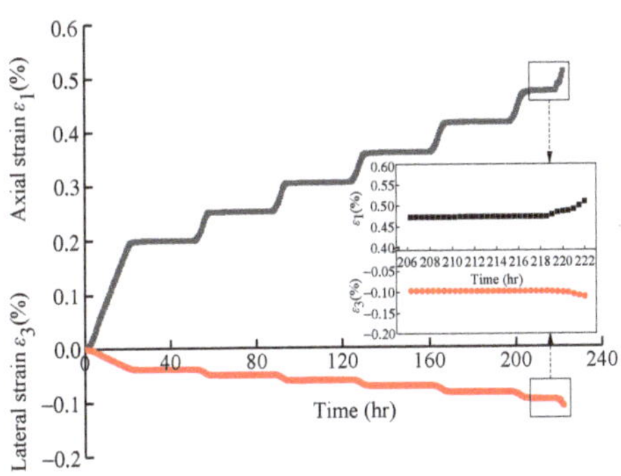

(**a**) Specimen 60°-0D

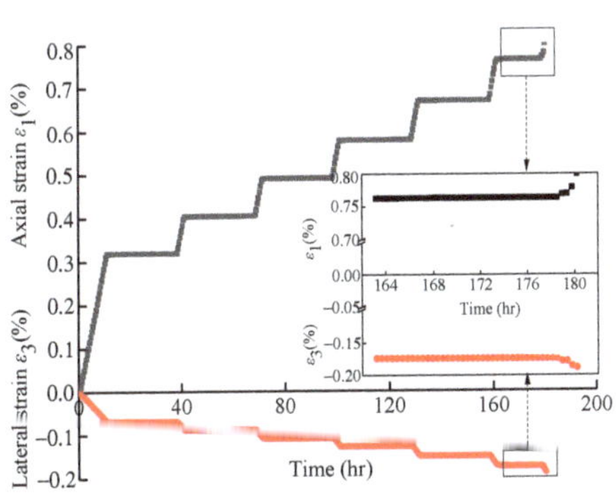

(**b**) Specimen 60°-30D

Figure 12. *Cont.*

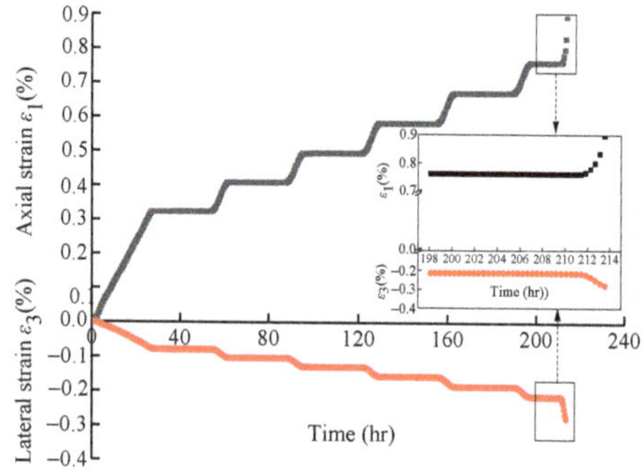

(**c**) Specimen 60°-60D

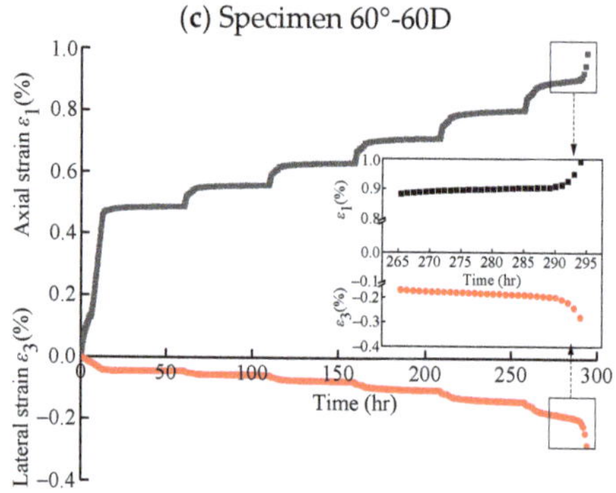

(**d**) Specimen 60°-90D

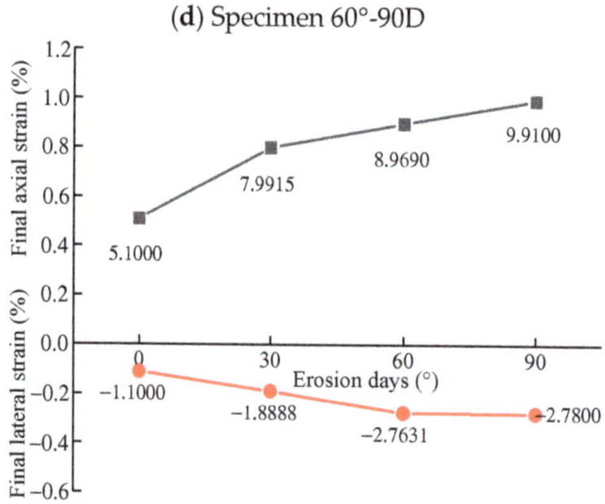

(**e**) Final strains of specimens under different erosion days

Figure 12. Creep curves of specimens with a bedding angle of 60° under different erosion days.

5.2. Creep Failure Modes

Analyzing the creep failure modes of the bedding's slate subjected to chemical erosion can provide references for predicting failure behavior and the service state of engineering the surrounding rock. In this section, the creep failure modes of specimens with different bedding angles and erosion days under creep conditions were analyzed and compared. Figure 13 shows the failure modes and failure sketches of specimens under creep tests.

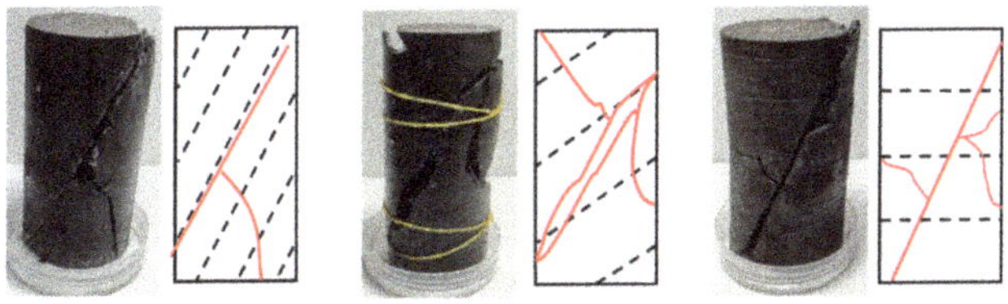

30°-90D 60°-90D 90°-90D

(**a**) Specimens with different bedding angles after 90 days of erosion

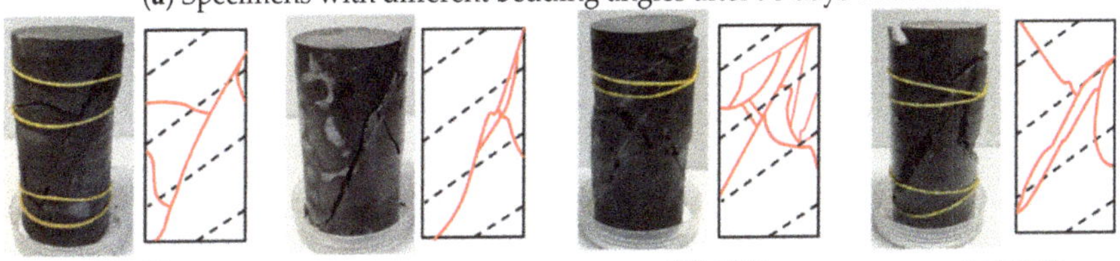

60°-0D 60°-30D 60°-60D 60°-90D

(**b**) Specimens with a bedding angle of 60° under different erosion days

Figure 13. Failure modes of specimens under creep tests.

As shown in Figure 13a, the failure modes of the specimens with different bedding angles are controlled by shear fractures, and there are some cavities along the shear planes. For the specimens of 60°-90D, the cavities along the shear planes are more obvious, inducing a small block separation, which is interpreted as an obvious increase of lateral deformation for the specimens of 60°-90D observed in Section 5.1. Therefore, this proves that in engineering practice, for engineering the surrounding rock of the bedding's slate under long-term chemical erosion conditions, more attention should be paid to prevent the surrounding rock's failure if the bedding angle is 60°.

As shown in Figure 13b, the failure modes of the specimens with different erosion days are controlled by shear fractures. Moreover, for the specimens under long-term erosion (60°-60D and 60°-90D), the failure is more serious, with more cavities and secondary cracks than that of the specimens under short-term erosion (60°-0D and 60°-30D). Thus, preventing the long-term erosion of the bedding's slate in engineering practice is meaningful for preventing serious damage to the surrounding rock under creep conditions.

6. Conclusions

In this paper, based on the experimental results, the influences of the coupling effect of the bedding structure and chemical erosion on the physical properties, conventional triaxial characteristics, and creep characteristics of carbonaceous slate specimens are analyzed for the first time. The following conclusions can be obtained:

(1) Chemical solutions containing SO_4^{2-}, Cl^-, and H^+ can damage the carbonaceous slate. According to the SEM results, the internal cracks and loose structures induced by chemical erosion are increasing with the increase of erosion days. Moreover, acoustic wave velocity results prove that specimens with smaller bedding angles suffer more serious erosion and induce more internal cracks;

(2) Under conventional triaxial compression conditions, for specimens with the same erosion days, the conventional triaxial compressive strength (σ_s) decreases with the decrease of the bedding angle. On the other hand, for specimens with the same bedding angle, the σ_s decreases with the increase of erosion days. The failure modes of the specimens are mainly controlled by oblique shear fractures and accompanied by the occurrence of slip dislocation fractures between the bedding inclination;

(3) Under creep conditions, for specimens with the same erosion days, the final axial deformations increase with the increase of the bedding angles, whereas the final lateral

deformation increases obviously for the specimens with bedding angles of 60°. Moreover, the final axial and final lateral deformations increase with the increase of erosion days, which means the longer the erosion days, the greater the deformations. The failure modes of the specimens with different bedding angles and different erosion days are controlled by shear fractures, and for the specimens with 60° bedding angles and long-term erosion, there are block separations and many cavities along the shear planes.

These conclusive remarks can provide useful information for the safety and operations of deep-buried tunnels with the surrounding rock of the bedding's carbonaceous slate. On the one hand, smaller bedding angles and longer erosion days will result in weakening the physical and mechanical properties. On the other hand, more attention should be paid to prevent serious failure and large deformation of the surrounding rock if the surrounding rock has a bedding angle of 60° or suffers long-term erosion. The drawback of the experiments herein is that the rock specimens were obtained from an integrated block rather than a filed borehole, which means the specimens were without any explicit natural fractures or flaws. However, in reality, the rock mass contains many natural fractures. Therefore, to further understand the damage and creep characteristics of carbonaceous slate rock masses, the effect of natural fractures should be involved, which will be considered in our follow-up studies.

Author Contributions: Conceptualization, Q.C.; Methodology, W.Z.; Formal analysis, Y.X.; Investigation, W.Z. and Z.C.; Data curation, Y.X.; Writing—original draft, W.Z.; Writing—review & editing, Z.C.; Funding acquisition, Z.C. All authors have read and agreed to the published version of the manuscript.

Funding: The National Natural Science Foundation of China (52004328), and the Natural Science Foundation of Hunan Province (No. 2022JJ40149). Research foundation of Education Bureau of Hunan Provincial (No. 22B0518).

Institutional Review Board Statement: Not applicable.

Informed Consent Statement: Not applicable.

Data Availability Statement: Data sharing is not applicable to this article.

Conflicts of Interest: The authors declare no conflict of interest.

References

1. Zhang, C.S.; Liu, N.; Chu, W.J. Key technologies and risk management of deep tunnel construction at Jinping II hydropower station. *Rock Mech. Rock Eng.* **2016**, *8*, 499–512. [CrossRef]
2. Chen, Z.H.; Chen, Q.N.; Li, X.B.; Wu, Q.H.; Huang, X.C. Stress distribution and crack propagation of high stress hard rock under excavation unloading-perforation fracturing. *Chin. J. Nonferrous Met.* **2023**, *33*, 952–968. (In Chinese)
3. Wang, S.F.; Li, X.B.; Yao, J.R.; Gong, F.Q.; Li, X.; Du, K.; Tao, M.; Huang, L.Q.; Du, S.L. Experimental investigation of rock breakage by a conical pick and its application to non-explosive mechanized mining in deep hard rock. *Int. J. Rock Mech. Min. Sci.* **2019**, *122*, 104063. [CrossRef]
4. Huang, L.Q.; Wu, X.; Li, X.B.; Wang, S.F. Influence of sensor array on MS/AE source location accuracy in rock mass. *Trans. Nonferrous Met. Soc. China* **2023**, *33*, 254–274. [CrossRef]
5. Chen, Y.F.; Wei, K.; Liu, W.; Hu, S.H.; Hu, R.; Zhou, C.B. Experimental Characterization and Micromechanical Modelling of Anisotropic Slates. *Rock Mech. Rock Eng.* **2016**, *49*, 3541–3557. [CrossRef]
6. Wu, C.Z.; Chen, Q.S.; Basck, S.; Karekal, S. Laboratory investigation on rheological properties of greenschist considering anisotropy under multi-stage compressive creep condition. *J. Struct. Geol.* **2018**, *114*, 111–120. [CrossRef]
7. Zhou, Y.Y.; Feng, X.T.; Xu, D.P.; Fan, Q.X. An enhanced equivalent continuum model for layered rock mass incorporating bedding structure and stress dependence. *Int. J. Rock Mech. Min. Sci.* **2017**, *97*, 75–98. [CrossRef]
8. Xie, Y.P.; Chen, Q.N.; He, Y.C.; Chen, Z.H.; Tian, W.Q. Creep model construction and application of carbonaceous slate with muscovite and graphite under deep high geostress environment. *J. Cent. South Univ.* **2021**, *52*, 568–578. (in Chinese).
9. He, M.C.; LI, J.Y.; Ren, F.Q.; Liu, D.Q. Experimental investigation on rockburst ejection velocity of unidirectional double-face unloading of sandstone with different bedding angles. *Rock Mech. Rock Eng.* **2021**, *40*, 433–447. (in Chinese).
10. Zhang, X.M.; Ou, X.; Gong, F.; YANG, J.S. Effects of Bedding on the Dynamic Compressive Properties of Low Anisotropy Slate. *Rock Mech. Rock Eng.* **2019**, *52*, 981–990. [CrossRef]
11. Liu, X.H.; Dai, F.; Zhang, R.; LIU, J.F. Static and dynamic uniaxial compression tests on coal rock considering the bedding directivity. *Environ. Earth Sci.* **2015**, *73*, 5933–5949. [CrossRef]

12. Hao, X.J.; Wang, S.H.; Xu, Q.S.; Yang, D.Q.; Zhang, Q.; Jin, D.X.; Wei, Y.N. Influences of confining pressure and bedding angles on the deformation, fracture and mechanical characteristics of slate. *Constr. Build. Mater.* **2020**, *243*, 118–255. [CrossRef]
13. Wasantha, P.L.P.; Ranjith, P.G.; Shao, S.S. Energy monitoring and analysis during deformation of bedded-sandstone: Use of acoustic emission. *Ultrasonics* **2014**, *54*, 217–226. [CrossRef] [PubMed]
14. Liu, Y.S.; Fu, H.L.; Rao, J.Y.; Dong, H.; Cao, Q. Research on Brazilian disc splitting tests for anisotropy of slate under influence of different bedding orientations. *Rock Mech. Rock Eng.* **2012**, *31*, 785–791.
15. Xia, L.; Zeng, Y.W.; Luo, R.; Liu, W. Influence of Bedding Planes on the Mechanical Characteristics and Fracture Pattern of Transversely Isotropic Rocks in Direct Shear Tests. *Shock Vib.* **2018**, *2018*, 6479031. [CrossRef]
16. Lin, Y.; Zhou, K.P.; Gao, R.G.; Li, J.L.; Zhang, J. Influence of Chemical Corrosion on Pore Structure and Mechanical Properties of Sandstone. *Geofluids* **2019**, *6*, 1–15. [CrossRef]
17. Miao, S.J.; Wang, H.; Cai, M.F.; Song, Y.F.; Ma, J.T. Damage constitutive model and variables of cracked rock in a hydro-chemical environment. *Arab J. Geosci.* **2018**, *11*, 19. [CrossRef]
18. Dragan, G.; Albert, G. The influence of different fluids on the static fatigue of a porous rock: Poro-mechanical coupling versus chemical effects. *Mech. Mater.* **2014**, *71*, 34–51.
19. Zhang, F.; Zhao, J.J.; Hu, D.w.; Sheng, Q.; Shao, J.F. Creep Strain and Permeability Evolution in Cracked Granite Subjected to Triaxial Stress and Reactive Flow. *Geofluids* **2018**, *2*, 4653973. [CrossRef]
20. Hu, D.H.; Zhou, H.; Zhang, F.; Shao, J.J. Modeling of Short- and Long-Term Chemomechanical Coupling Behavior of Cement-Based Materials. *J. Eng. Mech.* **2014**, *140*, 206–218. [CrossRef]
21. Callahan, O.A.; Eichhubl, P.; Olson, J.E.; Davatzes, N.C. Experimental investigation of chemically aided fracture growth in silicified fault rocks. *Geothermics* **2020**, *83*, 101724. [CrossRef]
22. Feucht, L.J.; Logan, J.M. Effects of chemically active solutions on shearing behavior of a sandstone. *Tectonophysics* **1990**, *175*, 159–176. [CrossRef]
23. Li, H.; Zhong, Z.L.; Liu, X.R.; Sheng, Y.; Yang, D.M. Micro-damage evolution and macro-mechanical property degradation of limestone due to chemical effect. *Int. J. Rock Mech. Min. Sci.* **2018**, *110*, 257–265. [CrossRef]
24. Xie, Y.P.; Chen, Q.N.; Huang, X.C.; Luo, P. An experimental study of microstructure and uniaxial compression test of carbonaceous slate in a deep buried tunnel. *Hydrogeol. Eng. Geol.* **2020**, *47*, 96–102. (In Chinese)
25. Ulusay, R. *The ISRM Suggested Methods for Rock Characterization, Testing and Monitoring: 2007–2014*; Springer: London, UK, 2015.
26. Naguib, H.M.; Zakil, E.G.; Abdelsattar, D.E.; Dhmees, A.S.; Azab, M.A.; Elsaeed, S.M.; Kandil, U.F. Environmentally Friendly Polymer Concrete: Polymer Treatment, Processing, and Investigating Carbon Footprint with Climate Change. *ACS Omega* **2023**, *8*, 8804–8814. [CrossRef] [PubMed]
27. Hou, G.H.; Yan, Z.W.; Sun, J.F.; Naguib, H.M.; Lu, B.; Zhang, Z.H. Microstructure and mechanical properties of CO_2-cured steel slag brick in pilot-scale. *Constr. Build. Mater.* **2021**, *271*, 121581. [CrossRef]
28. Varma, M.; Maji, V.B.; Boominathan, A. Influence of rock joints on longitudinal wave velocity using experimental and numerical techniques. *Int. J. Rock Mech. Min. Sci.* **2021**, *141*, 104699. [CrossRef]
29. Mohd-Nordin, M.M.; Song, K.I.; Cho, G.C.; Mohamed, Z. Long-Wavelength Elastic Wave Propagation Across Naturally Fractured Rock Masses. *Rock Mech. Rock Eng.* **2014**, *47*, 561–573. [CrossRef]
30. Hubschen, G. *Ultrasonic Techniques for Materials Characterization*; Woodhead Publishing: Cambridge, UK, 2016; pp. 177–224.
31. Wang, P.; Xu, J.Y.; Fang, X.Y.; Wang, P.X.; Zheng, G.H.; Wen, M. Ultrasonic time-frequency method to evaluate the deterioration properties of rock suffered from freeze-thaw weathering. *Cold Reg. Sci. Technol.* **2017**, *143*, 13–22. [CrossRef]

materials

MDPI

Article

Scaled Model Tests Investigating Deformation Characteristics of Geosynthetic Reinforced Soil (GRS) Abutments under Vertical Loads

Chao Xu [1], Qingming Wang [1], Panpan Shen [2,*], Geye Li [1], Qiushen Wang [1], Xiao Zhang [1] and Chongxi Zhao [1]

1 Department of Geotechnical Engineering, Tongji University, Shanghai 200092, China
2 Shanghai Investigation, Design & Research Institute Co., Ltd., Shanghai 200434, China
* Correspondence: hermit_shpp@hotmail.com

Abstract: This study conducted plane-strain scaled model tests to investigate the deformation characteristics of geosynthetic reinforced soil (GRS) abutments subjected to vertical loads. Setback distance, i.e., the distance between the back of the abutment facing and the front of the loading plate, was chosen as the investigated influencing factor since it is one of the most frequently used variables by engineers for the design of GRS abutments. This study analyzed the settlements at the top of the abutment, the lateral displacements of the abutment facing, and the volumetric deformations of the abutment under the applied vertical loads. Test results showed that increasing the setback distance could effectively reduce the deformations of the GRS abutment. There existed an optimum setback distance and further increasing the setback distance beyond this optimum value did not have a significant effect on reducing the abutment deformations. The vertical, lateral, and total volumetric deformations of the GRS abutment showed an approximately linear increase with the increase of the applied vertical loads. The lateral volumetric deformations of the GRS abutment were larger than its vertical volumetric deformations and therefore the total volumetric strains of the GRS abutment were not zero based on the test results. However, the theory of zero volume change may still be suitable for the deformation calculation of the GRS abutment since the values of the volumetric strains were minimal. The measured maximum lateral facing displacements were compared with the calculated values using the US Federal Highway Administration (FHWA) method, which assumes zero volume change of the GRS abutment under vertical loads. Comparison results indicated that the FHWA method overestimated the maximum lateral facing displacements of the GRS abutment under vertical loads. An improved method was proposed in this study to calculate the maximum lateral facing displacements under vertical loads based on the theory of zero volume change and a revised distribution of the settlements at the top of the GRS abutment. Results showed that the improved method could better predict the maximum lateral facing displacements as compared to the FHWA method.

Keywords: abutment; deformation; geosynthetics; geosynthetic reinforced soil; volumetric deformation

Citation: Xu, C.; Wang, Q.; Shen, P.; Li, G.; Wang, Q.; Zhang, X.; Zhao, C. Scaled Model Tests Investigating Deformation Characteristics of Geosynthetic Reinforced Soil (GRS) Abutments under Vertical Loads. *Materials* **2023**, *16*, 4601. https://doi.org/10.3390/ma16134601

Academic Editor: Oldrich Sucharda

Received: 17 May 2023
Revised: 10 June 2023
Accepted: 19 June 2023
Published: 26 June 2023

1. Introduction

For the last few decades, geosynthetic reinforcement has shown great economic benefits in geotechnical engineering structures and has been widely used in reinforced retaining walls, slopes, and embankments. In recent years, geosynthetic reinforced soil (GRS) with closely spaced geosynthetic layers (i.e., no larger than 0.3 m) has been proven to have the performance of a composite material through the interaction between reinforcement and soil mass, hence the recent emerging use of the GRS as load-carrying structures such as bridge abutments. Different from the traditional reinforced concrete pile-supported abutments illustrated by Fu et al. [1] and Ma et al. [2], GRS abutments have the advantages of reduced construction cost and time [3,4]. In recent years, the US Federal Highway Administration (FHWA) developed a specific bridge system called the Geosynthetic Reinforced

Soil–Integrated Bridge System (GRS-IBS). In 2005, Bowman Bridge, constructed in Ohio, USA, was the first application of the GRS-IBS [3]. Since then, more and more GRS-IBSs have been constructed across the USA. Researchers have conducted field monitoring to investigate the long-term performances of the GRS-IBSs under working stress conditions. Results showed that the GRS-IBSs have good service performances and significant advantages in eliminating bumps at the end of the bridge slabs [5–12].

Since the GRS abutment serves as the most important part of the GRS-IBS to directly support the bridge load, its bearing capacity and service performance (e.g., deformation) are extremely important to this specific technology. As an effective and convenient method, numerical analysis has been widely used to investigate the performances of the GRS abutment. Many studies have been published in literature by using finite element or finite difference methods to investigate different influencing factors on the performances of the GRS abutment [13–19]. Results show that the GRS abutment with close spacing has an excellent service performance and a high load-carrying capacity to meet the design requirements. However, the input parameters for constitutive models in the numerical analysis are based on a large number of laboratory soil tests, which may cause the uncertainty of the numerical results [20]. Raja et al. [21] proposed a novel hybrid artificial intelligence (AI)-based model to predict the load-settlement behavior of the GRS abutment. However, this model was still limited by its training range [21].

Scaled model test is another effective method to simulate engineering prototypes and has been carried out to investigate the performances of the GRS abutment. Mini-pier tests, which treat the reinforced soil mass as a composite material, were constructed to systematically investigate the influences of reinforcement spacing, tensile strength, and backfill properties on the service performances of the reinforced soil mass under vertical loads [22–25]. Wu et al. [26] evaluated the failure loads of two full-scale segmental-facing GRS abutments and pointed out that the design values were much smaller than the measured loads. In addition, several model tests published in literature [27–29] showed that different facing conditions have an influence on the performances of the GRS abutment. However, there are few studies considering the setback distance, i.e., the distance between the back of the abutment facing and the front of the loading plate, as an influencing factor on the performances of the GRS abutment. In the FHWA design guideline [30], the suggested setback distance value for design is based on empirical data and test results. Therefore, it is important for model tests to incorporate the setback distance as an influencing factor.

This study conducted three plane-strain scaled model tests to investigate the effects of the setback distance (a_b), which is the distance between the back of the abutment facing and the front of the loading plate, on the deformation characteristics of the GRS abutment subjected to vertical loads. Since the service performance of the GRS abutment greatly depends on its deformation, this study chose the deformation characteristics rather than the ultimate bearing capacity of the GRS abutment as the investigated subject. Setback distance was chosen as the investigated influencing factor since it is one of the most frequently used variables by engineers for the design of GRS abutments. The analyzed deformation of the GRS abutment included the settlements at the top of the abutment, the lateral displacements of the abutment facing, and the volumetric deformations of the abutment. The FHWA method was selected for the comparison of the calculated maximum lateral facing displacements with the measured data. An improved method was proposed in this study to calculate the maximum lateral facing displacements under vertical loads.

2. Model Test Geometry, Material, and Plan

2.1. Model Geometry

The Bowman Bridge constructed in Ohio, USA [3] with a height of 4.7 m was chosen as the prototype of the model tests conducted in this study. The model abutment was constructed in a test pit with dimensions of 4.8 m (long) × 3.4 m (wide) × 2.0 m (high). Considering the geometry of the test pit and the capacity of the loading device, a length scaling factor of 3 was adopted in this study. It should be noted that using even smaller scaling factors (e.g., 1.5 or 2) in 1-g model tests could better reflect the stress and strain levels of the prototype. However, the limited space of the test pit made it impossible, hence the choice of 3 as the scaling factor. Figure 1 shows the dimensions of the model GRS abutments. The whole model constructed in this study had a total length (L) of 1.95 m, a total height (H) of 1.91 m, and a total width (w) of 1.50 m in the out-of-plane direction. The height of the GRS abutment itself (h) was 1.46 m, the width of loading plate (b) was 0.40 m, and the cut slope of the retained soil was 1:1. The loading plate was placed on top of the GRS abutment at a setback distance (a_b) from the back of the abutment facing. The lengths of all primary reinforcement layers used in the GRS abutment were such that they reached the cut slope. The reinforcement vertical spacing (S_v) between primary reinforcement layers was 0.14 m. Two layers of bearing bed reinforcement underneath the loading plate were used in the model tests. According to the design guideline provided by the FHWA [30], the length of the bearing bed reinforcement layers should be no smaller than the sum of the width of the loading plate and twice the setback distance ($b + 2a_b$). Therefore, the length of the bearing bed reinforcement was 1.27 m in this study.

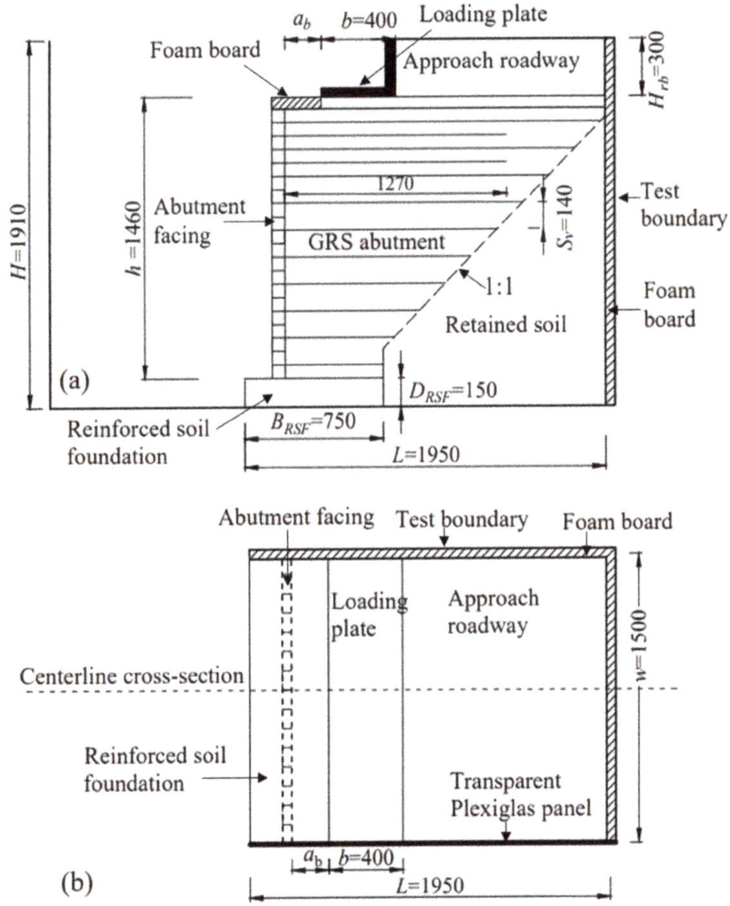

Figure 1. Test configuration of the model GRS abutment: (**a**) Front view and (**b**) top view (unit: mm).

2.2. Test Material

Open-graded quartz sand was used as backfill in the model tests. Figure 2 shows the gradation curves of both the prototype backfill used in the Bowman Bridge and the model backfill (i.e., open-graded quartz sand). The model backfill used in this study was scale-reduced from the prototype backfill so that the gradation curves of the prototype and the model backfill, as shown in Figure 2, satisfied the similitude relationship. According to the sieve analysis results shown in Figure 2, the particle size of the backfill ranged from 0.1 mm to 4.2 mm. The model backfill had a maximum dry density of 1860 kg/m^3. During construction, the controlled dry density of the model backfill was 1760 kg/m^3, which corresponded to a compaction degree of 95%. Triaxial test results showed that the model backfill had a friction angle of 48° and a cohesion of 0 kPa at the compaction degree of 95%.

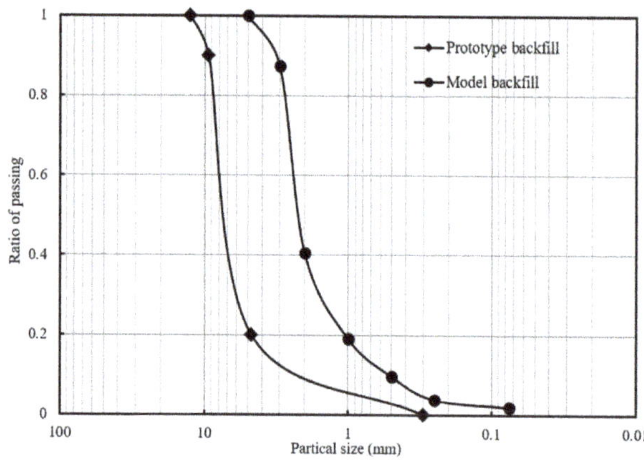

Figure 2. Sieve analysis results of both the prototype and the model backfill.

This study selected biaxial woven polypropylene geotextiles as the reinforcement material. Figure 3 shows the results of three wide-width tensile tests conducted on the geotextiles. The average value of the tensile strength at 10% reinforcement strain was 7.53 kN/m for the model geotextile, which corresponded to a tensile strength of 67.77 kN/m for the prototype geotextile at 10% reinforcement strain.

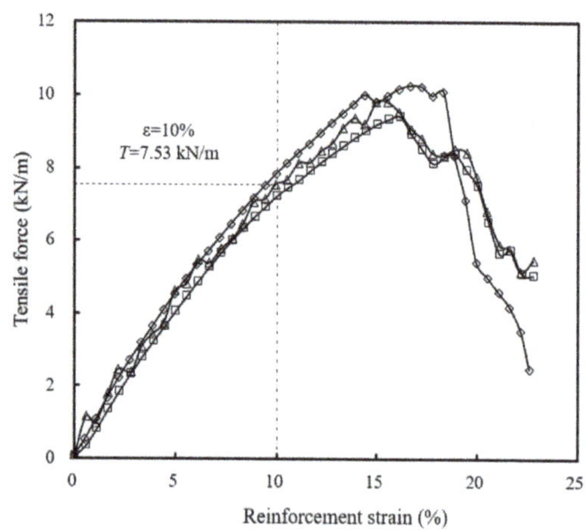

Figure 3. Wide-width tensile tests of model geotextiles used in the tests.

The abutment facing was simulated in the model tests using modular blocks with dimensions of 130 mm (long) × 70 mm (high) × 70 mm (wide). An L-shaped steel loading plate with the dimensions of 1.50 m (long) × 0.40 m (wide) × 0.04 m (thick) was used as the beam seat to simulate the interaction between the bridge girders and the GRS abutment. Concrete blocks were laid behind the loading plate to simulate the approach roadway.

2.3. Construction

Before construction, polytetrafluoroethylene membrane and lubricating oil were applied to the inner surface of the test pit and the Plexiglas panel, respectively, to reduce the negative effects of friction on the test results and to ensure the plane-strain condition. The reinforced soil foundation was first constructed using a mass-volume control method, followed by the construction of the GRS abutment in 20 lifts. In each lift, a layer of the modular facing blocks was placed, followed by a layer of the model backfill and the geotextile reinforcement. The top three geotextile layers were connected to the modular facing blocks using double-faced duct tape to simulate the mechanical connection used in the prototype abutment while the remaining geotextile layers were frictionally connected to the modular facing blocks. Finally, the steel loading plate and the concrete blocks simulating the approach roadway were put on top of the GRS abutment. Figure 4 shows the construction process of the model GRS abutment.

Figure 4. Construction of the model GRS abutment: (**a**) construction of the reinforced soil foundation; (**b**) application of the polytetrafluoroethylene membrane; (**c**) placement of the modular facing blocks; (**d**) backfill compaction; (**e**) placement of the loading device; (**f**) completion of the model GRS abutment.

2.4. Test Plan and Instrumentation

This study investigated the influence of the setback distance (a_b) on deformation characteristics of GRS abutments under vertical loads. Therefore, three tests with different setback distances were conducted in this study, as shown in Table 1.

Figure 5 shows the instrumentation layout when a_b was 0.2 m. The instrumentation layouts for the other two a_b were similar and omitted to save space. Six Linear Variable Differential Transformers (LVDTs) were installed at the top of the GRS abutment to monitor the settlements. Among these six LVDTs, only 2 (i.e., V2 and V3) were used to monitor the settlements of the loading plate. The loading plate is a rigid plate and steel ribs were used to reinforce the loading plate to ensure its rigidity. Therefore, it was assumed that no bending deformation of the loading plated occurred under vertical loading and the settlements of the loading plate were linearly distributed along its width. Using two LVDTs was considered sufficient to capture the linear distribution of the settlements. In addition, the limited area at the top surface of the loading plate resulted in difficulties in placing additional LVDTs since the loading device occupied most of the space. Lateral displacements of the abutment facing were not uniformly or linearly distributed along the abutment height due to the flexibility of the facing consisting of multiple modular blocks. Discretization of the lateral facing displacements was considered in this study by using seven draw-wire displacement sensors (DWDS) installed in front of the abutment facing to monitor the lateral displacements.

Table 1. Model test plan.

No.	Setback Distance a_b (m)	Reinforcement Spacing S_v (m)
T1	0.2	0.14
T2	0.3	0.14
T3	0.4	0.14

Note: a_b represented the setback distance, i.e., the distance between the back of the abutment facing and the front of the loading plate.

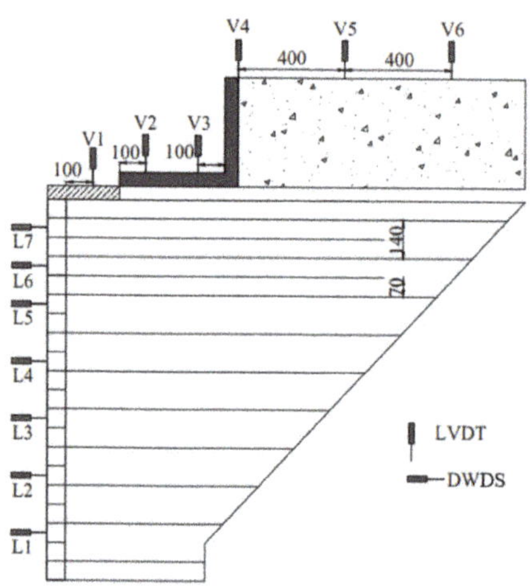

Figure 5. Instrumentation layout (Unit: mm).

It should be noted that all the sensors were installed at the centerline cross section along the out-of-plane direction of the GRS abutment, as shown in Figure 1b. The readings of all the sensors were zeroed out after the completion of construction prior to loading. In other words, the results of all these sensors represented the deformations of the GRS abutment induced by vertical loading only.

For each test, vertical loads were applied to the GRS abutment by multi-stage loading with increments of 10 kN. According to the method recommended by the FHWA design guideline [30], the ultimate bearing capacity of the model GRS abutment was 30 kN. Considering this calculated ultimate bearing capacity as well as the capacity of the loading device, this study adopted 90 kN as the maximum applied load, which was three times the calculated ultimate bearing capacity and should fully cover the range of the working stress condition of the GRS abutment. In other words, this 90 kN maximum applied load fully satisfied the purpose of this study, which was to investigate the deformation characteristics of the GRS abutment subjected to vertical loads. The stage loading was terminated when one of the following conditions occurred: (1) The global failure of the GRS abutment occurred, (2) the settlement of the top of the abutment reached 5% of the abutment height, or (3) the maximum lateral displacement of the abutment facing reached 10% of the abutment height.

3. Test Results

The model GRS abutment did not show significant deformation when the applied loads reached the predetermined maximum load of 90 kN. In order to obtain greater deformation values and better investigate the deformation characteristics of the GRS abutment, stage loading was not terminated and the applied load continued increasing to 100 kN, which was the maximum capacity of the loading device. No obvious failure phenomenon was observed in any of the three model tests, indicating that all the three model GRS abutments were under working stress conditions.

3.1. Settlement at the Top of the Abutment

Figure 6 shows the distributions of the normalized settlements at the top of the abutment subjected to different applied loads. The normalized settlement was calculated as the ratio of the settlement s to the abutment height h.

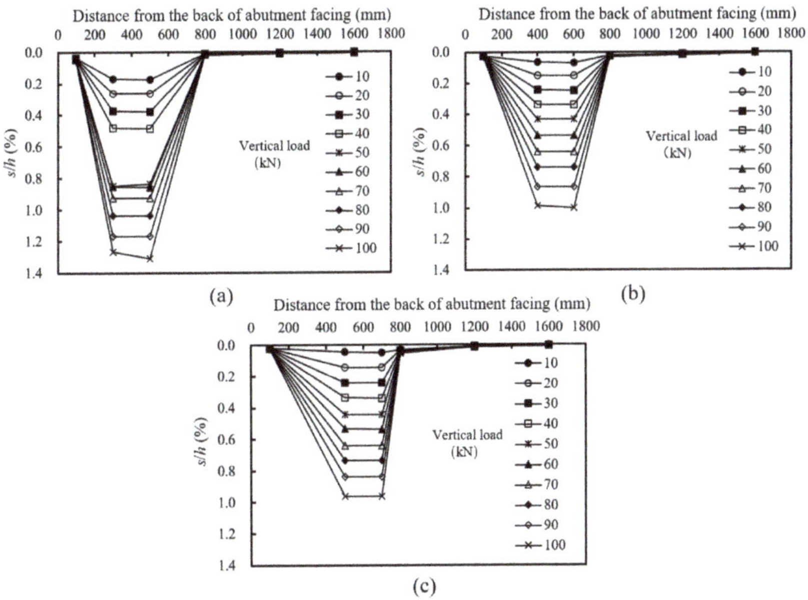

Figure 6. Distributions of normalized settlements at the top of the GRS abutment under loading: (**a**) T1: a_b = 0.20 m; (**b**) T2: a_b = 0.30 m; (**c**) T3: a_b = 0.40 m.

Figure 6 indicates that the change of the setback distance did not produce significant effects on the distributions of the normalized settlements at the top of the GRS abutment. The settlements under the loading plate were obviously larger than those in other areas and the settlements increased approximately linearly with the increase of the applied vertical load in all tests, indicating that the model abutments were still under working stress conditions. In T1, when the applied load increased from 40 kN to 50 kN, the settlement increased significantly. However, the increment of the settlement decreased in the subsequent loading stages. It was speculated that a malfunction of the loading device occurred at this loading stage, thus resulting in the significant increase of the normalized settlement. Figure 6 also shows that with the increase of the setback distance, the settlement under the loading plate decreased in the beginning and then remained approximately stable, indicating that increasing the setback distance could reduce the settlement at the top of the abutment, but there existed an optimum setback distance $(ab)_{opt}$ and further increasing the setback distance beyond this optimum value did not have a significant effect on reducing the abutment settlements.

3.2. Lateral Facing Displacement

Figure 7 shows the distributions of normalized lateral facing displacements along the abutment height under different applied loads. The normalized lateral facing displacement was calculated as the ratio of the lateral facing displacement δ to the abutment height h. As expected, with the increase of the applied loads, the lateral facing displacements increased. The maximum and minimum lateral facing displacements occurred near $1/3\ h$ from the top of the abutment and near the bottom of the abutment, respectively. Comparing T1, T2, and T3, it could be found that with the increase of the setback distance, the location of the maximum lateral facing displacement gradually moved from the top to the mid-height of the abutment.

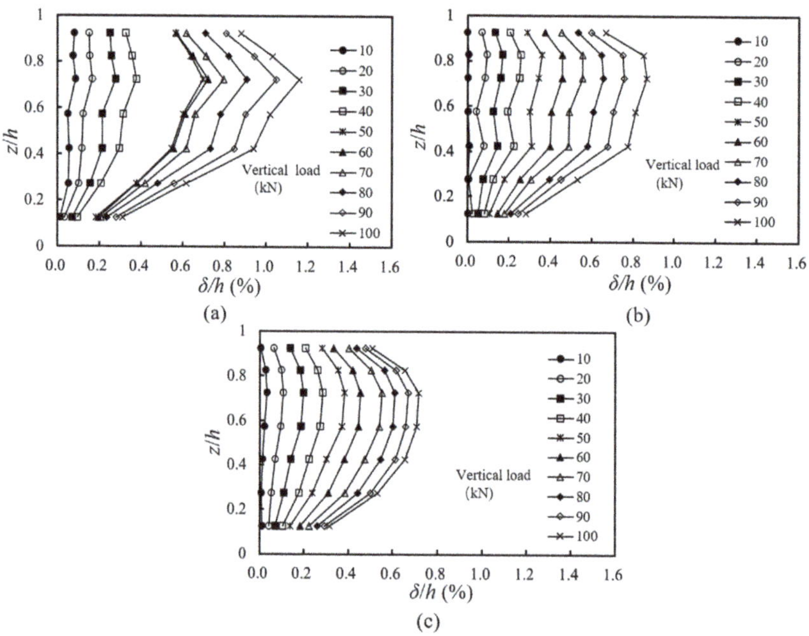

Figure 7. Distributions of normalized lateral facing displacements along the abutment height under loading: (**a**) T1: a_b = 0.20 m; (**b**) T2: a_b = 0.30 m; (**c**) T3: a_b = 0.40 m.

In addition, as compared with the other two cases, T1 with the smallest setback distance had a maximum lateral facing displacement of approximately 1.2%h under the applied load of 100 kN, which did not exceed the allowable lateral facing displacement of 2%h under working stress conditions according to the design guideline published by the FHWA [30]. In T1 with the smallest setback distance, when the vertical load was larger than 50 kN, the increments of the lateral facing displacements increased gradually with the increase of the applied load. In T2 when the setback distance increased to 0.30 m, the increments of the lateral facing displacements under each loading stage were approximately the same. In T3 when the setback distance further increased to 0.40 m, the increments of the lateral facing displacements decreased with the increase of the applied load.

Adams et al. [31] proposed a zero volume change assumption based on the mini-pier tests. This assumption was adopted by the FHWA design guidelines [30] to calculate the maximum lateral displacement of the GRS abutment. This assumption indicates that the volume lost at the top of abutment due to compression is equal to the volume gained at the facing due to lateral deformation under vertical loads. Few studies were published in literature analyzing the volumetric deformation of GRS abutments. Therefore, further research is needed to analyze the volumetric deformations of the GRS abutment under vertical loads.

3.3. Volumetric Deformation

Figure 8 shows the schematic of the deformed GRS abutment based on the measured lateral facing displacements as well as the top settlements. Under the applied vertical loads, the top of the GRS abutment settled and the abutment facing expanded laterally. The vertical volumetric deformation of the GRS abutment due to the applied load can be calculated as $\Delta V_v = -\Delta S_v \times w$, where w is the width of the GRS abutment in the out-of-plane direction (w = 1.5 m in this study) and ΔS_v is calculated by integrating the settlement at the top of the GRS abutment along its length direction. According to the settlement distributions of the GRS abutment as shown in Figure 6, the reading of LVDTs V1, V4, V5, and V6 were minimal under different applied loads. In other words, the settlement of the abutment occurred mainly underneath the L-shaped loading plate. Therefore, the calculation of the vertical volumetric deformation of the GRS abutment did not consider the settlements behind the loading plate (i.e., the area of the approach roadway) and the settlements close to the abutment facing (i.e., the area in front of the LVDT V1). The negative sign in front of ΔS_v indicated that the vertical compression was deemed as negative during the volumetric deformation calculation. The lateral volumetric deformation of the GRS abutment due to the applied load can be calculated as $\Delta V_l = \Delta S_l \times w$, where ΔS_l is calculated by integrating the lateral displacements of the abutment facing over the abutment height. During the calculation of ΔS_l, this study assumed that the lateral displacement at the top and the bottom of the abutment facing were the same as those measured by the DWDSs L7 and L1 (as shown in Figure 5), respectively.

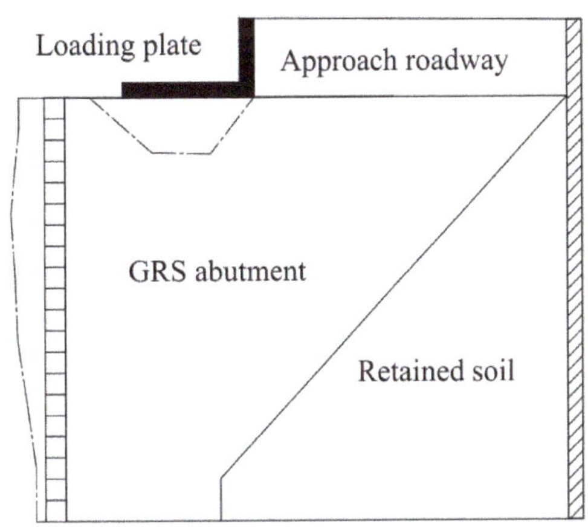

Figure 8. Schematic of the deformed GRS abutment based on the monitored results.

The vertical volumetric strain of the GRS abutment was defined as $\Delta V_v / V_0$, where V_0 was the original volume of the GRS abutment (V_0 = 2.40 m^3 in this study). Similarly, the lateral volumetric strain of the GRS abutment was defined as $\Delta V_l / V_0$. Therefore, the total volumetric strain of the GRS abutment was $\varepsilon_v = (\Delta V_v + \Delta V_l)/V_0$.

Figure 9 shows the effects of the setback distance on the volumetric strains of the GRS abutment under different vertical loads. The vertical, lateral, and total volumetric strains of the GRS abutment showed approximately linear increases with the increase of the applied vertical loads. Meanwhile, Figure 9 shows that the setback distance did have some influences on the volumetric deformation of the GRS abutment. The maximum volumetric strain of the GRS abutment was 0.52% under the 100 kN vertical load in T1. In T2 and T3, the volumetric strains of the GRS abutment were similar and both of them were smaller than that in T1. In other words, increasing the setback distance could reduce the vertical and lateral volumetric deformations of the GRS abutment. However, there existed an optimum setback distance $(a_b)_{opt}$ and further increasing the setback distance beyond this optimum value did not have a significant effect on reducing the volumetric deformations of the GRS abutment.

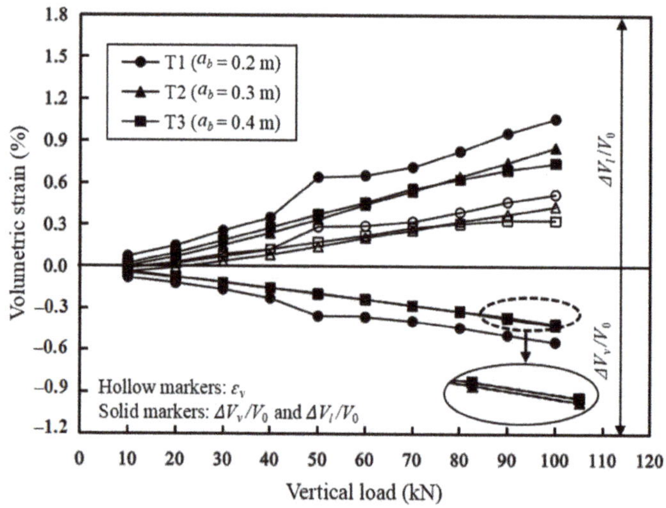

Figure 9. Effects of the setback distance on the volumetric deformations of the GRS abutment under vertical loads.

In addition, Figure 9 also shows that, despite the change of the setback distance, the lateral volumetric deformations of the GRS abutment were always larger than its vertical volumetric deformations and the differences between the lateral and vertical volumetric deformations increased with the increase of the vertical load. Therefore, the total volumetric strain of the GRS abutment was not zero based on the test results. It should be pointed out that the theory of zero volume change was based on the results of mini-pier tests. The axisymmetric boundary condition of the mini-pier tests was significantly different from the two-dimensional plane-strain boundary condition of the GRS abutment. The differences of the boundary conditions between mini-piers and abutments could be the reason causing the non-zero volumetric deformations of the GRS abutments in this study. However, Figure 9 also shows that the total volumetric strain of the GRS abutment was less than 0.5%. Therefore, the theory of zero volume change may still be suitable for the deformation calculation of GRS abutments.

3.4. Comparison with FHWA Method

The FHWA [30] proposed the following Equation (1) to predict the maximum lateral facing displacement (D_L) of the abutment based on the theory of zero volume change. This equation also assumed a triangular distribution of the lateral facing displacements along the abutment height and a uniform distribution of the settlements at the top of the abutment, as shown in Figure 10a:

$$D_L = \frac{2b_q D_v}{h} = \frac{2(a_b + b)D_v}{h} \quad (1)$$

where b_q is the width of the load area along the top of the abutment including the setback distance (i.e., $b_q = a_b + b$), D_v is the vertical settlement at the top of the GRS abutment, and h is the height of the GRS abutment. Figure 11a shows the measured lateral facing displacements from the tests and the calculated values using Equation (1) proposed by the FHWA. Due to the malfunction of the loading device which occurred in T1 when the applied load increased from 40 to 50 kN, the measured results of T1 under the applied load larger than 40 kN were not presented in Figure 11a. Figure 11a indicates that the FHWA method overestimated the lateral facing displacements under vertical loads under all three different setback distances, which was contrary to the results from filed monitoring by Saghebfar [6]. The model tests conducted in this study were under two-dimensional plane-strain condition, which was different from the three-dimensional GRS abutment conducted in the field [6]. Different boundary conditions between the model and the field GRS abutments may result in the different volumetric deformation results.

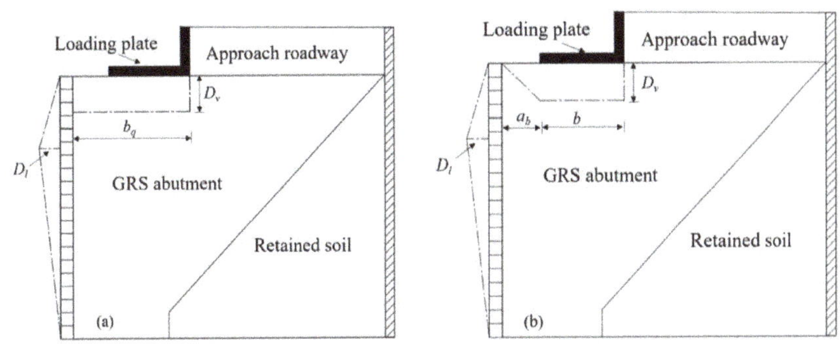

Figure 10. Assumed distributions of the volumetric deformation of the GRS abutment under vertical loading: (**a**) The FHWA method; (**b**) the improved method proposed in this study.

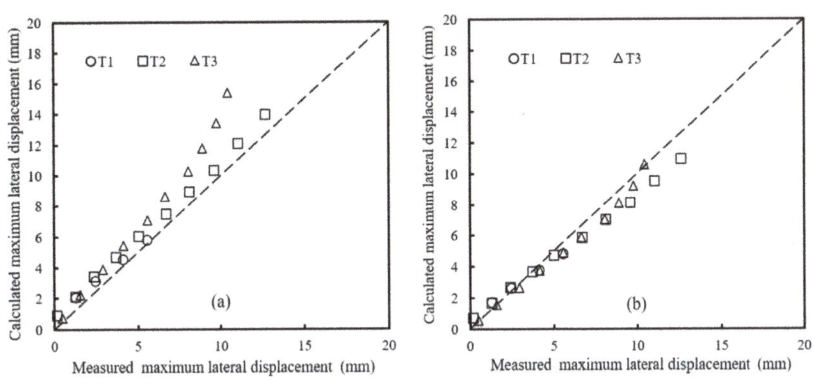

Figure 11. Comparison between the calculated and the measured maximum lateral facing displacements of the abutment facing under vertical loading: (**a**) calculated using the FHWA method; (**b**) calculated using the improved method proposed in this study.

In addition, Figure 11a also shows that, with the increase of the setback distance, the deviations of the calculated results from the measured ones became larger, indicating that the setback distance had a significant influence on the maximum lateral facing displacement. Test results in Figure 6 show that the settlements under the setback distance area were obviously different from those under the beam seat width, indicating that it is not reasonable for the FHWA method to assume a uniform distribution of the settlements at the top of the GRS abutment under vertical loads. In other words, it was speculated that FHWA's assumption of uniformly distributed settlements at the top of the GRS abutment could be the reason for overestimating the maximum lateral facing displacements under vertical loads.

A revision was made based on the test results in this study to the uniform distribution of the settlements at the top of the GRS abutment adopted by the FHWA. As shown in Figure 10b, the distributions of the settlements under the setback distance area and the beam seat width were assumed to be triangular and uniform, respectively. The triangular distribution of the lateral facing displacements along the abutment height was kept the same as that in the FHWA method. Under the theory of zero volume change and the revised settlement distribution, an improved method was proposed in this study to calculate the maximum lateral facing displacement of the GRS abutment under vertical loads, shown as the following Equation (2):

$$D_L = \frac{(a_b + 2b)D_v}{h} \tag{2}$$

Figure 11 shows the comparison between the FHWA method (i.e., Equation (1)) and the improved method proposed in this study (i.e., Equation (2)). The comparison between Figure 11a,b shows that the improved method could better predict the maximum lateral facing displacements as compared to the FHWA method. However, other parameters such as the beam seat width and the abutment height could also have influences on the calculation of the maximum lateral facing displacements. Further research is needed to investigate the distributions of the volumetric deformations of the GRS abutment under vertical loads.

4. Limitations

It is important to acknowledge the limitations of this study, which could be helpful for the improvements of future research. Due to the fact that constructing and loading the model GRS abutments were time- and labor-consuming as well as technique-challenging, only three model tests were conducted in this study investigating three different setback distances. The conclusions of this study were limited by the number of the model tests. The boundary conditions of the model tests conducted in this study (i.e., two-dimensional plane-strain conditions) are different from those in the field (i.e., three-dimensional conditions). The two-dimensional boundary condition generally results in more conservative results (e.g., larger deformations under the same vertical loads) as compared to the three-dimensional boundary condition. Although test results of this study indicated that the volumetric deformation of the GRS abutment under vertical loads was not zero, the zero volume change assumption made by the FHWA may still be suitable for the design of GRS abutments since the value of the volumetric strains were less than 0.5%. Therefore, both the FHWA method and the improved method proposed in this study were based on the zero volume change assumption. The improved method revised the distribution of the settlements at the top of the GRS abutment. Further studies may be necessary to investigate the volumetric distributions of the GRS abutment under vertical loads in order to give better predictions of the deformations. In addition, the humidity of the backfill was not considered in this study. Despite the above limitations, the results of this study provided valuable insights into the influence of the setback distance on the deformation characteristics of the GRS abutments under working stress conditions and gave important references for engineering applications.

5. Conclusions

In this study, two-dimensional plane-strain scaled model tests were conducted to investigate the deformation characteristics of geosynthetic reinforced soil (GRS) abutments subjected to vertical loads. Setback distance, which is the distance between the back of the abutment facing and the front of the loading plate, was chosen as the investigated influencing factor since it is one of the most frequently used variables by engineers for the design of GRS abutments. This study analyzed the settlements at the top of the abutment, the lateral displacements of the abutment facing, and the volumetric deformations of the GRS abutment under the applied vertical loads. The US Federal Highway Administration (FHWA) method was selected for the comparison of the calculated maximum lateral facing displacements with the measured data. An improved method was proposed in this study to calculate the maximum lateral displacements of the abutment facing under vertical loads. The following conclusions can be made from this study:

(1) Increasing the setback distance could effectively reduce the settlements at the top of the abutment, the lateral facing displacements, and the volumetric deformations of the GRS abutment. However, there existed an optimum setback distance $(a_b)_{\text{opt}}$ and further increasing the setback distance beyond this optimum value did not have a significant effect on reducing the abutment deformations.
(2) With the increase of the applied vertical loads, the vertical, lateral, and total volumetric deformations of the GRS abutment increased linearly. The lateral volumetric deformations of the GRS abutment were larger than its vertical volumetric deformations and therefore the total volumetric strains of the GRS abutment were not zero based on the test results. However, the theory of zero volume change may still be suitable for the deformation calculation of GRS abutments since the values of the volumetric strains were minimal.
(3) A comparison between the measured maximum lateral facing displacements and the calculated values using the method proposed by the FHWA showed that the FHWA method overestimated the lateral facing displacements under vertical loads. It was speculated that the FHWA's assumption of uniformly distributed settlements at the top of the GRS abutment could be the reason for overestimating the maximum lateral displacements.
(4) An improved method was proposed in this study to calculate the maximum lateral facing displacements under vertical loads based on the theory of zero volume change and a revised distribution of the settlements at the top of the GRS abutment. Results showed that the improved method could better predict the maximum lateral facing displacements as compared to the FHWA method.
(5) Results of this study could provide valuable insights into the influence of the setback distance on the deformation characteristics of the GRS abutment under working stress conditions and give important references for engineering applications of GRS abutments. A reasonable setback distance has the advantages of both reducing the bridge span and controlling the abutment deformations.

Author Contributions: Methodology, Q.W. (Qingming Wang), P.S. and C.X.; validation, P.S. and C.X.; formal analysis, Q.W. (Qingming Wang), X.Z. and G.L.; data curation, Q.W. (Qiushen Wang), X.Z. and C.Z.; writing—original draft preparation, Q.W. (Qingming Wang); writing—review and editing, P.S. and G.L.; supervision, P.S.; project administration, P.S. and C.X.; funding acquisition, P.S. and C.X. All authors have read and agreed to the published version of the manuscript.

Funding: This study was financially funded by the National Natural Science Foundation of China (grant number 42002270 and 41772284) and the Key Research and Development Project of the Chinese Ministry of Science and Technology (grant number 2016YFE0105800). The authors would like to acknowledge these supports.

Institutional Review Board Statement: Not applicable.

Informed Consent Statement: Not applicable.

Data Availability Statement: Not applicable.

Conflicts of Interest: The authors declare no conflict of interest.

References

1. Fu, R.H.; Briseghella, B.; Xue, J.Q.; Aloisio, A.; Lin, Y.B.; Nuti, C. Experimental and finite element analyses of laterally loaded RC piles with pre-hole filled by various filling materials in IABs. *Eng. Struct.* **2022**, *272*, 114991. [CrossRef]
2. Ma, H.B.; Zhuo, W.D.; Lavorato, D.; Nuti, C.; Fiorentino, G.; Marano, G.C.; Briseghella, B. Probabilistic seismic response and uncertainty analysis of continuous bridges under near-fault ground motions. *Front. Struct. Civ. Eng.* **2019**, *13*, 1510–1519. [CrossRef]

3. Adams, M.T.; Schlatter, W.; Stabile, T. Geosynthetic Reinforced Soil Integrated Abutments at the Bowman Road Bridge in Defiance County, Ohio. In *Geotechnical Special Publication*; ASCE: Reston, VA, USA, 2007.
4. Han, J.; Jiang, Y.; Xu, C. Recent advances in geosynthetic-reinforced retaining walls for highway applications. *Front. Struct. Civ. Eng.* **2017**, *12*, 239–247. [CrossRef]
5. Warren, K.A.; Whelan, M.J.; Hite, J.; Adams, M. Three-Year Evaluation of Thermally Induced Strain and Corresponding Lateral End Pressures for a GRS-IBS in Ohio. In *Geo-Congress 2014 Technical Papers*; ASCE: Reston, VA, USA, 2014; pp. 4238–4251.
6. Saghebfar, M.; Abu-Farsakh, M.; Ardah, A.; Chen, Q.; Fernandez, B.A. Performance monitoring of Geosynthetic Reinforced Soil Integrated Bridge System (GRS-IBS) in Louisiana. *Geotext. Geomembr.* **2017**, *45*, 34–47. [CrossRef]
7. Abu-Farsakh, M.; Saghebfar, M.; Ardah, A.; Chen, Q. A Case Study on Evaluating the Performance of a Geosynthetic Reinforced Soil Integrated Bridge System (GRS-IBS). *Geotech. Front. Walls Slopes* **2017**, *278*, 12–22.
8. Hogan, R.J.; Pirando, R.; Anderson, J.B.; Montgomery, J. Construction and Monitoring of Alabama's First Geosynthetic Reinforced Soil-Integrated Bridge System. In *Geo-Congress 2019: Earth Retaining Structures and Geosynthetics*; ASCE: Reston, VA, USA, 2019; pp. 222–235.
9. Talebi, M.; Meehan, C.L.; Leshchinsky, D. Applied bearing pressure beneath a reinforced soil foundation used in a geosynthetic reinforced soil integrated bridge system. *Geotext. Geomembr.* **2017**, *45*, 580–591. [CrossRef]
10. Jelušic, P.; Žlender, B. Experimental study of a geosynthetic-reinforced soil bridge abutment. *Geosynth. Int.* **2021**, *28*, 479–490. [CrossRef]
11. Zhang, J.; Guo, W.; Ji, M.; Zhao, J.; Xu, C.; Zheng, Y. Field monitoring of vertical stress distribution in GRS-IBS with full-height rigid facings. *Geosynth. Int.* **2022**, *29*, 610–621. [CrossRef]
12. Gebremariam, F.; Tanyu, B.F.; Christopher, B.; Leshchinsky, D.; Han, J.; Zornberg, J.G. Evaluation of vertical stress distribution in field monitored GRS-IBS structure. *Geosynth. Int.* **2020**, *27*, 414–431. [CrossRef]
13. Zheng, Y.; Fox, P.J. Numerical investigation ofthe geosynthetic reinforced soil–integrated bridge system under static loading. *J. Geotech. Geoenviron. Eng.* **2017**, *143*, 04017008. [CrossRef]
14. Zheng, Y.; Fox, P.J.; McCartney, J.S. Numerical simulation of deformation and failure behavior of geosynthetic reinforced soil bridge abutments. *J. Geotech. Geoenviron. Eng.* **2018**, *144*, 04018037. [CrossRef]
15. Ardah, A.; Abu-Farsakh, M.; Voyiadjis, G. Numerical evaluation of the performance of a geosynthetic reinforced soil-integrated bridge system (GRS-IBS) under different loading conditions. *Geotext. Geomembr.* **2017**, *45*, 558–569. [CrossRef]
16. Ardah, A.; Abu-Farsakh, M.Y.; Voyiadjis, G.Z. Numerical evaluation of the effect of differential settlement on the performance of GRS-IBS. *Geosynth. Int.* **2018**, *25*, 427–441. [CrossRef]
17. Abu-Farsakh, M.; Ardah, A.; Voyiadjis, G. 3D finite element analysis of the Geosynthetic Reinforced Soil-Integrated Bridge System (GRS-IBS) under different loading conditions. *Transp. Geotech.* **2018**, *15*, 70–83. [CrossRef]
18. Gebremariam, F.; Tanyu, B.F.; Güler, E.; Urgessa, G.S.; Shen, P. Numerical investigation of reinforced soil structures with GRS-IBS design features. *Geosynth. Int.* **2021**, *28*, 95–112. [CrossRef]
19. Shen, P.; Han, J.; Zornberg, J.; Tanyu, B.F.; Christopher, B.R.; Leshchinsky, D. Responses of geosynthetic-reinforced soil (GRS) abutments under bridge slab loading: Numerical investigation. *Comput. Geotech.* **2020**, *123*, 103566. [CrossRef]
20. Xu, C.; Liang, C.; Shen, P. Experimental and theoretical studies on the ultimate bearing capacity of geogrid-reinforced sand. *Geotext. Geomembr.* **2019**, *47*, 417–428. [CrossRef]
21. Raja MN, A.; Jaffar ST, A.; Bardhan, A.; Shukla, S.K. Predicting and validating the load-settlement behavior of large-scale geosynthetic-reinforced soil abutments using hybrid intelligent modeling. *J. Rock Mech. Geotech. Eng.* **2023**, *15*, 773–788. [CrossRef]
22. Adams, M.T.; Ketchart, K.; Wu, J.T.H. Mini Pier Experiments: Geosynthetic Reinforcement Spacing and Strength as Related to Performance. Geosynthetics in Reinforcement and Hydraulic Applications. In *Geotechnical Special Publication*; ASCE: Reston, VA, USA, 2007.
23. Adams, M.T.; Ooi, P.S.K.; Nicks, J.E. Mini-Pier Testing to Estimate Performance of Full-Scale Geosynthetic Reinforcement Soil Bridge Abutments. *J. Geotech. Test.* **2014**, *37*, 884–894. [CrossRef]
24. Nicks, J.; Esmaili, D.; Adams, M. Deformations of geosynthetic reinforced soil under bridge service loads. *Geotext. Geomembr.* **2016**, *44*, 641–653. [CrossRef]
25. Wu, J.T.H.; Ketchart, K.; Adams, M.T. Two full-scale loading experiments of geosynthetic-reinforced soil (GRS) abutment wall. *Int. J. Geotech. Eng.* **2008**, *2*, 305–317. [CrossRef]
26. Ahmadi, H.; Bezuijen, A. Full-scale mechanically stabilized earth (MSE) walls under strip footing load. *Geotext. Geomembr.* **2018**, *46*, 297–311. [CrossRef]
27. Doger, R.; Hatami, K. Influence of facing on the performance of GRS bridge abutments. *Int. J. Geosynth. Grou. Eng.* **2020**, *6*, 42. [CrossRef]
28. Hatami, K.; Doger, R. Load-bearing performance of model GRS bridge abutments with different facing and reinforcement spacing configurations. *Geotext. Geomembr.* **2021**, *49*, 1139–1148. [CrossRef]
29. Zhang, J.; Jia, Y.; Guo, W.; Zhao, J.; Ling, J.; Zheng, Y. Experimental study on the load bearing behavior of geosynthetic reinforced soil bridge abutments with different facing conditions. *Geotext. Geomembr.* **2022**, *50*, 632–643. [CrossRef]

30. Adams, M.; Nicks, J. *Design and Construction Guidelines for Geosynthetic Reinforced Soil Abutments and Integrated Bridge Systems*; FHWA-Hrt-17-080; U.S. Department of Transportation Federal Highway Administration: Washington, DC, USA, 2018.
31. Adams, M.T.; Lillis, C.P.; Wu, J.T.H.; Ketchart, K. Vegas Mini Pier Experiment and Postulate of Zero Volume Change. In Proceedings of the Seventh International Conference on Geosynthetics, Nice, France, 22–27 September 2002.

materials

MDPI

Article

Experimental Study on the Shear Strength and Failure Mechanism of Cemented Soil–Concrete Interface

Jie Zhou [1,2,*], Chao Ban [1], Huade Zhou [1], Junjie Ren [1] and Zhong Liu [3]

1 Department of Geotechnical Engineering, College of Civil Engineering, Tongji University, 1239 Siping Road, Shanghai 200092, China; 2230201@tongji.edu.cn (C.B.); 2210401@tongji.edu.cn (H.Z.); 1930175@tongji.edu.cn (J.R.)

2 Key Laboratory of Geotechnical and Underground Engineering, Ministry of Education, Tongji University, 1239 Siping Road, Shanghai 200092, China

3 Zhejiang Kunde Innovate Geotechnical Engineering Co., Ltd., Ningbo 315000, China; liuz@kundeyt.com

* Correspondence: zhoujie1001@tongji.edu.cn; Tel.: +86-136-3665-1710; Fax: +86-021-65987079

Abstract: Cement is always used in underground construction to reinforce and improve soft clay, resulting in the formation of a cemented soil–concrete interface. It is of great importance to study interface shear strength and failure mechanisms. So, in order to figure out the failure mechanism and characteristics of a cemented soil–concrete interface, a series of large-scale shear tests of a cemented soil–concrete interface, and corresponding unconfined compressive tests and direct shear tests of cemented soil, were carried out specifically under different impact factors. A kind of bounding strength was observed during large-scale interface shearing. Resultantly, three stages of the shear failure process of the cemented soil–concrete interface are proposed, and bonding strength, peak (shear) strength and residual strength are pointed out, respectively, in interface shear stress–strain development. Based on the analysis results of the impact factors, the shear strength of the cemented soil–concrete interface increases with age, the cement mixing ratio and normal stress, and decreases with the water–cement ratio. Additionally, the interface shear strength grows much more rapidly after 14 d to 28 d compared to the early stage (1~7 d). Additionally, the shear strength of the cemented soil–concrete interface is positively related to unconfined compressive strength and shear strength. However, the trends of the bonding strength and unconfined compressive strength or shear strength are much closer than those of the peak and residual strength. This is considered to be related to the cementation of cement hydration products and probably the particle arrangement of the interface. Particularly, the cemented soil–concrete interface shear strength is always smaller than the cemented soil's own shear strength at any age.

Keywords: cemented soil–concrete interface; large-scale interface shear test; interface shear strength; unconfined compressive strength

Citation: Zhou, J.; Ban, C.; Zhou, H.; Ren, J.; Liu, Z. Experimental Study on the Shear Strength and Failure Mechanism of Cemented Soil–Concrete Interface. *Materials* **2023**, *16*, 4222. https://doi.org/10.3390/ma16124222

Academic Editor: René de Borst

Received: 10 April 2023
Revised: 27 May 2023
Accepted: 29 May 2023
Published: 7 June 2023

1. Introduction

Interfaces are important in all kinds of geotechnical and underground engineering structures [1–4]. Since the mechanical properties of two materials in contact with each other are different, and the interface is also the main carrier for the interaction between them, large shear stress and differential deformation are easily generated at the interface. In such cases, this will most likely lead to structural failure and instability. Therefore, the study of the shear characteristics of interfaces has always been an important research topic in geotechnical and underground engineering.

In underground projects, an interface between the soil and structures (soil–concrete interface) is always formed. Additionally, many studies have been conducted on soil–concrete interfaces. Gong [5] studied the effect of normal-stress history on the shear characteristics of a clay–concrete interface by using a large direct shear system, and found that the shear stress would be larger at the same shear displacement with increasing initial normal stress. Additionally, the larger the initial normal stress was, the larger the maximum shear stress at the interface would be. Liu [6] conducted an experimental study on the shear properties of a silty clay–concrete interface under freeze–thaw cycles, and found that the shear strength of the interface was positively related to normal stress and negatively related to the number of freeze–thaw cycles and the initial water content of the soil. Namdar [7] found that the differential settlement of soil depends on the soil–concrete foundation interaction through a numerical investigation on soil–concrete foundation interactions.

However, in coastal areas such as Shanghai, Tianjin and Ningbo, soft clay is widely distributed. Their engineering properties of low strength, high compressibility and low permeability substantially increase the difficulty of engineering construction. In order to effectively improve the properties of soft clay, cement is usually used in practical construction to reinforce and improve the soil, such as cement mixing pile [8,9], high-pressure rotary jet grouting pile, concrete-cored DCM pile [10], etc. Even in the process of the artificial ground freezing method (AGF), in order to reduce the impact of frost heave and thaw collapse, the soil will also be grouted to improve it. In these projects, there will be a cemented soil–concrete interface formed between the cemented soil and some underground structures, as shown in Figure 1.

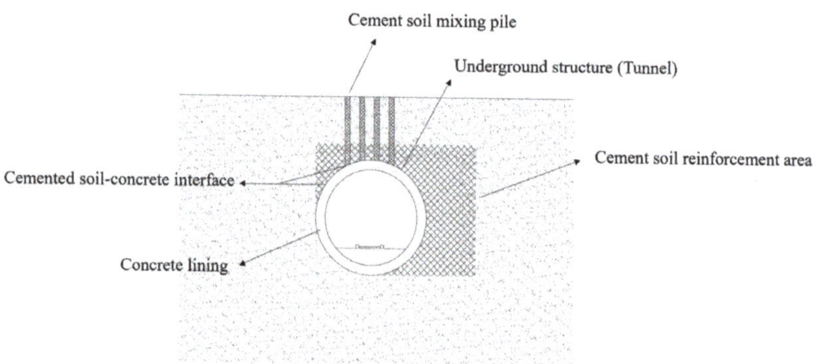

Figure 1. Diagram of the cemented soil–concrete interface of underground structures.

The cemented soil–concrete interface plays a significant role in the structure's overall structural stability. However, the cemented soil is often cast in situ. Its strength is generally influenced by its age, the cement mixing ratio, the water–cement ratio, etc., which will definitely further influence the performance of the shear strength of the cemented soil–concrete interface. Zhou [11] studied the behavior of pre-bored grouting planted piles under compression and tension, and found that the frictional capacity of the concrete–cemented soil interface was mainly controlled by the properties of the cemented soil. Correia [12] and Horpibulsuk [13], respectively, studied chemically stabilized soft soils and cement-admixed high-water-content clays. It was found that the strength of cemented soil was related to the binder content, the liquidity index and clay-water/cement ratio. Considering the existence of concrete–cemented soil interfaces in underground constructions, there is no doubt that the study of the characteristics of these interfaces cannot be neglected because of the bond forces generated by cement. Tanchaisawat [14] studied the characteristics of cemented soil–concrete interfaces through shear testing. The results showed that the interface shear strength increased linearly with the unconfined compressive strength of the cemented soil. Wu [15] conducted laboratory tests on the interface of cemented soil–concrete and found that the value of interface shear strength was about 0.194 times the unconfined compressive strength of the cemented soil specimen. Peng [16] concluded that the interfacial shear strength of cement-treated soil and concrete was about 0.188 times the unconfined compressive strength of cement-treated soil. Jamsawang [17] conducted a pullout test of a concrete core pile in the field of a stiff composite pile, and the results showed that the interfacial shear strength could be 0.4 times the unconfined compressive strength of the cement pile. Li [18] took the effect of normal stress into consideration, and established an empirical equation between the shear strength of a cemented soil–concrete interface and normal stress. Yu [19] conducted an experimental study on the frictional capacity of the concrete–cemented soil interface of a concrete-cored cemented soil column and found that the relationship between the ultimate lateral friction and the unconfined compressive strength of the cemented soil was approximately linear. Zhou [20] analyzed the interface shear characteristics of concrete pile body and cemented soil by examining the pile axial force and load–settlement curve in a model test of pre-bored pile. Additionally, it was concluded that the concrete-cemented soil interface strength was much greater than the cemented soil–soil interface strength. The authors of [21] then conducted an experimental study on the strength characteristics of a concrete–cemented soil interface. The experimental results showed that the interface strength had a positive correlation with the strength of cemented soil.

However, the above studies mainly concentrate on analyzing the relationship between the shear strength of the cemented soil–concrete interface and the unconfined compressive strength of the cemented soil. The failure mechanism and shear characteristics of cemented soil–concrete interfaces are not clarified. At the same time, in some codes and specifications, such as the "Technical specification for strength composite piles" (DGJ32/TJ 151-2013) [22], the bearing capacity of the pile is determined by the frictional force at the cemented soil–concrete and cemented soil–soil interface. However, the friction at the interface is always determined by local experience. The influence of the real interface characteristics on the bearing capacity of the pile is not reflected in such a specification. Additionally, the mechanism of interaction between cemented soil and underground structures is not clear. Therefore, this paper discusses the effects of cemented soil strength and normal stress on the shear strength of a cemented soil–concrete interface through a series of large-scale shear tests. From the perspective of cemented soil strength, the influences of age, the cement mixing ratio and the water–cement ratio are also investigated. In order to establish a more practical method for estimating the shear strength of the cemented soil–concrete interface later, the shear strength of cement–soil is also considered, and a comprehensive analysis of the relationship between the shear strength of the cemented soil–concrete interface, the unconfined compressive strength and the shear strength of the cemented soil is carried out. The research results of this paper could provide references for future studies about the shear characteristics of cemented soil–concrete interfaces and a better understanding of the interface failure mechanism. Additionally, it will provide a basis for the design of underground construction involving cemented soil–concrete interfaces.

2. Materials and Methods

2.1. Materials

2.1.1. Soil In Situ

The lengths of cement soil mixing piles and stiffened deep cement mixing piles used in engineering practice normally range from 10 m to 20 m [23,24]. Additionally, underground structures such as subway tunnels are also distributed from 10 m to 30 m underground. According to the geologic investigation data of Ningbo [25], a large amount of mucky soft clay is distributed along the piles. Therefore, the soil used in this research was mucky soft clay of the second layer in Ningbo, from an approximate depth of 12 m. The grain gradation curves of the soil samples are shown in Figure 2 and their basic physical properties are shown in Table 1.

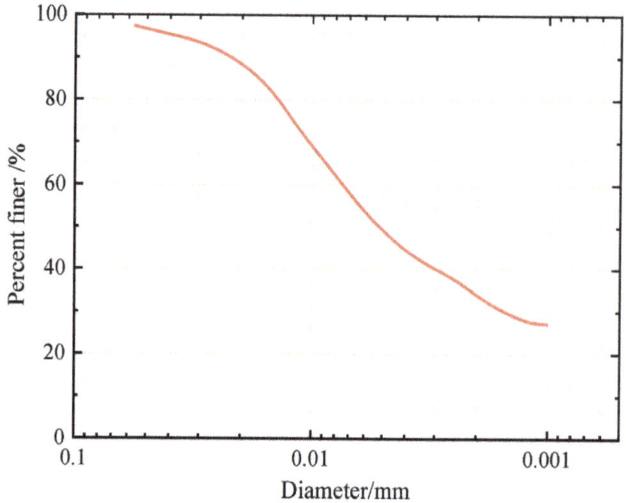

Figure 2. Gradation of soil sample.

Table 1. Basic properties of soil sample.

Water Content ω/%	Density ρ/g cm^{-3}	Void Ratio e	Liquidity Index I_L	Plasticity Index I_P	Liquid Limit ω_L/%	Plastic Limit ω_P/%	Shear Strength *	
							Cohesion c/kPa	Friction Angle φ/°
49.3	1.73	1.36	1.35	19.0	42.7	23.7	11.90	8.00

Shear strength *: from the direct shear test.

2.1.2. Cemented Soil

Considering that the large-scale shear tests needed a large quantity of the cemented soil sample, the soil sample used in this paper was remolded soil. According to the "Specification for mix proportion design of cement soil" (JGJ/T 233-2011) [26], the air-dried soil sample was crushed and sieved. Then, the water content of air-dried soil sample was measured as w_0 and the mass ratio of cement to soil was measured as a_w. Additionally, a_a represented the mass ratio of admixture to cement. The mass of air-dried soil, cement, water and admixture needed to configure cemented soil was calculated using Equations (1)–(4) while referring to the "Specification for mix proportion design of cement soil" (JGJ/T 233-2011) [26]. For Equation (1), according to the equal mass of soil particles in wet soil and air-dried soil, the mass of air-dried soil could be calculated. Additionally, the mass of cement and admixture could be obtained based on the definition of cement mixing ratio and admixture content in Equations (2) and (4). For Equation (3), the mass of water was calculated in two parts. The mass of water required in air-dried soil was calculated in the first part according to the difference in water content. At the same time, the required water in cement was calculated according to the definition of the water–cement ratio.

$$m_0 = \frac{1+0.01w_0}{1+0.01w} m_s \tag{1}$$

$$m_c = \frac{1+0.01w}{1+0.01w_0} 0.01a_w m_0 \tag{2}$$

$$m_w = \frac{0.01w - 0.01w_0}{1+0.01w_0} m_0 + 0.01\mu a_w m_s \tag{3}$$

$$m_a = 0.01 a_a m_c \tag{4}$$

where w_0 is the water content of air-dried soil, %; w is the natural water content, %; m_s is the mass of wet soil, kg; m_0 is the mass of air-dried soil, kg; a_w is the cement mixing ratio, %; m_c is the mass of cement, kg; m_w is the mass of water, kg; μ is the water–cement ratio; a_a is the admixture content, %; and m_a is the mass of admixture, kg.

Additionally, the cement type used was P42.5 ordinary Portland cement, which is commonly used in construction. Its properties are shown in Table 2.

Table 2. Properties of the cement.

Dry Density/g·cm^{-3}	Fineness/%	Initial Setting Time/min	Final Setting Time/min	Compressive Strength (28 d)/MPa	Flexural Strength (28 d)/MPa
3.1	1.1	130	210	43.5	7.8

2.1.3. Concrete Slab

C80 high-strength concrete is always used in underground engineering. According to "Specification for mix proportion design of ordinary concrete" (JGJ 55-2011) [27] and "Technical specification for high performance concrete" (GB/T 41054-2021) [28], concretes of different proportions were test-matched. Hence, the concrete samples were configured as presented in Table 3. In addition to the samples to be tested with sizes of 600 × 400 × 100 mm, three cubes with sizes of 70.7 × 70.7 × 70.7 mm were also prepared for unconfined compression tests. Through unconfined compression tests on concrete, the average compressive strength of cubes was found to be 85.7 MPa. Steel molds with sizes of 600 × 400 × 100 mm were used to cast the concrete samples.

Table 3. Concrete proportion design.

Water–Cement Ratio	Cement/kg	Ground Sand/kg	Sand/kg	Stone/kg	Water Reducing Agent/kg
0.3	301	129	697	1250	14

2.2. Experimental Program and Sample Preparation

2.2.1. Large-Scale Shear Tests of Cemented Soil–Concrete Interface

(1) Experimental apparatus and program

In order to decrease the size effect, large-scale shear tests were carried out to measure the shear strength of the cemented soil–concrete interface. According to the "Standard for test methods of engineering rock mass" (GB/T 50266-2013) [29], the test apparatus used was designed by the Department of Geotechnical Engineering College of Tongji University, and is shown in Figure 3. Firstly, the cemented soil–concrete sample was placed in a shear box with the concrete on the bottom and the cemented soil on the top. After the vertical load was applied, the strain control method was employed during the shear process at a rate of 1.0 mm/min. When the shear stress was almost stable, the shear test was completed. The performance parameters of this apparatus were a maximum normal load of 100 kN with precision of 0.1 N; the largest shear displacement of ±75 mm with precision of 0.1 mm; and a shear rate of 0.1~10 mm/min.

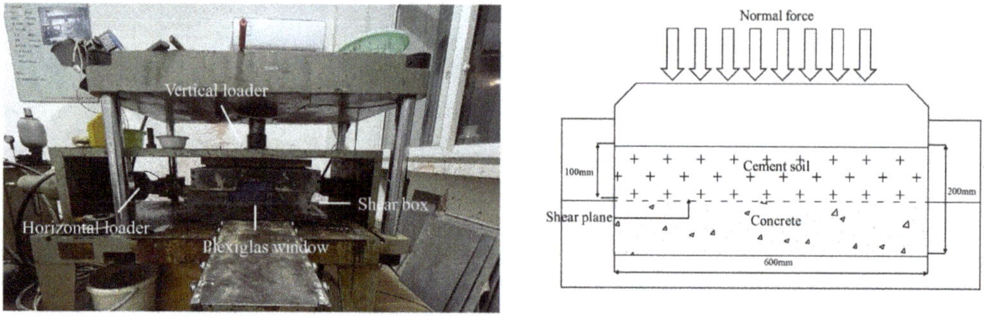

(**a**) (**b**)

Figure 3. Large-scale interface shear apparatus. (**a**) Large-scale interface shear apparatus; (**b**) Schematic diagram of loading mechanism.

Taking the soil in situ properties into account, normal stresses of 50 kPa, 150 kPa, and 250 kPa were considered in the experimental design. Considering the influencing factors of cemented soil strength, we set age, cement mixing ratio, water–cement ratio and normal stress as impact factors to discuss the interface shear characteristics. Since the increase in cement strength grows greatly in the early stage and becomes much more stable after 28 d, in this experiment, the age division was short initially and became longer afterwards. So, 1 d, 2 d, 3 d, 7 d, 14 d and 28 d were, respectively, set as the age factors. The cement mixing ratio and water–cement ratio were set based on both specifications, technical codes (shown in Table 4) and local engineering experience. In the Ningbo area, the common values of the cement mixing ratio and the water–cement ratio in engineering are 15% and 0.5, respectively. At the same time, in order to increase the fluidity of cemented soil, 0.01% polycarboxylate superplasticizer is also commonly added in engineering practices. Therefore, in this experiment, 15% was used as the reference value of the cement mixing ratio, and it fluctuated by one level smaller and larger (13%, 18%), respectively (based on the smallest 13% in the cement mixing pile in the specifications, as shown in Table 4), to analyze the influence of the cement mixing ratio on the cemented soil strength parameters. Additionally, the fluidity of the cemented soil was poor when the water–cement ratio was 0.5. If the water–cement ratio decreased, it would lead to a decrease in fluidity, making it difficult to ensure the uniformity of the cemented soil sample. Thus, combined with the recommended value range in Table 4, 0.5 was used as the reference value, and two levels (0.8, 1.0) were designed to analyze the influence of the water–cement ratio on cemented soil strength parameters. Considering that the preparation of large-scale cemented soil–concrete interface samples is time- and labor-consuming, and the water–cement ratios commonly used in engineering practice are between 0.5 and 0.8, in the large-scale interface shear tests, only two levels of 0.5 and 0.8 were set for the water–cement ratio factor. In order to investigate the effect of age, cement mixing ratio, water–cement ratio and normal stress on the shear strength of the cemented soil–concrete interface, a detailed experimental program of large-scale shear tests of the cemented soil–concrete interface was designed and is shown in Table 5, in which 28 d, 15%, 0.5 and 150 kPa are the basic reference parameter values.

Table 4. Recommendations for cement mixing ratio and water–cement ratio in specifications.

Specification	Cement Mixing Ratio/%	Water–Cement Ratio
Specification for mixed proportion design of cement soil (JCJ/T 233-2011) [26]	3~25	0.45~2.0
Technical code for composite foundation (GBT50783-2012) [23]	10~20	-
Technical specification for pile foundation of pipe pile embedded in cemented soil (JGJ/T 330-2014) [24]	≥20	0.8~1.5
Technical specification for strength composite piles (DGJ32/TJ 151-2013) [22]	15~25	0.8~1.2
Technical code for excavation engineering (DG/TJ 08-61-2010) [30]	Biaxial cement mixing pile: 13~15 Triaxial cement mixing pile: 20~22	0.5–0.6

Table 5. Experimental program of large-scale interface shear tests.

Age/d	Cement Mixing Ratio/%	Water–Cement Ratio	Normal Stress/kPa
1	15	0.5	150
2	15	0.5	150
3	15	0.5	150
7	15	0.5	150
14	15	0.5	150
28	15	0.5	150
28	13	0.5	150
28	18	0.5	150
28	15	0.8	150
28	15	0.5	50
28	15	0.5	250

(2) Preparation of cemented soil–concrete interface sample

The size of the shear box used in the large-scale shear test was 600 × 400 × 200 mm, and molds of same size were also prepared. Firstly, the concrete slab, which had reached curing age, was placed on the bottom of the mold, and we applied a thin layer of Vaseline around the mold. Then, the configured cemented soil was filled into the mold in layers and covered the concrete slab. Because the sample was relatively large, a vibrating rod was used to ensure the cemented soil and concrete were well compacted and avoid the formation of large bubbles inside. Finally, the surface of the cemented soil was scraped, and the sample was cured for 24 h before demolding. The entire set of cemented soil–concrete standard samples (Figure 4) were put in a curing room for a specified amount of time.

(a) (b)

Figure 4. Cemented soil–concrete interface sample. (**a**) Interface sample before curing, after preparation; (**b**) Interface sample before testing, after curing.

(3) Specimen quality control

It was necessary for the specimen quality to be strictly controlled throughout the experiments. Density control was conducted and best uniformity determined for sample preparation. Firstly, the soils were sampled in situ (Figure 5a) and totally dried in the air (Figure 5b); then, they were smashed and sieved into a uniform soil powder (Figure 5c). Most importantly, according to the water content and soil density of field soil (Table 1) and the specifically designed cement mixing ratio and water–cement ratio (Table 5), the total mass of air-dried soil, cement, water and admixture was calculated based on Equations (1)–(4) for the cemented soil–concrete interface sample volume. Additionally, we then mixed these materials together (Figure 5d). In this step, uniformity was very important, as well, for specimen quality control. A large high-speed motor stirrer was used for initial cement mixing. Additionally, then, the mass of each sublayer was determined and filled into a mold overlying a concrete slab (Figure 5e). Here, it was necessary for the density to be strictly controlled, i.e., a certain mass of cement should be weighed and totally filled into a predesigned sublayer volume. During this process, a shaking machine was also used to remove air bubbles in the cemented soil via overall shaking. In addition, a high-speed vibrating tube was used to help all the cement to be poured without any voids, vibrating it locally piece by piece, especially ambient walls. After all the layers were sub-filled, the specimen was scraped and subsequently maintained in a curing chamber for a specific amount of time, as shown in Table 5. Finally, small specimens were kept for strength comparison with the field samples during cement mixing pile construction.

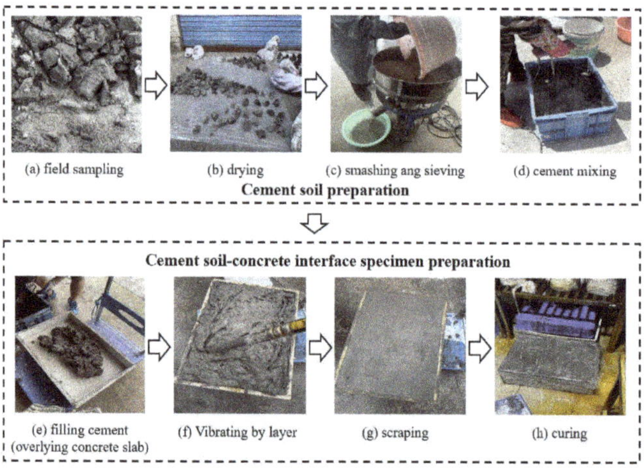

Figure 5. Large-scale cemented soil–concrete interface specimen preparation and quality control procedures.

2.2.2. Unconfined Compression Test and Direct Shear Test of Cemented Soil

Corresponding to the large-scale interface shear tests, unconfined compression tests and direct shear tests of the cemented soil were also designed and conducted under the same impact factors of age, cement content, water–cement ratio and normal stress, according to the "Standard for geotechnical testing method" (GB/T 50123-2019) [31]. The only difference was that the water–cement ratio was set at three levels of 0.5, 0.8 and 1.0 since the preparation was much easier, and the normal stress could be applied in stages. Therefore, the whole experimental program of the unconfined compression test and the direct shear test of cemented soil is shown in Table 6.

Table 6. Experimental program of unconfined compression test and direct shear test of cemented soil.

Experiments	Age/d	Cement Ratio/%	Water–Cement Ratio	Normal Stress/kPa
Unconfined compression test	1, 2, 3, 7, 14, 28	15	0.5	-
	28	13	0.5	-
	28	18	0.5	-
	28	15	0.8	-
	28	15	1.0	-
Direct shear test	1, 2, 3, 7, 14, 28	15	0.5	50, 150, 250
	28	13	0.5	50, 150, 250
	28	18	0.5	50, 150, 250
	28	15	0.8	50, 150, 250
	28	15	1.0	50, 150, 250

A rock mechanics testing machine (Figure 6a) was used for the unconfined compression tests of the cemented soil. The testing specimens were cubes with sizes of 70.7 × 70.7 × 70.7 mm. The direct shear apparatus (Figure 6b) was used to measure the shear strength of the cemented soil samples. Thirty groups of cemented soil cutting ring specimens with sizes of 61.8 × 20 mm (diameter × height) were prepared.

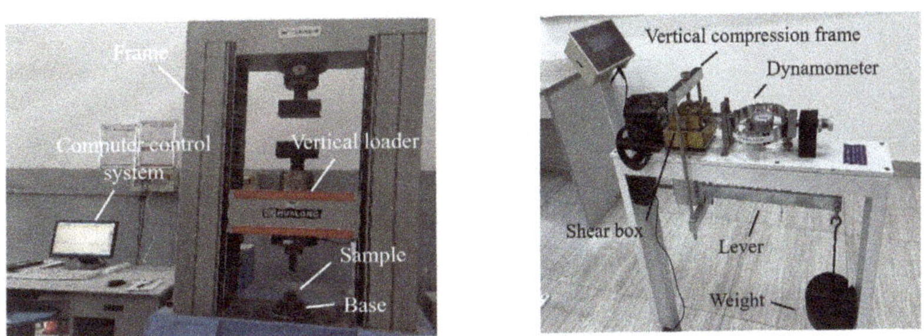

(a) (b)

Figure 6. Experimental apparatus for unconfined compression test and direct shear test of cemented soil. (**a**) Rock mechanics testing machine; (**b**) Direct shear apparatus.

Similarly, specimen quality control was very important for the experimental results. All the small cemented soil specimens for the unconfined compression test and the direct shear test were prepared during the preparation of the large-scale interface shear test specimens for consistency, as shown in Figure 7.

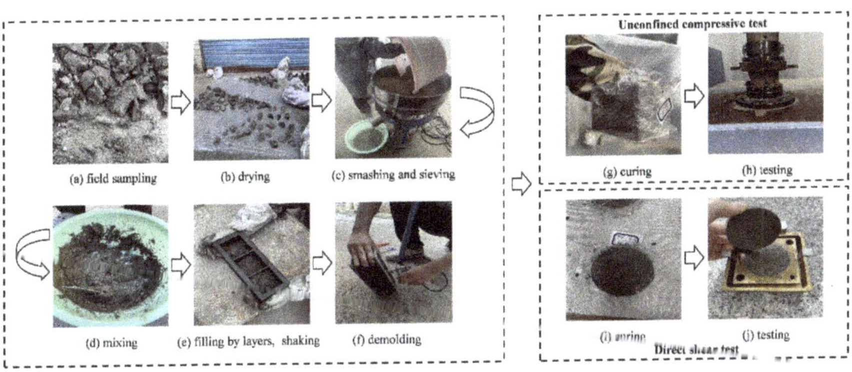

Figure 7. Specimen preparation procedures of unconfined compressive and direct shear tests of cemented soil.

3. Results and Discussion

3.1. Shear Strength of Cemented Soil–Concrete Interface

The shear stress–displacement development characteristics of the concrete-cemented soil samples aged 1 d and 28 d are shown in Figure 8. From the shear stress–displacement curve, they both show a typical kind of brittle failure mode. The maximum shear stress could be taken as the shear strength of the cemented soil–concrete interface, as shown in Figure 8, while the shear stress–displacement curves of the interface have a small peak before the maximum value of shear stress. This is presented at the end of the linear stage of the curve. All the shear stress–displacement data of the interface at different ages of 1~28 d are all presented in Figure 9. It can be found that all the curves present a small peak strength. Additionally, with a longer curing age, this small peak of interface shear strength is larger. The occurrence of this shear strength before the peak shear strength is due to the bonding effect of cemented soil and concrete during sample preparation, i.e., cemented soil was cast in place on a precast concrete mold, and not prepared separately. Actually, the preparation of casting-in-place was much more consistent with real engineering construction. Therefore, this small peak in shear strength is regarded as bonding shear strength (Figure 8). It is important to understand the shearing mechanism of a cemented soil–concrete interface when small displacement is encountered, which is also the main difference with previous studies on cemented soil–concrete interface shear strength, where only peak shear strength occurs. It is considered that the small peak of bonding strength presents a certain degree of unloading caused by the generation of small cracks on the interface. This indicates that it is due to the previous destruction of the bonding effect, not the interface itself. Aa shown in Figure 10, under different ages of cemented soil, i.e., different bonding effects, the failure occurrence at the interface after shearing is totally different, as observed in the cracks aged 1 d and 28 d.

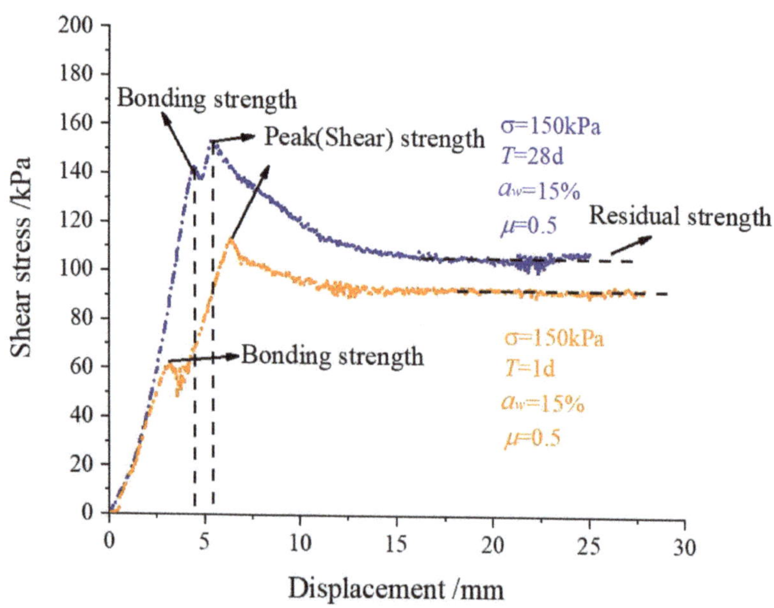

Figure 8. Shear stress–displacement curve of cemented soil–concrete interface with basic reference parameter values from large-scale shear tests.

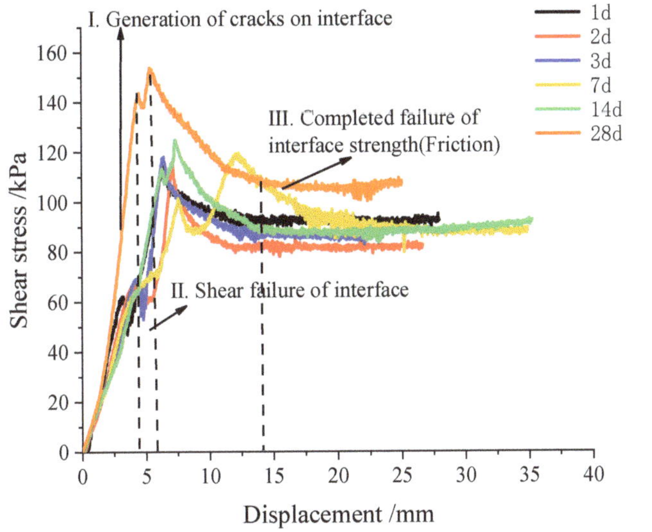

Figure 9. Shear stress–displacement curve of concrete–cemented soil interface from large-scale shear tests.

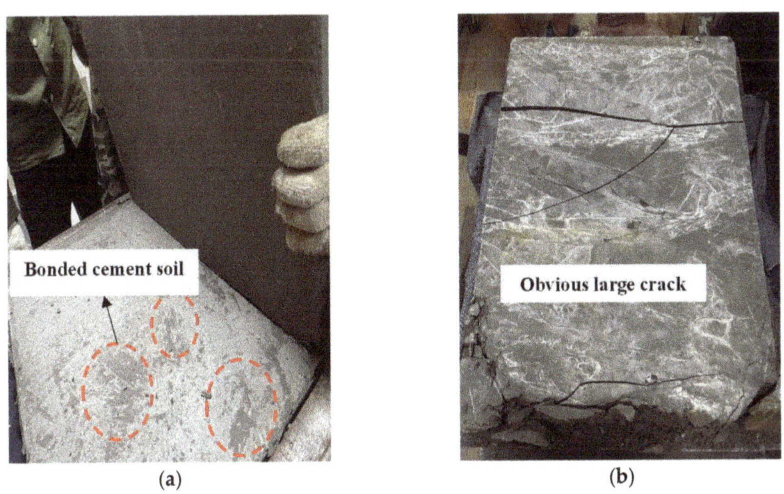

Figure 10. Different occurrences of interfaces after shear stress at difference ages. (**a**) Only broken bonding surface, no obvious internal cracks (age of 1 d). (**b**) Interface totally destroyed, obvious cracks (age of 28 d).

When the displacement continues to increase, the cemented soil and concrete are compacted again, and the shear stress continues to increase. When the shear stress reaches the maximum value, the interface is destroyed (Figure 10b). The shear stress becomes stable when the interface is completely destroyed and the rough surface is smoothed via shearing. This stable value can be defined as residual shear strength (Figure 8). Friction plays a main role in residual strength.

Generally, based on our observations and the above analysis of the shear stress–displacement characteristics, the failure process of the cemented soil–concrete interface can be divided into three parts: (1) the generation of small cracks at the interface; (2) the interface is destroyed and shear failure occurs; and (3) the interface is completely cracked, and the residual strength mainly consists of friction. Correspondingly, the bonding shear strength, peak shear strength and residual shear strength of the cemented soil–concrete interface are defined, and all the specific values of the cemented soil–concrete interface at different ages, the cement mixing ratio, the water–cement ratio and normal stress are presented in Table 7. It is found that the shear strength of the interface increases with the age, the cement mixing ratio and normal stress, and decreases with an increase in the water–cement ratio. At the same time, the interface shear strength grows rapidly during at 14 d to 28 d. Additionally, from Figure 11, it can be seen that the peak strength and residual strength have similar trends. They all increase slowly in the early stage before 14 d, but have rapid growth at 14 d to 28 d. On the contrary, the bonding strength increases rapidly from 1 d to 7 d. Additionally, the rate of growth becomes slower from 7 d to 28 d.

Table 7. Shear strength of concrete–cemented soil interface.

Age/d	Cement Mixing Ratio/%	Water Cement Ratio	Normal Stress/kPa	Bonding Strength/kPa	Shear (Peak) Strength/kPa	Residual Strength/kPa
1	15	0.5	150	62.158	114.216	92.5
2	15	0.5	150	66.667	115.908	82.0
3	15	0.5	150	69.382	117.828	85.5
7	15	0.5	150	101.522	119.449	88.2
14	15	0.5	150	113.359	124.793	88.0
28	15	0.5	150	143.665	153.786	105
28	13	0.5	150	75.193	140.292	90.5
28	18	0.5	150	81.703	163.978	118.8
28	15	0.8	150	110.090	135.825	101.0
28	15	0.5	50	23.450	62.296	39.6
28	15	0.5	250	203.693	238.815	198.5

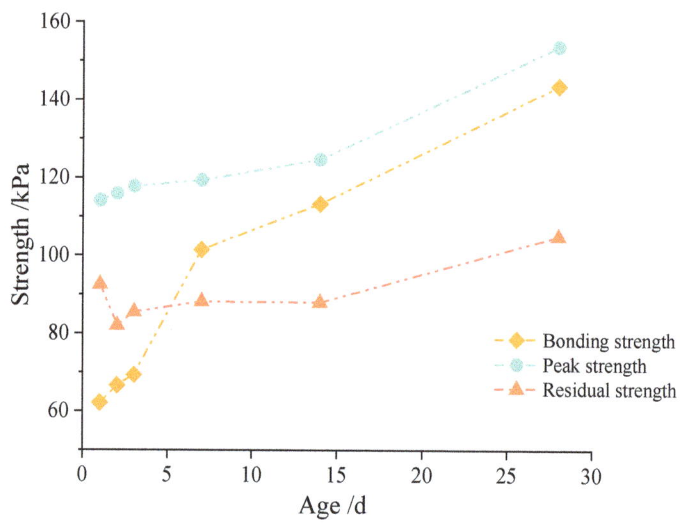

Figure 11. Bonding strength, peak strength and residual strength of the interface at different ages from large-scale shear tests.

3.2. Unconfined Compressive Strength and Shear Strength of Cemented Soil

The unconfined compressive strength and shear strength of the cemented soil samples are shown in Table 8.

Table 8. Unconfined compressive strength and shear strength of cemented soil.

Age/d	Cement Mixing Ratio/%	Water-Cement Ratio	Normal Stress/kPa	Unconfined Compressive Strength/kPa	Shear Strength/kPa	Cohesion *c*/kPa	Friction Angle *φ*/°
1	15	0.5	150	586.18	169.83	65.93	34.68
2	15	0.5	150	676.28	239.43	135.57	34.56
3	15	0.5	150	883.35	286.29	173.70	35.90
7	15	0.5	150	1076.31	367.92	257.19	38.16
14	15	0.5	150	1489.29	494.49	366.04	44.63
28	15	0.5	150	1675.95	569.75	408.03	45.59
28	13	0.5	150	1392.14	548.56	363.63	46.83
28	18	0.5	150	2043.90	594.37	450.47	43.67
28	15	0.8	150	1259.98	535.78	362.80	49.05
28	15	1.0	150	938.08	501.68	341.88	50.04
28	15	0.5	50	1675.95	454.76	408.03	45.59
28	15	0.5	250	1675.95	658.91	408.03	45.59

The results of Table 8 indicate that the unconfined compressive strength and shear strength are positively correlated, and both of them increase with age. They also increase with increases in the cement mixing ratio and decrease with increases in the water–cement ratio. Additionally, the shear strength increases with normal stress. From the perspective of age, the ratio of unconfined compressive strength to shear strength is around 3.0, as shown in Figure 12a. To some extent, age is not the key factor that influences the relationship between unconfined compressive strength and shear strength. For different types of cemented soil samples at the same age of 28 d, there is a positive linear correlation between unconfined compressive strength and shear strength, as shown in Figure 12b. For the shear strength index of cemented soil, it is obvious that cohesion is positively correlated with unconfined compressive strength and shear strength. The friction angle increases relative to age and the water–cement ratio, and decreases with the cement mixing ratio.

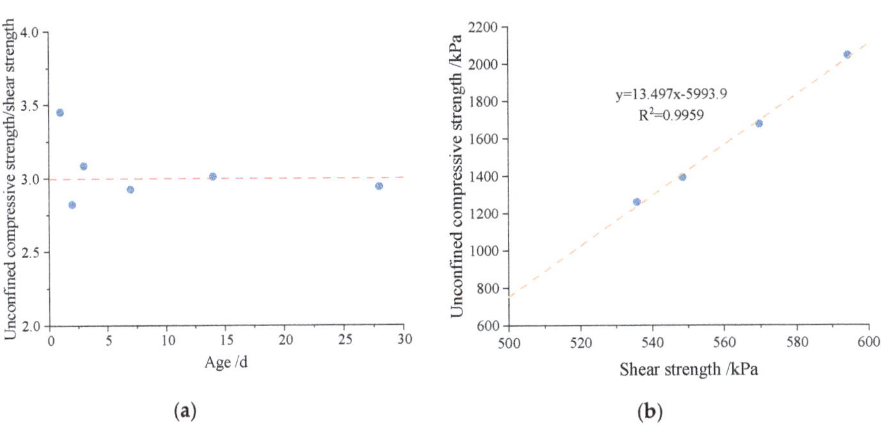

Figure 12. Relationship between unconfined compressive strength and shear strength of cemented soil. (**a**) Ratio of unconfined compressive strength to shear strength of cemented soil at different ages. (**b**) Relationship between unconfined compressive strength and shear strength of cemented soil at the same age of 28 d.

3.3. Failure Mechanism of Cemented Soil–Concrete Interface

To better understand the failure mechanism of the cemented soil–concrete interface, the relationship between the cemented soil–concrete interface shear strength and the unconfined compressive strength of the cemented soil, the cemented soil–concrete interface shear strength and the shear strength of the cemented soil was analyzed, under a same normal stress of 150 kPa, at different ages. The comparison indicates that the cemented soil–concrete interface shear strength at any stage of curing is always smaller than the cemented soil's own shear strength, as shown in Table 9. This is probably to say that the bonding effect of the cemented soil–concrete interface is weaker than the strength of the cemented soil particles. During the shear process, the interface of cemented soil–concrete must be the weak surface, instead of the cemented soil. So, when the cemented soil–concrete interface that forms in some underground constructions is subjected to shear action, the interface will be destroyed before the cemented soil or concrete. Additionally, the value of the unconfined compressive strength is much greater than that of the shear strength of the interface and cemented soil under the same condition. Additionally, the shear strength and unconfined compressive strength of cemented soil increase more quickly than the interface shear strength with curing duration.

Table 9. Strength of cemented soil–concrete interface, unconfined compressive strength and shear strength of cemented soil at different times (under a same normal stress of 150 kPa).

Age/d	Cemented Soil–Concrete Interface			Cemented Soil	
	Interface Shear (Peak) Strength/kPa	Bonding Strength/kPa	Residual Strength/kPa	Shear Strength/kPa	Unconfined Compressive Strength/kPa
1	114.216	62.158	92.5	169.83	586.18
2	115.908	66.667	82.0	239.43	676.28
3	117.828	69.382	85.5	286.29	883.35
7	119.449	101.522	88.2	367.92	1076.31
14	124.793	113.359	88.0	494.49	1489.29
28	153.786	143.665	105.0	569.75	1675.95

From Table 9, it can be figured out that the shear strength of the cemented soil–concrete interface is positively related to the unconfined compressive strength and shear strength of the cemented soil. With an increase in unconfined compressive strength and shear strength, the shear strength of the cemented soil–concrete interface will increase correspondingly. As shown in the above analysis of Figure 11, it can be easily seen that the trends of bonding strength and peak and residual strength are different, although all of them are positively related to unconfined compressive strength and shear strength. Upon looking further, the trends of bonding strength and unconfined compressive strength or shear strength are much closer than those of peak and residual strength, as shown in Figure 13. That is, it grows fast in the early age and slows down from 7 d to 28 d. In principle, the increase in the unconfined compressive strength and shear strength of cemented soil is mainly related to the cementation of cement hydration products. In the initial stage of curing, the hydration of cement can proceed quickly and react adequately due to sufficient materials, water, etc. The hydration products form cementation with other components and the strength of the cemented soil grows rapidly. With the continuous development of time, the water and reaction materials decrease. So, hydration slows down, and the strength growth of the cemented soil slows down as well. At the same time, the bonding strength of the interface also represents the degree of cementation. So, the relationship between the bonding strength and the unconfined compressive strength or shear strength of cemented soil is closer. On the other hand, after reaching the bonding strength, the cementation of the cemented soil gradually disappears. So, the peak and residual strength relate more closely to the properties of the interface itself rather than cemented soil material. At this point, the particle arrangement and friction characteristics at the interface may affect the peak (shear) strength and residual strength even more. This allows both to have the same rapid growth from 14 to 28 days. As a result, the curing influence on interface shear strength and unconfined compressive strength or shear strength are not close, although there is positive relationship between peak and residual strength and unconfined compressive strength or shear strength.

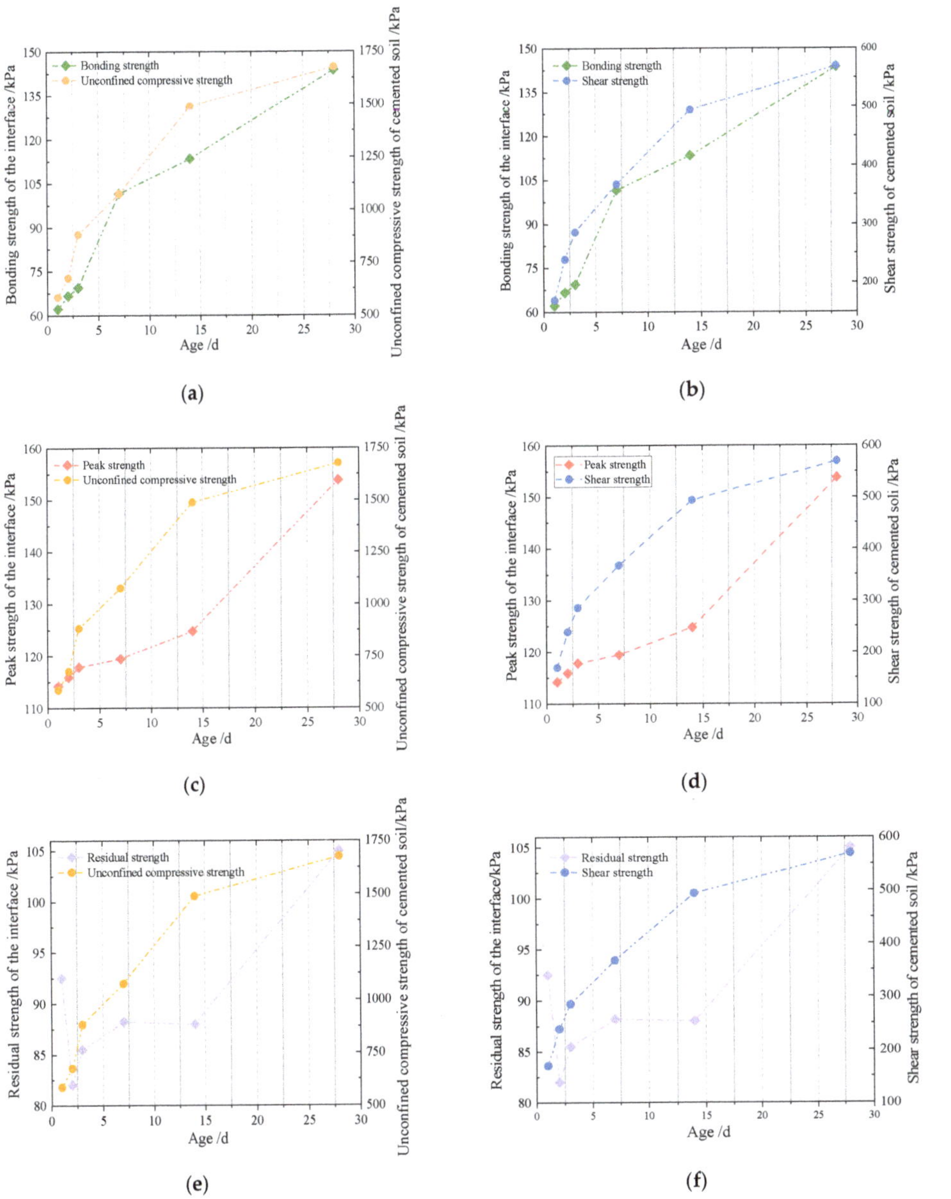

Figure 13. Relationship between bonding strength, peak strength and residual strength of the interface and unconfined compressive strength or shear strength of cemented soil. (**a**) Bonding strength of the interface and unconfined compressive strength of cemented soil. (**b**) Bonding strength of the interface and shear strength of cemented soil. (**c**) Peak strength of the interface and unconfined compressive strength of cemented soil. (**d**) Peak strength of the interface and shear strength of cemented soil. (**e**) Residual strength of the interface and unconfined compressive strength of cemented soil. (**f**) Residual strength of the interface and shear strength of cemented soil.

4. Conclusions

To figure out the failure mechanism and characteristics of cemented a soil–concrete interface, a series of large-scale contact surface shear tests and corresponding unconfined compressive tests and direct shear tests of cemented soil were carried out. The following important conclusions can be drawn:

1. Three stages of the failure process of the cemented soil–concrete interface can be observed: (1) the generation of cracks at the interface; (2) the interface is destroyed and shear failure occurs; (3) the interface completely fails, and the residual strength mainly consists of friction. The bonding strength, peak (shear) strength and residual shear strength of the cemented soil–concrete interface are correspondingly proposed and analyzed.
2. The shear strength of the interface increases with the age, cement mixing ratio and normal stress, and decreases with an increase in the water–cement ratio. Additionally, the interface shear strength grows rapidly at 14 d to 28 d. Hence, the cemented soil used in some practical projects needs to be cured for a sufficient amount of time, such that the strength of the cemented soil–concrete interface can be improved sufficiently.
3. The results of the unconfined compressive test and direct shear test on the cemented soil indicate that the unconfined compressive strength and shear strength are positively correlated. Additionally, both of them increase with age and the cement mixing ratio and decrease with an increase in the water–cement ratio. There is a positive correlation between interface shear strength and the unconfined compressive strength or shear strength of cemented soil. However, the trends of the interface bonding strength and unconfined compressive strength or shear strength of cemented soil are much closer than those of the interface peak strength and unconfined compressive strength or shear strength, and the interface residual strength and unconfined compressive strength or shear strength. This is considered to be related to the cementation of cement hydration products and probably the particle arrangement of the interface and its friction. Additionally, interface strength is always smaller than the unconfined compressive strength or shear strength of the cemented soil itself.

This research is of great significance to the study of the contribution of materials' strength to interface evolution at different stages. However, the effects of age, the cement mixing ratio, the water–cement ratio and normal stress on the cementation and shear properties of cemented soil–concrete interfaces have not been analyzed at the microscopic level. That is to say, the essence is not clear. So, further studies on the inherent mechanism for interface strength improvement, and tests of more influencing factors on the cemented soil–concrete interface, are expected.

Author Contributions: Conceptualization, J.Z.; methodology, C.B. and H.Z.; software, C.B., H.Z. and J.R.; validation, C.B., H.Z. and J.R.; formal analysis, C.B., H.Z. and J.R.; investigation, J.Z., C.B. and Z.L.; resources, J.Z., J.R. and Z.L.; data curation, C.B. and J.R.; writing—original draft preparation, C.B., J.R. and J.Z.; writing—review and editing, J.Z., C.B. and H.Z.; visualization, J.Z.; supervision, J.Z.; project administration, J.Z. and Z.L.; funding acquisition, J.Z. and Z.L. All authors have read and agreed to the published version of the manuscript.

Funding: The research work herein was supported by the National Natural Science Foundation of China (No. 41702299) and the Tongji consulting service program for Zhejiang Kunde Innovate geotechnical engineering Co., Ltd. (kh0023020210725).

Institutional Review Board Statement: Not applicable.

Informed Consent Statement: Not applicable.

Data Availability Statement: Not applicable.

Conflicts of Interest: The authors declare no conflict of interest.

References

1. Xia, H.; Zhou, G.; Du, Z. Experimental study of the soil-underground structure interfacial layer effect. *J. China Univ. Min. Technol.* **2011**, *40*, 846–851.
2. He, Y.X.; Wu, H.B. Ground Liner Interaction Analysis in Underground Opening. In Proceedings of the International Conference on Civil Engineering and Building Materials (CEBM), Kunming, China, 29–31 July 2011.
3. Veerla, V.K.; Srinivas, K.A. Study on soil interface interaction; proceedings of the International Conference on Advances in Materials Science (ICAMS). *Mater. Today Proc.* **2022**, *51*, 2561–2568. [CrossRef]
4. Isaev, O.N.; Sharafutdinov, R.F. Soil Shear Strength at the Structure Interface. *Soil. Mech. Found. Eng.* **2020**, *57*, 139–146. [CrossRef]
5. Gong, H.; Zhao, C.F.; Tao, G.X.; Zhao, C. Research on effect of stress history on shear behavior of interface between clay and concrete. *Chin. J. Rock. Mech. Eng.* **2011**, *30*, 1712–1719.
6. Liu, J.Q. Experimental Study on the Shear Performance of Silty Clay-CONCRETE interface under Freeze-Thaw Cycles. Master's Thesis, Jilin University, Jilin, China, 2022.
7. Namdar, A. A numerical investigation on soil-concrete foundation interaction. *Procedia Struct. Integr.* **2016**, *2*, 2803–2809. [CrossRef]
8. Bergado, D.T.; Long, P.V.; Jamsawang, P.; Lampun, C.N.; Balasubramaniam, A.S. On Prefabricated Vertical Drain (PVD) and Deep Cement Mixing (DCM)/Stiffened DCM (SDCM) Techniques for Soft Ground Improvement. *Geotech. Eng.* **2018**, *49*, 164–181. [CrossRef]
9. Zhou, S.Q.; Zhang, Y.F.; Zhou, D.W.; Wang, W.J.; Ke, Z.B.; Halder, S. Fly ash foundation reinforced by cement-soil mixing piles. *DYNA-Ing. E Ind.* **2020**, *95*, 198–204.
10. Dong, P.; Chen, Z.; Qin, R. Use of concrete-cored DCM pile in soft ground. *Chin. J. Geotech. Eng.* **2002**, *24*, 204–207.
11. Zhou, J.J.; Gong, X.N.; Wang, K.H.; Zhang, R.H.; Yan, J.J. Testing and modeling the behavior of pre-bored grouting planted piles under compression and tension. *Acta Geotech.* **2017**, *12*, 1061–1075. [CrossRef]
12. Correia, A.A.S.; Oliveira, P.J.V.; Lemos, L.J.L. Strength assessment of chemically stabilised soft soils. *Proc. Inst. Civ. Eng.-Geotech. Eng.* **2019**, *172*, 218–227. [CrossRef]
13. Horpibulsuk, S.; Miura, N.; Nagaraj, T.S. Assessment of strength development in cement-admixed high water content clays with Abrams' law as a basis. *Geotechnique* **2003**, *53*, 439–444. [CrossRef]
14. Tanchaisawat, T.; Suriyavanagul, P.; Jamsawang, P. Stiffened deep cement mixing (SDCM) pile: Laboratory investigation. In Proceedings of the International Conference on Concrete Construction, London, UK, 9–10 September 2008.
15. Wu, M.; Zhao, X.; Dou, Y.M.; Wang, E.Y. Experimental study on stiffened DCM pile in laboratory. *Ind. Constr.* **2004**, *11*, 45–48.
16. Peng, T.; Li, L.Y.; Zhang, D.W. Experimental research on interfacial performance between cement treated soil and concrete. *Mod. Transp. Technol.* **2017**, *14*, 19–23.
17. Jamsawang, P.; Bergado, D.T.; Voottipruex, P. Field behaviour of stiffened deep cement mixing piles. *Proc. Inst. Civ. Eng.-Ground Improv.* **2011**, *164*, 33–49. [CrossRef]
18. Li, Y.H.; Sun, L.Q.; Li, X.; Huang, M.S. Experimental study on the shear mechanical properties of the cemented soil-concrete interface. *Eur. J. Environ. Civil. Eng.* **2022**, *26*, 4725–4739. [CrossRef]
19. Yu, J.L.; Xu, J.C.; Zhou, J.J.; Gong, X.N. Experimental study on frictional capacity of concrete-cemented soil interface of concrete-cored cemented soil column. *Chin. Civ. Eng. J.* **2022**, *55*, 93–104, 117.
20. Zhou, J.J.; Yu, J.L.; Gong, X.N.; Naggar, M.H.; Zhang, R.H. The effect of cemented soil strength on the frictional capacity of precast concrete pile-cemented soil interface. *Acta Geotech.* **2020**, *15*, 3271–3282. [CrossRef]
21. Zhou, J.J.; Gong, X.N.; Zhang, R.H. Model tests comparing the behavior of pre-bored grouted planted piles and a wished-in-place concrete pile in dense sand. *Soils Found.* **2019**, *59*, 84–96. [CrossRef]
22. *DGJ32/TJ 151–2013*; Technical Specification for Strength Composite Piles. Department of Housing and Urban-Rural Development of Jiangsu Province: Nanjing, China; Jiangsu Science Press: Nanjing, China, 2013.
23. *GB/T 50783–2012*; Technical Code for Composite Foundation. Ministry of Housing and Urban-Rural Development of the People's Republic of China: Beijing, China; China Planning Press: Beijing, China, 2012.
24. *JGJ/T 330-2014*; Technical Specification for Pile Foundation of Pipe Pile Embedded in Cement Soil. Ministry of Housing and Urban-Rural Development of the People's Republic of China: Beijing, China; China Architecture and Building Press: Beijing, China, 2014.
25. Yi, K.J. Analysis on Engineering Property of Soft Soils in Shanghai and Ningbo Area. *Zhejiang Constr.* **2012**, *29*, 30–32.
26. *JCJ/T 233-2011*; Specification for Mix Proportion Design of Cement Soil. Ministry of Housing and Urban-Rural Development of the People's Republic of China: Beijing, China; China Architecture and Building Press: Beijing, China, 2011.
27. *JGJ 55-2011*; Specification for Mix Proportion Design of Ordinary Concrete. Ministry of Housing and Urban-Rural Development of the People's Republic of China: Beijing, China; China Architecture and Building Press: Beijing, China, 2011.
28. *GB/T 41054-2021*; Technical Specification for High Performance Concrete. State Administration for Market Regulation, Standardization Administration: Beijing, China, 2021.
29. *GB/T 50266-2013*; Standard for Test Methods of Engineering Rock Mass. Ministry of Housing and Urban-Rural Development of the People's Republic of China: Beijing, China; China Planning Press: Beijing, China, 2013.

30. *DG/TJ 08-61-2010*; Technical Code for Excavation Engineering. Shanghai Urban-Rural Construction and Municipal Transportation Commission: Shanghai, China, 2010.
31. *GB/T 50123-2019*; Standard for Geotechnical Testing Method. Ministry of Housing and Urban-Rural Development of the People's Republic of China, State Administration for Market Regulation: Beijing, China; China Planning Press: Beijing, China, 2019.

materials

MDPI

Article

Influence of In-Situ Stress on Cut Blasting of One-Step Raise Excavation Using Numerical Analysis Based on a Modified Holmquist-Johnson-Cook Model

Kai Liu [1], Qiyue Li [1], Chengqing Wu [2], Xibing Li [1,*] and Wei Zhu [1,*]

1 School of Resources and Safety Engineering, Central South University, Changsha 410083, China; kailiu30@csu.edu.cn (K.L.)
2 School of Civil and Environmental Engineering, University of Technology Sydney, Sydney, NSW 2007, Australia
* Correspondence: xbli@csu.edu.cn (X.L.); csuzhuwei@csu.edu.cn (W.Z.)

Abstract: Due to different tensile and compressive properties of rock material, the corresponding tensile and compressive damage evolution show major differences. To investigate the tensile and compressive damage evolution in deep cut blasting with different in-situ stresses, an improved Holmquist-Johnson-Cook (HJC) material model considers the tensile and compressive damage separately is developed. The improved HJC model is implemented into LS-DYNA via a user-defined subroutine in this study. Then, a numerical model with different in-situ stresses loading schemes is modelled. Numerical simulation results show that in-situ stress can inhibit the development of tensile damage evolution, while promote the development of compressive damage evolution. The overall damage zone presents a decreasing trend with the increase of in-situ stress, because the tensile damage is more sensitive than the compressive damage for rock material. In addition, the maximum principal stress can determine the development of the direction of damage. Further, for a field test of blind cut raise in deep, the actual in-situ stress values are loaded on the numerical model. Then, in order to overcome the difficulties caused by in-situ stress, the cut blasting design is optimized by reducing hole spacing. Subsequently, the optimized cut parameters are applied in the blind cut raise. However, the one-step raise excavation method is adjusted to two steps to ensure success due to a serious borehole deviation between drilling and design drawing. After these steps, the formation of the blind cut raise with 8.7 m depth is met the requirements of design.

Keywords: cut blasting; damage model; in-situ stress; numerical simulation; field test

Citation: Liu, K.; Li, Q.; Wu, C.; Li, X.; Zhu, W. Influence of In-Situ Stress on Cut Blasting of One-Step Raise Excavation Using Numerical Analysis Based on a Modified Holmquist-Johnson-Cook Model. *Materials* **2023**, *16*, 3415. https://doi.org/10.3390/ma16093415

Academic Editor: Pradeep Menezes

Received: 10 February 2023
Revised: 14 April 2023
Accepted: 20 April 2023
Published: 27 April 2023

1. Introduction

Raises such as ventilating raise, cut raise and service raise is widely used in underground space engineering [1–3]. However, conventional raise excavation methods, which need workers to gain access to the raise heading face and drill 3~4 m per excavation cycle, are insecure and inefficient. Compared with the conventional raise excavation methods, the one-step raise excavation technique has huge advantages in security, cost and efficiency [4]. Thus, it has been gradually applied in the field of underground engineering. The one-step raise excavation technique can be divided into two styles based on charging structure: spherical cartridge blasting mode (SCBM) and burn cuts blasting mode (BCBM) [5]. Compared to SCBM, the BCBM is more flexible in underground engineering, especially in deep mines with narrow space. Meanwhile, BCBM has more advantages in excavation of small section raise. The BCBM needs a cut cavity, which is formed by initiation of a series of cut holes in sequence, to serve as the free surface and swelling space for the boreholes [6]. Thus, it is crucial to investigate the damage evolution process of cut-blasting. When cut blasting is performed in a deep mine, damage induced by blasting load is affected by not only cut parameters but also high in-situ stress which will cause difficulties to cut blasting [7–9]. As a result, the application of one-step raise excavation is limited in deep mines. Therefore, it is necessary to study damage evolution and failure mechanisms of cut blasting under high in-situ stress.

In recent years, many studies on damage evolution of rock mass under blasting loading have been conducted [10,11]. Liu and Katsabanis [12] described a constitutive model for predicting rock damage and fragment size distribution based on continuum mechanics and statistical fracture mechanics. Zhang et al. [13] proposed an anisotropic continuum damage model to study rock damage induced by stress wave. In addition, some damage models are used in the numerical simulation of cut blasting. Yang et al. [14] implemented a statistical damage evolution law into LS-DYNA to simulate rock damage evolution of deep tunnel excavation. Xie et al. [15] developed a tension and compression-shear damage model to investigate the damage mechanics of cut blasting under high in-situ stress. However, little work on one-step raise excavation has been done except that covering tunneling excavation. The cut model, excavation depth, section and direction are different between raises and tunnels, which will result in differences in the damage evolution of rock mass during cut blasting [4].

It is well-known that rock material exhibits different damage evolution mechanisms under tension and compression loading. However, some damage constitutive models cannot describe well the differences between tensile damage and compressive damage [16–18]. For example, the Holmquist-Johnson-Cook (HJC) model can express compression damage behavior via its damage model, which defines damage accumulates from equivalent plastic strain and plastic volumetric strain, but the tension damage behavior is not taken into account in the HJC model [19]. Due to the tensile strength of rock mass which is much less than its compressive strength, the tensile damage is the main cause for rock failure. Therefore, many damage-constitutive models have been proposed and implemented to describe both the compressive and tensile damage evolution mechanisms of rock mass. Hao et al. [20] present a numerical model, in which both the tensile and compressive damage are involved, to simulate the damage zone around the charge hole. Li and Shi [21] established a tensile and compressive damage model based on the extended Drucker-Prager strength criterion and the Johnson–Cook material model. Considering that the compression behavior of materials was described in detail and a few parameters needed to be determined, the original HJC model is modified in this study.

The objective of this study is to investigate the damage evolution mechanism of cut blasting in the BCBM under different in-situ stresses. Firstly, a modified HJC damage model which considers tension and compression damage behaviors is presented and implemented into LS-DYNA software (R 11.0, Sydney, Australia) via user subroutines. After this, for deep raise excavated by the BCBM, a simplified cut blasting numerical model is built to analyze the behaviors of rock damage under different in-situ stress load schemes. Further, the actual in-situ stresses in −665 m sublevel of a gold mine are applied to the numerical model to obtain the optimized cut blasting parameters. Finally, the optimization parameters are applied to the field test of BCBM.

2. Engineering Background

2.1. General

Xincheng gold mine is located in Laizhou city, Shandong province, China. The mine area is about 35 km southwest of Laizhou city (37°25′ N, 120°08′ E). Via the G206 national road which runs through the mine area, the mine is connected with the Weifang railway station about 135 km to the southwest, as shown in Figure 1a. Xincheng gold mine is a medium mechanized underground mine with a production capacity of over 3600 t/day. The ore deposit is mainly composed of I# and V# ore bodies in Xincheng gold mine. The I# ore body, which is located at a depth from −20 m to −630 m, has been exhausted. Therefore, the burden of the main ore output is carried by the V# ore body which is located at a depth from −430 m to −900 m, as shown in Figure 1b. In order to improve the production capacity of Xincheng gold mine, a new mining method should be determined for the V#

ore body. Because the cut and fill stoping method has been used in the thin ore body I#, it is no longer suitable for the large and thick ore body V#. Sublevel fill stoping, which is a mass mining method and can meet the production requirement [22,23], will be adopted for the V# ore body. For sublevel fill stoping, a vertical cut raise should be excavated as the free surface for production holes blasting. Therefore, the BCBM is used in construction of the cut raise to improve excavation efficiency and ensure operational safety.

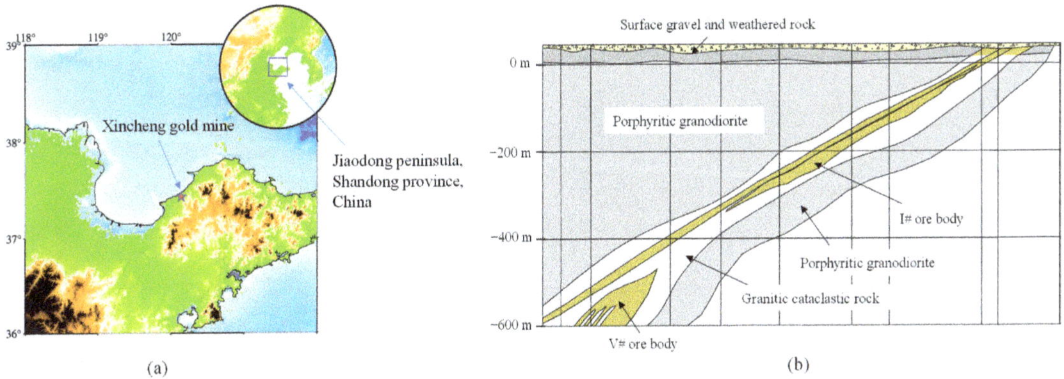

Figure 1. Basic information of Xincheng gold mine: (**a**) the location of Xincheng gold mine; (**b**) the engineering geology profile of Xincheng gold mine.

2.2. Design of Cut Blasting for One-Step Raise Excavation

According to the surrounding rock mass stability and sublevel height, a blind cut raise with a height of 9 m and a cross section of 2.5×2.5 m^2 is designed. The BCBM is used to excavate the deep cut raise in this study. For the application of the BCBM, cut model and parameters should be determined firstly, especially for the prime cut hole which detonates firstly and uses empty holes as the free surface and swelling space in the process of cut blasting. However, the constriction of burden rock in cut-blasting is much higher than that in bench blasting. Moreover, high in-situ stress will cause more difficulties for cut blasting in deep mines. According to the research of Xie et al. [15], the empty hole not only plays the role of free surface but can also transfer the blasting energy to the free surface. Consequently, more empty holes applied in the cut model means more blasting energy will be used to break rock mass. Thus, a burn-cut-with-four-empty-holes model is adopted for the BCBM in a deep mine, as shown in Figure 2, which includes one prime cut hole (No. 1), four empty holes (No. 2, 3, 4, and 5) and four secondary cut holes (No. 6, 7, 8, and 9) in the cut zone, and four supplementary holes (No. 10, 11, 12, and 13) and four peripheral holes (No. 14, 15, 16, and 17) in the stoping and contour zones. It can be seen that as the empty holes are arranged around the prime cut hole, more blasting energy will be shifted to the direction of empty hole and spent in the creation of the cut cavity after the prime cut hole detonation. The sizes of boreholes are determined based on the existing drilling equipment in Xincheng gold mine. The diameters of charge holes and empty holes are 70 mm, 130 mm respectively. According to the requirement of swelling coefficient k, the hole spacing L between the prime cut hole and empty holes is set as 250 mm to ensure $k > 1.45$ [24].

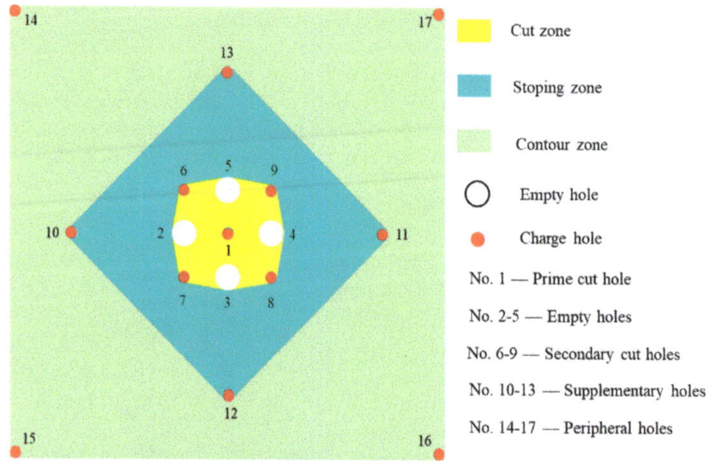

Figure 2. Layout of boreholes for burn cuts with four empty holes.

3. Material Model for Numerical

3.1. Blasting Damage Model for Rock Material

As we all know, the explosive energy can convert into two main forms: stress waves, and detonation gas after it is initiated. The rock damage induced by blasting also results from the two forms. Nevertheless, it is difficult to quantify the effects of the two forms and describe the two damage mechanisms in a single damage model. Thus, the effects of stress waves and detonation gas on rock damage are generally independently modeled in most studies [25,26]. The damage evolution models produced by detonation gas cannot give a reasonable prediction in numerical simulation and have not been further pursued [27]. Meanwhile, the damage models only consider the effect of stress waves can predict the rock respond to blasting load. Thus, the blasting damage model only induced by stress waves is modeled in this study.

3.1.1. Damage Evolution Model

For brittle material, such as rock or concrete, its compressive strength is much greater than its tensile strength. In the process of explosion stress wave propagation, the compressive component derived by shock wave produces compressive-shear damage in the vicinity of charging hole. After this, a tensile damage zone will appear outside the compressive-shear damage zone with the propagation of stress wave. Due to different mechanical properties under tensile and compression loads, the rock damage evolution model can be divided into two parts: tensile damage function and compressive damage function. When the rock material is in a tensile state, its damage evolution should be expressed by the tensile damage function. Conversely, its damage evolution can be expressed by the compressive damage function.

In major reported studies on tensile fracture of the brittle material [28,29], the stress-strain curve under uniaxial tension is similar to the exponential form in the tensile softening stage. Therefore, referring to the study of Weerheijm and Doormaal [29], an exponential softening expression in which the tensile damage D_t is accumulated with plastic strain is adopted, as following,

$$D_t = \left[1 + \left(c_1 \frac{\bar{\varepsilon}_p}{\varepsilon_{frac}}\right)^3\right] \exp\left(-c_2 \frac{\bar{\varepsilon}_p}{\varepsilon_{frac}}\right) - \frac{\bar{\varepsilon}_p}{\varepsilon_{frac}}\left(1 + c_1^3\right) \exp(-c_2) \tag{1}$$

where, $\bar{\varepsilon}_p = \sum \Delta\varepsilon_p$ is the effective plastic strain, ε_{frac} is the fracture strain which is depended on the element size in the numerical model. The constants c_1 = 3 and c_2 = 6.93. ε_{frac} is the fracture strain, and its value depends on the size of the element in the numerical model. It can be calculated by

$$\varepsilon_{\text{frac}} = \frac{5.136G_{\text{f}}}{h_{\text{c}}T} \tag{2}$$

where G_{f} is the fracture energy and is taken as 80 Nm/m^2 from Kong et al. [19]. h_{c} is the characteristic length of the element, which may be approximated by the cube root of the volume of the element in a 3D analysis. T is the tensile strength of material.

For compressive damage evolution, not only the plastic strain which is caused by plastic shear deformation but also the plastic volume strain which is induced by plastic crushing of the pores in the rock mass should be considered in compression damage accumulation. According to the definition in the HJC model, the compressive damage D_{c} is proposed as follows,

$$D_{\text{c}} = \sum \frac{\Delta\varepsilon_{\text{p}} + \Delta\mu_{\text{p}}}{\varepsilon_{\text{p}}^{\text{f}} + \mu_{\text{p}}^{\text{f}}} \tag{3}$$

where, $\Delta\varepsilon_{\text{p}}$, $\Delta\mu_{\text{p}}$ are the effective plastic strain increment and plastic volumetric strain, respectively, during a cycle of integration. $\varepsilon_{\text{p}}^{\text{f}} + \mu_{\text{p}}^{\text{f}}$ is the total plastic strain under a constant pressure until fracture, which is expressed as,

$$\varepsilon_{\text{p}}^{\text{f}} + \mu_{\text{p}}^{\text{f}} = D_1(p^* + T^*)^{D_2} \geq EF_{\text{min}} \tag{4}$$

where, D_1 and D_2 are the damage constants. $p^* = \frac{p}{f_c}$, $T^* = \frac{T}{f_c}$ denote the normalized pressure and the normalized hydrostatic tension respectively, and p is the actual pressure, T is the maximum hydrostatic tension, f_{c} is the quasi-static uniaxial compressive strength. EF_{min} is used to suppress fracture from low tensile stress waves.

In order to show which kind of damage predominates in the failure of rock mass, the maximum function is applied. Thus, the final damage variable can be expressed as,

$$D = \text{Max}\,(D_{\text{c}}, D_{\text{t}}) \tag{5}$$

3.1.2. Yield Strength Model

Just like many previous studies, the shape of the yield surface of the brittle material in the deviatoric plane exhibits a triangular shape at low pressure and transforms to a circular shape at high pressure. However, many material models such as the classic Drucker–Prager model [21] and HJC model [19] have assumed that the shape of yield surface remains a circle the whole time in the deviatoric plane. That means there are not differences between the tension meridian and the compression meridian, which result in the consequence that the capacity of rock subjected to tension loading will be overestimated in the above models. Therefore, the Lode-angle function [30] depending on Lode-angle (θ) and e should be introduced to the yield strength surface, which is expressed as,

$$R(\theta, e) = \frac{2(1-e^2)\cos\theta + (2e-1)\left[4(1-e^2)\cos^2\theta + 5e^2 - 4e\right]^{\frac{1}{2}}}{4(1-e^2)\cos^2\theta + (2e-1)^2} \tag{6}$$

where, the Lode-angle θ is determined by,

$$\theta = \frac{1}{3}\cos^{-1}\left(\frac{27J_3}{2\sigma_{\text{eq}}^3}\right) \tag{7}$$

with J_3 being the third invariant of the deviatoric part of the stress tensor, σ_{eq} being the equivalent stress. e is the ratio of the tensile meridian to the compressive meridian, which can describe the transition of deviatoric plane shape from triangle to circle. As discussed by Polanco-Loria et al. [30], the e which is sensitive to the biaxial-compression strength is simply defined by a linear interpolation of the discrete points, as follows

$$e = \begin{cases} 0.65 & p^* < 0 \\ 0.65 + (1 - 0.65)\frac{p^*}{p^*_{\text{ref}}} & 0 \le p^* \le p^*_{\text{ref}} \\ 1.0 & p^* > p^*_{\text{ref}} \end{cases} \tag{8}$$

where, p^*_{ref} is the reference normalized pressure and $p^*_{\text{ref}} = 10$ is adopted in this study.

3.1.3. Strain Rate Effect Model

The strain rate effect is used to describe the phenomenon that material strength increases with strain rate under dynamic loading. In order to describe the strain rate effect in the constitutive model, a simplified approach in which the static yield surface is multiplied by the dynamic increase factor (*DIF*) as the dynamic yield surface is implemented. However, the *DIF* should be defined separately for the tension and compression loading, since dynamic experimental results show that the *DIF* for tension is much higher than that for compression. Many tension-compression strain rate effect expressions have been put forward recently [31]. However, most of the expressions are of a exponential relationship, which will lead to an overestimate of *DIF* with the strain rate increases to the range of $10^4\ \text{s}^{-1}$~$10^6\ \text{s}^{-1}$. Gebbeken and Greulich [32] suggested a hyperbolic function to describe the relationship between strain rate and *DIF*, which sets a cut-off of *DIF* in high strain rate region and shows a good agreement with the experiment data of brittle material.

$$DIF_{\text{t}} = \{[\tanh((\lg(\dot{\varepsilon}/\dot{\varepsilon}_0) - W_x)S)][\frac{F_m}{W_y} - 1] + 1\}W_y \tag{9}$$

$$DIF_{\text{c}} = (DIF_{\text{t}} - 1)(T/f_{\text{c}}) + 1 \tag{10}$$

in which, DIF_{t} and DIF_{c} are the dynamic increase factor for tension and compression, respectively. W_x, S, F_m, W_y are the fitting constants, the values of W_x = 1.6, S = 0.8, F_m = 10, W_y = 5.5 are determined referred the research of Tedesco et al. [33], $\dot{\varepsilon}_0 = 1\text{s}^{-1}$ is the reference strain rate. In this study, the above hyperbolic function for strain rate effect is adopt. Thus, according to the form of continuous yield surface function proposed by Polanco-Loria et al. [30], it can be written as follows,

$$\sigma^* = \begin{cases} B[T^*(1-D) + p^*]^N R(\theta, e) DIF, & P^* \ge -T^*(1-D) \\ 0 & , P^* < -T^*(1-D) \end{cases} \tag{11}$$

where, B is the pressure hardening coefficient, N is the pressure hardening exponent, and the other parameters are the same as above. In addition, a typical three-stage equation of state for brittle material is used to describe the relationship between pressure and volume strain [19,21]. The three-stage equation of state contains bulk modulus K, pressure parameter K_1, K_2 and K_3, hydrostatic pressure p_{c} and volume strain μ_{c} at the elastic limit, p_{l} and μ_{l} at the crushing region. A detailed description of this modified HJC model and its validation can be found in Ref. [34], which is a previous study of the authors.

3.2. JWL EOS for Explosive

The borehole pressure profiles can be approximated using two main methods: pressure-decay functions and Equation-of-State (EOS), while the EOS can describe the process of rock-explosive interaction and is convenient for application in LS-DYNA. Therefore, the JWL EOS is used to simulate detonation products of high explosives in the present study. The JWL EOS defines the pressure as:

$$P_J = A_J(1 - \frac{\omega}{R_1 V})e^{-R_1 V} + B_J(1 - \frac{\omega}{R_2 V})e_1{}^{-R_2 V} + \frac{\omega E}{V} \tag{12}$$

where P_J is the pressure of the detonation products, V is the relative volume of detonation products, E is the special internal energy with an initial value of E_0, A_J, B_J, R_1, R_2 and ω are material constants.

3.3. Air Material Model

In addition, the radial decoupling charge is usually applied on cut blasting to control the blasting damage zone in practice. In this study, the radial air-decoupling charge technique is implemented. As for air, material type 9 of LS-DYNA (*MAT_NULL) is used to calculate the pressure from a specified EOS, which is expressed as:

$$P_A = C_0 + C_1\delta + C_2\delta^2 + C_3\delta^3 + (C_4 + C_5\delta + C_6\delta^2)e_2 \tag{13}$$

where P_A is the pressure, e_2 is the internal energy per volume, δ is dynamic viscosity coefficient, C_0, C_1, C_2, C_3, C_4, C_5 and C_6 are material constants. In addition, the air is modeled as an ideal gas by setting $C_0 = C_1 = C_2 = C_3 = C_6 = 0$ and $C_4 = C_5 = 0.401$, and air mass density and initial internal energy are 1.255 kg/m^3 and 0.25 J/cm^3, respectively [35].

4. Numerical Simulation and Analysis

4.1. Research Methodology and Steps

The numerical software ANSYS-LSDYNA is used to simulate and reproduce the whole process of cut blasting under in-situ stress [36]. In order to load in-situ stress at the boundary of the numerical model, the implicit-explicit sequential solution method is involved. First, the implicit module is called to analyze the static in-situ stress state of the cut blasting model. Then the implicit calculation result contains strains, displacements and stresses is output as a *drelax* file which will be imported to the explicit module during the explicit calculation. In the explicit stage, the boundary conditions and element type should be updated. The unnecessary constrains are deleted from the implicit analysis and the new boundary conditions are set up for explicit solution. Then the numerical model is initialized for the explicit solution via the *drelax* file, as shown in Figure 3. The element type should be transformed from solid 185 in implicit to solid 164 in explicit. Besides, in the explicit analysis stage, there are basic materials such as rock, explosive to be modeled. In the process of high-speed explosion, the explosives can be treated as fluid and modeled with Euler mesh, while the rock material is modeled with Lagrangian mesh in the ALE algorithm. The interaction between explosive and rock is achieved by the keyword "*CONSTRAINED_LAGRANGE_IN_SOLID".

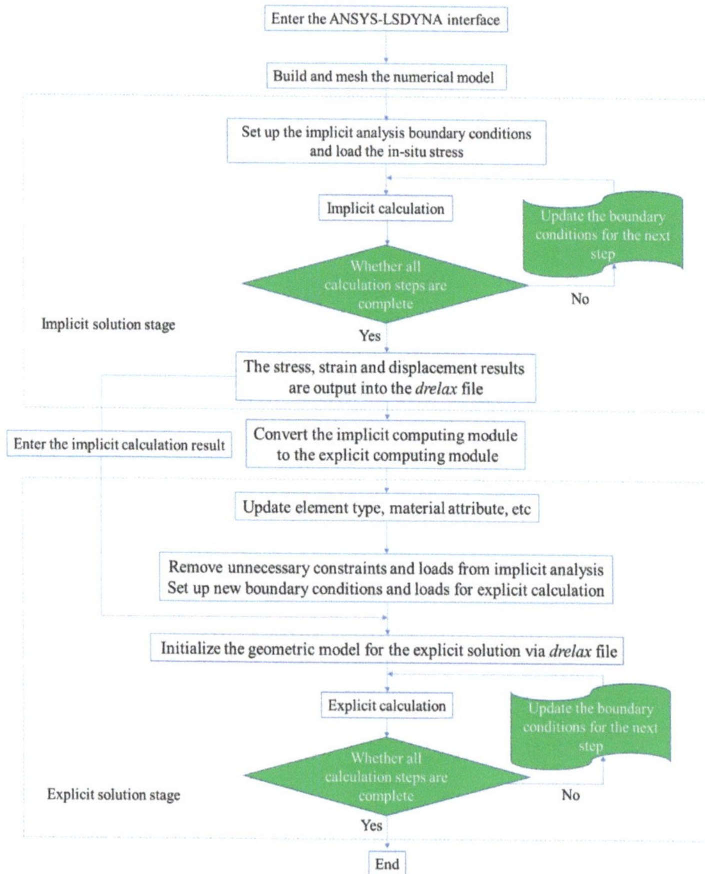

Figure 3. Flow chart of the implicit-explicit sequential solution.

4.2. Numerical Model and Material Paramaters

The damage evolution process of the prime cut hole is only taken into account in the numerical simulation. Because it is the most critical and difficult step in the cut blasting in which the empty holes serve as the only free surface for the prime cut hole [6]. Therefore, a prime cut numerical model with the dimension of 20 × 20 m^2 (length × width) is modelled based on the cut blasting design, as shown in Figure 4. According to the influence of stress wave reflection and the crack zone of rock blasting [37,38], the numerical model size is set to 20 × 20 m^2 (length × width). To simplify the problem, the numerical model is assumed as a plane strain problem, provided that the mechanical characteristics are the same in every cross section along the cut raise's centerline. Thus, the numerical model can set as a three-dimensional single layer mesh model, which can improve the accuracy of the simulated results and reduce the computation time [39,40]. In addition, it can be seen that the parameters of hole diameters and hole spacing are consistent with the design. The radial air decoupling charge structure with 10 mm air layer is also used for the prime cut hole. In the implicit solution stage, initial horizontal in-situ stresses σ_{Hx} and σ_{Hz} are applied to the x and z direction faces of the numerical model, and the other two opposite surfaces are fixed boundaries. In the explicit solution stage, the unnecessary loads and constrains are deleted, and the non-reflection boundary conditions are set up in the four outerrounded faces to reduce the influence of stress wave reflection.

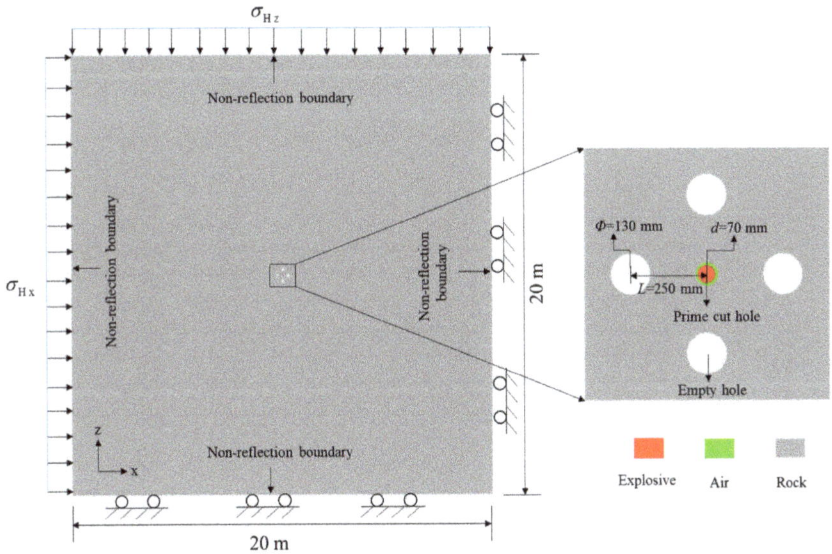

Figure 4. Geometric model and boundary conditions of the prime cut hole for the BCBM.

For the rock material, the basic parameters, such as density, uniaxial compressive strength, uniaxial tensile strength and shear modulus, are obtained by rock mechanics experiments. The other parameters such as yield surface strength parameters, equation-of-state parameters and damage parameters can be determined by referring to the author's previous research [34]. The all-parameter values for the modified HJC model that need to be inputted into the LS-DYNA are presented in Table 1. Besides this, the JWL EOS parameters of explosive are listed in Table 2.

Table 1. Parameters of the improved damage model.

ρ/(kg/m^3)	G/GPa	K/GPa	B	N	f_c/MPa	f_t/MPa	D_1	D_2
2810	17.2	33.2	2.57	0.62	133.5	6.76	0.04	1.0
p_c/MPa	μ_c	p_1/GPa	μ_1	K_1/GPa	K_2/GPa	K_3/GPa	EF_{min}	
44.5	0.0013	1.89	0.110	18.8	15.3	94.1	0.03	

Table 2. Parameters of explosive material and JWL EOS.

ρ_e (kg/m^3)	VoD (m/s)	P_{CJ} (GPa)	A_J (GPa)	B_J (GPa)	R_1	R_2	ω	E_0 (GPa)
1210	5660	9.7	214.4	0.182	4.2	0.9	0.15	4.192

4.3. Damage Evolution under Different In-Situ Stresses

In order to investigate the effect of in-situ stress on rock damage evolution mechanisms of cut blasting during deep raise excavation, the corresponding loading schemes of in-situ stress are established. Since the current mining depth of Xincheng gold mine is 1200 m, the corresponding horizontal maximum in-situ stress is about 60 MPa, so the numerical simulation research scheme sets the values of in-situ stress within the range of 0–60 MPa, as shown in Table 3. Schemes 1–4 are used to study the effect of the hydrostatic pressure, while Schemes 5–8 are used to study the effect of the anisotropic in-situ stress.

Table 3. In-situ stress loading stress for the cut blasting.

Loading Schemes	1	2	3	4	5	6	7	8
Horizontal stress in the x direction (MPa)	0	10	30	50	30	30	30	30
Horizontal stress in the z direction (MPa)	0	10	30	50	10	20	40	60

The numerical simulation results of Schemes 1–4 are presented in Figure 5. It can be seen that with the increase of in-situ stress, the area of the overall damage tends to decrease gradually. That indicates that in-situ stress can restrain the development of rock damage evolution in cut blasting. Further, the tensile damage and compressive damage are separately analyzed. It is obvious that the area of the tensile damage drops, especially in the stress waves superposition zone behind the empty holes, with the increase of in-situ stress, while that of the compressive damage appears to rise. It is demonstrated that in-situ stress can promote the development of compressive damage, while inhibit the development of tensile damage. The main reason is that the tensile component of stress wave is suppressed by in-situ stress, which causes the tensile damage reduced significantly. When the stress wave propagates to the superposition zone behind the empty holes, the attenuation is more serious. Thus, the tensile damage area of the region decreases most obviously. In addition, when in-situ stress is below 30 MPa, the tensile damage zone drops sharply with the increase of in-situ stress. Nevertheless, when in-situ stress is higher than 30 MPa, the gradient of tensile damage zone drops is lower than the previous. The probably reason for this phenomenon is that when in-situ stress is higher than 30 MPa, the tensile component of stress wave in the superposition zone is reduced by in-situ stress and is lower than the dynamic tensile strength of rock. Therefore, in the mining industry, the linear charge density in deep mines is greater than that in shallow mines, to rise the peak value of shock wave and overcome the difficulties from in-situ stress.

Figure 6 shows the rock damage evolution of cut blasting under the action of anisotropic in-situ stresses. The results demonstrate that both the compressive damage and tensile damage are aligned with the direction of the maximum principal stress, and the phenomenon becomes more and more obvious with the increase of difference between the two principal stresses. The compressive damage shows a slight increase with the increase of in-situ stress, while the tensile damage drops sharply. The main reason for this is that the tensile stress component perpendicular to the maximum principal stress is greatly suppressed, while the tensile stress component perpendicular to the minor principal stress is slightly suppressed. In contrast, the compressive stress component is promoted by in-situ stress.

Therefore, for the cut blasting in deep mines, in order to overcome the suppressive effect of in-situ stress on tensile damage evolution, the simple and effective method is to reduce the hole spacing between the prime cut hole and empty holes [7], that is, to reduce the burden. According to the numerical simulation results, the burden can be reduced slightly in the direction of the maximum principal stress, while it needs to be reduced to a greater extent in the direction of the minor principal stress. Similarly, the promoting effect of in-situ stress on compressive damage evolution can be applied. If in-situ stress is high enough, the compressive component will be greatly promoted and the burden can be increased appropriately. For a project to determine the blasting parameters, such factors needs to be considered as the magnitude of in-situ stress and rock mechanical properties, and then using these to optimize the parameters through numerical simulation.

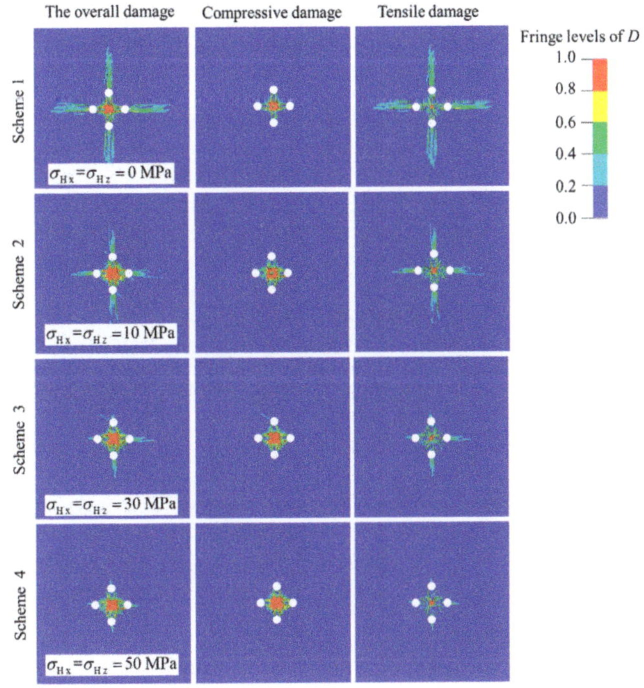

Figure 5. Damage distribution of prime cut blasting under the action of hydrostatic pressure.

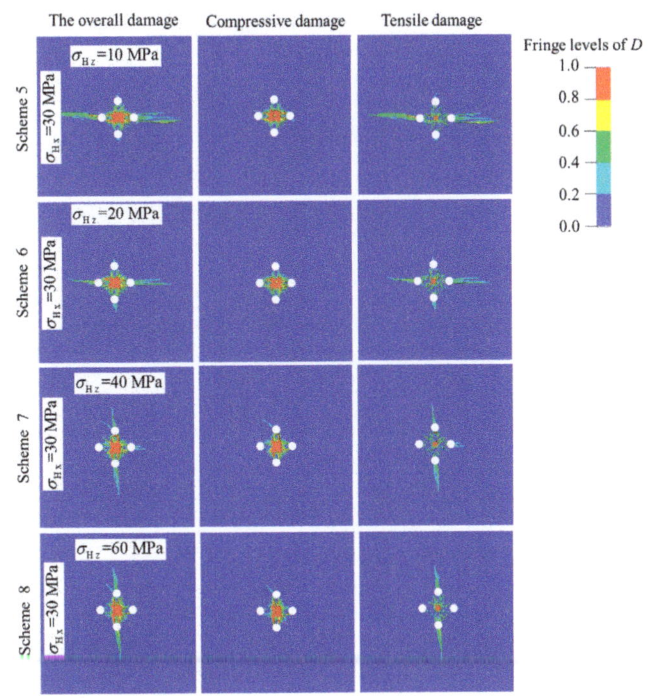

Figure 6. Damage distribution of prime cut blasting under the action of anisotropic in-situ stress.

4.4. Case Study of Cut Parameters Optimization for Deep Raise

The experimental stope is located in −665 m sublevel, V# ore body, Xincheng gold mine. According to the literature [41], a study of in-situ stress distribution in Xincheng gold mine was conducted by Cai et al. [41]. It was found that in-situ stress field is dominated by horizontal tectonic stress rather than gravity stress in Xincheng gold mine. The measured results show that the two principal stresses in the horizontal direction are the maximum and minimum principal stresses, respectively, and the vertical stress is the intermediate principal stress. In addition, the vertical stress is approximately equal to the gravity stress. If we assume that the vertical stress is equal to the gravity stress, and the average unit weight γ is 26.50 kN/m^3 for the rock mass, the vertical stress σ_V= 17.6 MPa can be calculated by $\sigma_V = H\gamma$. According to relevant research [42], R_1 = 1.9 and R_2 = 0.85, which are the ratio of horizontal maximum principal stress σ_{Hmax} and vertical stress, and minimum principal stress σ_{Hmin} and vertical stress. respectively, are determined in this study. Thus, the $\sigma_{Hmax} = \sigma_V \times R_1$= 33.4 MPa, $\sigma_{Hmin} = \sigma_V \times R_2$= 14.8 MPa can be obtained.

The numerical model in which the blasting parameters, the boundary conditions and the model size are consistent with Section 4.2 is built in ANSYS-LSDYNA. Only in-situ stress loading changes to the real values of the experimental stope in implicit analysis. In addition, the blasting processes of the four secondary cut holes are added to investigate the damage evolution mechanism of the overall cut cavity. The numerical results are shown in Figure 7. As we can see, a complete cut cavity is formed by a series of cut holes denotation in sequence. However, it can be found from Figure 7a that the damage zones between the prime cut hole and the empty holes are not interconnected, and that the burden rock did not break completely in the direction of the minimum principal stress. In practical engineering, this may lead to a failure of formation of the prime cut cavity, as it cannot provide the free surface and swelling space for the subsequent secondary cut holes. Therefore, the hole spacing needs to be optimized.

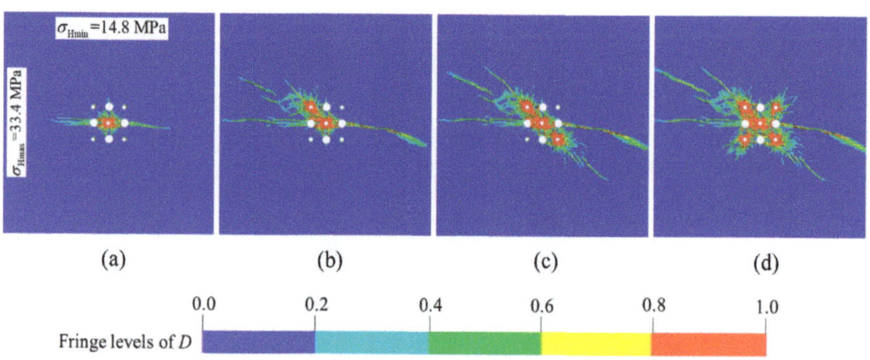

Figure 7. Damage evolution processes of burn cuts with the actual in-situ stress in Xincheng gold mine: (**a**) rock damage after the prime cut hole detonation; (**b**) rock damage after the No. 6 secondary cut holes detonation; (**c**) rock damage after the No. 8 secondary cut holes detonation; (**d**) rock damage after the No. 7 and 9 secondary cut holes detonation.

The hole spacings L_{max} are kept a constant with 250 mm in the direction of the maximum principal stress, while, the hole spacings L_{min} are reduced to 240 mm in the direction of the minimum principal stress. Meanwhile, the distance components of the corresponding secondary cut holes in the direction of the minimum principal stress are also reduced to 240 mm. The simulation results of cut model optimization are shown in Figure 8. It can be seen from Figure 8a that the damage zones are interconnected and that the cavity is in a "lotus" shape after the prime cut hole denotation. From the comparison of the final damage clouds between Figures 7d and 8d, the area of D = 1 is larger, and the radial tensile cracks which extend to outside of the cut cavity are shorter in the optimization scheme than that in the original scheme. That means more blasting energy is spent in the creation of the cut cavity, and that the size of the rock fragments is smaller and they more easily fall out of the cut cavity in the optimization scheme. According to the above analysis, the hole spacings L_{max} = 250 mm and L_{min} = 240 mm for one-step raise excavation in the −665 m sublevel of Xincheng gold mine are determined.

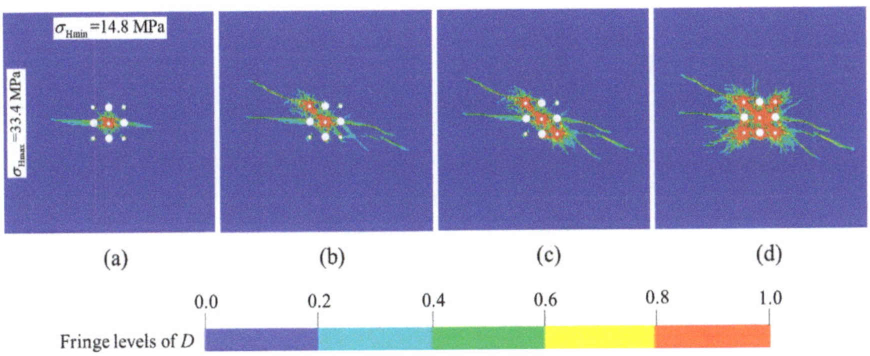

Figure 8. Damage evolution processes of the optimized burn cuts design by reducing hole spacing: (**a**) rock damage after the prime cut hole detonation; (**b**) rock damage after the No. 6 secondary cut holes detonation; (**c**) rock damage after the No. 8 secondary cut holes detonation; (**d**) rock damage after the No. 7 and 9 secondary cut holes detonation.

5. Field Test

According to the drilling equipment selection and numerical simulation results, the cut parameters—such as hole diameters d = 70 mm, Φ = 130 mm and hole spacing L_{max} = 250 mm, L_{min} = 240 mm in the burn cuts with four empty holes—are determined. The in-situ stress measurement results show that the distribution of the maximum principal stress in Xincheng gold mine is close to the east-west trend per Cai et al. [41], while the trend of the V# ore body is north-south trend and the stope is arranged vertically, so that the direction of the maximum principal stress is relatively consistent with the length direction of the stope. Therefore, the hole spacing L_{max} is aligned with the length direction of the stope, while L_{min} is aligned with the width direction of stope. Moreover, the raise section is changed into a rectangle with a size of 2.4 × 2.6 m^2 according to the in-situ stress distribution, as shown in Figure 9a. The layout of boreholes is consistent with Figure 1, and the parameters of hole spacings and diameters of boreholes are marked in detail. In addition, the depth of the empty holes is increased to 9.5 m to reduce the constriction at the top of the blind cut raise, which only has one free surface at the bottom.

The Sandvik drill rig with a drilling rate of 20 m/h is used to drill the boreholes based on the design drawing of the blind cut raise. However, it is difficult to locate accurately the point of boreholes due to dimly light and drilling straight up. Thereby, a serious borehole deviation appears between drilling and design drawing, as shown in Figure 9b. It can be seen that the distance between No. 3 (empty hole) and No. 1 (prime cut hole) is 498 mm, which is much larger than the design hole spacing of 250 mm. The corresponding burden rock may only perform a plastic deformation and not break into fragments. No. 3 empty hole cannot serve as the free surface and swelling space when the prime cut hole is initiated. Fortunately, although the distances from No. 2, 4, 5 to the prime cut hole have a deviation from the design (202 mm, 246 mm and 267 mm, respectively), it is still under control (swelling coefficient k > 1.45). In addition, it is found that there are two extra boreholes next to No. 7 and 8 secondary cut holes, which are named as No. 7′ and 8′, respectively.

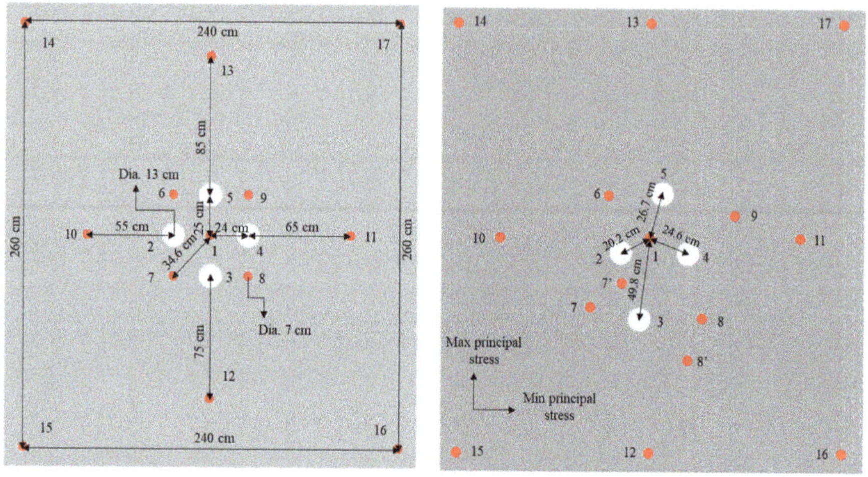

(a) (b)

Figure 9. Borehole layout of burn cuts for the blind cut raise: (**a**) design scheme; (**b**) drilling result.

The deviation of boreholes increases the difficulty of cut blasting in the BCBM. In order to solve this problem, some adjustments should be conducted. First, the charge density of cut holes is increased through reducing the air deck length. For cut holes No. 1, 7, 7′, 6, 9, 8 and 8′, the charge structure is set as 0.6 m explosive with 0.05 m air deck. Besides this, the stemming length is 0.2 m. The charge of each cut hole is 24.5 kg. Then a continuous charge structure is implemented in the supplementary holes and peripheral holes. For supplementary holes No. 10, 11, 12 and 13, the stemming length is 0.7 m and the charge is 38.3 kg. For peripheral holes No. 14, 15, 16 and 17, the stemming length is 1.0 m and the charge is 36.9 kg. Meanwhile, the bottom of the three empty holes, No. 2, 4, and 5, are filled with 0.2 kg explosive and 0.05 m stemming, respectively, to accelerate the removal of rock fragments from the cavity.

After this, the BCBM is split into two steps: the cut holes are detonated first, after the cut cavity is formed, and then the supplementary holes and peripheral holes are detonated in sequence, as shown in Figure 10. According to the swelling space theory, the initiation sequence is determined as: No. 1→No. 2, 4 and 5→No. 7′→No. 7→No. 6→No. 9→No. 8 and 8′ at the first initiation, No. 10→No. 11→No. 12→No. 13→No. 14 and 16→No. 15 and 17 at the second initiation. The decay time of two adjacent boreholes $t \geq 0.338$ s is calculated; the detail calculation procedure refers to Liu et al. [4]. There are non-electric millisecond decay detonators and half-second decay detonators in the mine, and the decay time of the adjacent half-second detonator is 500 ms, so the half-second detonator is used to fire the blastholes, and the initiation circuit is determined based on delay times and initiation sequence, as shown in Figure 11.

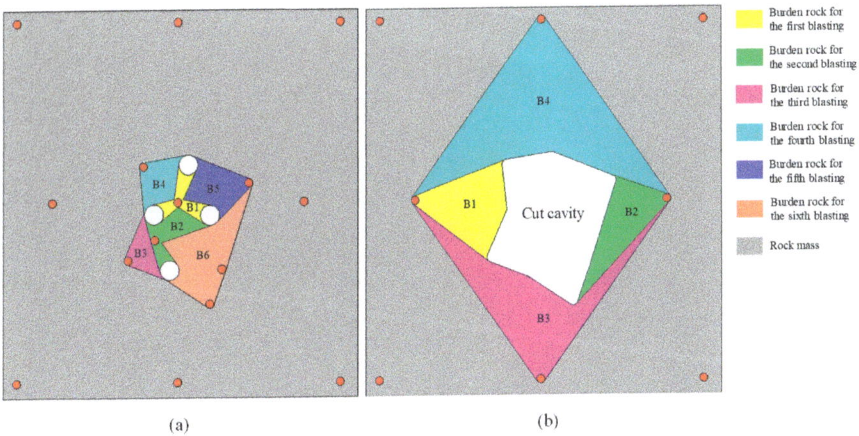

Figure 10. Adjustment for burn cuts with two steps: (**a**) detonation of cut holes in the first step; (**b**) detonation of supplementary and peripheral holes in the second step.

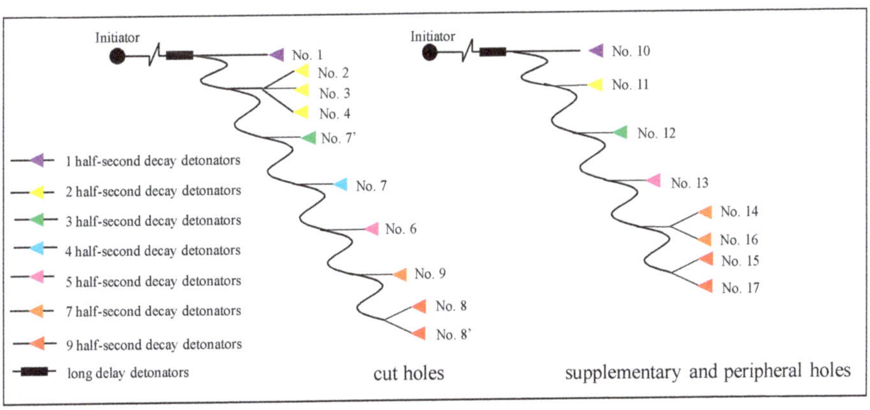

Figure 11. Initiation circuit of the 9 m blind cut raise with two steps.

The blasting effect of BCBM for the blind cut raise in a deep mine is shown in Figure 12. As we can see, the result of the first detonation is presented in the Figure 12a, the cut cavity is formed completely with a crater at the mouth. Moreover, there is no blockage in the cavity, which can used as the free surface and swelling space for the subsequence boreholes. The Figure 12b shows the complete blind cut raise formed after the second detonation. It can be seen that the raise section is rectangular and the crater is expanded. Further, the cavity explorer of C-ALS is applied to scan the profile of the blind cut raise, as shown in Figure 13. The raise is smooth and free of clogging. The depth of 8.7 m basically reaches the design requirement. Meanwhile, the crater with 5.1 × 2.1 m^2 (diameter × height) is formed at the bottom of the raise. The diameter of middle section of the raise is 2.4 m, which is consistent with the design. However, the diameter of the section at the top of the raise is only 2.0 m, less than the design value. The main reasons for this result is that there is a free surface at the bottom of the raise, which reduces the constriction of the burden rock, so a blasting crater is formed at the bottom of the raise. For the burden rock at the top of the raise, it is subjected to a high degree of constriction due to there being only empty holes as the free surface and swelling space. In order to overcome the high degree of constriction caused by a narrow free surface and high in-situ stress, explosive consumption can be increased at the top of the raise to form a more complete raise.

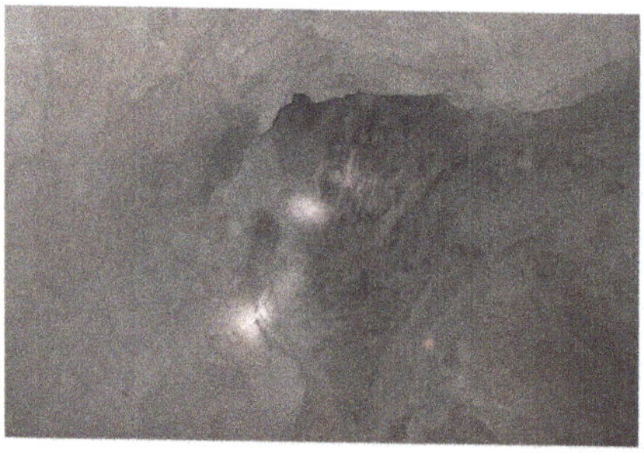

(a) (b)

Figure 12. Blasting effect of the BCBM for the blind cut raise in deep: (**a**) cut cavity after the first initiation; (**b**) blind cut raise after the second initiation.

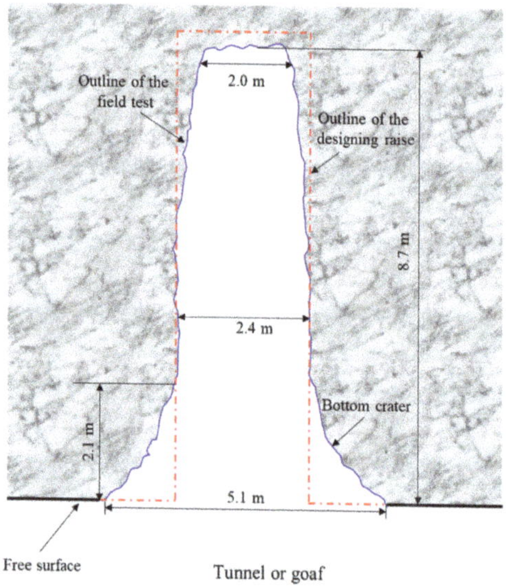

Figure 13. Profile of blind cut raise via cavity explorer scanned.

6. Discussion

The objective of this study is to investigate the damage evolution mechanism of deep cut blasting by using a tension-compression damage model. For cut blasting in deep mines, high in-situ stress has a very serious influence on rock damage evolution induced by blasting. Moreover, the rock constriction in cut blasting is higher than that in bench blasting [43]. Therefore, discovering how to solve the problem of cut blasting under high in-situ stress is urgent. Due to the efficiency, flexibility and low cost of computer algorithms, a numerical simulation method is used to study damage evolution mechanism of rock in this paper. Many damage models have been proposed to predict the damage evolution of rock in recent years [44,45]. However, most of the models, such as HJC or TCK, have a problem: the differences between tensile and compressive properties of rock materials cannot be described well. Thus, a damage model which considers the tensile and compressive damage separately is presented in this investigation. Meanwhile, two formulas are involved to describe strain rate effect under tension and compression, respectively. Moreover, the J_3 is taken into account through introduction of the Lode-angle function.

The numerical results of cut blasting under different in-situ stresses show that the overall damage tends to decrease with the increase of in-situ stress, which is consistent with many researches. Separately, the tensile damage is reduced sharply while the compressive damage is increased under the action of in-situ stress. The trend of tensile and compressive damage can be intuitively seen by the new damage mode, which is conductive to analyze the effect of in-situ stress on tensile and compressive damage of rock. Further, the optimization cut parameters by numerical simulation of the new damage model are applied in a blind cut raise in deep mine. In the process of drilling, a deviation of boreholes between site operation and design is occurred. Therefore, a temporary adjustment was made to the design scheme: the one-step raise excavation is divided into two steps, and a better effect is obtained. In practical engineering, borehole deviation is a frequent phenomenon due to drilling operation or joints in rock mass. To ensure the success rate of one-step raise excavation, the field application should be flexible and changeable based on the actual situation.

In addition, the in-situ stress can enhance the compressive component of stress wave, which will result in the compressive damage increasing according to the above analysis. However, most of the papers are devoted to how to solve the problem of rock blasting suppression induced by in-situ stress in deep mines [7,14,26,46]. Some papers suggest using in-situ stress to expand the compressive damage and improve blasting effect. For large-scale rock blasting, the rock fragments are mainly produced by radial tensile cracks, so it is greatly affected by the negative effect of in-situ stress. As for the cut blasting in rock excavation, due to its small zone, the burden rock can be crushed by the shock wave induced by the explosion. Under the appropriate blasting parameters, the in-situ stress can promote the rock breaking in cut blasting. Therefore, it provides a direction for us to investigate how to use the promotion effect of in-situ stress on compressive damage in the future.

7. Conclusions

Due to the differences between tensile and compressive properties of rock material, an improved tensile-compressive damage blasting model based on the original HJC model is proposed to investigate cut blasting in deep raise. Then a numerical simulation of cut blasting under different in-situ stresses is carried out through the improved damage model, which is implemented into LS-DYNA software via user subroutines. The numerical results showed that the in-situ stress can promote the development of compressive damage while inhibiting the development of tensile damage. Subsequently, the cut parameters are optimized for blind cut raise in a specific project in Xincheng gold mine. Further, the optimization cut parameters are applied in the field test. The research conclusions are as follows:

1. An improved damage blasting model is proposed to describe the tensile and compressive damage separately, and then implemented into LS-DYNA to simulate the rock damage evolution in cut blasting for one-step raise excavation under different in-situ stresses.
2. In-situ stress causes resistance on the overall rock damage evolution induced by blasting stress wave. For compressive damage, the damage increases with the increase of in-situ stress, while for tensile damage, it decreases with the increasing in-situ stress. The underlying reason is that in-situ stress can superimpose the peak pressure of the compressive stress component, but weaken that of the tensile stress component.
3. The damage zones, including tensile and compressive damage, tend to develop along the direction of the maximum principal stress under the action of anisotropic in-situ stress. The phenomenon becomes distinctive as the difference between the two principal stresses increases, which is the main problem which should be solved in the deep cut blasting.

4. For a blind cut raise in −665 m of the Xincheng gold mine, the optimal cut parameters are obtained by numerical simulation results of the improved damage model. The blind cut raise with 8.7 m depth is successfully formed, which demonstrates that the optimization parameters are valid for one-step excavation in deep mines, and the adjustments of design are necessary for the actual situation.

Author Contributions: Conceptualization, X.L.; Methodology, C.W.; Software, K.L.; Formal analysis, Q.L.; Investigation, K.L.; Data curation, Q.L.; Writing—original draft, K.L.; Writing—review & editing, W.Z.; Funding acquisition, W.Z. All authors have read and agreed to the published version of the manuscript.

Funding: The National Natural Science Foundation of China (52204116, 52204263).

Institutional Review Board Statement: Not applicable.

Informed Consent Statement: Not applicable.

Data Availability Statement: No new data were created or analyzed in this study. Data sharing is not applicable to this article.

Conflicts of Interest: The authors declare no conflict of interest.

References

1. Liu, K.W.; Yang, J.C.; Li, X.B.; Hao, H.; Li, Q.Y.; Liu, Z.X.; Wang, C.Y. Study on the long-hole raising technique using one blast based on vertical crater retreat multiple deck shots. *Int. J. Rock Mech. Mining Sci.* **2018**, *109*, 52–67. [CrossRef]
2. Rao, Z.B.; Cai, S.J. The Blasting Test and Blasting Vibration Monitoring of Vertical Crater Retreat Mining Method in the Luohe Iron Mine. *Geotech. Geolog. Eng.* **2016**, *34*, 1047–1056. [CrossRef]
3. Balland, C.; Morel, J.; Armand, G.; Pettitt, W. Ultrasonic velocity survey in Callovo-Oxfordian argillaceous rock during shaft excavation. *Int. J. Rock Mech. Min. Sci.* **2009**, *46*, 69–79. [CrossRef]
4. Liu, K.; Li, Q.Y.; Wu, C.Q.; Li, X.B.; Li, J. A study of cut blasting for one-step raise excavation based on numerical simulation and field blast tests. *Int. J. Rock Mech. Min. Sci.* **2018**, *109*, 91–104. [CrossRef]
5. Li, Q.Y.; Li, X.B.; Fan, Z.P.; Zhang, R.H. One time deep hole raise blasting technology and case study. *Chin. J. Rock Mech. Eng.* **2013**, *32*, 664–670. (In Chinese)
6. Li, Q.Y.; Liu, K.; Li, X.B.; Wang, Z.W.; Weng, L. Cutting parameter optimization for one-step shaft excavation technique based on parallel cutting method. *Trans. Nonfer. Metals Soc. China* **2018**, *28*, 1413–1423. [CrossRef]
7. Xie, L.X.; Lu, W.B.; Zhang, Q.B.; Jiang, Q.H.; Chen, M.; Zhao, J. Analysis of damage mechanisms and optimization of cut basting model under high in-situ stresses. *Tunn. Undergr. Space Technol.* **2017**, *66*, 19–33. [CrossRef]
8. Wang, S.F.; Li, X.B.; Rao, J.R.; Gong, F.Q.; Li, X.; Du, K.; Tao, M.; Huang, L.Q.; Du, S.L. Experimental investigation of rock breakage by a conical pick and its application to non-explosive mechanized mining in deep hard rock. *Int. J. Rock Mech. Min. Sci.* **2019**, *122*, 104063. [CrossRef]
9. Wang, S.F.; Sun, L.C.; Tang, Y.; Jing, Y.; Li, X.B.; Rao, J.R. Field application of non-blasting mechanized mining using high-frequency impact hammer in deep hard rock mine. *Trans. Nonfer. Metals Soc. China* **2022**, *32*, 3051–3064. [CrossRef]
10. Hong, Z.X.; Tao, M.; Cui, X.J.; Wu, C.Q.; Zhao, M.S. Experimental and numerical studies of the blast-induced overbreak and underbreak in underground roadways. *Undergr. Space* **2023**, *8*, 61–79. [CrossRef]
11. Taylor, L. Microcrack-induced damage accumulation in brittle rock under dynamic loading. *Comp. Meth. Applied Mech. Eng.* **1985**, *55*, 301–320. [CrossRef]
12. Liu, L.Q.; Katsabanis, P.D. Development of a continuum damage model for blasting analysis. *Int. J. Rock Mech. Mining Sci.* **1997**, *34*, 217–231. [CrossRef]
13. Zhang, Y.Q.; Hao, H.; Lu, Y. Anisotropic dynamic damage and fragmentation of rock materials under explosive loading. *Int. J. Eng. Sci.* **2003**, *41*, 917–929. [CrossRef]
14. Yang, J.H.; Lu, W.B.; Hu, Y.G.; Chen, M.; Yan, P. Numerical simulation of rock mass damage evolution during deep-buried tunnel excavation by drill and blast. *Rock Mech. Rock Eng.* **2015**, *48*, 2045–2059. [CrossRef]
15. Xie, L.X.; Lu, W.B.; Zhang, Q.B.; Jiang, Q.H.; Wang, G.H.; Zhao, J. Damage evolution mechanisms of rock in deep tunnels induced by cut blasting. *Tunn. Undergr. Space Technol.* **2016**, *58*, 257–270. [CrossRef]
16. Tao, M.; Li, X.B.; Wu, C.Q. Characteristics of the unloading process of rocks under high initial stress. *Comp. Geotech.* **2012**, *45*, 83–92. [CrossRef]
17. Brannon, R.M.; Leelavanichkul, S. *Survey of Four Damage Models for Concrete*; Sandia Report, Sand 2009-5544; Sandia National Laboratories: Albuquerque, NM, USA; Livermore, CA, USA, 2009; pp. 1–82.
18. Riedel, W.; Thoma, K.; Hiermaier, S.; Schmolinske, E. Penetration of reinforced concrete by BETA-B-500 numerical analysis using a new macroscopic concrete model for hydrocedes. In Proceedings of the 9th International Symposium on the Effects of Munitions with Structures, Berlin, Germany, 3–7 May 1999.

19. Kong, X.Z.; Qin, F.; Hao, W.; Peng, Y. Numerical predictions of cratering and scabbing in concrete slabs subjected to projectile impact using a modified version of HJC material model. *Int. J. Impact Eng.* **2016**, *95*, 61–71. [CrossRef]
20. Hao, H.; Wu, C.Q.; Seah, C.C. Numerical Analysis of Blast-Induced Stress Waves in a Rock Mass with Anisotropic Continuum Damage Models Part 2: Stochastic Approach. *Rock Mech. Rock Eng.* **2002**, *35*, 95–108. [CrossRef]
21. Li, H.Y.; Shi, G.Y. A dynamic material model for rock materials under conditions of high confining pressures and high strain rates. *Int. J. Impact Eng.* **2014**, *89*, 38–48. [CrossRef]
22. Yi, C.P.; Sjöberg, J.; Johansson, D. Numerical modelling for blast-induced fragmentation in sublevel caving mines. *Tunn. Undergr. Space Technol.* **2017**, *68*, 167–173. [CrossRef]
23. Wang, S.F.; Liu, K.H.; Wang, S.Y.; Liang, Y.T.; Tian, F.C. Three-dimensional stochastic distribution characteristics of void fraction in longwall mining-disturbed overburden. *Bull. Eng. Geol. Environ.* **2022**, *81*, 414. [CrossRef]
24. Zhang, Z.X. Rock blasting in open cut and tunneling. In *Rock Fracture and Blasting*; Elsevier: Amsterdam, The Netherlands, 2016; pp. 334–352.
25. Hao, H.; Wu, C.Q.; Zhou, Y. Numerical Analysis of Blast-Induced Stress Waves in a Rock Mass with Anisotropic Continuum Damage Models Part 1: Equivalent Material Property Approach. *Rock Mech. Rock Eng.* **2002**, *35*, 79–94. [CrossRef]
26. Ma, G.W.; An, X.M. Numerical simulation of blasting-induced rock fractures. *Int. J. Rock Mech. Min. Sci.* **2008**, *45*, 966–975. [CrossRef]
27. Yilmaz, O.; Unlu, T. Three dimensional numerical rock damage analysis under blasting load. *Tunn. Undergr. Space Technol.* **2013**, *38*, 266–278. [CrossRef]
28. Hartmann, T.; Pietzsch, A.; Gebbeken, N. A Hydrocode Material Model for Concrete. *Int. J. Protective Struc.* **2010**, *1*, 443. [CrossRef]
29. Weerheijm, J.; Van Doormaal, J.C.A.M. Tensile failure of concrete at high loading rates: New test data on strength and fracture energy from instrumented spalling tests. *Int. J. Impact Eng.* **2007**, *34*, 609–626. [CrossRef]
30. Polanco-Loria, M.; Hopperstad, O.S.; Børvik, T.; Berstad, T. Numerical predictions of ballistic limits for concrete slabs using a modified version of the HJC concrete model. *Int. J. Impact Eng.* **2008**, *35*, 290–303. [CrossRef]
31. Zhang, Q.B.; Zhao, J. A Review of Dynamic Experimental Techniques and Mechanical Behaviour of Rock Materials. *Rock Mech. Rock Eng.* **2014**, *47*, 1411–1478. [CrossRef]
32. Gebbeken, N.; Greulich, S. A new material model for SFRC under high dynamic loadings. In Proceedings of the 1th international Symposium on Interaction of the Effects of Munitions with Structures (ISIEMS), Mannheim, Germany, 5–9 May 2003.
33. Tedesco, J.W.; Powell, J.C.; Ross, C.A.; Hughes, M.L. A strain-rate-dependent concrete material model for ADINA. *Comp. Struct.* **1997**, *64*, 1053–1067. [CrossRef]
34. Liu, K.; Wu, C.Q.; Li, X.B.; Li, Q.Y.; Fang, J.G.; Liu, J. A modified HJC model for improved dynamic response of brittle materials under blasting loads. *Comp. Geotech.* **2020**, *123*, 103584. [CrossRef]
35. Hu, Y.G.; Lu, W.B.; Chen, M.; Yan, P.; Yang, J.H. Comparison of blast-induced damage between presplit and smooth blasting of high rock slope. *Rock Mech. Rock Eng.* **2014**, *47*, 1307–1320. [CrossRef]
36. Hallquist, J.O. *Ls-Dyna Theory Manual*; Livermore Software Technology Corporation: Livermore, CA, USA, 2006.
37. Liu, K.; Qiu, J.D. Investigation of Burn Cut Parameters and Model for One-Step Raise Excavation Based on Damage Evolution Mechanisms. *Geofluids* **2020**, *2020*, 8879477. [CrossRef]
38. Kononenko, M.; Khomenko, O. New theory for the rock mass destruction by blasting. *Min. Miner. Depos.* **2021**, *15*, 111–123. [CrossRef]
39. Li, X.B.; Weng, L. Numerical investigation on fracturing behaviors of deep-buried opening under dynamic disturbance. *Tunn. Undergr. Space Technol.* **2016**, *54*, 61–72. [CrossRef]
40. Weng, L.; Huang, L.Q.; Taheri, A.; Li, X.B. Rockburst characteristics and numerical simulation based on a strain energy density index: A case study of a roadway in Linglong gold mine, China. *Tunn. Undergr. Space Technol.* **2017**, *69*, 223–232. [CrossRef]
41. Cai, M.F.; Qiao, L.; Li, C.H. Measuring results and regularity of in situ stress in Xincheng gold mine. *Nonfer. Metals.* **2000**, *52*, 1–6. (In Chinese)
42. Xie, H.P.; Gao, F.; Ju, Y. Research and development of rock mechanics in deep ground engineering. *Chin. J. Rock Mech. Eng.* **2015**, *34*, 2161–2178. (In Chinese)
43. Zare, S.; Bruland, A. Comparison of tunnel blast design models. *Tunn. Undergr. Space Technol.* **2006**, *21*, 533–541. [CrossRef]
44. Holmquist, T.J.; Johnson, G.R.; Cook, W.H. A computational constitutive model for concrete subjected to large strains, high strain rates, and high pressures. In Proceedings of the 14th International Symposium on Ballistics, Quebec, QC, Canada, 26–29 September 1993; pp. 591–600.
45. Shahzamanian, M.M. Implementation of a rate dependent tensile failure model for brittle materials in ABAQUS. *Int. J. Impact Eng.* **2016**, *97*, 127–147. [CrossRef]
46. Zhu, Z.M.; Xie, H.P.; Mohanty, B.H. Numerical investigation of blasting-induced damage in cylindrical rocks. *Int. J. Rock Mech. Min. Sci.* **2008**, *45*, 111–121.

materials

MDPI

Article

Analysis of Electrical Resistivity Characteristics and Damage Evolution of Soil–Rock Mixture under Triaxial Shear

Mingjie Zhao [1,2], Songlin Chen [1,*], Kui Wang [1] and Gang Liu [1]

1 Engineering Research Center of Diagnosis Technology and Instruments of Hydro-Construction, Chongqing Jiaotong University, Chongqing 400074, China; m.j.zhao@163.com (M.Z.); anhuiwk@163.com (K.W.); cqjtulg@163.com (G.L.)
2 School of Civil Engineering and Architecture, Chongqing University of Science and Technology, Chongqing 401331, China
* Correspondence: cqchensl@163.com

Abstract: Construction of engineering structures in geomaterials with soil–rock mixture (S-RM) is often a challenging task for engineers. When analyzing the stability of the engineering structures, the mechanical properties of S-RM often receive the most attention. To study the mechanical damage evolution characteristics of S-RM under triaxial loading conditions, a modified triaxial apparatus was used to conduct shear test on S-RM, and the change of electrical resistivity was measured simultaneously. The stress–strain–electrical resistivity curve and stress–strain characteristics under different confining pressures were obtained and analyzed. Based on the electrical resistivity, a mechanical damage model was established and verified to analyze the damage evolution regularities of S-RM during shearing. The results show that the electrical resistivity of S-RM decreases with increasing axial strain and that the differences in decrease rates correspond to the different deformation stages of the samples. With the increase in loading confining pressure, the stress–strain curve characteristics change from a slight strain softening to a strong strain hardening. Additionally, an increase in rock content and confining pressure can enhance the bearing capacity of S-RM. Moreover, the derived damage evolution model based on electrical resistivity can accurately characterize the mechanical behavior of S-RM under triaxial shear. Based on the damage variable *D*, it is found that the damage evolution process of S-RM can be divided into a non-damage stage, a rapid damage stage and a stable damage stage. Furthermore, the structure enhancement factor, which is a model modification parameter for the effect of rock content difference, can accurately predict the stress–strain curves of S-RMs with different rock contents. This study sets the stage for an electrical-resistivity-based monitoring method for studying the evolution of internal damage in S-RM.

Keywords: soil–rock mixture (S-RM); electrical resistivity; damage model; triaxial shear; mechanical behavior

Citation: Zhao, M.; Chen, S.; Wang, K.; Liu, G. Analysis of Electrical Resistivity Characteristics and Damage Evolution of Soil–Rock Mixture under Triaxial Shear. *Materials* **2023**, *16*, 3698. https://doi.org/10.3390/ma16103698

Academic Editor: Krzysztof Schabowicz

Received: 5 April 2023
Revised: 28 April 2023
Accepted: 4 May 2023
Published: 12 May 2023

1. Introduction

Soil–rock mixture (S-RM) is widely distributed around the world to meet a wide range of architectural demands; "material origin" and "geological dynamic action formed by accumulation" jointly determine S-RMs in nature. Xu et al. [1] concluded that the origins of S-RM are very complex and mainly include the origin types of remnant slope accumulation, collapse accumulation, scour accumulation, glacial accumulation and artificial accumulation. Due to the complexity of geological origin and formation process, S-RM is generally different from traditional soil types. Xu et al. [2] and Afifipour et al. [3] distinguish S-RM from traditional geotechnical mass from the perspective of key physical indices and divide it into an emerging geotechnical medium system. Wang et al. [4] concluded that the main characteristics of S-RM include a complex composition of components, the mixing of different blocks with a range of shape and size, a stochastic distribution of rock blocks and an evident scale effect. In recent years, S-RM has been used in various applications such as earth–rockfill dams, high-fill works, subgrade works or foundation aggregates for construction work [5].

Among many factors affecting the mechanical properties of S-RM, the rock block content is always the first factor to be considered. Yao et al. [6] investigated the effect of rock content and shape distribution on the shear strength of S-RM under different stress conditions. The results showed that the contribution of contact force within gravel particles increased with increasing rock content. Before loading, the rock and soil in S-RM have a good coupling state; after being loaded, the breakage of coarse particles occurs, leading to the formation and expansion of microcracks. These anomalous changes are often regarded as the process of cumulative damage of S-RM and determine the magnitude of the load-bearing capacity of material and the destabilization damage process. In this paper, localized deterioration caused by the formation, convergence and propagation of microcracks inside the material is defined as damage. Therefore, accurate characterization of the damage evolution process of S-RM has gradually become a focus of interest for researchers.

Electrical resistivity is a physical quality of a material that characterizes the conductive properties of a material. For porous media such as soil and rock, their electrical resistivity is often considered as the representation of the spatial and temporal variability of many physical properties of soil (i.e., internal structure, water content or fluid composition) [7]. Electrical resistivity has been used by many researchers in various studies on soil [8,9], rock [10], coal [11,12] and concrete [13] to interpret the various properties of the research object. Khurshid and Afgan [14] investigated the effects of the injection of engineered water into carbonate reservoirs on electrical conductivity, ion mobility, electrical double layer thickness and the related oil recovery. Some quantitative and qualitative correlations between electrical resistivity and soil parameters such as cohesion, friction angle and plasticity index have been obtained. Moreover, the electrical resistivity characteristics of S-RM have also been initially investigated. According to the experimental results from Zhao et al. [15], variables such as rock content, compaction and water content all have an effect on the electrical resistivity variations of S-RMs. Moreover, electrical resistivity models, such as the series–parallel connection model [16], were established to characterize the relationship between the electrical resistivity and the physical property parameters of S-RM. Wang et al. [17] obtained the response regularity of the electrical resistivity of S-RMs with different rock contents and compactness during the water absorption–saturation process. Wang et al. [18] studied the variations in matrix suction and electrical resistivity in S-RM and found that there was a correlation between the matrix suction and the change in the electrical resistivity of S-RM.

Currently, most of the investigations conducted on the damage evolution of different materials are generally focused on brittle materials such as rock, concrete and coal. In a study on rock damage, Wang et al. [19] determined the damage variable of rock under loading conditions based on acoustic emission measurements, and the results showed that the damage variable increased gradually with the increasing deformation process. Zhang et al. [20] studied the variations in acoustic emission and infrared radiation of granite under uniaxial compression; they found that the complementary damage model based on sound–heat could accurately characterize the damage evolution characteristics of rock. Khurshid and Fujii [21] examined the influence of low-temperature CO_2 on the decrease of formation breakdown pressure and the associated reservoir rock damage from a geomechanical prospective. Chung et al. [22] found that electrical resistivity had distinct responses to the damage state of concrete such as the freeze–thaw cycle and drying. Cao et al. [23] studied the variation in the electrical resistance of concrete under loading conditions and found that there was a synchronous relationship between the compressive damage and the change in concrete electrical resistance. When studying the relationship between concrete electrical resistivity and compressive damage, Zeng et al. [24] reported that the derived relationship between the damage variable and electrical resistivity is well correlated with the experimental data, in which electrical resistivity showed a trend of decreasing first and then increasing. Li et al. [25] applied acoustic emission to monitor the damage evolution process of coal mass induced by multi-stage loading and pointed out that the damage evolution in coal mass can be revealed and characterized by the acoustic emission. Similar conclusions had been drawn by Jia et al. [26] through a triaxial compression test, where the different damage evolution regularities of coal mass at different depths can be effectively comprehended by the acoustic emission. Based on the electrical resistivity testing data, the resistivity damage model of residual soil under uniaxial loading was proposed [27]; however, the general applicability of the damage model needs to be further explored. These investigations are capable of providing beneficial quantitative information for the damage assessment but are insufficient, as most of them rarely focus on the research of S-RM.

In summary, the electrical resistivity is closely related to the traditional geotechnical parameters of S-RM. Although many significant efforts have been made to investigate the electrical properties of S-RM, almost all the experiments mainly focus on the static measurement of indoor samples using the four-electrode or two-electrode methods. However, there is a lack of an in-depth study on the mechanical properties of S-RM based on the electrical resistivity method. Moreover, the quantitative study of the damage evolution of S-RM based on electrical resistivity in the process of shearing has not been reported, and the damage evolution characteristics of S-RM are the basis for understanding the formation and propagation of internal cracks. Therefore, it is of crucial significance to develop a method for characterizing the internal damage evolution of S-RM.

This study aims to explore the influence of triaxial shear behavior on the electrical response to applied voltage and the damage evaluation method based on electrical resistivity monitoring. In this paper, a modified triaxial apparatus was used to measure the electrical resistivity of S-RM subjected to triaxial shear, and the electrical resistivity method was used to evaluate the change in damage variable D, which quantifies the degree of damage. The relationship between the electrical resistivity and the stress–strain of S-RM was derived. A mechanical damage model based on the electrical resistivity of S-RM was established, and the calculated results of the model were compared with the measured results to verify the reliability of the damage model. Additionally, the damage evolution regularity of S-RM was discussed from the perspective of electrical resistivity, which enriches the application of the electrical resistivity method in the study of S-RM damage. Based on the electrical resistivity results, we modified the established damage model to meet the mechanical damage analysis of S-RM with different rock contents.

2. Materials and Methods

2.1. Experimental Setup

In order to measure the electrical resistivity of S-RM samples under triaxial shear, the conventional triaxial apparatus was modified. The schematic drawing and experimental setup of the modified triaxial apparatus are illustrated in Figure 1. The apparatus can measure the change in the electrical resistivity of the sample during shearing simultaneously. The modified triaxial apparatus was manufactured by the Yongchang Science and Education Instrument Manufacturing Co., Ltd., Liyang, Jiangsu, China. It consists of four components: axial loading equipment, pressure chamber, confining pressure and in–out water control system and data acquisition system. The inner dimensions of the cylindrical sample are 100 mm (diameter) × 200 mm (height). The maximum confining pressure that the pressure chamber can bear and the maximum axial force that the axial loading system can apply are 2000 kPa and 250 kN, respectively. Various test variables such as confining pressure, axial force, electrical resistivity, displacement, pore pressure, etc., are recorded and stored by the data acquisition control system; the confining pressure and axial loading can be controlled automatically according to their target values. To measure the electrical resistivity, two metal electrodes are attached to the top cap and base pedestal of the triaxial apparatus. The electrodes are made of porous stainless-steel discs with a diameter of 100 mm. Two thin copper wires wrapped with insulating material lead out from the base pedestal of the triaxial apparatus. The copper wires in the pressure chamber are connected to the upper and lower electrodes, and the copper wires outside the pressure chamber are connected to the electrical resistivity measurement device.

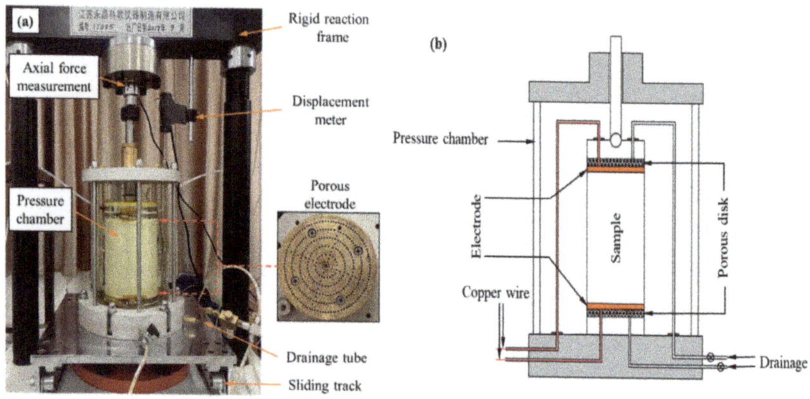

Figure 1. (**a**) Experimental setup. (**b**) Schematic drawing of the modified triaxial apparatus.

To date, the two-electrode and four-electrode methods have been adopted by most researchers for the laboratory electrical resistivity measurement. In the four-electrode method the most important consideration is related to the disturbance of test sample introduced by the insertion of electrical probes. Moreover, the four-electrode method is difficult to adopt in triaxial tests due to the difficulty in determining the distance between the electrodes [28]. In this study, the two-electrode method is adopted for the electrical resistivity measurement. For the two-electrode method, since the same pair of electrodes are used for both the injection of current and the measurement of potential difference, a polarization effect usually occurs. In the present work, the polarization effect can be avoided by changing the polarity of the current electrode. In order to reduce the contact resistance between the two ends of the sample and the metal electrodes, both ends of the sample were evenly coated with conductive graphite before loading. The power supply provides an output voltage of 9 V.

2.2. Materials

The formation lithology of the Chongqing region, China, consists mainly of silty clay, sandstone, mudstone and sandy mudstone. These materials are widely used to construct levees and earthen dams in China and are internally unstable in essence. In order to facilitate the local selection of the experimental materials, a mixture of mudstone and clay from this region was selected as the raw material for this study. The raw materials retrieved from the site were sieved and then secondly proportioned. The particles with a diameter of 0–20 mm were then selected as the experimental materials (see Figure 2a). The soil–rock threshold (S-RT) can be used to distinguish the particle size limits of "soil matrix" and "rock block" in S-RM; it is a very significant physical property index of S-RM. According to some previous studies [29,30], the method of determining the S-RT has been given. Medley proposed that the S-RT is not a fixed value but should be a variable value related to the area of the S-RM, denoted as $0.05L_c$, where L_c is the scale of engineering features such as the tunnel diameter, slope height and laboratory sample diameter. Taking into account the geomechanical contribution of the blocks, Zhang et al. [30] and Tu et al. [31] gave the S-RT criterion as 5 mm. Meanwhile, in this paper, considering the size of the laboratory test samples, the S-RT criterion of 5 mm is adopted. This means that the S-RM could be defined as "soil" if the particle size less than 5 mm, whereas it could be defined as "rock" if the particle size greater than 5 mm. The main physical property indexes of S-RM retrieved from the site are summarized in Table 1. Based on X-ray diffraction, the specific mineral composition of the rock is obtained, as shown in Table 2. Generally, the mechanical properties of S-RM depend on the joint action of soil and rock only when the rock content is between 25% and 75%. Considering the rock content of the raw material, four rock contents of 20%, 30%, 40% and 50% are set, and the corresponding particle size distribution of each rock content is shown in Figure 2b. Figure 3 shows an SEM image of the soil particles' surfaces. We found that the surfaces of the particles had an irregular scaly stacking structure with different sizes, and a small number of holes presented after local amplification.

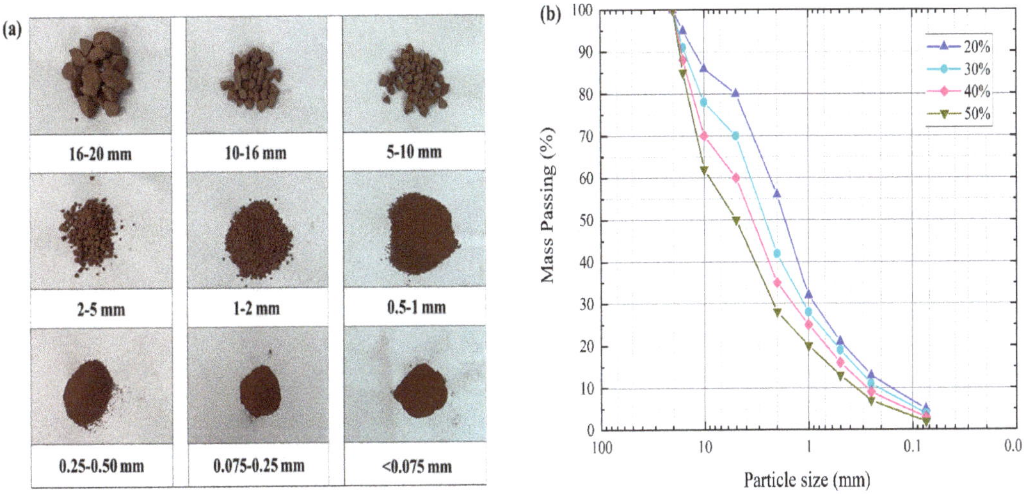

Figure 2. Material used for S-RM samples. (**a**) Photographs of each particle group. (**b**) Particle size distribution.

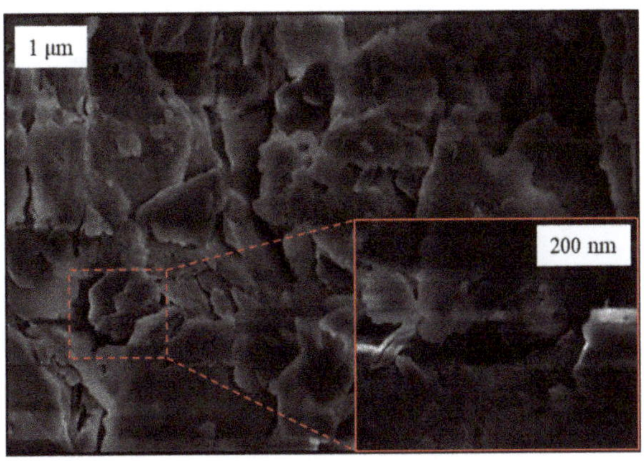

Figure 3. SEM image of particles' surfaces.

Table 1. Main physical property indexes of S-RM.

R/%	w_{na}/%	G_s	w_{op}/%	$\rho_{d\,max}$/(g·cm^{-3})
47.29	2.57	2.72	7.94	1.91

R = rock content; w_{na} = natural water content; G_s = specific gravity; w_{op} = optimum water content; $\rho_{d\,max}$ = maximum dry density.

Table 2. Mineral composition of rock (in %).

Quartz	Illite	Albite	Kaolinite	Chlorite	Calcite	Hematite
48.8	22.0	17.9	2.7	5.5	1.8	1.2

In addition, the maximum dry density at four rock contents was determined through the proctor compaction test. The target dry density is determined by considering the maximum dry density of raw materials and S-RM with different rock contents, that is, the target dry density is 1.95 g/cm^3 with 50% rock content and the target dry density of remaining rock content decreases in order. The water content of the sample preparation is 5%. The values of the main physical parameters of the S-RM samples with different rock contents are shown in Table 3.

Table 3. Main physical parameters of S-RM samples.

R/%	*n*	*w*/%	$\rho_{d\,max}$/(g·cm^{-3})	ρ_t/(g·cm^{-3})
20	0.353	5	1.71	1.65
30	0.318	5	1.78	1.75
40	0.283	5	1.87	1.85
50	0.249	5	1.98	1.95

R = rock content; *n* = porosity; *w* = water content; $\rho_{d\,max}$ = maximum dry density; ρ_t = target dry density.

2.3. Test Procedure

Triaxial compression tests were conducted under the conventional path, which consists of three basic steps. The first step is the preparation of the S-RM samples. Firstly, soil and rock are oven-dried at 103 °C for 24 h; the amounts of soil and rock required to achieve the target dry density are weighed according to the particle size distribution curve with different rock contents. Secondly, the required amount of water (gravimetric water content of 5%) is weighed and mixed evenly with the S-RM. Finally, the prepared S-RM is put into a cylindrical mold (which has a height of 200 mm and a diameter of 100 mm) in three layers and compacted in a stratified manner. In order to enhance the interlayer connectivity of the samples, artificial chiseling is applied to the interlayer contact surfaces. Double rubber membranes are utilized to prevent the membrane from being punctured by the sharp corners of irregular-shaped rock blocks.

The second step was to saturate and consolidate the S-RM samples. The prepared S-RM samples were installed on the base pedestal of the triaxial apparatus. In the whole process of installation, the disturbance to the sample was minimized. After the installation, in order to speed up the saturation process, the backpressure saturation method was adopted to saturate the S-RM samples. After the S-RM samples were saturated, the confining pressure was adjusted to the design value and then consolidation tests were conducted. During consolidation, pore water was allowed to be discharged and the confining pressure remained stable. When the drainage volume changes very little and almost no longer changes, the consolidation of the S-RM samples was considered to be completed. In order to capture the electrical resistivity data of tested sample more accurately, the consolidation tests started when the data were stable.

The third step was to conduct a shear test on the S-RM samples. Given the long time required for the shearing test, the time interval for the electrical resistivity data acquisition was set to 120 s. After the consolidation, the sample underwent a slight deformation, resulting in a decrease in sample height. Therefore, it was necessary to make full contact between the pressure shaft and the top cap before starting the shearing test. Once the electrical resistivity data remained stable, drained triaxial shear tests were conducted on the S-RM samples under different confining pressure levels (250 kPa, 400 kPa, 550 kPa and 700 kPa). The shear mode is strain control, and the shear rate is set to 0.13 mm/min. When the axial strain reached 15% (axial displacement of 30 mm), the shearing process was terminated. As a result, the deviatoric stress, axial strain and electrical resistivity were recorded and stored by the data acquisition system in the shearing process. In addition, we noted that the contact resistance in this experiment was diminished since the polishing treatment of the electrode surfaces was conducted after each measurement. Figure 4 shows all the S-RM samples after the triaxial shear was completed.

Figure 4. Photograph of S-RM samples after the shear.

3. Experimental Results and Discussion

3.1. Stress–Strain–Electrical Resistivity Curve

Through the triaxial test results, it is found that the stress–strain–resistivity variation regularities of the S-RM samples with the same rock contents under four confining pressures are similar. Therefore, some typical resistivity variation curves are selected for response regularity analysis. Figure 5 plots the typical response of electrical resistivity with increasing deviator stress. It can be observed that the electrical resistivity varies continuously with increasing deviator stress, which also provides the possibility for the establishment of a subsequent electrical resistivity damage model.

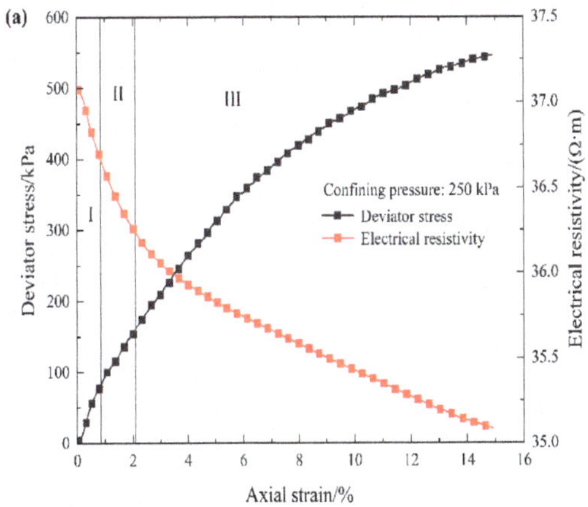

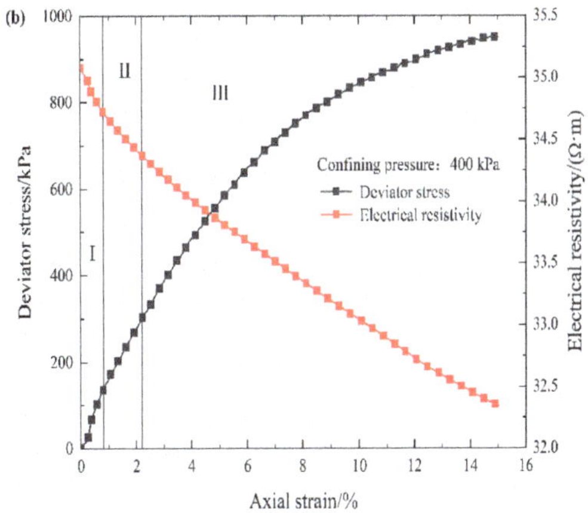

Figure 5. *Cont.*

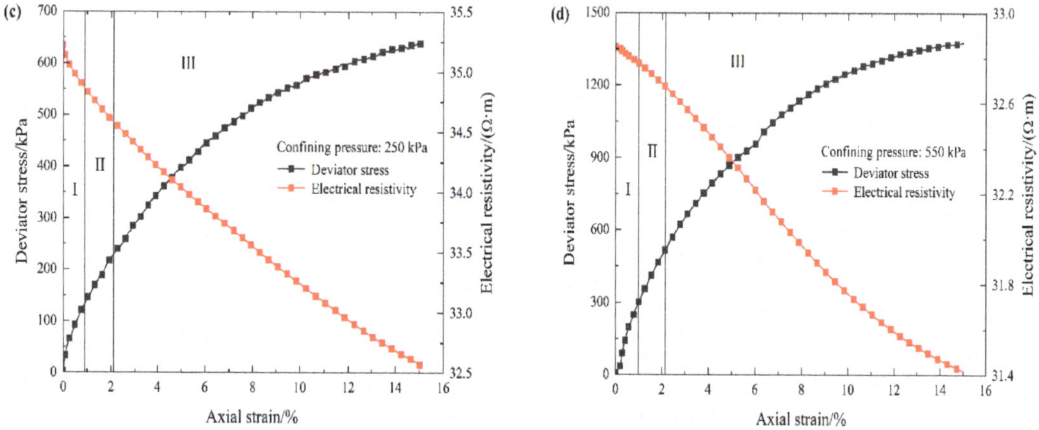

Figure 5. Stress–strain–electrical resistivity curves of S-RM samples. (**a**) Rock content 20%. (**b**) Rock content 30%. (**c**) Rock content 40%. (**d**) Rock content 50%.

In general, in the process of complete strain, the electrical resistivity curve of the S-RM samples shows a gradually decreasing trend that presents an irregular "slightly concave shaped". The volumetric strain curves of the loaded coal and rock mass have obvious inflection points; based on this, for convenience, their stress–strain curves are commonly divided into different deformation stages in existing experiments to study the electrical resistivity variation properties [10–12]. In this paper, there is no obvious inflection point in the volumetric strain curve of S-RM, as shown in Figure 6. Therefore, combining the electrical resistivity evolution characteristics of S-RM samples and the classification methods of different deformation stages of coal and rock mass, the deformation process of samples is roughly divided into three distinct deformation stages, and there are differences in the reduction rate of electrical resistivity under each stage. In the initial compaction stage (I), there is a short and rapid decrease in electrical resistivity; in the elastic deformation stage (II), the reduction rate of the electrical resistivity value diminishes with increasing axial strain; in the yield deformation stage (III), the reduction rate further decreases. The electrical resistivity of S-RM samples generally follows the following variation regularity: it initially decreases rapidly, then declines slowly and finally reduces gently. As reported in some triaxial shear test studies [28,32], the obtained stress–strain curves are also divided into several distinct deformation stages.

However, the electrical resistivity response mechanism of S-RM is distinct in different deformation stages. It can be surmised that the decrease in electrical resistivity under the initial compaction stage is dependent upon the closure of pores and microcracks in the sample and the discharge of high-resistance gas. The deceleration of electrical resistivity under the elastic deformation stage is mainly influenced by the increase in sample saturation and the electron conductivity. The slow decrease in electrical resistivity during the yield stage can mainly be attributed to the progressive discharge of pore water in the S-RM sample and the qualitative change in its internal structure. There are similar assumptions to explain the change in soil electrical resistivity under freeze–thaw cycles in the field of soil damage evaluation [33].

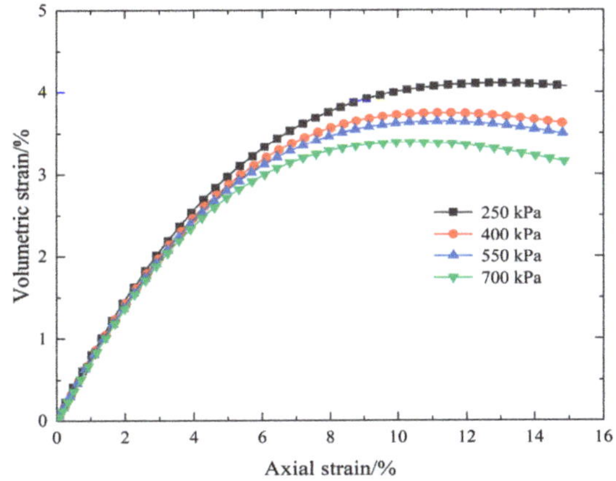

Figure 6. Volumetric strain curves of S-RM (rock content 40%).

3.2. *Stress–Strain Characteristics of S-RM*

The stress–strain curves of each rock content under four confining pressures are plotted in Figure 7. By analyzing and summarizing the stress–strain curves, three different variation regularities can be derived, as shown in Figure 8a. Such variation characteristics are consistent with the results obtained by Zhang et al. [34]. This implies that the stress–strain characteristics of S-RM gradually change from a slight strain softening to a strong strain hardening as the confining pressure increases. The increasing rate of the deviator stress is fast at the beginning and then slows down soon after. Unlike rock mass, the stress–strain curve of S-RM has no obvious peak stress phenomenon. This indicates that S-RM has the strength property of being able to withstand a certain loading even in the late stage of strain. Therefore, it can be predicted that if the confining pressure is further increased, then the strain hardening characteristics of S-RM will be further enhanced. All of the S-RM samples show radial dilatation deformation that is more prominent in the middle of the samples, which can be attributed to the bottom layer of the samples withstanding the major compaction energy during the sample preparation and the pressure shaft acting directly on the top of the samples. Furthermore, in the late stage of strain, the energy dissipation caused by the partial coarse particle breakage is compensated for by adjusting the internal structure. Therefore, the apparent particle breakage does not result in a corresponding reduction in the deviator stress.

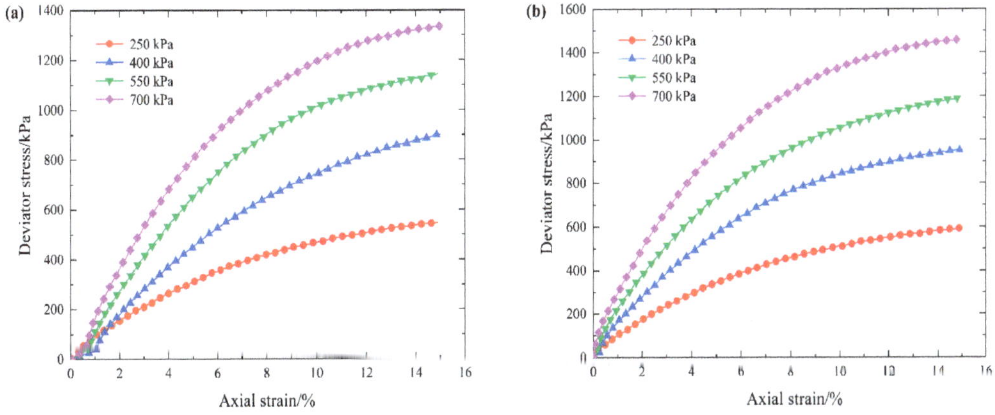

Figure 7. *Cont.*

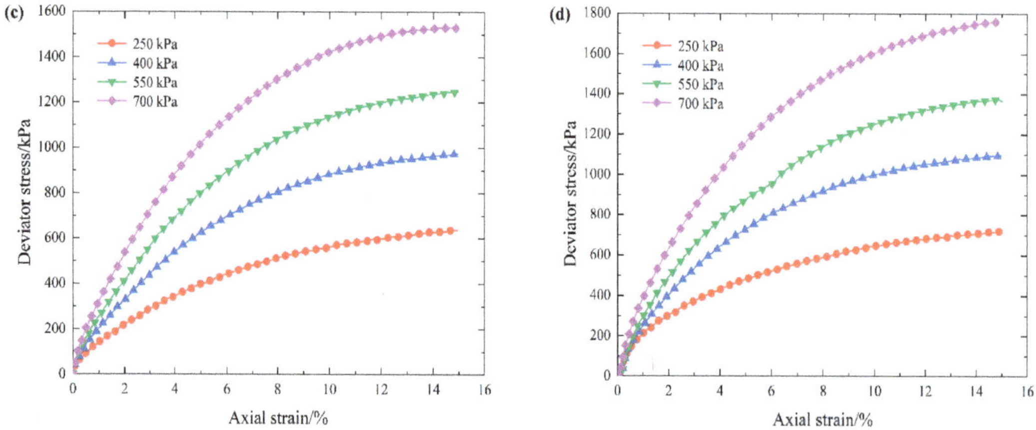

Figure 7. Stress–strain curves of S-RM for different confining pressures. (**a**) Rock content 20%. (**b**) Rock content 30%. (**c**) Rock content 40%. (**d**) Rock content 50%. The red circle symbols, the blue triangle symbols, the green inverted triangle symbols and the purple rhombus symbols represent confining pressures of 250 kPa, 400 kPa, 550 kPa and 700 kPa, respectively.

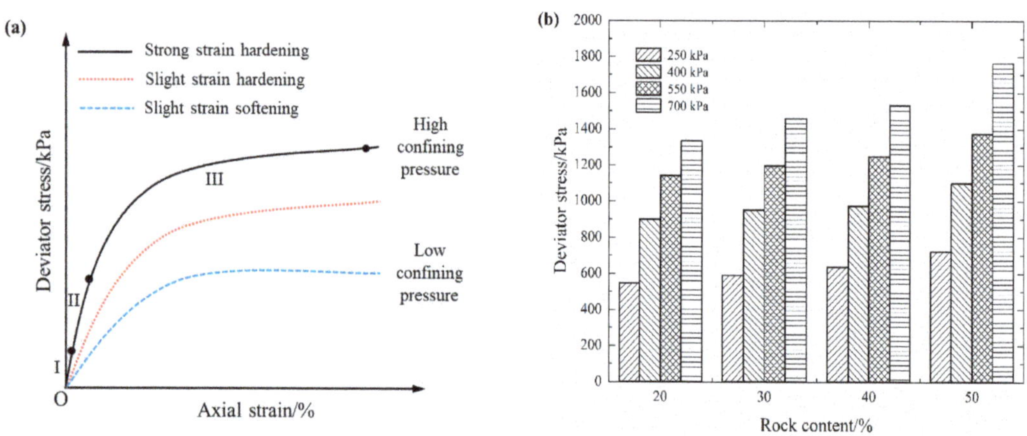

Figure 8. Variation regularities of deviator stress. (**a**) Variation of deviator stress with axial strain. (**b**) Variation of deviator stress with rock content.

Figure 8b shows the maximum deviator stress that S-RM samples can withstand under different confining pressures. When the confining pressure is constant, the maximum deviator stress increases with the increase in rock content. We refer to the difference in shear strength as the "enhancement effect of rock block". The cause of such a phenomenon is that the skeleton structure of the rock blocks has been generated inside the S-RM samples and has the capacity to resist deformation under external loading. Although the samples with different rock contents have different initial compactions, it is undeniable that increasing rock content, skeleton structure formation and the interlock effect among the coarse particles play a significant role in enhancing the strength characteristics of S-RM. Moreover, with the increase in confining pressure, the corresponding deviator stress of the samples with the same rock content also presents an incremental trend.

3.3. Establishment of Mechanical Damage Model

On the basis of the original electrical resistivity data, we further establish the damage model of S-RM. As mentioned earlier, the variation characteristics for electrical resistivity and deviator stress with axial strain provide the possibility for the establishment of a damage model. Specifically, if the sample is in the initial compaction stage, there is almost no structure damage inside it and the elastic deformation can be restored to the initial state when no external loading is applied. In this case, we suppose that no damage occurs inside the sample during the compaction and elastic deformation stages. From the test results (see Figure 5 for part), it can be seen that the yield deformation for the studied sample starts to occur when the axial strain is about 2%. Therefore, the axial strain of 2% is taken as the initial strain for damage propagation, which is denoted as ε_2; the corresponding electrical resistivity is denoted as ρ_{ini}.

For the convenience of analysis, it is assumed that the damaged sample consists of the middle section with structural damage and the two ends with structural integrity [27]. The structural damage mainly includes the changes in the failure morphology, deformation behavior and other characteristics, such as shear dilatation and strain hardening, which are closely related to the evolution of the meso-structure of S-RM. Assuming that the cross-sectional area and the height of the S-RM sample in the initial stage of damage propagation are S_0 and L_0, respectively, the heights of two ends with intact structure are L_0k_1 and L_0k_2, where k_1 and k_2 are the proportion coefficients of the height of the two ends to the height of the entire sample, respectively. The height of the intermediate damage section is $L_0\beta$, where β is the proportion coefficient of the middle section to the height of the entire sample. The damage mode of the S-RM sample is shown in Figure 9. k_1, k_2 and β satisfy the following formula:

$$k_1 + k_2 + \beta = 1 \tag{1}$$

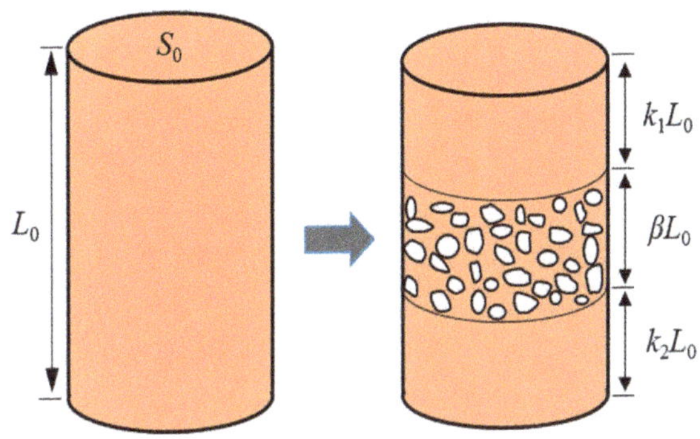

Figure 9. The schematic diagram of the S-RM damage mode.

Assuming that the S-RM sample is a uniform conductive cylinder when the strain is 2%, then the resistance can be written as follows:

$$R = \frac{\rho L_0}{S_0} = \frac{\rho L_0^2}{V_0} \tag{2}$$

where R is the resistance of the S-RM sample, Ω; ρ is the measured electrical resistivity, $\Omega{\cdot}$m; V_0 is the volume of the S-RM sample in the initial stage of damage propagation, m^3.

The sample height will gradually change during the shearing process, so the measured electrical resistivity may be different from the actual resistivity. Considering the effect of the dimensional change, the amended electrical resistivity value is obtained as:

$$\rho_{\text{r}} = \frac{\rho}{(1-\varepsilon)^2} \tag{3}$$

where ρ_r is the realistic electrical resistivity, Ω·m; ε is the axial strain, %.

Equation (2) can be written as follows:

$$R = \frac{\rho_r (1-\varepsilon_2)^2 L_0^2}{V_0} \tag{4}$$

When the axial strain of the S-RM sample is ε, the resistance can be written as follows:

$$R_\phi = k_1 \frac{\rho_{r,ini} L_0^2 (1-\varepsilon)^2}{V_0} + k_2 \frac{\rho_{r,ini} L_0^2 (1-\varepsilon)^2}{V_0} + \beta \frac{\rho_{r,ini} L_0^2 (1-\varepsilon)^2}{\phi_v V_0} = \frac{\rho_\phi L_0^2}{V_0} \tag{5}$$

where ϕ is the extent of damage development; ρ_ϕ is the damage electrical resistivity (measured value); $\rho_{r,ini}$ is the realistic electrical resistivity under initial damage, which is expressed as $\rho_{r,ini} = \frac{\rho_{ini}}{(1-\varepsilon_2)^2}$; ϕ_v is the damage volume coefficient of the sample.

Equation (5) can be further simplified:

$$\frac{\rho_\phi (1-\varepsilon_2)^2}{\rho_{ini} (1-\varepsilon)^2} = \beta(\frac{1}{\phi_v} - 1) + 1 \tag{6}$$

The damage volume coefficient ϕ_v can be further expressed as:

$$\phi_v = \frac{V_a}{V_0} = \frac{L_0 (1-\varepsilon) S_a}{L_0 S_0} = (1-\varepsilon)\phi_s \tag{7}$$

where V_a is the volume of the S-RM sample; S_a is the average cross-sectional area of the S-RM sample; ϕ_s is the damage coefficient. Substituting Equation (7) into Equation (6):

$$\frac{\rho_\phi (1-\varepsilon_2)^2}{\rho_{ini} (1-\varepsilon)^2} = \beta\left(\frac{1}{(1-\varepsilon)\phi_s} - 1\right) + 1 \tag{8}$$

The effective conductivity coefficient ϕ_ρ is introduced, and its expression is:

$$\phi_\rho = \frac{\rho_\phi (1-\varepsilon_2)^2}{\rho_{ini} (1-\varepsilon)^2} \tag{9}$$

Therefore, the damage variable D can be derived from the damage coefficient ϕ_s, which can be written as follows:

$$D = 1 - \phi_s = 1 - \frac{1}{(1-\varepsilon)(1 + \frac{\phi_\rho - 1}{\beta})} \tag{10}$$

Neglecting the geometric effects due to deformation, Equation (10) can be simplified:

$$D = 1 - \phi_s = 1 - \frac{1}{(1 + \frac{\phi_\rho - 1}{\beta})} \tag{11}$$

According to the effective stress principle, the axial strain of the damaged material under stress σ is equal to that of undamaged material under effective stress σ'. We obtain

$$\varepsilon = \frac{\sigma}{E'} = \frac{\sigma'}{E} \tag{12}$$

where E and E' are the elastic modulus of the S-RM sample in the undamaged and damaged states, respectively.

The effective stress σ' and the damage variable D satisfy the following expression:

$$\sigma' = \frac{\sigma}{1-D} \tag{13}$$

Substituting Equation (12) into Equation (13):

$$\sigma = (1-D)E\varepsilon \tag{14}$$

Substituting Equation (11) into Equation (14), the mechanical damage model of S-RM based on electrical resistivity can be written as follows:

$$\sigma = \frac{1}{(\frac{\phi_\rho - 1}{\beta} + 1)} E\varepsilon \tag{15}$$

The coefficient β represents the proportional magnitude of the damaged section in the entire sample, which changes gradually with increasing axial strain. Previous research had shown that the total crack area in S-RM samples under triaxial deformation gradually tends to a certain value with the increase in strain, which follows the following variation regularity: it initially increases rapidly and then increases gently; such changes in microcracks always occur in the range of 6~8% axial strain [35]. Therefore, in this paper, it is assumed that the coefficient β hardly varies in the late stage of axial strain and is a fixed value. The critical value of the abovementioned axial strain is taken as 6%, which is denoted as ε_6.

When the axial strain $\varepsilon < \varepsilon_2$, β = 0; when the axial strain ε is in the range of $\varepsilon_2 \sim \varepsilon_6$, β increases monotonically from 0 to the maximum value $\beta_{\max}$; when the axial strain $\varepsilon > \varepsilon_6$, β = $\beta_{\max}$. Therefore, a piecewise function is used to describe the variation characteristics of coefficient β:

$$\beta(\varepsilon) = \begin{cases} 0 & (0 \le \varepsilon \le \varepsilon_2) \\ \beta_{\max}\sqrt{\frac{\varepsilon - \varepsilon_2}{\varepsilon_6 - \varepsilon_2}} & (\varepsilon_2 < \varepsilon \le \varepsilon_6) \\ \beta_{\max} & (\varepsilon > \varepsilon_6) \end{cases} \tag{16}$$

The maximum value $\beta_{\max}$ can be obtained as follows:

$$\beta_{\max} = \phi_{\rho,\max} - 1 \tag{17}$$

where $\phi_{\rho,\max}$ can be derived from Equation (9) by taking the maximum value of ε, ε_{15}.

In order to verify the reliability of the established damage model, the stress–strain curves of S-RM samples with 20% rock content under four confining pressures obtained from the test are compared with the calculated results from the damage model proposed in this paper, and the results are shown in Figure 10.

As expected, the stress–strain results calculated by the electrical resistivity damage model are in good agreement with the experimental results. Only when the confining pressure is 400 kPa is there a large difference between the above two results under large axial strain. The difference between the calculated value and the measured result is perhaps due to the sharp increase in deviator stress at 400 kPa and the limitation of the measurement accuracy caused by the test. This indicates that the reliability of the damage model still needs to be further refined due to the presence of partial assumptions during the model derivation. On balance, we believe that the damage model based on electrical resistivity obtained in this study can not only better characterize the internal structure damage of S-RM but also further validate the general applicability of the damage model proposed in the literature [24].

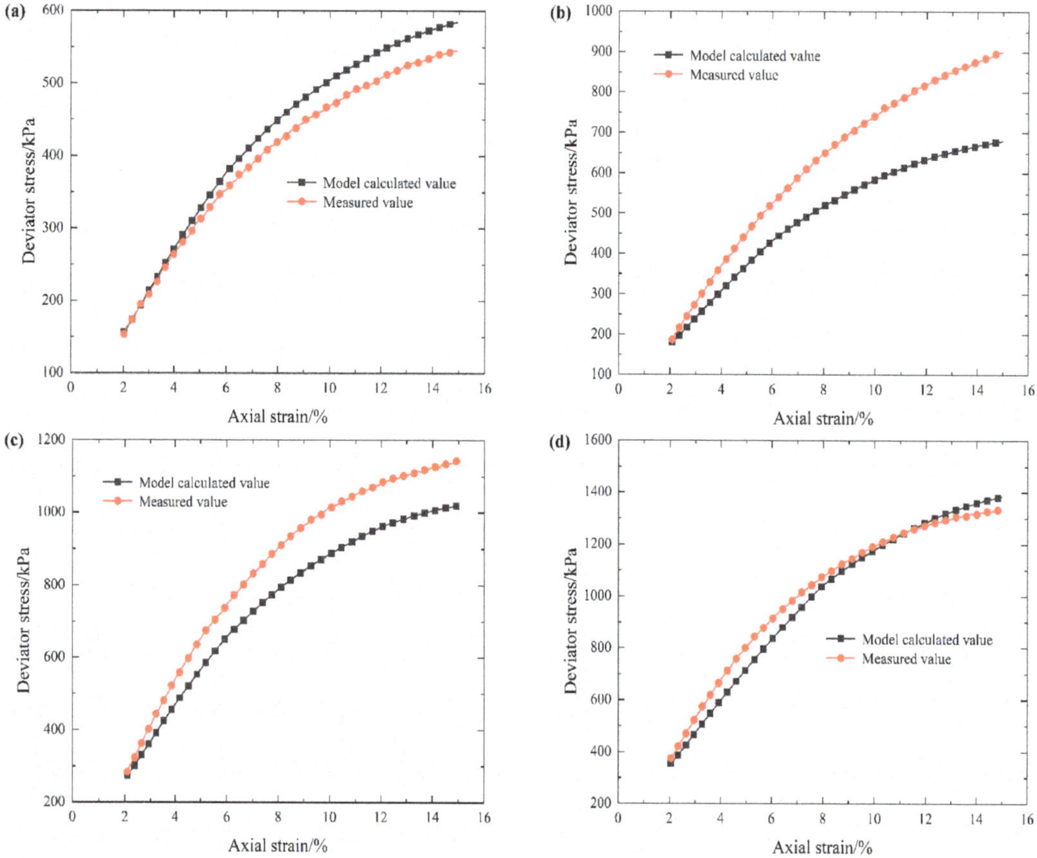

Figure 10. Measured results and calculated values from the damage model under different confining pressures. (**a**) 250 kPa. (**b**) 400 kPa. (**c**) 550 kPa. (**d**) 700 kPa.

3.4. Damage Evolution Analysis of S-RM

Electrical resistivity is an indicator of the generation of structural damage inside the geomaterials until the occurrence of macroscopic damage by monitoring the closure of pores and the generation and propagation of microcracks, which can effectively evaluate the damage degree of geomaterials. Moreover, this experiment seems to prove the conclusion in most of the literature that electrical resistivity can accurately characterize the mechanical behavior of soil. Through the variation curve of damage variable D, the damage evolution characteristics of S-RM are analyzed from the perspective of the changes in internal microstructure. Figure 11 plots the variation curve of damage variable D during triaxial shear (confining pressure 700 kPa) for S-RM samples with 20% rock content. The maximum value of D is about 0.483. The stress–strain curve and damage variable–strain curve in Figure 11 are obtained from the experimental data in this paper and Equation (11), respectively.

As shown in Figure 11, the damage evolution of S-RM under triaxial shear can be divided into three stages: the non-damage stage, the rapid damage stage and the stable damage stage. As deformation grows, the damage of the S-RM sample also means that it is gradually compressed. During the non-damage stage, it can be seen that the damage variable D is 0 and that the damage–strain curve varies horizontally, which may be due to the fact that there is no structural damage such as particle breakage and microcrack formation in the sample when the sample is in the initial compaction and elastic deformation stages.

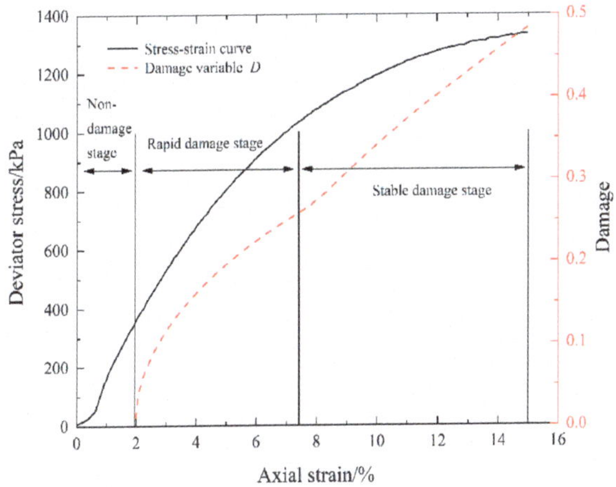

Figure 11. Variation curve of damage variable D during triaxial shear.

In the stage of rapid damage, the sample has entered the yield deformation stage, the damage variable D exhibits a trend of rapid increase and the damage–strain curve is a convex arc. It can be seen that the increasing rate of damage variable D decreases gradually. The interparticle contact forces increase further in the yield deformation stage, and coarse particles with their own defects in the S-RM start to break due to the stress concentration, so the damage can occur rapidly. With the increase in axial strain, the greater inter-particle contact force makes the existing defects of the particles further develop or causes new defects and the particle breakage rate increases, accompanied by the gradual propagation of microcracks. Compared with coarse particles, the fine particles formed after particle breakage have a lower breakage potential and are more difficult to break.

In the stage of stable damage, the damage variable D exhibits a steadily increasing trend and the gradient of the damage–strain curve tends to be stable. Because the coarse and fine particles have been broken to different degrees, a more uniform inter-particle contact force is formed, which leads to a decreasing possibility for massive breakage. Meanwhile, the internal microcracks of S-RM gradually increase until the formation of shear localized bands. Although there is still the particle breakage phenomenon at this stage, it seems that the propagation of the crack has a more significant effect on the damage. Under the combined effect of particle breakage and microcracks, the corresponding damage variable D presents an incremental trend.

Compared with that in the S-RM with a confining pressure of 700 kPa, the maximum damage variable of S-RM with lower confining pressure changes, which exhibits a slight increase with the decrease in confining pressure. The maximum damage variables of the sample with 20% rock content under the confining pressures of 250 kPa, 400 kPa and 550 kPa are 0.512, 0.498 and 0.490, respectively. For the S-RM, the confining pressure level can not only enhance the compactness of the sample but also may improve the overall strength effectively, so the degree of damage varies. Due to the difference in the continuity of the material composition between rock and S-RM, the damage evolution regularity of rock is significantly distinct from that of S-RM. The damage of rock under load was investigated in the literature [18]. They found that there is a turning point in the curve for the damage variable as a function of strain. There is no significant change in the damage variable before the turning point, but above this point the damage variable rapidly rises to the critical value.

3.5. Modification of Effect of Rock Content Difference

The results of previous studies indicate that the existence of rock blocks not only results in the enhancement of the compressive strength but also affects the difference in the electrical resistivity for S-RM samples [15,31]. Table 4 summarizes the statistics of the electrical resistivity values measured after the saturation of S-RM samples with four rock contents. It can be found that the range (difference between the maximum and minimum), standard deviation and coefficient of variation of each rock content are small, suggesting that all electrical resistivity values are relatively stable and have no significant fluctuations. Thus, the mean is chosen as the electrical resistivity value for each rock content after saturation, which is the comprehensive performance of both rock content and compaction.

Table 4. Statistics of the electrical resistivity values.

R/%	*ER*/(Ω·m)	Range/(Ω·m)	Mean/(Ω·m)	*SD*/(Ω·m)	*CV*/%
20	47.18, 46.85, 47.25, 47.56	0.71	47.21	0.25	0.53
30	44.07, 44.28, 43.92, 44.45	0.53	44.18	0.20	0.46
40	39.90, 40.47, 40.55, 39.88	0.67	40.20	0.31	0.77
50	38.42, 38.16, 37.87, 38.07	0.55	38.13	0.19	0.50

R = rock content; *ER* = electrical resistivity (from four measurements); *SD* = standard deviation; *CV* = coefficient of variation.

According to the electrical resistivity values in Table 4, the structure enhancement factor of S-RM is introduced, which is used as a model modification parameter to characterize the effect of the rock content difference, denoted as $S(R)$, and its expression is:

$$S(R) = \frac{ER(R)}{ER(20)} \tag{18}$$

where $ER(20)$ is the electrical resistivity of S-RM with rock content 20% after saturation; $ER(R)$ is the electrical resistivity of S-RM with different rock contents of 30%, 40% and 50% after saturation.

Based on the derived structure enhancement factor $S(R)$ and the established damage model, the stress–strain curves of S-RM with different rock contents (30%, 40% and 50%) can be obtained. One confining pressure is selected for analysis for each rock content, as shown in Figure 12.

From Figure 12, it can be seen that the calculated values for the damage model after the introduction of the modification parameter have a good correlation with the experimental data for S-RM. Through simultaneous measurement of electrical resistivity in the triaxial shear test, the stress–strain curves of S-RM with different rock contents can be accurately predicted. Therefore, it can be considered that the electrical resistivity signals can effectively characterize and evaluate the internal damage degree of S-RM.

It should be noted that the coefficient β, which characterizes the magnitude of the middle-damaged section of the S-RM sample, is a piecewise function that has been simplified in the calculation process. In the actual shear process, the damage of the S-RM sample exists all the time, which increases gradually from early to late shear, and is far more complex than the monotonic increase assumed in this work. To well reveal the role of rock block in affecting damage behavior, the damage mechanism of the S-RM samples with various rock contents needs to be further discussed.

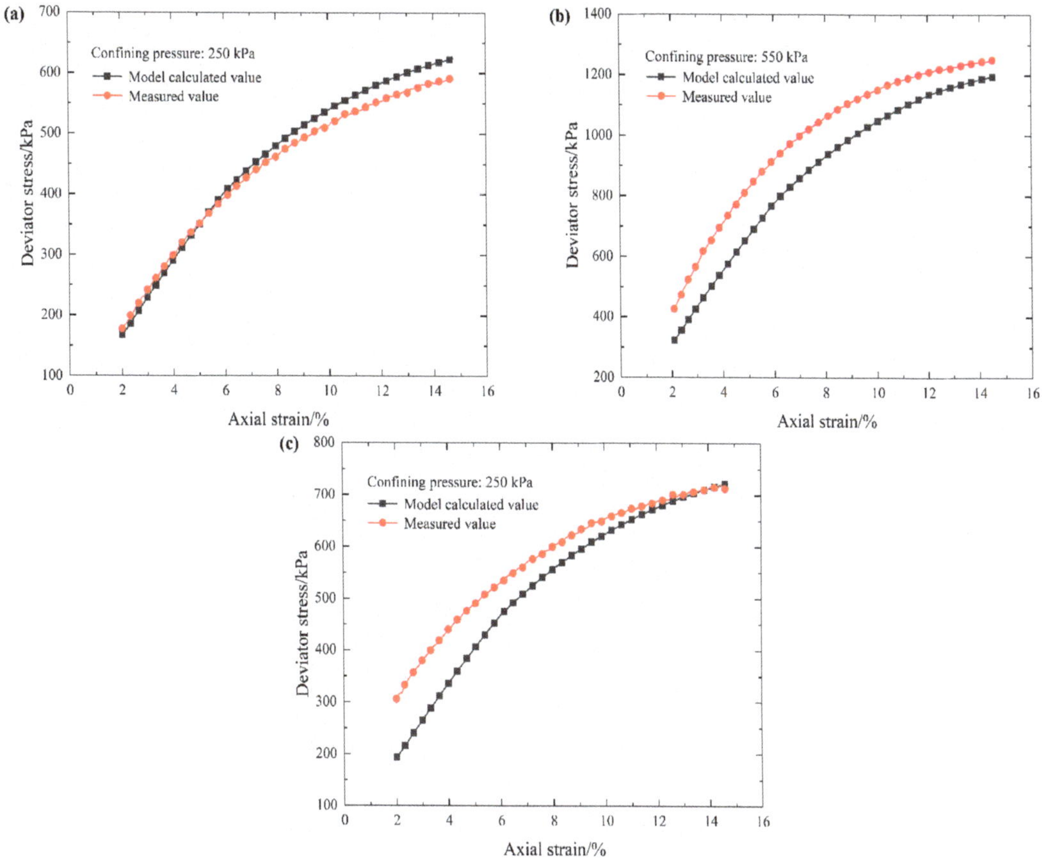

Figure 12. Measured results and calculated values of the damage model under partial confining pressures. (**a**) Rock content 30%. (**b**) Rock content 40%. (**c**) Rock content 50%.

In summary, the experimental results show that the electrical resistivity is highly correlated with internal damage. Both the existence and the development of structural damage can be reflected by the electrical resistivity measurement of the S-RM models, which provides implications for developing an electrical-resistivity-based method for monitoring the structural failure in earth–rockfill dams. Therefore, compared with acoustic emission, infrared radiation and ultrasonic measurement, the electrical resistivity technique is more suitable for in situ testing in the field and has prospects for application in the study of geotechnical mechanics.

4. Conclusions

In order to explore the feasibility of the electrical resistivity method to characterize the damage behavior of S-RM under triaxial shear, the relationship between the change in electrical resistivity and the damage evolution of S-RM was studied. Under the applied voltage, the electrical resistivity response of S-RM samples was collected during triaxial deformation. The electrical resistivity variation regularities and stress–strain characteristics of S-RM under stress were explored using modified triaxial apparatus. The mechanical damage model, the damage evolution process and the modification of the effect of rock content difference were investigated and discussed. The main conclusions are as follows:

(1) During the triaxial loading process the electrical resistivity of S-RM varies in stages with the axial strain, which follows the following variation regularity: it initially decreases rapidly, then declines slowly and finally reduces gently. The response mechanisms of electrical resistivity under different deformation stages are distinct.
(2) The stress–strain characteristics of S-RM gradually change from a slight strain softening to a strong strain hardening as the confining pressure increases. Under the same confining pressure, the deviator stress that S-RM samples can withstand is not the same; the more the rock content, the higher deviator stress is. With the increase in confining pressure, the corresponding deviator stress of the samples with the same rock contents increases gradually.
(3) The electrical-resistivity-based mechanical damage model for S-RM can accurately characterize the degree of structure damage. This mathematical model agrees with the experimental results. The damage evolution of S-RM under triaxial shear can be divided into three stages: a non-damage stage, a rapid damage stage and a stable damage stage. In addition, the modified model can accurately predict the mechanical behavior of S-RM with different rock contents. These discoveries can provide a basis for further research on the application of electrical resistivity.

Author Contributions: Methodology, M.Z. and K.W.; formal analysis, S.C. and G.L.; investigation, S.C. and G.L.; data curation, S.C.; writing—original draft preparation, S.C.; writing—review and editing, M.Z.; project administration, M.Z. and K.W.; funding acquisition, M.Z. All authors have read and agreed to the published version of the manuscript.

Funding: This research was funded by National Natural Science Foundation of China, grant number 51879017 and Scientific and Technological Research Program of Chongqing Municipal Education Commission, grant number KJZD-K202100705.

Institutional Review Board Statement: Not applicable.

Informed Consent Statement: Not applicable.

Data Availability Statement: Not applicable.

Conflicts of Interest: The authors declare no conflict of interest.

References

1. Xu, W.J.; Yue, Z.Q.; Hu, R.L. Study on the mesostructure and mesomechanical characteristics of the soil–rock mixture using digital image processing based finite element method. *Int. J. Rock Mech. Min.* **2008**, *45*, 749–762. [CrossRef]
2. Xu, W.J.; Xu, Q.; Hu, R.L. Study on the shear strength of soil-rock mixture by large scale direct shear test. *Int. J. Rock Mech. Min.* **2011**, *48*, 1235–1247. [CrossRef]
3. Afifipour, M.; Moarefvand, P. Mechanical behavior of bimrocks having high rock block proportion. *Int. J. Rock Mech. Min.* **2014**, *65*, 40–48. [CrossRef]
4. Wang, Y.; Li, X.; Wu, Y.F. Damage evolution analysis of SRM under compression using X-ray tomography and numerical simulation. *Eur. J. Environ. Civ. Eng.* **2014**, *19*, 400–417. [CrossRef]
5. Liu, Z.R.; Ye, W.M.; Zhang, Z.; Wang, Q.; Chen, Y.G.; Cui, Y.J. A nonlinear particle packing model for multi-sized granular soils. *Constr. Build. Mater.* **2019**, *221*, 274–282. [CrossRef]
6. Yao, Y.S.; Li, J.; Ni, J.J.; Liang, C.H.; Zhang, A.S. Effects of gravel content and shape on shear behaviour of soil-rock mixture: Experiment and DEM modelling. *Comput. Geotech.* **2022**, *141*, 104476. [CrossRef]
7. Samouëlian, A.; Cousin, I.; Tabbagh, A.; Bruand, A.; Richard, G. Electrical resistivity survey in soil science: A review. *Soil Tillage Res.* **2005**, *83*, 173–193. [CrossRef]
8. Wayal, V.; Sitharam, T.G. Geo-electric assessment of the compacted sand-bentonite mixes. *Int. J. Geotech. Eng.* **2019**, *16*, 74–89. [CrossRef]
9. Kibria, G.; Hossain, M.S. Investigation of geotechnical parameters affecting electrical resistivity of compacted clays. *J. Geotech. Geoenviron. Eng.* **2012**, *138*, 1520–1529. [CrossRef]
10. Sun, Q.; Zhu, S.Y.; Xue, L. Electrical resistivity variation in uniaxial rock compression. *Arab. J. Geosci.* **2015**, *8*, 1869–1880. [CrossRef]
11. Li, X.C.; Zhang, Q.; An, Z.X.; Chen, X.L.; Zhang, F. Experimental research on electrical resistivity variation of coal under different loading modes. *Arab. J. Geosci.* **2020**, *13*, 1068. [CrossRef]

12. Chen, P.; Yang, T.; Chen, X.X.; Liu, Y.J.; Li, X.L.; Zhang, K.X. Experimental study on electrical characteristics of gassy coal during extrusion process in different stage. *Arab. J. Geosci.* **2019**, *12*, 430–438. [CrossRef]
13. Zhou, C.S.; Li, K.F.; Han, J.G. Characterizing the effect of compressive damage on transport properties of cracked concretes. *Mater. Struct.* **2012**, *45*, 381–392. [CrossRef]
14. Khurshid, I.; Afgan, I. Geochemical investigation of electrical conductivity and electrical double layer based wettability alteration during engineered water injection in carbonates. *J. Petrol. Sci. Eng.* **2022**, *215*, 110627. [CrossRef]
15. Zhao, M.J.; Li, G.; Huang, W.D.; Li, J. Experiment on electrical resistivity properties of polyphase soil-stone mediums. *J. Chongqing Jiaotong Univ. (Nat. Sci.)* **2010**, *29*, 928–933.
16. Wang, K.; Zhao, M.J. Characteristics and application of electrical resistivity of soil-stone composite medium. *J. Chongqing Jiaotong Univ. (Nat. Sci.)* **2014**, *33*, 90–94. Available online: http://xbzk.cqjtu.edu.cn/EN/abstract/abstract510.shtml (accessed on 3 May 2023).
17. Wang, R.S.; Zhao, M.J.; Wang, R.Q. Experimental research on variation of resistivity in soil rock composite media. *J. Build. Mater.* **2019**, *22*, 94–100. [CrossRef]
18. Wang, K.; Xia, Z.T.; Li, X. Matrix suction evaluation of soil-rock mixture based on electrical resistivity. *Water* **2021**, *13*, 2937. [CrossRef]
19. Wang, Y.H.; Liu, Y.F.; Ma, H.T. Changing regularity of rock damage variable and resistivity under loading condition. *Safe. Sci.* **2012**, *50*, 718–722. [CrossRef]
20. Zhang, Y.B.; Wu, W.R.; Yao, X.L.; Liang, P.; Tian, B.Z.; Huang, Y.L.; Liang, J.L. Acoustic emission, infrared characteristics and damage evolution of granite under uniaxial compression. *Rock Soil Mech.* **2020**, *41*, 139–146. [CrossRef]
21. Khurshid, I.; Fujii, Y. Geomechanical analysis of formation deformation and permeability enhancement due to low-temperature CO_2 injection in subsurface oil reservoirs. *J. Pet. Explor. Prod.* **2021**, *11*, 1915–1923. [CrossRef]
22. Chung, D.D.L. Damage in cement-based materials, studied by electrical resistance measurement. *Mater. Sci. Eng.* **2003**, *42*, 1–40. [CrossRef]
23. Cao, J.Y.; Chung, D.D.L. Defect dynamics and damage of concrete under repeated compression, studied by electrical resistance measurement. *Cem. Concr. Res.* **2001**, *31*, 1639–1642. [CrossRef]
24. Zeng, X.H.; Liu, H.C.; Zhu, H.S.; Ling, C.B.; Liang, K.; Umar, H.A.; Xie, Y.J.; Long, G.C.; Ma, C. Study on damage of concrete under uniaxial compression based on electrical resistivity method. *Constr. Build. Mater.* **2020**, *254*, 119270. [CrossRef]
25. Li, X.C.; Zhang, Q. Study on damage evolution and resistivity variation regularities of coal mass under multi-stage loading. *Appl. Sci.* **2019**, *9*, 4124. [CrossRef]
26. Jia, Z.Q.; Xie, H.P.; Zhang, R.; Li, C.B.; Wang, M.; Gao, M.Z.; Zhang, Z.P.; Zhang, Z.T. Acoustic emission characteristics and damage evolution of coal at different depths under triaxial compression. *Rock Mech. Rock Eng.* **2020**, *53*, 2063–2076. [CrossRef]
27. An, R.; Kong, L.W.; Bai, W.; Li, C.S. The resistivity damage model of residual soil under uniaxial load and the law of drying-wetting effects. *Chin. J. Rock Mech. Eng.* **2020**, *39*, 3159–3167. [CrossRef]
28. Chen, Y.L.; Wei, Z.A.; Irfan, M.; Xu, J.J.; Yang, Y.H. Laboratory investigation of the relationship between electrical resistivity and geotechnical properties of phosphate tailings. *Measurement* **2018**, *126*, 289–298. [CrossRef]
29. Medley, E.W. *The Engineering Characterization of Melanges and Similar Block-in-Matrix Rocks (Bimrocks)*; University of California: Berkeley, CA, USA, 1994.
30. Zhang, H.Y.; Xu, W.J.; Yu, Y.Z. Triaxial tests of soil-rock mixtures with different rock block distributions. *Soils Found.* **2016**, *56*, 44–56. [CrossRef]
31. Tu, Y.L.; Chai, H.J.; Liu, X.R.; Wang, J.B.; Zeng, B.; Fu, X.; Yu, J.Y. An experimental investigation on the particle breakage and strength properties of soil-rock mixture. *Arab. J. Geosci.* **2021**, *14*, 840. [CrossRef]
32. Wayal, V.; Sitharam, T.G.; Anjali, M. Geo-electrical characterization of physical and mechanical properties of zinc tailing. *J. Appl. Geophys.* **2021**, *188*, 104315. [CrossRef]
33. Yuan, G.L.; Che, A.L.; Tang, H. Evaluation of soil damage degree under freeze–thaw cycles through electrical measurements. *Eng. Geol.* **2021**, *293*, 106297. [CrossRef]
34. Zhang, Y.Q.; Jing, H.J.; Dai, J.Y. Experimental analysis on the mechanical properties of soil-rock mixed fillers with different rock contents on a high embankment. *DYNA-Ing. E Ind.* **2021**, *96*, 478–483. [CrossRef] [PubMed]
35. Wang, Y.; Li, C.H.; Hu, Y.Z. Use of X-ray computed tomography to investigate the effect of rock blocks on meso-structural changes in soil-rock mixture under triaxial deformation. *Constr. Build. Mater.* **2018**, *164*, 386–399. [CrossRef]

materials

MDPI

Article

Measurement and Classification Criteria of Strength Decrease Rate and Brittleness Indicator Index for Rockburst Proneness Evaluation of Hard Rocks

Kun Du [1], Songge Yang [1], Jian Zhou [1] and Lichang Wang [2,*]

[1] School of Resources and Safety Engineering, Central South University, Changsha 410083, China
[2] School of Geosciences and Info-Physics, Central South University, Changsha 410083, China
* Correspondence: wlccsu@csu.edu.cn

Abstract: Rockburst is one of the common geological hazards. It is of great significance to study the evaluation indexes and classification criteria of the bursting liability of hard rocks, which is important for the prediction and prevention of rockbursts in hard rocks. In this study, the evaluation of the rockburst tendency was conducted using two indoor non-energy indexes, namely the brittleness indicator (*B*2) and the strength decrease rate (*SDR*). The measuring methods of *B* and *SDR* as well as the classification criteria were analyzed. Firstly, the most rational calculation formulas for *B* and *SDR* were selected based on previous studies. The *B*2 equaled to the ratio between the difference and sum of uniaxial compressive strength and Brazilian tensile strength of rocks. The *SDR* was the average stress decrease rate of the post-peak stage in uniaxial compression tests and equaled the uniaxial compressive strength dividing the duration time of post-peak rock failure in uniaxial compression tests. Secondly, the uniaxial compression tests of different rock types were designed and carried out, and the change trend of *B* and *SDR* with the increase of loading rate in uniaxial compression tests were studied in detail. The results showed that after the loading rate was greater than 5 mm/min or 100 kN/min, the *B* value was affected, limited by the loading rate, while the *SDR* value was more affected by the strain rate. The displacement control, with a loading rate of 0.1–0.7 mm/min, was recommended for the measurement of *B* and *SDR*. The classification criteria of *B*2 and *SDR* were proposed, and four grades of rockburst tendency were defined for *SDR* and *B*2 according to the test results.

Keywords: hard rock; rock burst tendency; lab testing; brittleness indicator; strength decrease rate

Citation: Du, K.; Yang, S.; Zhou, J.; Wang, L. Measurement and Classification Criteria of Strength Decrease Rate and Brittleness Indicator Index for Rockburst Proneness Evaluation of Hard Rocks. *Materials* **2023**, *16*, 3101. https://doi.org/10.3390/ma16083101

Academic Editor: Evangelos J. Sapountzakis

Received: 22 February 2023
Revised: 5 April 2023
Accepted: 11 April 2023
Published: 14 April 2023

1. Introduction

With the development of society and the economy, the needs for space and mineral resources continue to increase, and the buried depths of underground engineering or mines continue to expand [1]. Taking underground mines as an example, the shallow mineral resources have been exhausted gradually, and the deeply buried geo-resources should be mined out urgently [2,3]. The depth means high geo-stress and elastic energy stored in rocks, which makes it easier to induce rockburst. Rockburst has occurred many times in Germany, Canada, South Africa, India, Japan, and other countries [4–9]. At present, rockburst is a global underground engineering problem; it is difficult to predict accurately, and to prevent and control effectively [10]. Once rockburst occurs, it may lead to more casualties and economic losses than other disasters [2,11,12].

With in-depth study and analysis on the phenomena of rockburst, the mechanism of rockburst has been understood gradually. The occurrence of rockburst usually requires three key conditions: one is that the rock itself has high rockburst tendency; then, that the rocks are subjected to high geo-stress and possess a large amount of elastic energy; and the last is that the elastic energy stored in the rocks is induced to release by excavation, dynamic stress, or other external factors [13–15]. In order to accurately predict and evaluate rockburst, many scholars have introduced field tests and theoretical analysis into the research field of rockburst [16]. At present, in order to study the local strain variation and stability of rock masses, researchers have conducted rockburst mechanism studies based on capacity theory, strength theory, blast reliability analysis, and chaos theory [17,18]. However, it is difficult to establish universal and suitable criteria for rockburst prediction by simply using theoretical methods from classical mechanics. Many scholars have applied some new scientific methods to rockburst prediction, such as cloud models [19], fuzzy evaluation [20], neural networks [21], extension methods [22,23], Bayes discriminant analysis [24], and cluster analysis [25].

However, machine learning also has some defects, such as the large sample base required, unclear parameter calibration, and over-simplified models. In order to control rockburst and minimize the risk of rockburst, the evaluation of the rockburst tendency in laboratory tests has received more and more attention in recent years. The measurement methods and grade classification criteria of different bursting indexes have become the hot research field [26]. Most bursting indexes are defined from the perspective of energy, such as the strain energy storage index W_{ET} [27], the peak-strength strain energy storage index W_{ET}^{P}, the peak-strength energy impact index A'_{CF} [28] and the residual elastic energy index A_{EF} [29,30]. Likewise, there are also many non-energy bursting indexes, such as the modified brittleness index *BIM*, the deformation brittleness index K_u [31], the decrease module index *DMI* [32], the brittleness indicator index *B* [33], and the strength decrease rate index *SDR* [34].

A multi-index bursting liability evaluation system has been formed for coal at present [35], but there is no suitable standard used for the rockburst proneness evaluation of hard rocks, especially in the aspects of the bursting indexes and classification criteria. In general, it directly adopts the standards of coal to predict the bursting proneness of hard rocks, and the prediction results are inconsistent with the field phenomena of rockburst. In fact, there are great differences between the mechanical parameters of coal and hard rocks, and the bursting proneness indexes and their criterion used in coal are not suitable for hard rocks [34]. In addition, the bursting classification criteria in hard rocks is often divided into four grades, i.e., none, weak, medium, and strong, by many researchers [26,29,30], which is a three-grade criterion for coal, i.e., none, weak, and strong. Figure 1 shows the statistics of rockburst tendency degrees of coal and non-coal rocks in underground engineering with different excavation depths in China [36–62]. The excavation depth of non-coal rock engineering is deeper than that of coal mines, and the intensity of rockburst in deep hard rock engineering is much stronger. Therefore, the rockburst of hard rocks should be paid more attention, accordingly.

In this study, the influence of the loading mode and rate on the *SDR*, *B*1, and *B*2 results is analyzed, and then the loading mode and rate consistent with the determination of the basic mechanical parameters of the rock are recommended. Combined with the determination results of *SDR* and *B*2 and the classification method, the rockburst tendency of different rocks is evaluated.

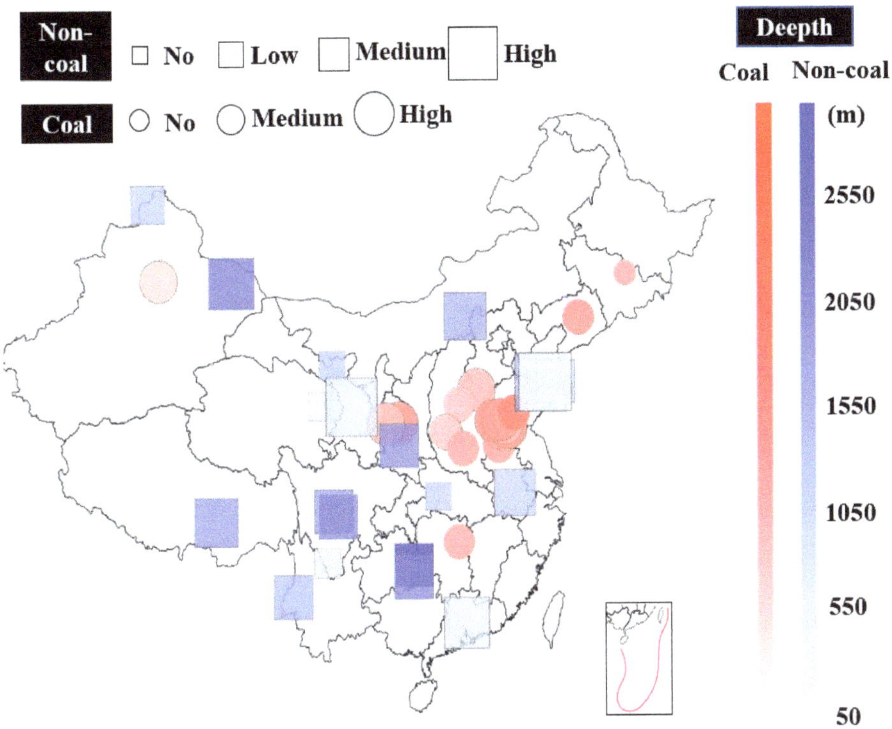

Figure 1. Distribution of rockburst tendency degrees of rocks in China (the darker the color, the deeper the buried depth, and the bigger the symbols, the stronger the rockburst).

2. Definitions of *B* and *SDR*

Several non-energy indexes of rockburst tendency in laboratory tests were put forward, e.g., brittleness indicator index *B*, decrease module index *DMI*, deformation brittleness index K_u, and strength decrease rate index *SDR* [12], as shown in Table 1. The K_u is defined as the ratio of strain corresponding to the stress at the unloading point and the terminal strain at the unloading stage, but the elastic strain at the peak is impossible to obtain, so it cannot accurately reflect the actual bursting tendency of rocks. The *DMI* is defined as the the ratio between pre-peak modulus and post-peak modulus, in which the pre-peak modulus reflects the rate of energy accumulation in rocks, while the post-peak modulus reflects the speed of energy release during the failure process of rocks. However, the post-peak modulus is also difficult to test and calculate. Therefore, this study focuses on the discussion of brittleness indicator index *B* and strength decrease rate index *SDR*.

The brittleness indicator is one of the important indexes for rockburst prediction, which can also be used to analyze the stability of surrounding rocks in underground engineering. The brittleness indicator index of rocks is closely related to the mineral composition, toughness, and hardness of rocks [63–65]. The strength decrease rate is closely related to the mechanical properties and post-failure state of rocks, can denote the release rate of elastic energy stored in rocks, and is important for the study of mechanical behaviors of rocks in the post-peak stage [66–68].

Table 1. Brittleness indicator and strength decrease rate for evaluation of rockburst tendency [69].

Indicators		Formula	Parameter
SDR		$SDR = \frac{\sigma_c}{D_t}$	σ_c and D_t represent the uniaxial compressive strength and dynamic failure time after peak strength, respectively
B	$B1$	$B1 = \frac{\sigma_c}{\sigma_t}$	σ_c and σ_t represent the uniaxial compressive strength and Brazilian tensile strength, respectively
	$B2$	$B2 = \frac{\sigma_c - \sigma_t}{\sigma_c + \sigma_t}$	
K_u		$K_u = \frac{\varepsilon^L}{\varepsilon_p}$	ε^L and ε_p represent the strain corresponding to the stress at the unloading point and the terminal strain at the unloading stage, respectively
DMI		$DMI = \frac{E_G}{\lvert E_M \rvert}$	E_G and E_M represent the pre-peak elastic modulus and post-peak elastic modulus, respectively

The non-energy indexes of rockburst tendency were easier to be obtain and calculate than the energy indexes, e.g., strain energy storage index W_{ET}, energy impact index A_{CF}, and residual elastic energy index A_{EF}. This is because the peak elastic energy cannot be obtained directly through the test. The energy indexes must be obtained indirectly through the cyclic loading and unloading test, and the results of energy indexes under different unloading points have large gaps between each other. Non-energy indexes of rockburst tendency are complementary to energy indexes. The energy indexes are influenced by the energy storage and dissipation ability of rocks. The brittleness indicator index in the non-energy index is affected by the natural characteristics of rocks. The SDR is affected by the energy release rate of rocks which can impact the intensity of rockburst. Many other bursting indexes ignore the effect of energy release rate on rockburst. The non-energy index and energy index can reflect the rockburst tendency from different views, which can keep the evaluation results of rockburst tendency more comprehensively.

Different brittleness indicator indexes have different application scenarios [33]. The initial stage of rockburst is the process of crack generation and propagation. Based on Griffith strength criterion and strength parameters, the brittleness B of rocks was put forward to evaluate the rockburst proneness of rocks [70]. One of the calculation methods of B is expressed by uniaxial compressive strength and Brazilian tensile strength as Formula (1):

$$B1 = \frac{\sigma_c}{\sigma_t} \tag{1}$$

The rockburst proneness criterion of $B1$ was as follows [41], which was negatively correlated with rockburst tendency.

$B1 > 40$ None;
$40 \geq B1 \geq 26.7$ Weak;
$26.7 \geq B1 \geq 14.5$ Medium;
$B1 < 14.5$ Strong.

However, the other rockburst proneness criterion of $B1$ was as follows [71], which was positively correlated with rockburst tendency.

$B1 < 15$ None;
$15 \leq B1 \leq 18$ Weak;
$18 \leq B1 \leq 22$ Medium;
$B1 > 22$ Strong.

The classification criteria of $B1$ were not unified, and the two categories of criteria of $B1$ disagreed with each other from the perspective of regularity [70]. Based on uniaxial compressive strength and Brazilian tensile strength, Singh [72] proposed a new method for calculating the brittleness indicator index $B2$, as expressed by Formula (2). Both the calculation Formulas (1) and (2) of brittleness indicator indexes B take two basic mechanical parameters, i.e., uniaxial compressive strength and Brazilian tensile strength, into account, which were affected by the fundamental mechanical properties of rocks [73,74]. There is no widely used classification criterion for $B2$ at present. Rocks with high compressive strength often have high Brazilian tensile strength, which may lead to a large dispersion of the results of $B1$, and it is difficult to judge the rockburst proneness using $B1$. According to the actual indoor rockburst situation, the rockburst tendency obtained by $B2$ is more consistent with the actual situation. Therefore, $B2$ is more suitable to use as an index of rockburst tendency.

$$B2 = \frac{\sigma_c - \sigma_t}{\sigma_c + \sigma_t} \tag{2}$$

The strength decrease rate SDR is a new index put forward by Du et al., and is defined as the ratio of uniaxial compressive strength and dynamic failure time after peak strength [22]. When the high elastic energy stored inside the rocks is induced, releasing quickly by external action, the ejection phenomenon of rock fragments is more obvious. The SDR can reflect the energy release rate of rocks to a certain extent. It is considered that among the non-energy evaluation indexes of rockburst tendency for rocks, the brittleness indicator index B and the strength decrease rate SDR can reflect the fundamental properties and energy release rate of rocks, respectively.

3. Loading Mode and Rate Effect on B and SDR

3.1. Testing Design

In order to investigate the rockburst tendency of different rocks, 4 kinds of rocks, including granite, marble, red sandstone, and coal, were selected to explore the influence of loading conditions on the brittleness indicator index B and strength decrease rate SDR. The rock specimens were prepared as cylindrical specimens, and all rock specimens were shown in Figure 2.

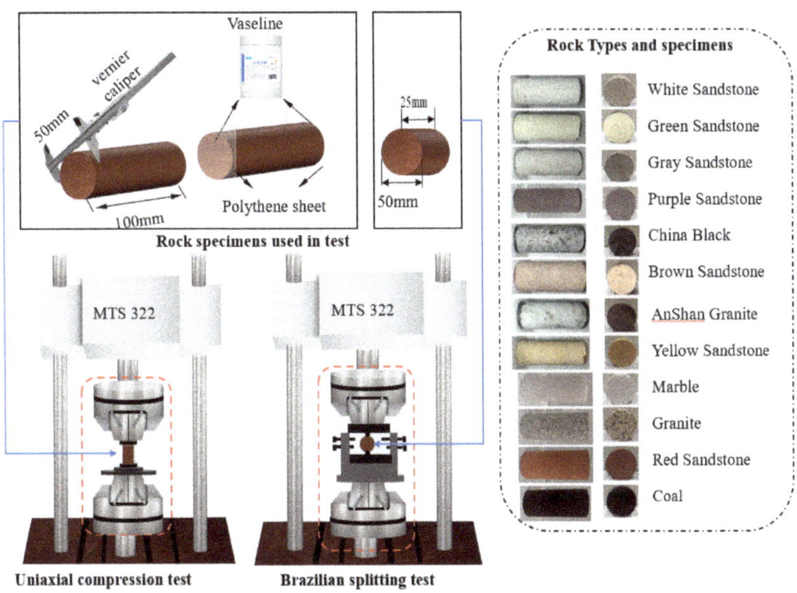

Figure 2. Testing machine and rock specimens used in lab test.

All specimens were drilled from intact rock blocks. The non-parallelism error of two end faces of rock specimens was not more than 0.05 mm. Along the height of the rock specimens, the diameter error was not more than 0.3 mm. The end face was perpendicular to the axis of the rock specimens, and the deviation was not greater than 0.25°. In the process of taking, transporting, and preparing rock specimens, it is necessary to avoid colliding with each other or accidentally falling to produce cracks.

Two series of testing, i.e., uniaxial compression tests and Brazilian splitting tests, were designed in the study. Before testing, rock specimens were grouped and numbered, and basic parameters, e.g., wave velocity and mass, were measured. The specimens had no defects, good uniformity and integrity, and were all in the state of natural water content. In uniaxial compression tests, the specimens were selected as cylindrical rock specimens with diameters of 50 mm, the diameters of the rock specimens were 10 times greater than the size of the largest mineral particle of rocks, and the height to diameter ratio of rock specimens was set as 2.0. In Brazil splitting test, the cylindrical rock specimens with diameter of 50 mm were used, and the ratio of height to diameter of specimens was designed as 0.5 [75].

3.2. Determination of *SDR*

The values of *SDR* were calculated by the results of uniaxial compression tests (Figure 3a) which were carried out on an MTS 322 T shaped material testing machine in Central South University (Figure 2); the maximum axial force applied by the testing machine was ±500 kN.

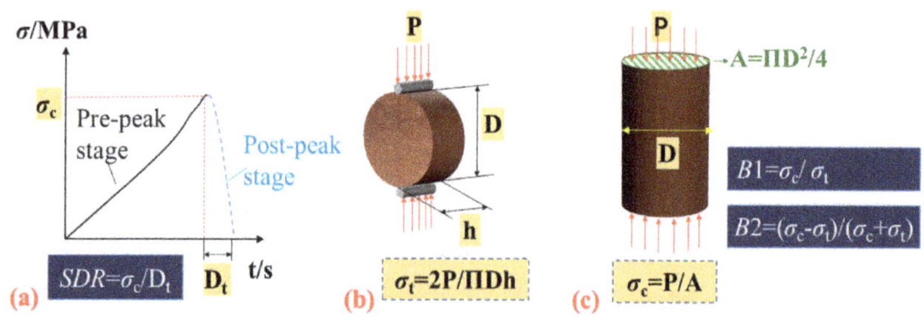

Figure 3. Schematic diagram of the *SDR* and *B* calculation process: (**a**) stress-time curve of rocks in uniaxial compression test and determination of *SDR*, (**b**) determination of σ_t; (**c**) determination of σ_c, *B*1, and *B*2. *SDR* denotes the strength decrease rate (MPa/s); D_t denotes the dynamic failure time in uniaxial compression test (s); σ_c denotes the uniaxial compressive strength of rock specimens (MPa); σ_t denotes the Brazilian tensile strength of rock specimens (MPa); *P* denotes the peak load in uniaxial compression test (kN); *A* denotes the area of the bearing section of a rock specimen (mm^2); *D* denotes the cross sectional diameter of a rock specimen (mm); *h* denotes the height of the rock specimen (mm).

In order to avoid the loss of rock fragments and ensure security in uniaxial compression tests, a cylindrical acrylic cylinder slightly larger than the specimens was used to cover the specimens during testing. The testing machine was stopped when the specimens lost their carrying capacity.

3.3. Determination of *B*

The values of the brittleness indicator index *B* were calculated by the results of uniaxial compression tests and Brazilian splitting tests. The specimens in Brazil splitting tests were destroyed by radial tensile stress which was applied by a pair of steel rods with diameters less than 3 mm along the diameter of the disk specimens (Figure 3b,c). In order to make sure that the steel rods, up and down, were parallel to each other and located along the diameter with the direction of vertical loading, a novel Brazil splitting testing device was designed, machined, and applied. The detailed usage flows are shown in Figure 4.

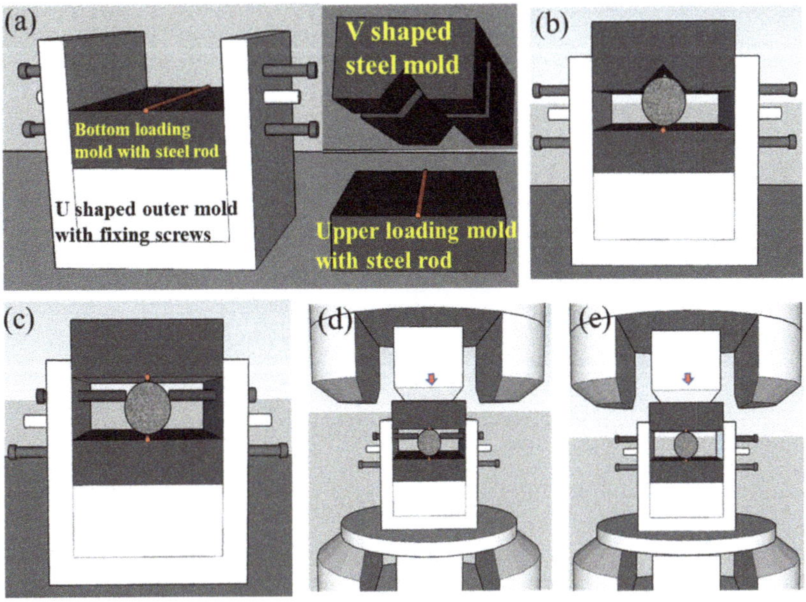

Figure 4. Composition and usage flow of a novel Brazil splitting testing device: (**a**) U Shaped outer steel mold with fixing screws, (**b**) V Shaped steel mold and upper loading mold with steel rod; (**c**) Install rock specimen, (**c**) Change V shaped steel mold as upper loading mode, (**d**) Install the entire mold, (**e**) Withdraw fixing screws and loading.

The novel device contains four parts, i.e., V shaped steel mold, upper loading mold with steel rod, bottom loading mold with steel rod, and U-shaped outer mold with fixing screws (Figure 4a). The detailed usage flows are as follows. Firstly, the bottom loading mold with steel rod was fixed in the U-shaped outer mold. The V-shaped steel mold was used to keep the rock specimens in the middle of the U-shaped outer mold, and the fixing screws were used to fix the rock specimens (Figure 4b). Secondly, the V-shaped steel mold was removed, and the upper loading mold with steel rode was placed at the location of the V-shaped steel mold (Figure 4c). Then, the whole mold and specimen was placed on the testing machine, and the pre-loads with small values were applied on the mold (Figure 4d). At last, the fixing screws moved far away from the specimens, and test was started (Figure 4e).

3.4. Data Processing

According to the testing data, the stress-time curves of rock specimens (Figure 3a) were drawn, and the uniaxial compressive strength and Brazilian tensile strength of rocks were calculated (Figure 3b,c), respectively. The dynamic failure time Dt is the time from the peak strength of rock to the time when the axial stress applied on specimens dropped to zero. According to the formulas in Table 1, the strength decrease rate SDR and the brittleness indicator index B of rocks were calculated directly.

3.5. Influence of Loading Terms on B and SDR

In order to explore the influence of loading rates on the brittleness indicator index B and strength decrease rate SDR, the displacement control mode and load control mode were selected in uniaxial compression tests. Under displacement control mode, the loading rates were pre-designed as 0.005 mm/min, 0.05 mm/min, 0.5 mm/min, 5 mm/min, 50 mm/min, and the maximum displacement loading rate of testing machine for each rock, respectively.

While under load control mode, the loading rate was pre-designed as in Table 2, and each rock was divided into 6 groups in load controlled uniaxial compression tests.

Table 2. Loading rate in load controlled uniaxial compression tests.

Rock Lithology	Loading Rate of Each Group								
	Group 1			Group 2			Group 3		
	kN/min	MPa/s	Sample Number	kN/min	MPa/s	Sample Number	kN/min	MPa/s	Sample Number
Granite	1.06	0.009	1	8.72	0.074	1	93.48	0.7935	1
Marble	0.64	0.006	1	6.66	0.057	1	66.67	0.5659	1
Red sandstone	0.07	0.0006	1	4.49	0.038	1	28.17	0.2391	1
Coal	0.22	0.002	1	1.50	0.013	1	14.87	0.1262	1
Rock Lithology	Group 4			Group 5			Group 6		
	kN/min	MPa/s		kN/min	MPa/s		kN/min	MPa/s	
Granite	1101.60	9.3507	1	8799.93	78.383	1	46,683.47	778.0578	1
Marble	693.32	5.8857	1	5082.66	43.1430	1	48,380.29	410.6647	1
Red sandstone	397.46	3.37377	1	3540.87	30.0558	1	27,172.56	230.6478	1
Coal	168.14	1.4273	1	2135.74	18.1287	1	7590.34	64.4288	1

In Brazilian splitting tests, the values of Brazilian tensile strength are small, and the influence of loading mode and rate on Brazilian tensile strength is ignored in this study. The load control mode with a loading rate of 0.3–0.5 MPa/s was selected in Brazilian splitting tests, which is recommended by the standard for test methods of engineering rock mass (GB/T 50266-2013) in China. The calculation results of *B* and *SDR* are shown in Tables 3 and 4.

Table 3. Calculation results of *B*1, *B*2, and *SDR* under different loading rates in displacement controlled uniaxial compression tests.

Rock Lithology	Loading Rate (mm/min)	*B*1 Formula (1)	*B*2 Formula (2)	*SDR* (MPa/s)
Granite	0.005	14.35	0.87	1.57
Granite	0.05	14.14	0.87	7.72
Granite	0.49	15.42	0.88	114.25
Granite	4.94	15.20	0.88	795.80
Granite	47.18	18.12	0.90	4299.82
Granite	281.23	15.95	0.88	8109.59
Marble	0.005	22.83	0.92	0.06
Marble	0.05	21.78	0.91	0.18
Marble	0.50	22.08	0.91	1.61
Marble	4.93	24.59	0.92	11.08
Marble	46.52	24.75	0.92	250.26
Marble	288.25	29.69	0.93	362.04
Red sandstone	0.001	13.27	0.86	0.07
Red sandstone	0.05	12.86	0.86	1.69
Red sandstone	0.50	12.54	0.85	16.25
Red sandstone	4.94	15.41	0.88	108.82
Red sandstone	46.63	14.80	0.87	932.70
Red sandstone	316.67	15.41	0.88	4627.39
Coal	0.005	23.95	0.92	0.22
Coal	0.05	19.30	0.90	1.16
Coal	0.50	26.95	0.93	8.47
Coal	4.93	30.12	0.94	32.00
Coal	46.04	44.38	0.96	676.21
Coal	298.95	32.50	0.94	2320.13

Table 4. Calculation results of *B*1, *B*2, and *SDR* under different loading rates in load controlled uniaxial compression tests.

Rock Type	Loading Rate (kN/min)	*B*1 Formula (1)	*B*2 Formula (2)	*SDR* (MPa/s)
Granite	1.06	12.19	0.85	/
Granite	8.72	13.20	0.86	305.62
Granite	93.48	15.06	0.88	697.20
Granite	1101.60	15.53	0.88	1534.98
Granite	8799.93	14.08	0.87	2251.52
Granite	46,683.47	14.78	0.87	-
Marble	0.64	22.81	0.92	23.38
Marble	6.66	23.66	0.92	51.82
Marble	66.67	23.24	0.92	184.78
Marble	693.32	23.71	0.92	492.96
Marble	5082.66	27.70	0.93	973.80
Marble	48,380.29	30.08	0.94	1904.80
Red sandstone	0.28	11.84	0.84	2.75
Red sandstone	4.49	15.77	0.88	36.57
Red sandstone	28.17	13.28	0.86	148.54
Red sandstone	397.46	13.89	0.87	466.09
Red sandstone	3540.87	15.31	0.88	646.69
Red sandstone	27,172.56	13.94	0.87	837.16
Coal	0.22	11.91	0.85	-
Coal	1.50	16.51	0.89	5.54
Coal	14.87	29.66	0.93	33.20
Coal	168.14	24.33	0.92	71.36
Coal	2135.74	19.77	0.90	230.81
Coal	7590.34	24.74	0.92	835.25

3.5.1. Impact of Loading Rate

As shown in Figure 5a–d, with the increase of loading rate, *B*1 (calculated using Formula (1)) and *B*2 (calculated using Formula (2)) show a slightly increasing trend, and the values of *B*1 and *B*2 of coal and marble are larger than that of the other two rocks under load and displacement control modes.

The increase of *B*2 on granite, marble, and red sandstone is lower than 0.05, and the increase of coal rock is slightly larger (i.e., lower than 0.08). Therefore, the loading rate effects on *B*2 can be ignored. The increasements of the 4 kinds of rocks are 3–13 for *B*1, and the loading rate effects on *B*1 are clearer than on *B*2, which cannot be ignored.

The influence of loading rates on the values of the *SDR* is more significant, as shown in Figure 5e,f. Different types of rocks have different sensitivities to loading rate. Among the 4 kinds of rocks, the loading rate has the least influence on the *SDR* values of marble, which shows a linear change trend in the scope of the loading rate in this study. Under displacement control mode, the values of the *SDR* of granite and red sandstone increased sharply after the displacement loading rate was larger than 5 mm/min. Under load control mode, the values of the *SDR* of all rock types increased sharply after the loading rate was larger than 100 kN/min. The influence degrees of loading rate on the *SDR* values in order from the largest to the smallest are as follows: granite, marble red sandstone, and coal, which is quite different from that under displacement control mode. Thus, when the *SDR* is used as the evaluation index of rockburst tendency, special attention should be paid to the test within a certain range, otherwise there is no comparability of *SDR* obtained under different loading rates.

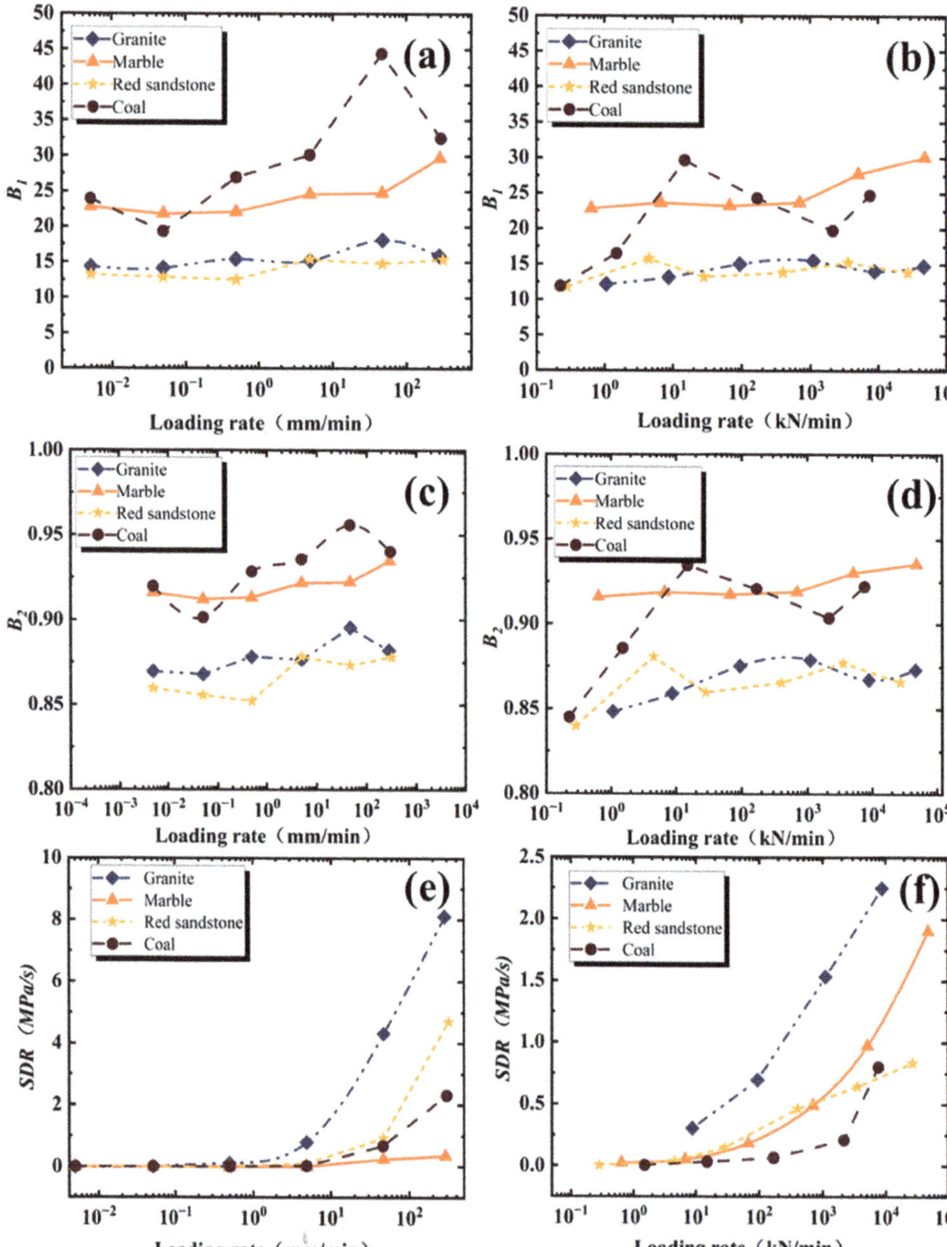

Figure 5. Effect of loading rate on *B*1, *B*2, and *SDR*: (**a**) *B*1 under displacement control mold, (**b**) *B*1 under load control mold, (**c**) *B*2 under displacement control mold, (**d**) *B*2 under load control mold, (**e**) *SDR* under displacement control mold, (**f**) *SDR* under load control mold.

3.5.2. Effect of Loading Mode

As shown in Figures 6–8, There are certain differences between the values of *B*1, *B*2, and *SDR* under displacement control mode and load control mode.

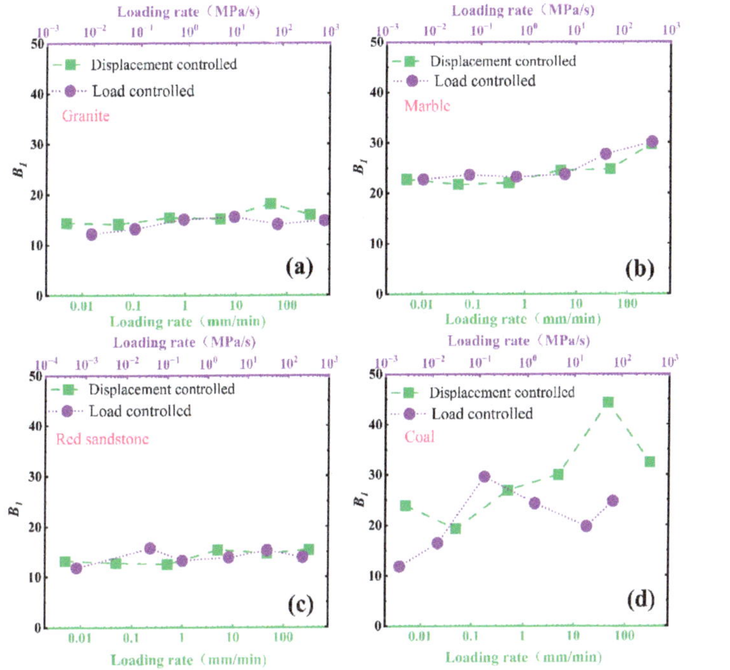

Figure 6. Effect of loading mode on *B*1: (**a**) granite, (**b**) marble, (**c**) red sandstone, (**d**) coal.

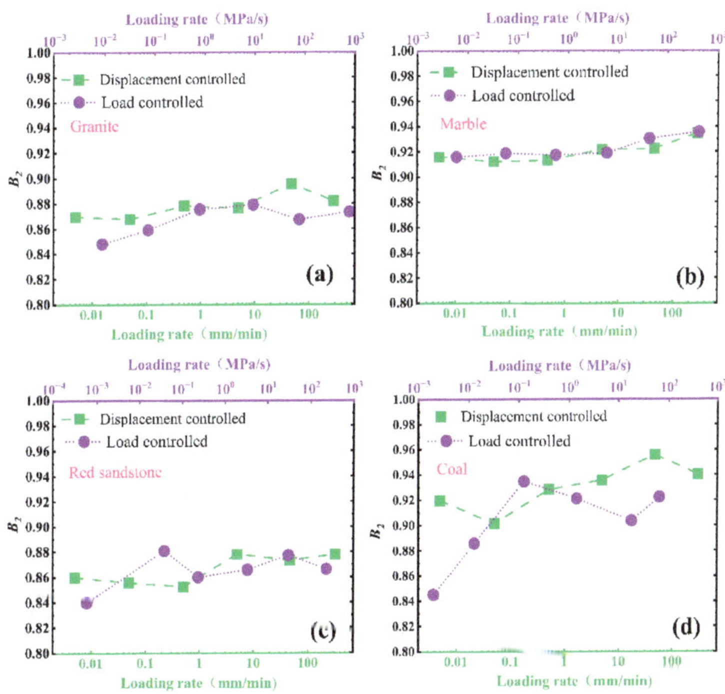

Figure 7. Effect of loading mode on *B*2: (**a**) granite, (**b**) marble, (**c**) red sandstone, (**d**) coal.

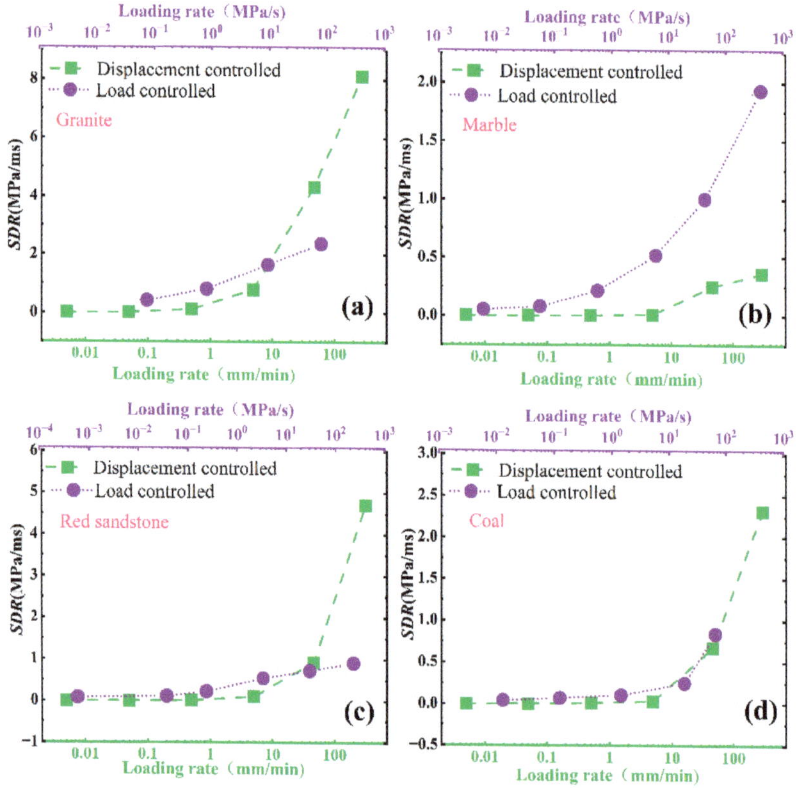

Figure 8. Effect of loading mode on *SDR*: (**a**) granite, (**b**) marble, (**c**) red sandstone, (**d**) coal.

For *B*1 and *B*2, the loading mode effect can be ignored, and the gaps of *B*1 and *B*2 under different loading modes fluctuate to a certain extent, especially for coals. The main factor inducing the phenomena is that coal is a typical heterogeneous anisotropic sedimentary rock, and the anisotropic properties of coal specimens induce the gaps of *B*1 and *B*2 under different loading modes. Therefore, the greater the anisotropy of specimens, the more pronounced the fluctuation of *B*1 and *B*2. It is suggested that more specimens should be prepared to measure the values of *B*1 and *B*2 (Figures 6 and 7).

For *SDR*, the control mode has great influence on the *SDR*. For granite, marble and coal, the values of the *SDR* under the load control mode are bigger than that of the *SDR* under displacement control mode. While for marble, the values of the *SDR* under displacement control mode are bigger than that of the *SDR* under loads control mode (Figure 8).

The loads applied on the specimens increase at first and then decrease, while the deformation of specimens always increases before total failure in uniaxial compression tests, so it is difficult to obtain the post-peak stage of rocks in load controlled uniaxial compression tests. When the rock specimens lose their carrying capacity after peak strength, the actuator of the testing machine under load control mode may trigger accelerated loading and shorten the duration time of the post- peak stage. The typical pre-peak and post -peak stress-time curves are shown in Figure 9.

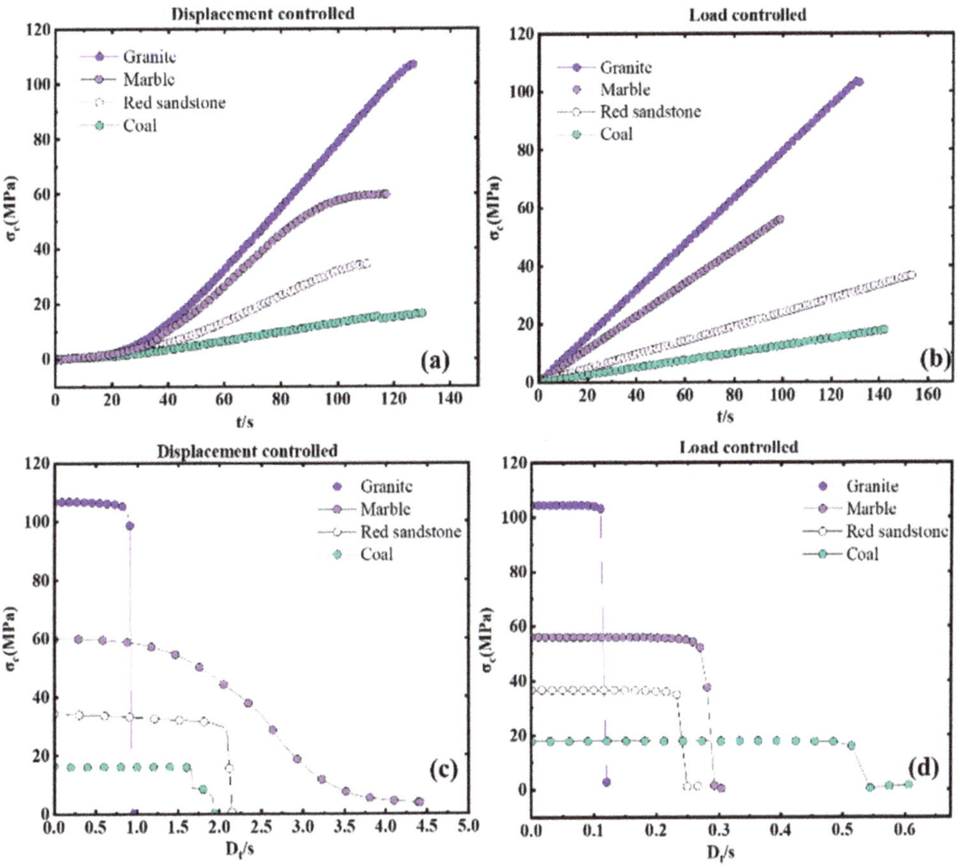

Figure 9. Stress-time curves of rocks in uniaxial compression tests (**a**,**c**) with a loading rate of 0.5 mm/min; (**b**,**d**) with a loading rate of 93 kN/min.

In this study, the main aim is to determine the rockburst proneness of hard rocks by the basic mechanical parameter tests of rocks, i.e., uniaxial compression tests and Brazilian tensile tests. According to GB/T 50266-2013 "Engineering Rock Mass Test Method Standard" [75], it is recommended that the loading terms are load control mode with loading rate of 0.5–1 MPa/s, while the displacement control mode is more suitable to test the rockburst proneness index of rocks in our study. Based on the strain rate effect of rocks, the difference between the testing results of basic mechanical parameters under load control mode with loading rate of 0.5–1 MPa/s and under displacement control mode with loading rate of 0.1–0.7 mm/min is not great, and both can be used in the measurement of brittleness indicators and the strength decrease rate.

4. Classification Criteria of *B* and *SDR*

Twelve kinds of rocks, including white sandstone, green sandstone, grey sandstone, purple sandstone, marble, granite, Chinese black granite and brown sandstone, Andesite, yellow sandstone, red sandstone, and coal (Figure 2), were selected to determine the rockburst criterion of the brittleness indicator index *B* and the strength decrease rate *SDR*. The displacement control mode with a loading rate of 0.12 mm/min was chosen in the uniaxial compression test designed in this part, and the load control mode with a loading rate of 0.3 MPa/s was chosen in the Brazil splitting test. The obtained results of *B*1, *B*2, and *SDR* were listed in Table 5.

Table 5. Calculation results of *B*1, *B*2, and *SDR*.

Rock Lithology	Sample Number	Brazilian Tensile Strength (MPa)	Uniaxial Compressive Strength (MPa)	*B*1	*B*2	*SDR* (MPa/s)
Purple sandstone	1	4.27	95.04	22.26	0.91	31.98
Yellow sandstone	1	1.16	43.72	37.69	0.95	4.48
Brown sandstone	1	3.38	64.92	19.21	0.90	5.86
Grey sandstone	1	2.99	70.45	23.56	0.92	9.3
Andesite	1	12.85	200.59	15.61	0.88	29.91
White sandstone	1	1.64	36.80	22.44	0.91	2.29
Green sandstone	1	2.60	67.15	25.83	0.93	48.77
Chinese black granite	1	8.97	150.98	16.83	0.89	24
Granite	1	6.68	112.65	16.86	0.89	17.03
Marble	1	2.41	109.54	45.45	0.96	1.63
Red sandstone	1	2.79	74.90	26.85	0.93	4.41
Coal	1	0.61	33.33	54.64	0.96	2.45

The rockburst tendency degrees of the rocks chosen in this study were determined by the failure phenomena. Different types of rocks will have different failure phenomena when rockburst occurs. Therefore, the classification criteria of failure phenomena are judged from two main aspects. First, the response degree of sound when the rock is about to destabilize and is damaged. The more intense or active the sound emitted near failure, the higher the rockburst tendency (an acoustic emission system is used to monitor the intensity of the sound, which is not described in detail in this manuscript due to space limitations). The second is the degree of fracture of the rock damage and the amount of debris after rock failure. The more cracks and debris produced by cracked rock, the higher the degree of rockburst tendency (debris is screened through the screen mesh). The detailed classification criterion of failure phenomena was summarized in Table 6.

Table 6. Rockburst classification criterion based on the failure phenomena of hard rocks.

Category	I (None)	II (Weak)	III (Medium)	IV (Strong)
Phenomenon	One or two penetrating cracks appeared on the specimen surface after failure. Few fragments formed with slight sound.	Multiple cracks appeared on the specimen surface after failure. A small number of fragments dropped slowly with a slight sound.	Multiple cracks appeared on the specimen surface after failure. Several fragments ejected with a clear failure sound.	Multiple cracks appeared on the specimen surface after failure. A large number of fragments ejected with great failure sound.
Classification results	 Marble	 Yellow sandstone Other rocks: Red sandstone, White sandstone, Coal	 Grey sandstone Other rocks: Brown sandstone, Green sandstone Purple sandstone Granite	 Chinese black granite Other rocks: Andesite

In this study, the classification of rocks under different rockburst tendency indexes was based on the results of the classification criteria in Table 6. The rockburst tendency degrees of hard rocks were also divided into four grades, i.e., none, weak, medium, and strong, as shown in Figure 10. The rock samples with the same category were grouped into the same interval as far as possible, thus the dividing line under different rockburst indicators was obtained. However, the values of green sandstone calculated by the two rockburst indicators had large deviations, and its rockburst tendency evaluation was fuzzy. Therefore, we will pay more attention to and explore methods to evaluate the rockburst tendency of green sandstone in future studies.

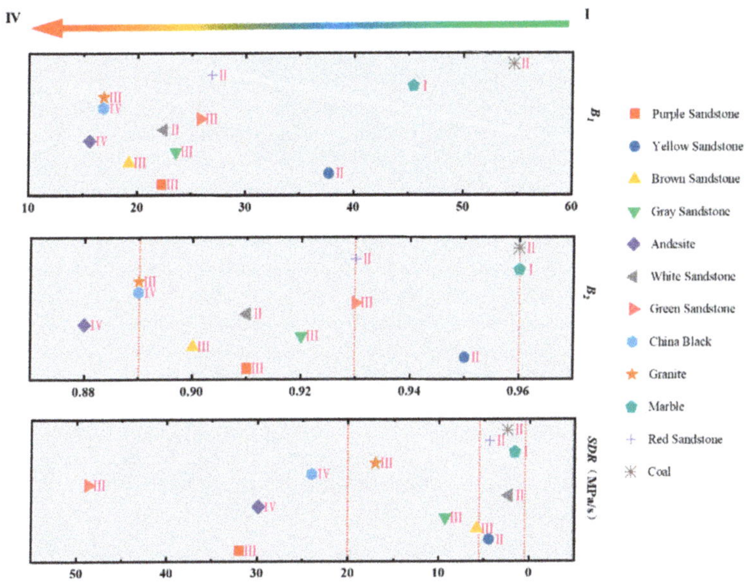

Figure 10. Classification criteria of the rockburst proneness of rocks in this study.

The values of *B*1 were concentrated in the range of 5–30, and the different bursting tendency degree results of rocks based on failure phenomena criteria were cross-distributed. It was different to formulate rational classification criterion of *B*1 with a crosswise distribution trend. In this study, *B*1 was not suitable for the evaluation of rockburst tendency. According to the bursting tendency degree results of rocks based on failure phenomena criteria, the results calculated by the rockburst tendency index (i.e., *B*2 and *SDR*) are basically consistent with the experimental phenomenon classification. Therefore. the *B*2 and *SDR* can be used for the evaluation of rockburst tendency. The detailed classification criteria of *B*2 and *SDR* are as follows:

$B2 > 0.96$ None;
$0.96 \geq B2 > 0.93$ Weak;
$0.93 \geq B2 > 0.89$ Medium;
$B2 \leq 0.89$ Strong;
$SDR < 2$ MPa/s None;
2 MPa/s $\leq SDR < 6$ MPa/s Weak;
6 MPa/s $< SDR < 20$ MPa/s Medium;
$SDR \geq 20$ MPa/s Strong.

5. Discussion

5.1. Rationality of Classification Criteria

In hard rock engineering, related academic studies at home and abroad showed that rockburst often occurs in hard and brittle rocks [76]. In our study, the brittleness indicator index *B*, i.e., *B*1 and *B*2, reflects the rockburst tendency degree of rocks from aspects of the basic mechanical parameters of rocks, i.e., uniaxial compressive strength and Brazilian tensile strength. The strength decrease rate *SDR* focuses on post-peak phase of complete stress-strain curves in uniaxial compression tests, using the average stress decrease velocity to reflect the rockburst tendency degree of rocks. Compared with other non-energy indicators, these two indicators have the advantage of accurate calculation, and reflect the rockburst tendency from two different aspects. Other redundant indexes are no longer needed. In addition, the non-energy index and the energy index complement each other, and the calculation of the non-energy index is more convenient. In this study, the bursting tendency grade of white sandstone was different between the classification grade on failure phenomena criteria and that on *B*2 criterion, while the bursting tendency grades of purple sandstone and green sandstone were different between the classification grade on failure phenomena criteria and that on *SDR* criteria, as shown in Figure 10. For other rocks close to the critical value, the evaluation results are considered to be accurate. The burst tendency grade of the rocks mentioned above on *B*2 or *SDR* criteria all raised by one level, which was conducive to field rockburst protection.

5.2. Strain Rate Effect on Rockburst Index

It is proved that the larger the loading rate is, the bigger the rockburst index is, and the more intense the rockburst is [77,78], which is reflected by *SDR* in this study. The loading rate and mode effects can be unified by study of the strain rate effect. The strain rates of specimens under different displacement and load loading rates were discussed. As shown in Figure 11, the strain rate was positively correlated with the displacement loading rate, and the strain rate was calculated as Formula (3). The strain rates of specimens with the same height value had the same value under the same displacement loading rate.

$$\varepsilon_{td} = \frac{d\varepsilon}{dt} = \frac{1}{H_0}\frac{dH}{dt} = \frac{D_{lr}}{H_0} \quad (3)$$

where ε_{td} is the strain rate of specimens under displacement control mode; H_0 is the height of the specimen, 100 mm in this study; H is the real time height of specimens during testing; and D_{lr} is the displacement loading rate.

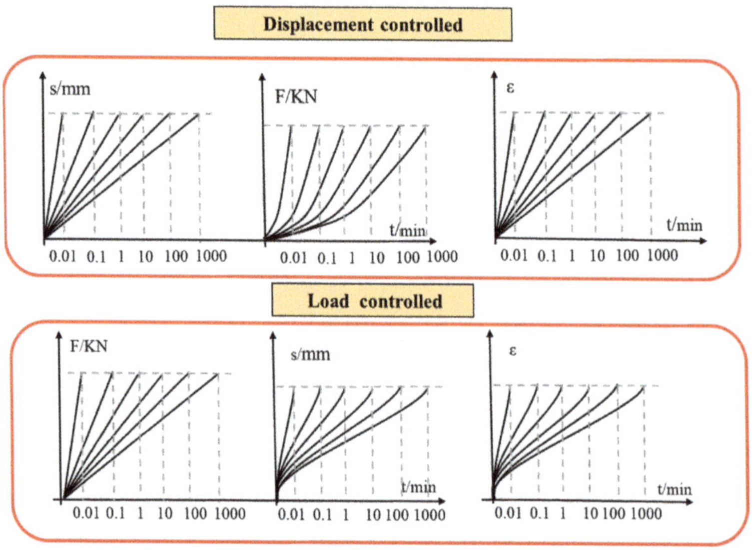

Figure 11. Relationship between strain rate under displacement controlled and load control mode.

Under loading control mode, if the same loading rate was applied for different rocks, the strain rates of different rocks were also different. The strain rate of rocks under displacement control mode showed a linear change trend (Figure 11 top), while that under load control mode had a non-linear change trend (Figure 11 bottom).

If the loading time of the pre-peak stage had a same value, the ε_{td} equal the average value of the strain rate ε_{tl} of specimens under load control mode. The ε_{td} and ε_{tl} of different rocks in uniaxial compression tests with varied loading rates are shown in Figure 12.

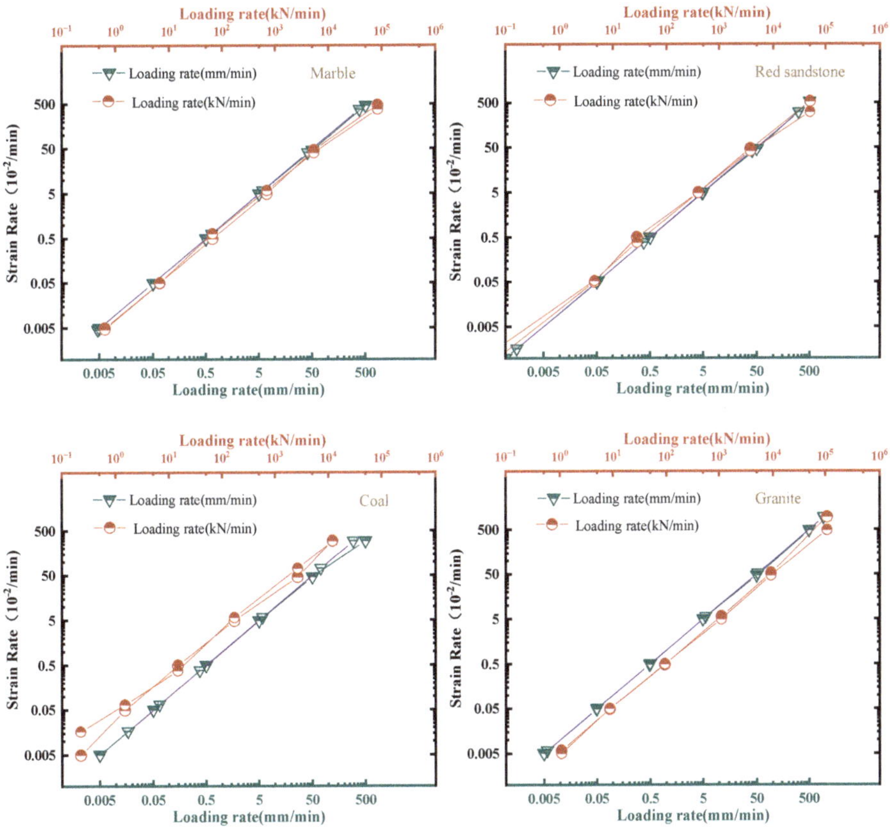

Figure 12. Relationship between experimental rate and strain rate of different rocks.

In this study, the *SDR* was affected greatly by the strain rate of rocks. The uniaxial compressive strength σ_c increased about 1.2 times, while *Dt* increase much more than σ_c, as shown in Figure 13. As can be seen from the figures, the Dt starts to increase rapidly when it is larger than 0.7 mm/min. Therefore, for the displacement loading rate, 0.1–0.7 mm/min is recommended.

The classification and evaluation criteria of *B* and *SDR* in non-energy rockburst indexes are mainly discussed, while some energy-based rockburst indexes are relatively ignored. It is also found that the evaluation results of a kind of rock under different classification criteria are different. Therefore, future research will pay more attention to the comprehensive criterion of rockburst and the quantitative research of risk assessment under the aggregation of multiple indicators, such as uncertainty theory, machine learning, and numerical simulation.

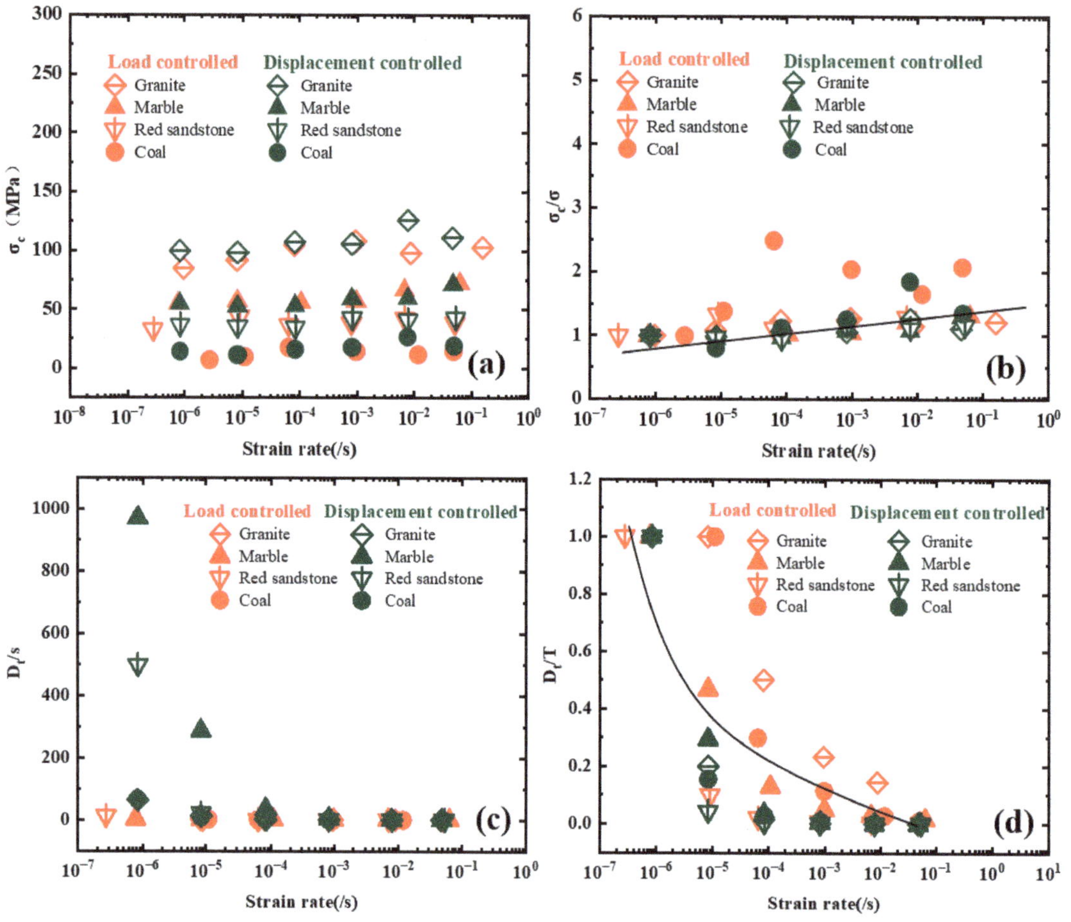

Figure 13. Relationships between (**a**) uniaxial compressive strength vs. time, (**b**) normalization uniaxial compressive strength vs. time, (**c**) Dt vs. time, and (**d**) normalization Dt vs. time.

6. Conclusions

In order to establish appropriate criteria for hard rock bursting liability, especially in terms of bursting indicators and classification criteria, the brittleness indicator index (*B*2) and strength decrease rate (*SDR*) index and their classification criteria in rockburst tendency evaluation were studied based on the results of lab testing. The research results provide reference for establishing the indoor evaluation standard of the rockburst tendency of hard rock. The conclusions of this paper are as follows:

1. The *B*2 and *SDR* were selected as the basic indexes for the rockburst tendency evaluation of hard rocks. It is considered that the strength decrease rate *SDR* reflects the speed of stress drop speed and energy release rate in rock failure stage, and the brittleness indicator index *B*2 value reflects the fundamental properties of rocks. The combination of the two indexes can describe rockburst tendency more comprehensively.
2. The reasonable loading mode and rate in rockburst tendency tests were determined. After studying the influence of loading rate and loading mode on rockburst tendency, it is suggested to adopt the displacement control mode of 0.1–0.7 mm/min for rockburst tendency assessment for *B*2 and *SDR*.

3. The classification criteria of *B*2 and *SDR* were put forward. According to the results in this study, *SDR* and *B*2 were divided into four grades. It was found that there is a strong consistency between the *B*2 index and *SDR*, which proves a rationality of the classification criteria based on the failure phenomena of rocks in uniaxial compression tests.

Author Contributions: Conceptualization, K.D.; Software, S.Y.; Investigation, S.Y.; Resources, L.W.; Data curation, J.Z.; Writing—original draft, S.Y.; Visualization, J.Z.; Project administration, K.D. All authors have read and agreed to the published version of the manuscript.

Funding: This research is supported by the National Natural Science Foundation of China (Grant No. 52104112), the Science and Technology Innovation Program of Hunan province (2021RC3007), the Natural Science Foundation of Hunan Province (2023JJ20062) and Research Foundation of the Department of Natural Resources of Hunan Province (Grant No. 2023-01).

Institutional Review Board Statement: Not applicable.

Informed Consent Statement: Not applicable.

Data Availability Statement: The data that support the findings of this study are available on request from the corresponding author, [wlccsu@csu.edu.cn], upon reasonable request.

Conflicts of Interest: The authors declare no conflict of interest.

References

1. Du, K.; Yang, C.; Su, R.; Tao, M.; Wang, S. Failure properties of cubic granite, marble, and sandstone specimens under true triaxial stress. *Int. J. Rock Mech. Min. Sci.* **2020**, *130*, 104309. [CrossRef]
2. Li, X.B.; Gong, F.Q.; Tao, M.; Dong, L.J.; Du, K.; Ma, C.D.; Zhou, Z.L.; Yin, T.B. Failure mechanism and coupled static-dynamic loading theory in deep hard rock mining: A review. *J. Rock Mech. Geotech. Eng.* **2017**, *9*, 767–782. [CrossRef]
3. Cai, M.; Brown, E.T. Challenges in the Mining and Utilization of Deep Mineral Resources. *Engineering* **2017**, *3*, 432–433. [CrossRef]
4. Wu, M.; Ye, Y.; Wang, Q.; Hu, N. Development of Rockburst Research: A Comprehensive Review. *Appl. Sci.* **2022**, *12*, 974. [CrossRef]
5. Solak, T. Ground behavior evaluation for tunnels in blocky rock masses. *Tunn. Undergr. Space Technol.* **2009**, *24*, 323–330. [CrossRef]
6. Farhadian, H. A new empirical chart for rockburst analysis in tunnelling: Tunnel rockburst classification (TRC). *Int. J. Min. Sci. Technol.* **2021**, *31*, 603–610. [CrossRef]
7. Keneti, A.; Sainsbury, B.-A. Review of published rockburst events and their contributing factors. *Eng. Geol.* **2018**, *246*, 361–373. [CrossRef]
8. Malan, D.; Napier, J. A limit equilibrium fracture zone model to investigate seismicity in coal mines. *Int. J. Min. Sci. Technol.* **2018**, *28*, 745–753. [CrossRef]
9. Li, X.; Jian, Z.; Wang, S.; Bing, L. Review and practice of deep mining for solid mineral resources. *Zhongguo Youse Jinshu Xuebao/Chin. J. Nonferr. Met.* **2017**, *27*, 1236–1262.
10. Ma, T.-H.; Tang, C.-A.; Liu, F.; Zhang, S.-C.; Feng, Z.-Q. Microseismic monitoring, analysis and early warning of rockburst. *Geomatics, Nat. Hazards Risk* **2021**, *12*, 2956–2983. [CrossRef]
11. Li, S.; Feng, X.-T.; Li, Z.; Chen, B.; Zhang, C.; Zhou, H. In situ monitoring of rockburst nucleation and evolution in the deeply buried tunnels of Jinping II hydropower station. *Eng. Geol.* **2012**, *137–138*, 85–96. [CrossRef]
12. Voza, A.; Valguarnera, L.; Marrazzo, R.; Ascari, G.; Boldini, D. A New In Situ Test for the Assessment of the Rock-Burst Alarm Threshold During Tunnelling. *Rock Mech. Rock Eng.* **2022**, *56*, 1645–1661. [CrossRef]
13. Wang, W.W. Testing study on rockburst pronounces of serpentine under a deep mine. *J. Wuhan Polytech. Univ.* **2017**, *36*, 55–58.
14. Wang, S.-F.; Sun, L.-C.; Tang, Y.; Jing, Y.; Li, X.-B.; Yao, J.-R. Field application of non-blasting mechanized mining using high-frequency impact hammer in deep hard rock mine. *Trans. Nonferr. Met. Soc. China* **2022**, *32*, 3051–3064. [CrossRef]
15. Wang, S.; Li, X.; Yao, J.; Gong, F.; Li, X.; Du, K.; Tao, M.; Huang, L.; Du, S. Experimental investigation of rock breakage by a conical pick and its application to non-explosive mechanized mining in deep hard rock. *Int. J. Rock Mech. Min. Sci.* **2019**, *122*, 104063. [CrossRef]
16. Sun, Y.; Tian, B.; Liu, X.; Chen, D.; Wu, Y. A Kind of Prediction Based on SOM Neural Clustering and Combination Weighting Evaluation Method. In Proceedings of the WWW'21: Companion Proceedings of the Web Conference 2021, Ljubljana, Slovenia, 19–23 April 2021; pp. 90–96.
17. Zhou, J.; Li, X.; Shi, X. Long-term prediction model of rockburst in underground openings using heuristic algorithms and support vector machines. *Saf. Sci.* **2012**, *50*, 629–644. [CrossRef]

18. Tang, C.; Wang, J.; Zhang, J. Preliminary engineering application of microseismic monitoring technique to rockburst prediction in tunneling of Jinping II project. *J. Rock Mech. Geotech. Eng.* **2010**, *2*, 193–208. [CrossRef]
19. Liu, H. Classified prediction model of rockburst using rough sets-normal cloud. *Neur. Comput. Appl.* **2019**, *31*, 8185–8193. [CrossRef]
20. Liang, W.; Zhao, G.; Hao, W.; Bing, D. Risk assessment of rockburst via an extended MABAC method under fuzzy environment. *Tunn. Undergr Space Technol.* **2019**, *83*, 533–544. [CrossRef]
21. Xuan, Z.; Bu, X. The Forecasting of Rockburst in Deep-buried Tunnel with Adaptive Neural Network. In Proceedings of the 2009 International Conference on Industrial and Information Systems, Haikou, China, 24–25 April 2009; pp. 3–6.
22. Chen, J.; Chen, Y.; Yang, S.; Zhong, X.; Han, X. A prediction model on rockburst intensity grade based on variable weight and matter-element extension. *PLoS ONE* **2019**, *14*, e0218525. [CrossRef]
23. Zhang, L.; Zhang, X.; Wu, J.; Zhao, D.; Fu, H. Rockburst prediction model based on comprehensive weight and extension methods and its engineering application. *Bull. Eng. Geol. Environ.* **2020**, *79*, 4891–4903. [CrossRef]
24. Li, D.; Liu, Z.; Armaghani, D.J.; Xiao, P.; Zhou, J. Novel ensemble intelligence methodologies for rockburst assessment in complex and variable environments. *Sci. Rep.* **2022**, *12*, 1844. [CrossRef]
25. Faradonbeh, R.S.; Taheri, A.; Ribeiro e Sousa, L.; Karakus, M. Rockburst assessment in deep geotechnical conditions using true-triaxial tests and data-driven approaches. *Int. J. Rock Mech. Min. Sci.* **2020**, *128*, 104279. [CrossRef]
26. Cui, Z.G. Rock burst tendency evaluation and analysis based on laboratory test. *Northeast Water Resour. Hydropower* **2019**, *37*, 3.
27. Gong, F.; Luo, S.; Jiang, Q.; Xu, L. Theoretical verification of the rationality of strain energy storage index as rockburst criterion based on linear energy storage law. *J. Rock Mech. Geotech. Eng.* **2022**, *14*, 1737–1746. [CrossRef]
28. Liu, S.; Xiao, F.; Li, T.; Zhang, B. Analysis of Impact Tendency and Sensitivity of Fractured Rock with Different Crack Arrest Measures. *Sustainability* **2022**, *14*, 13833. [CrossRef]
29. Gong, F.; Yan, J.; Li, X.; Luo, S. A peak-strength strain energy storage index for rock burst proneness of rock materials. *Int. J. Rock Mech. Min. Sci.* **2019**, *117*, 76–89. [CrossRef]
30. Gong, F.Q.; Yan, J.Y.; Li, X.B. A new criterion of rock burst proneness based on the linear energy storage law and the residual elastic energy index. *Chin. J. Rock Mech. Eng.* **2018**, *37*, 1993–2014.
31. Kang, Q.; Xia, Y.; Shi, M.; Zhang, W.; Wang, W.; Kong, D.; Wang, Y. Evaluation of Rock Burst Propensity and Rock Burst Mechanism in Deep Phosphate Mines: A Case Study of Sujiapo Phosphate Mine, Hubei Province, China. *Adv. Mater. Sci. Eng.* **2022**, *2022*, 7874016. [CrossRef]
32. Liang, W.Z.; Zhao, G.Y. Long and short term rockburst risk assessment of deep hard rock. *Chin. J. Rock Mech. Eng.* **2022**, *41*, 21.
33. Xia, Y.; Zhou, H.; Zhang, C.; He, S.; Gao, Y.; Wang, P. The evaluation of rock brittleness and its application: A review study. *Eur. J. Environ. Civ. Eng.* **2019**, *26*, 239–279. [CrossRef]
34. Du, K.; Sun, Y.; Yang, S.; Lv, S.; Wang, S. Differences of Mechanical Parameters and Rockburst Tendency Indices between Coal and Non-Coal Rocks and Modified Rockburst Tendency Classification Criteria for Non-Coal Rocks. *Appl. Sci.* **2021**, *11*, 2641. [CrossRef]
35. Methods for Determination, Monitoring and Control of Rock Burst—Part 2: Classification and Index Determination of Coal Bursting Liability. 2010. Available online: https://kns.cnki.net/kcms2/article/abstract?v=kxaUMs6x75KB1kF296ShrYo_9ZoYIWW4VfT4bfQP25xUu13oQ_hU9IEAfQx8mV-5auDdJFuN3RSHbQ9f9FWvw%3d%3d&uniplatform=NZKPT (accessed on 10 April 2023).
36. Jian, M.X.; Lin, Y.; Ma, C.D.; Xu, M.S.; He, Z.G.; Wang, Y.Q. Evaluation and prevention of rockburst tendency in deep mining of Malu Ping Mine. *Min. Technol.* **2011**, *11*, 52–54.
37. Qin, M.; Liu, C. Application of Comprehensive Analysis of Rockburst Tendency in Deep Mining of Ashele Copper Mine. *Min. Metall. Eng.* **2015**, *35*, 39–45.
38. Huang, B.; Li, S.L.; Gao, Z.P.; Hu, G.Y.; Wu, X.H.; Xu, G. Comprehensive Evaluation of Rockburst on Deep and Large-scale Mining in Dahongshan Iron Mine. *Ment. Mine* **2016**, *45*, 19–25.
39. Zhou, H.; Chen, S.K.; Zhang, G.Z.; Wang, H.L.; He, H.D.; Feng, J. Rockburst Prediction in Deep Lying and Long Tunnel Based on Efficiency Coefficient Method and Ground Stress Field Inversion. *J. Eng. Geol.* **2019**. [CrossRef]
40. Yi, Y.L.; Cao, P.; Pu, C.Z. Multi-factorial Comprehensive Estimation for Jinchuan's Deep Typical Rockburst Tendency. *Sci. Technol. Rev.* **2010**, *28*, 76–80.
41. Qiao, C.S.; Tian, Z.Y. Study on the possibility of rock burst in Dongguashan Copper Deposit. *Chin. J. Rock Mech. Eng.* **1998**, *17*, 917–921.
42. Liu, R.T.; Zhang, X.M.; Liu, J.X. Experimental study on rock burst tendency of granite in Linglong Gold Deposit. *ShanXi Archit.* **2014**, *40*, 57–59.
43. Jiang, C.L.; Jiang, Z.Q.; Zhang, R. Experimental Study of Impact Proneness of Deep Coal Seam and Roof Rock. *Coal Technol.* **2010**, *29*, 108–110.
44. Guo, D.M.; Zhang, T.; Li, Y.; Wang, G.D.; Li, G.E.; Guo, A.M. Research on rockburst tendency and its preventive measures of 800m deep surrounding-rock in Huxi coal mine. *China Min. Mag.* **2008**, *17*, 50–63.
45. Chen, S.J.; Guo, W.J.; Yang, Y.J.; Zou, J. Experimental study on bursting liability of coal seam. *Min. Saf. Environ. Prot.* **2007**, *34*, 10–15.
46. Guo, D.M.; Hou, J.F.; Yan, H.X.; Wang, J.; Cheng, H.Y.; Wang, Y.X. Study on bursting liability of deep surrounding rock of Suncun Coal Mine -1 300 m and countermeasures. *J. Jiaozuo Inst. Technol. (Nat. Sci.)* **2003**, *22*, 9–11.

47. Tian, S.F. The Risk Assessment and Prediction-Prevention of Rockburst in PengZhuang Mine. 2011. Available online: https://kns.cnki.net/kcms2/article/abstract?v=3uoqIhG8C475KOm_zrgu4lQARvep2SAkWGEmc0QetxDh64Dt3veMp_smSqBMOO34GtmQPt4ZcHD-metONYmBhxwPqQEl0d7G&uniplatform=NZKPT&src=copy (accessed on 10 April 2023).
48. Wang, G.F.; Li, Q.; Zhao, Z.L.; Pang, Y.H. The Impact Tendentiousness Testing of Working Faces and Roadways with Strong Rock burst and Fore Support System. *J. Shandong Univ. Sci. Technol. (Nat. Sci.)* **2011**, *30*, 1–8.
49. Sun, Q.J. Experimental study on bursting liability of Longjiabao No. 2 coal seam. *Shanxi Coking Coal Sci. Technol.* **2010**, 23–26. [CrossRef]
50. Guan, W. The Risk Assessment and Prediction of Rock Burst in the No. 9 District of No. 2 Jining Coal Mine. 2014. Available online: https://kns.cnki.net/kcms2/article/abstract?v=3uoqIhG8C475KOm_zrgu4lQARvep2SAkkyu7xrzFWukWIylgpWWcEnphZzgD0g6QolGV8xVCW1j7aGuBkkqKhZFS6E685Xmi&uniplatform=NZKPT&src=copy (accessed on 10 April 2023).
51. Liu, Z.P.; Liu, W.B.; Wu, L.Q.; Lu, S.B. Experiment Study of Burst Tendency of Coal and Rock Layers at Tangkou Coal Mine. *Saf. Coal Min.* **2009**, *40*, 8–10.
52. Li, Y.S.; Hu, Y.C. A study on the coal seam pressure bump tendency in Ding6 coal seam in Pingdingshan Coal Group No. 1 Coal Mine. *Saf. Coal Min.* **2009**, *36*, 99–101.
53. Meng, H.D.; Hou, X.Z.; Li, H.Y.; Zhao, Q. Monitoring and Prevention Research of Working Face Rock Burst in Deep Coal Well with Large Dip Angle. *Coal Technol.* **2014**, *33*, 63–65.
54. Wang, H.W.; Jiang, Y.D.; Deng, D.X.; Zhang, D.Q.; Lv, J.G. Investigation on the inducing factors of coal bursts under complicated geological environment in Yima mining area. *Chin. J. Rock Mech. Eng.* **2017**, *36*, 4085–4092.
55. Zhang, P.D.; Cui, J.S.; Wang, H.; Cai, F.; Gao, C. Experimental study on bursting liability of No.7 coal and roof in Yaoqiao Coal Mine. *Coal Sci. Technol. Mag.* **2015**, 47–50.
56. Zheng, S.D. Exploration on comprehensive control technology of rock burst in Gengcun Coal Mine. *Inn. Mong. Coal Econ.* **2014**, 2.
57. Wang, J.; Cui, J.S.; Hao, Y.X.; Zhang, Y.L.; Meng, Z.G.; Wang, H. Experement study on bursting Liability of 12# coal and roof rock of Hongyang coal mine. *China Min. Mag.* **2016**, *25*, 70–101.
58. Hu, S.Y.; Zhang, H. Research on rockburst risk assessment and prevention techniques in Changcun Coal Mine. *China Energy Environ. Prot.* **2019**, *41*, 15–23.
59. Wang, L.F.; Zhong, Z.L.; Liu, Y.L.; Xin, C.R. Evaluation and Control of Rock Burst Hazard of 2315 Working Face of Anju Coal Mine. *Coal Min. Technol.* **2019**, *24*, 120–123.
60. Su, C.D.; Yuan, R.F.; Zhai, X.X. Experimental Research on Bursting Liability Index of Coal Samples of ChengJiao Coal Mine. *Chin. J. Rock Mech. Eng.* **2013**, *32*, 3696–3704.
61. Jiang, F.X.; Zhu, S.T.; Wang, X.Y.; Jiang, Y.W.; Ning, T.Z.; Sun, S.H. Energy accumulation characteristics and rockburst mechanism of surrounding rock at heading face of extra-thick coal seam. *Chin. J. Geotech. Eng.* **2019**, *41*, 2071–2078.
62. Pan, J.F.; Jian, F.J.; Liu, S.H.; Zhao, W.G.; Yuan, D.S.; Wang, H.T.; Liu, H. Geological Characteristic and Control of Rock Burst of Huanglong Jurassic Coal Mine Field. *Coal Min. Technol.* **2019**, *24*, 110–115.
63. Kim, J.; Moridis, G.J. Numerical analysis of fracture propagation during hydraulic fracturing operations in shale gas systems. *Int. J. Rock Mech. Min. Sci.* **2015**, *76*, 127–137. [CrossRef]
64. Hajiabdolmajid, V.; Kaiser, P. Brittleness of rock and stability assessment in hard rock tunneling. *Tunn. Undergr. Space Technol.* **2003**, *18*, 35–48. [CrossRef]
65. Kahraman, S. Correlation of TBM and drilling machine performances with rock brittleness. *Eng. Geol.* **2002**, *65*, 269–283. [CrossRef]
66. Guo, F.; Zhang, D.; Su, J.; Niu, X. Experimental study of influences of confining pressures and ground water on residual strength and post-peak volumetric change of soft rock. *Chin. J. Rock Mech. Eng.* **2009**, *28*, 2644–2650.
67. Jing, H.W. Experimental study on dilatancy nature of the rock post-peak behavior. *Rock Soil Mech.* **2003**, *24*, 93–96.
68. Wong, L.N.Y.; Meng, F.; Guo, T.; Shi, X. The Role of Load Control Modes in Determination of Mechanical Properties of Granite. *Rock Mech. Rock Eng.* **2019**, *53*, 539–552. [CrossRef]
69. Gong, F.-q.; Wang, Y.-l.; Luo, S. Rockburst proneness criteria for rock materials: Review and new insights. *J. Cent. South Univ.* **2020**, *27*, 2793–2821. [CrossRef]
70. Peng, Z.; Wang, Y.H.; Li, T.J. Griffith theory and criterion of rockburst. *J. Rock Mech. Eng.* **1996**, *15*, 491–495.
71. Zhang, J.; Fu, B.; Li, Z.; Song, S.; Shang, Y. Criterion and classification for strain mode rockbursts based on five-factor comprehensive method. *Harmonis. Rock Eng. Environ.* **2012**, 1435–1440.
72. Singh, S.P. The influence of rock properties on the occurrence and control of rockbursts. *Min. Sci. Technol.* **1987**, *5*, 11–18. [CrossRef]
73. Du, K.; Li, X.; Su, R.; Tao, M.; Lv, S.; Luo, J.; Zhou, J. Shape ratio effects on the mechanical characteristics of rectangular prism rocks and isolated pillars under uniaxial compression. *Int. J. Min. Sci. Technol.* **2022**, *32*, 347–362. [CrossRef]
74. Du, K.; Sun, Y.; Zhou, J.; Khandelwal, M.; Gong, F. Mineral Composition and Grain Size Effects on the Fracture and Acoustic Emission (AE) Characteristics of Rocks Under Compressive and Tensile Stress. *Rock Mech. Rock Eng.* **2022**, *55*, 6445–6474. [CrossRef]

75. Ministry of Housing and Urban-Rural Development of the People's Republic of China. State Administration for Market Regulation, Standard for Test Methods of Engineering Rock Mass. 2013. Available online: https://kns.cnki.net/kcms2/article/abstract?v=kxaUMs6x75KB1kF296ShrYo_9ZoYIWW4VfT4bfQP25xUu13oQ_hU422dDazkYSspThIUfafUZKejgo7g1IOMg%3d%3d&uniplatform=NZKPT&src=copy (accessed on 10 April 2023).
76. Cheng, Y.; Zhang, Y.L.; Ma, Y.L. The Mechanism of Rockburst and the Control Countermeasure in Deep-Well Tough Rock, Controlling Seismic Hazard and Sustainable Development of Deep Mines. In Proceedings of the 7th International Symposium on Rockburst and Seismicity in Mines (Rasim7), Dalian, China, 21–23 August 2009; Volume 1 and 2, pp. 313–318.
77. Si, X.F.; Huang, L.Q.; Gong, F.Q.; Liu, X.L.; Li, X.B. Experimental investigation on influence of loading rate on rockburst in deep circular tunnel under true-triaxial stress condition. *J. Cent. South Univ.* **2020**, *27*, 2914–2929. [CrossRef]
78. Hu, F.; Cheng, J.Z.; Li, C.Y. Experiment study on rock burst of granite roadway under the loading rate effect. *Min. Metall.* **2019**, 28.

materials

MDPI

Article

Influence of Measurement Uncertainty in the Determination of Selected Rock Parameters—A Realistic Approach

Klaus Voit [1], Oliver Zeman [2,*], Peter Gappmaier [1], Karin Wriessnig [1] and Renata Adamcova [3]

1 Institute of Applied Geology, University of Natural Resources and Life Sciences, Vienna, Peter Jordan-Straße 82, 1190 Vienna, Austria
2 Institute of Structural Engineering, University of Natural Resources and Life Sciences, Vienna, Peter Jordan-Straße 82, 1190 Vienna, Austria
3 Faculty of Natural Sciences, Comenius University in Bratislava, Ilkovicova 6, 842 15 Bratislava, Slovakia
* Correspondence: oliver.zeman@boku.ac.at; Tel.: +43-14-765487541; Fax: +43-14-765487509

Abstract: To determine and compare the measurement uncertainty of different geological-geotechnical testing methods, numerous test locations were selected in a hard rock quarry. Measurements were carried out along two vertical measurement lines perpendicular to the mining levels of an existing exploration. Along these lines, the rock quality basically varies due to weathering (decreasing influence with increasing distance from the original ground surface), but also due to the influence of the geological-tectonic conditions on site. The mining conditions (blasting) are identical over the considered area. The rock quality was examined as follows: as field tests, the rock compressive strength was determined by means of point load test and rebound hammer, as laboratory method the Los Angeles test (standard laboratory test for the determination of the mechanical rock quality) was used to identify the impact abrasion resistance. The statistical evaluation and comparison of the results allowed conclusions to be drawn about the contribution of the individual test methods to the measurement uncertainty whereas, in practice, a priori information can be applied complementarily. It shows that the influence on the combined measurement uncertainty u of the different methods due to the geological variability in horizontal direction reaches values between 17 and 32%, whereby the rebound hammer method shows the highest values. However, the highest influences on the measurement uncertainties are a result of the vertical direction due to weathering phenomena with percentages of 55 to 70%. For the point load test, the vertical direction shows the highest significance with an influence of approximately 70%. This leads to the conclusion that a higher weathering degree of the rock mass shows an increasing effect on the measurement uncertainty which needs to be considered using a priori information in measurements.

Keywords: hard rock characterization; rock properties; field tests; rebound/Schmidt hammer; point load test; Los Angeles test; aleatoric and epistemic measurement uncertainty

Citation: Voit, K.; Zeman, O.; Gappmaier, P.; Wriessnig, K.; Adamcova, R. Influence of Measurement Uncertainty in the Determination of Selected Rock Parameters—A Realistic Approach. *Materials* **2023**, *16*, 3045. https://doi.org/10.3390/ma16083045

Academic Editors: Shaofeng Wang, Tianshou Ma, Linqi Huang, Jie Zhou and Changjie Zheng

Received: 28 February 2023
Revised: 4 April 2023
Accepted: 10 April 2023
Published: 12 April 2023

1. General

1.1. Intent of Research

Rock classifications to assess the stability of rock in tunneling or slope construction are among the top disciplines in engineering geology [1–3]. In this context, rock strength plays a decisive role—in addition to the interface characteristics. The same applies to the estimation and monitoring of rock quality in quarries during the production of high-quality aggregates: methods for a basic characterization as well as for a continuous quality addressing are essential.

In principal, a basic distinction concerning different test methods for the determination of mechanical rock parameters—with different time and equipment requirements concerning the execution of the tests—can be made as follows [4]:

- field tests vs. laboratory tests;
- destructive vs. non-destructive methods;
- direct vs. indirect methods.

Field tests are portable and can be carried out quickly on site. As far as destructive methods are concerned, the rock or a rock fragment is required to test the mechanical strength and is destroyed in the course of the test (loading up to the breaking strength). By applying non-destructive testing methods, geophysical methods are used to determine the rock strength indirectly without destroying a sample. The distinction between direct and indirect methods again concerns the determination of a specific strength parameter via another rock parameter that can be determined more easily (e.g., the determination of the tensile strength and conversion to the rock compressive strength) [4,5].

There are numerous standard geological and geotechnical investigation methods that have proven themselves over several decades, such as the rebound (Schmidt) hammer (RH), point load test (PLT), and Los Angeles test (LAT) compared in the present research paper, e.g., [4–6].

There is also current research into the development of new methods for determining the mechanical properties and quality of rock mass [6–8]. In addition, further development is carried out via computer-based studies using distinct-element methods such as PFC when modeling without pre-defined discontinuities, e.g., [9–11] or UDEC [12–14], if the rock mass is mainly characterized by preestablished discontinuities and joints.

Simple, quick, and cheap on-site standard methods for the indirect characterization of the mechanical rock quality and estimation of the rock strength are still the rebound (Schmidt) hammer [15,16] and the point load test [17]. These sometimes differ considerably in terms of the methodological approach. The same applies to the conversions of the obtained basic values (e.g., the rebound value R or the corrected point load index $Is_{(50)}$), into an unconfined compressive strength (UCS) based on empirical tests. Therefore, according to ISRM [18], it is necessary to speak of an estimated value (UCS*), if the uniaxial compressive strength determination (UCS) was not carried out on drill cores [19].

Nevertheless, the results of the different methods are often used synonymously to assess or describe the mechanical rock quality. The possible uncertainties regarding the reliability and comparability of the results are often underestimated, although these have already been published as implementation guidelines [18]. Therefore, this article deals with the determination of the "rock strength" via the common testing methods rebound hammer [15,16], point load test [17], and Los Angeles (*LA*) test [20] with regard to the determination of the measurement uncertainty for the individual testing methods. Due to the intense fracturing of the rock, an extraction of drill cores for the direct determination of the UCS was not possible.

The above-mentioned investigation methods, rebound hammer, point load test, and Los Angeles test, were applied simultaneously in numerous test locations in a hard rock quarry in Austria. The investigated rock is a granulite from the family of metamorphic rocks. Results are available in different numbers for the respective test. The measurements were carried out along two vertical measurement lines perpendicular to the mining levels. Along these lines, the rock quality basically varies due to weathering (decreasing influence with increasing distance from the original ground surface), but also due to the influence of the geological-tectonic conditions on site expressed as a horizontal variability. The mining conditions (blasting) are identical over the considered area. The objective is to assess the suitability for the intended use of the rock as aggregates and to consider the measurement uncertainties in the significance of the results.

1.2. Influences on Rock Material Parameters

From an engineering-geological point of view, a fundamental distinction is made between rock strength and rock mass strength. The rock strength is basically dependent on the type of rock (magmatic, metamorphic, sedimentary) and the minerals present, which depends on the weathering degree as well, e.g., [4]. The bigger the scale of observation, the more important are the discontinuities (e.g., joints) and weathering phenomena with regard to the rock mass strength [1,2]. For the evaluation of an aggregate (grain group or grain mixture) as in the present case, weathering as well as smaller scale foliation or stratification planes have a great influence on the rock strength. In contrast, dividing planes with a larger separation distance—if larger than the final grain diameter—play almost no significant role due to the processing of the rock material (blasting, crushing, sieving, washing). Basically, the following characteristics and influences affect the strength parameters of rock aggregates (compare [21]):

- rock type and mineralogy (for example, an igneous granite composed of quartz, feldspar and mica is usually stronger than limestone composed of calcite minerals);
- rock formation and genesis: tectonic stress during ductile deformation may lead to formation of a foliation, while a brittle deformation leads to the formation of joints and fissures [22];
- weathering: whereby the weathering starts preferentially at weak zones (e.g., interfaces), where agents such as water and air can penetrate the rock;
- type of material excavation and processing: possible damage due to microcracks (could not be observed for the examined rock by means of thin section analysis).

These parameters can vary spatially (horizontally and vertically) depending on the geological-tectonic conditions. As a result, the rock strengths can vary significantly within a very small scale. This circumstance can mainly be attributed to the presence of interfaces and a varying degree of weathering [23–25]. Weathering can affect both the interface itself and the surrounding rock. In terms of aggregate production, weathering of the interface surface is mostly unproblematic, but the progress of weathering into the surrounding rock results in a reduction in the rock (aggregate) strength. Starting from the surface level, weathering must always be considered as an influencing factor.

1.3. Contributions on Measurement Uncertainty

Measurement uncertainty can be described as the trustworthiness of numerical values/measurement results. This uncertainty or inaccuracy of the measured value is negatively influenced by errors. Uncertainty of measurement, therefore, is defined as "a parameter, which is associated with the result of a measurement, characterizing the dispersion of the values that could reasonably be attributed to the measurand" [26]. Therefore, the uncertainty of measurement expresses the estimation of the "error" of the measurement result differing from the "true value" of the measurement value.

According to JCGM 100:2008 [26], JCGM 104:2009 [27], and JCGM 200:2008 [28], two components are distinguished concerning the measurement error: (1) the 'systematic error' and (2) the 'random error'. Systematic measurement errors can be quantified and are significant in terms of accuracy and a correction factor can be applied to identify this effect in the overall measurement scope. If performed this way, the systematic error reaches a value of zero. In contrast, a random error is the result of an unpredictable stochastic variation without the possibility of correction or compensation. This is only reduced by an increasing number of attempts [26].

When dealing with natural materials, the terms 'aleatoric' and 'epistemic' are used instead of 'random error' and 'systematic error'. This is due to the fact, that the variability of the properties of natural complex materials, such as rock masses, defines by its origin an initial level of uncertainty. This is in contrast, for example, to building materials (e.g., concrete), whose production takes place under controlled conditions and which are basically homogeneous. This type of uncertainty cannot be removed due to it being a feature of the material itself and therefore leads to an 'aleatoric' uncertainty. 'Epistemic' uncertainty, on the other hand, refers to uncertainty caused by a lack of knowledge, for example the measures of a given material property. This type of uncertainty can be reduced by expanding the quantity and quality of data describing the considered property, e.g., [29].

1.3.1. Uncertainty of Measurement

In [26], a concept of contributions to measurement uncertainty, covering both random/aleatoric and systematic/epistemic errors, is provided. However, the concept does not make a strict distinction between these two. Generally, the contribution is used in the way of a standard uncertainty format and is calculated as the result of the standard deviation or the coefficients of variation considering the number of samples.

1.3.2. Standard Uncertainty

In [26], the contributing factors influencing the uncertainty of measurement are identified as type A and type B. These can be differentiated as follows: while type A is identified by statistical analysis of observation series—like the standard deviation (coefficient of variation) of the arithmetic mean value under the assumption of a normal distribution; type B is acquired by means of other methods than statistical analysis of observation series [26]. For type B, each measured value must be subject to a statistical distribution which allows the statistical parameters to be defined on the basis of the assumed distribution. Specifications for the distribution assumption can be preceding measurement data, experience or expertise in the behaviour. Information regarding the assumption of the distribution type may be: previous measurement data, experience or knowledge of the behaviour observed, specifications from manufacturers, literature or calibration procedure.

1.3.3. Modelling the Measured Values

To characterize the input quantities on the measurement result according to [26], a modelling of measurement is re-quired, whereby the output quantity Y for the mathematical model is the measured value itself as stated in Equation (1):

$$Y = f(X_1, \ldots, X_N) \tag{1}$$

If the true values of the quantities $X_1, \ldots, X_N$ of the measurement are unknown, the usage of estimated input values $x_1, \ldots, x_N$ is necessary (Equation (2)).

$$y = f(x_1, \ldots, x_N) \tag{2}$$

According to [30], the modelling of the measurement should include the entire measurement chain, i.e., the entire measurement process, which obeys the principle of cause and effect.

1.3.4. Combined Standard Uncertainty

Entry values ca may correlate or may not correlate to each other. Uncorrelated quantities are independent and have no relation with each other. The combined standard uncertainty is derived from the different single standard uncertainties (Equation (3)).

$$u_c^2(y) = \sum_{i=1}^{n} \left(\frac{\partial f}{\partial x_i} \right)^2 u^2(x_i) \tag{3}$$

Thereby, a sensitivity coefficient c accounting the influence of a single quantity on the combined measurement uncertainty is included and calculated according to Equation (4).

$$c = \frac{\partial f}{\partial x_i} \Rightarrow \frac{\Delta f}{\Delta x_i} \tag{4}$$

1.3.5. Expanded Uncertainty of Measurement

The expanded uncertainty of measurement U is derived by the multiplication of the combined standard uncertainty $u_c(y)$ by a coverage factor k (Equation (5)), the latter defining the level of confidence for the measurement uncertainty based on a probabilistic approach. In [26] a coverage factor $2 \leq k \leq 3$ is recommended, whereby $k = 2$ defines a level of confidence of approximately 95% and $k = 3$ of approximately 99%.

$$U = k \cdot u_c(y) \quad (5)$$

2. Materials and Methods

2.1. Tested Rock Structure

The investigated rock site lies in the Bohemian Massif in Austria. The rock at present is a granulite from the Moldanubicum with an age of approximately 340 Ma, which underwent metamorphism during the Variscan Orogenesis at about 800 °C and a pressure of about 1800 MPa. The granulite was subjected to mylonitic foliation in the course of exhumation, as well as to scarping of the "bench-like" metamorphic structures [31,32]. The structure is mylonitic and is characterized by recrystallization processes, whereby an intensive interlocking of the individual minerals develops, which finally also contributes to the high strength of this rock. In the course of cooling (retrograde branch of metamorphism), a very subordinate chloritization of the garnet minerals occurs. Ores are present as accessory minerals.

The slightly formed foliation generally dips very steeply to the north, and the dividing planes formed in the course of emplacement and tectonic overprinting also stand very steeply. From a mineralogical point of view, the rocks in the area consist of quartz, feldspars (plagioclase and potassium feldspar), and mica, occasionally also accompanied by accessory minerals such as kyanite [33]. The rock is exposed to weathering in the near-surface area especially along interfaces and fractures, which in turn influences the rock properties. This leads to mineral transformations, which generally reduce the rock strength.

Lithologically, the area shows a heterogeneous structure, both in terms of horizontal extent and depth (excavation levels), and is roughly composed of the following different granulite types: (1) the very light-colored unit with a bright sound when crushed ("vitreous granulite"), (2) a brown granulite unit showing stronger weathering signs (highest, weathered area), and (3) a dark, more deeply lying granulite unit. Additionally, serpentinite lenses are situated north and south of the investigated granulite area. Photos of areas (1), (2), and (3) are shown in Figures 1–3.

Figure 1. Light-colored, vitreous granulite, corresponding to (1).

Figure 2. Weathered, strongly jointed granulite, corresponding to (2).

Figure 3. Darker, more deeply lying granulite, corresponding to (3).

The highest quality granulite rocks (light-colored, "vitreous granulite") are extracted for the production of aggregates for railroad gravel, concrete, and asphalt. These are characterized by very high compressive strength (≥200 MPa), high impact abrasion and frost resistance. The other granulite rocks are mainly used for the production of base course and fill material. For the quality classification, the regular assessment of the rock properties—primarily by means of the Los-Angeles test—plays an essential role in ensuring the high-quality requirements of the aggregate.

In order to systematically demonstrate the variability of the rock properties, sampling and examination of the granulite bedrock at different depth levels (a total of seven depth levels with generally 15 to 20 m height difference were investigated) took place along two vertical profiles (Figure 4). As a reference, three depth levels in the area of the glassy granulite were also sampled and surveyed (reference testing profile, Figure 4). A partial section of the sampling area is shown in Figure 5. The light coloured, glassy vitreous granulite can be seen in the background, the darker granulite in the front, with the alteration (and thus the brownish discoloration) increasing towards the top of the terrain. The vitreous granulite was used as a reference profile due to its very high and desired strength properties and the values determined were then standardized accordingly.

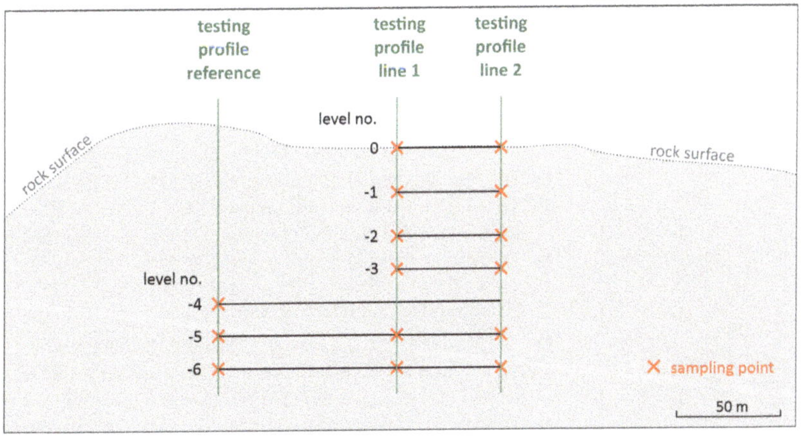

Figure 4. Schematic illustration of the sampling along two vertical measurement lines and of the reference rock along a third vertical measurement line.

Figure 5. Section of the investigated quarry wall (oblique view); compare Figure 4.

2.2. Testing Methods

In order to record and compare the mechanical rock properties at different sampling positions, the following methods were used:

- the rebound hammer (RH)—or also referred as Schmidt hammer—was used as an indirect non-destructive method for determining the rock strength (derivation of the unconfined compressive strength) [15,16,34];
- the point load test (PLT) was applied as a destructive method to determine the rock strength (convertible to unconfined rock compressive strength) using rock hand-pieces [17,18,35];
- the Los Angeles test (LAT) as a standard industrial test procedure for the recording of the impact-abrasion-resistance of rock aggregate [5,20,36].

These methods offer the possibility of estimating the uniaxial compressive strength (UCS) as a substitute for the uncon-fined compression test on drill cores. Especially in highly fractured rock, it is not possible to obtain a suitable drill core for a compression test on the rock cylinder. The methods mentioned above offer an indirect possibility of rock strength estimation. On the basis of empirical comparative studies, it is possible to deduce the UCS from the parameters (re-bound value, point load resistance, LA class) obtained by these methods [17,37–39].

The methods are presented in detail in the following subchapters. Additionally, X-ray diffraction (using a PANanalytical X'pert PRO from Malvern Panalytical, Malvern, United Kingdom) and thin section analysis (via Leica DM4500P polarization microscope from Leica Microsystems GmbH, Wetzlar, Germany) were applied to exclude mineralogical and structural sources of error.

2.2.1. Schmidt/Rebound Hammer (RH)

The Schmidt rebound hammer is an indirect method for determining the strength of rock or concrete [15,16,40]. Originally designed for the non-destructive testing of concrete hardness, the RH is an index apparatus to measure the surface hardness very quickly and inexpensively. Nowadays the RH represents the most frequently used index method to estimate the uniaxial compressive strength (UCS) of rock and concrete in laboratory as well as in situ [41,42].

Regarding the testing mechanism, the RH is a spring-loaded piston, which is manually prestressed and automatically releases when the tip of the piston is pressed against the surface to be measured (Figure 6). The energy of the spring is released and is then mainly consumed by the plastic deformation of the material under the tip. A part of the energy is rebounded depending on the impact penetration resistance. The harder or stronger the test subsurface, the smaller is the distance travelled by the piston tip and the lower the spring extension. The result of the measurement is a rebound value, measured as a percentage of the initial extension of the spring to the spring extension after testing. This ratio represents an index for surface hardness. This index can be converted into an UCS via empirical relationships for both rock and concrete [15,16]. For this study, however, the index value was used without conversion to a UCS to avoid introducing an uncertainty component via empirical variables.

Figure 6. Schmidt hammer when testing the repeatability on a 30 cm thick concrete slab (used as reference material).

The Schmidt rebound hardness ("rebound value" R) was determined as a non-destructive field test method via the rebound hammer "Rock Schmidt" model N of the company "proceq" (impact energy 2.2 Nm). Testing was carried out according to the specifications of [34]. A total of 20 impact tests were carried out in a narrow rock area at the different sampling points (Figure 4), all measured values of a standard measurement run were included in the statistical evaluation. The same was done comparatively on a high strength concrete slab to check the repeatability (Figure 6).

2.2.2. Point Load Test (PLT)

During the PLT, a rock sample is clamped between two standardized test tips with 60° test point angle and loaded until failure (point load test device type "Wille Geotechnik"), i.a. [17,18,38,39]. The contact area between the cone tips and the specimen are called load application points. The breaking load obtained is called the point load index *Is* [6]. This corresponds to the strength of a rock specimen that breaks under a concentrated applied load. Similar to the Schmidt hammer, there are also empirically determined correlations between point load index and UCS depending on the rock type [43–45]. For this study, the initial values (point load index *Is*) were used for the correlation calculations.

Fist-sized rock specimens were used to perform the PLT [17,34]. The point load index Is was calculated according to Equation (6):

$$I_s = \frac{P}{D_e^2} \tag{6}$$

whereby:

Is—point load index [MPa],
P—load at failure [N],
D_e—equivalent core diameter [mm].

A correction of the sample size to calculate the corrected point load Index $Is_{(50)}$ [MPa] was carried out as follows (Equation (7)):

$$I_{s(50)} = \left(\frac{D_e}{50}\right)^{0.45} \cdot I_s \tag{7}$$

At each sampling point (Figure 4), 30 individual measurements were carried out to account for the high scatter of this measurement method. Comparative tests on concrete mortar specimens were carried out for comparative purposes (Figure 7).

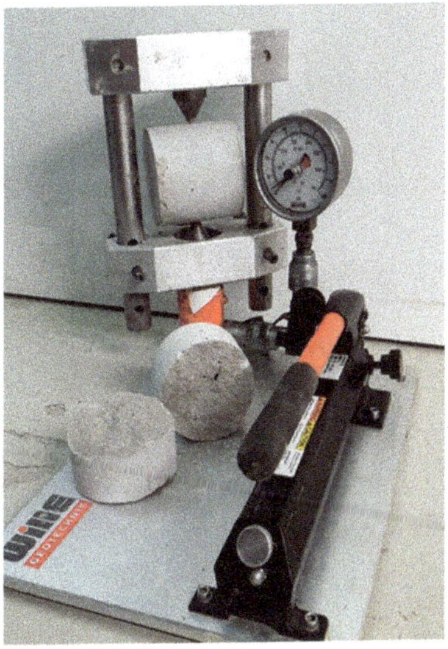

Figure 7. Point load testing on concrete mortar cylinders (used as reference material).

2.2.3. Los Angeles Test (LAT)

To evaluate the impact abrasion resistance, the LAT was performed according to the standard EN 1097-2 [20] using a Los-Angeles testing device from the manufacturer 'testing' (Figure 8). At the beginning of the test, a 5 kg (±5 g) rock sample with a grain size of 10 to 14 mm is placed in a rotatable steel drum together with eleven standardized steel balls. The subsequent rotation (32 ± 1 min^{-1}) of the drum crushes or abrades the rock sample during 500 rotations. The LA value corresponds to the passing of the tested material through a 1.6 mm sieve [5,20]. Accordingly, the higher the percentage of sieve passage smaller than 1.6 mm, the less abrasion resistant the rock sample is as an expression of mechanical resistance to impact and abrasion exposure [20]. The UCS of a rock can also be derived from the LA value by means of empirically determined correlations [46].

Figure 8. Los Angeles testing machine with steel balls in the front (in soundproof enclosure).

2.3. *Contributions on the Budget of Measurement Uncertainty*

In accordance with the general explanations provided in Section 1.3, a model for the influences on the uncertainty for the determination of rock material parameters can be established. As target value the material parameter of rebound value (R) as well as the point load index ($Is_{(50)}$) and the Los-Angeles-value (LA-value) from the Los Angeles fragmentation test are considered. The determination of this model and the influences on the model are based on a discussion process. Figure 9 shows the investigated influences resp. error causes on the measurement on the left side by means of a cause-and-effect diagram, as well as the obtained effects on the measurement in accordance to [47].

The following influences on the uncertainty of the measurement have been identified in Figure 9 and are discussed below.

- Geospatial variability: large-scale variability considering the spatial variability of the rock as a result of:
  - (a) the weathering processes progressing vertically downwards from the top level which is exposed to the environment;
  - (b) the weathering processes caused by exposed surfaces as a result of mining;
  - (c) the geological variability, which can vary vertically and horizontally in consequence of the rock formation processes.

- Small scale variability: variation in the rock in the small range as a result of the different (micro)structural conditions: dividing planes (e.g., joints) and the resulting inherent uncertainties.
- Methodology: the applied methods such as rebound hammer method, point load test, and fragmentation test, as well as the uncertainties coming from the used method itself including the repeatability.
- Test setup: the quality and traceability of the used measurement devices, correct technical use and in the appropriate measurement range influencing the reproducibility.

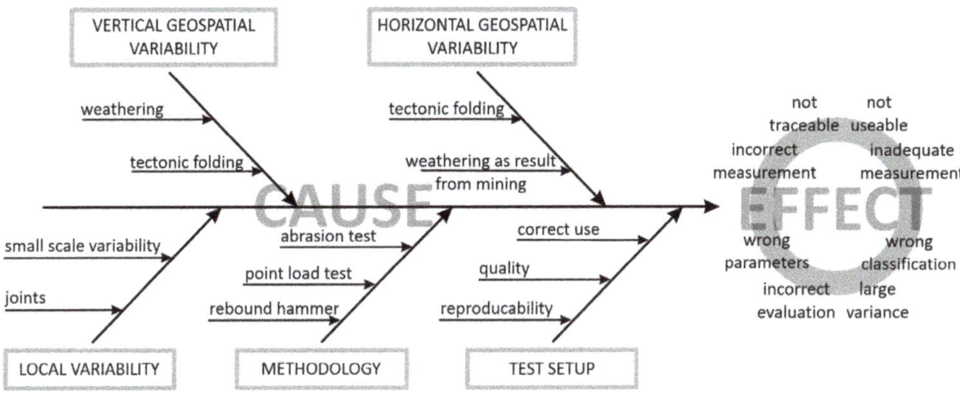

Figure 9. Cause-and-effect diagram for the determination of the influences on the measurement uncertainty of rock material parameters.

Hence, as a result of the findings above, the model of the influences on the measurement uncertainty might be defined in analytical form as following (Equation (8)).

$$Y = f\left(u_{var,hor}, u_{var,vert}, u_{method}, u_{repeatability}, u_{reproducability}\right) \quad (8)$$

in which:

$u_{var,hor}$—uncertainty due to geological variability in horizontal direction of the rock mass (as a result of the site only relevant in this direction),
$u_{var,ver}$—uncertainty due to weathering processes progressing vertically downwards from the top level resulting from environmental influences,
u_{method}—uncertainty due to different testing methodologies,
$u_{repeatability}$—uncertainty considering the repeatability of test results under ideal conditions,
$u_{reproducability}$—uncertainty considering the reproducibility of test results under real conditions.

The model from Equation (8) is considered as a basic theoretical consideration. Subsequently, individual components are quantified on the basis of available data and evaluated as to whether they should be considered or not.

3. Results and Evaluation

3.1. General

This section presents the results of the conducted investigations as described above and their evaluation. The geological formation presented in Section 2.1 was analyzed regarding its mechanical rock parameters using the testing methods according to Section 2.2. Section 3.2 provides the test results, and Section 3.3 a detailed approach for the estimation of the measurement uncertainty of the determination of the rock properties.

3.2. Test Results

The investigation area consists of a total of seven exposed levels over a total depth of approximately 100 m, which means 15 m to 20 m between each horizon (Figure 4). An overview of the performed tests is shown in Table 1.

Table 1. Overview of the considered tests for rock material parameters.

Testing Method	Line	Testing Horizons Resp. Levels	n = Number of Tests at Each Location
Rebound hammer	1	0/−1/−2/−3/−4/−5/−6	3 × 10
	2	0/−1/−2/−3/−4/−5/−6	3 × 10
	reference	−4/−5/−6	3 × 10
Point load test	1	0/−1/−2/−3/−4/−5/−6	1 × 30
	2	0/−1/−2/−3/−4/−5/−6	1 × 30 [1]
	reference	−4/−5/−6	1 × 30
LA-test	1	0/−1/−2/−3/−4/−5/−6	1 × 1
	2	0/−1/−2/−3/−4/−5/−6	1 × 1
	reference	−4/−5/−6	1 × 1

[1] in horizon −6 for comparison reasons two independent test series were performed.

The results of the performed tests as the rebound value (*R*), the $Is_{(50)}$ from the PLT and the *LA* value are shown in Figure 5. The tests were performed in two vertically aligned measurement lines (line 1 and line 2). Additionally, the reference values were determined in more or less undisturbed rock at measuring points in the measuring horizons −4 to −6, which were considered to be unweathered a priori due to their geological conditions (Figure 1). Level −4 showed the highest rock quality with the highest material properties and therefore was set as the reference value to which all measured values of the particular testing methods in the presentation are referred. Accordingly, the values shown in Figure 10 are to be regarded as percentages of the respective reference value, whereby the *LA* value behaves reciprocally. Figure 10 shows the normalized ratio of the respective test results (test respective test results were normalized to the values of the reference rock) for the different test profiles (cf. Figure 1) and the reference rock material for the respective levels. As reference value, the maximum value of the reference measurement (level −4) has been considered. In Figure 10, a structured overview of the distribution of the various parameters of the rock over the depth of the measured lines is presented.

As shown in Figure 10, there are considerable differences in the test results, which are evident as follows:

- the top horizon or the two top horizons show the lowest values for both the RH and the PLT, i.e., material damage is evident here. In the case of the LAT, the highest values are reached in these levels, which also indicates increasing material damage, which causes increased fragmentation. This is clearly due to the influences of the environmental conditions (weathering processes).
- The two vertical measurement lines do not indicate any clearly identifiable differences in the corresponding levels, so it can be assumed that the two measurement lines are comparable.
- The reference measurements in levels −4 to −6 (high value grain of the quarry) show values that indicate a high-quality rock material as these are within the range of the reference value.

3.3. Estimation of Measurement Uncertainty for the Determination of Rock Material Parameters

The general principles for determining the contributions to the measurement uncertainty budget are stated in Section 1.3. Following the determination of the measurement uncertainty, this is derived for the three test methods that were performed.

(a)

(b)

(c)

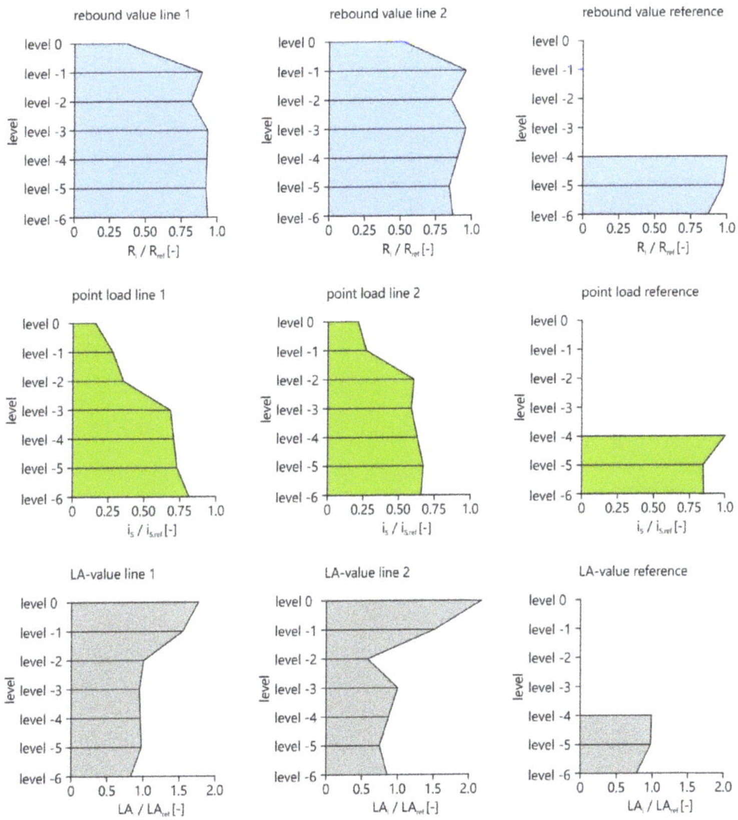

Figure 10. Overview of the test results of (**a**) rebound values, (**b**) point load tests and (**c**) Los Angeles test referred to the maximum value of level −4 for each considered material parameter.

3.3.1. Measurement Uncertainty for Determination of Rebound Value

For the determination of the rebound value R, the contributions for the measurement uncertainty have been considered as listed in Equation (8). The estimation of the individual quantities is conducted as follows (Table 2):

The expanded uncertainty of measurement for the rebound value has a value of approximately 20% to 30% depending on whether the weathering effects in the two upper horizons are considered or not.

3.3.2. Measurement Uncertainty for Determination of Uniaxial Compressive Strength from Point Load Test

For the determination of the corrected point load index $Is_{(50)}$ from point load test, the contributions for the measurement uncertainty have been considered as listed in Equation (8) with estimation of the quantities as follows (Table 3):

The value for the expanded uncertainty of measurement for the corrected point load index $Is_{(50)}$ ranges from approximately 40% to 116% depending on whether the weathering effects in the two upper horizons are considered or not.

3.3.3. Measurement Uncertainty for Determination of LA Value from Fragmentation Test

For the determination of the LA value from fragmentation testing, the contributions for the measurement uncertainty are as follows (Table 4):

Table 2. Estimation of measurement uncertainty for determination of rebound value.

Uncertainty	Description	Calculation	Valid for	n	u_{mean}	u_{median}	Min.Value u	Max.Value u
			[-]	[-]	[-]	[-]	[-]	[-]
$u_{var,hor}$	uncertainty due to geological variability of the rock mass in horizontal direction	$\left(\frac{\lvert R_{ref,m}-R_{i,m}\rvert}{R_{ref,m}}\right)$	all levels	3	0.07	0.07	0.03	0.10
$u_{var,ver}$	uncertainty due to weathering processes progressing vertically downwards from the top level resulting from environmental conditions	$\left(\frac{\lvert R_{level\ i}-R_{level\ j}\rvert}{R_{level\ i}}\right)$	all levels level −1/−6 level −2/−6	30 × 10 20 × 10 10 × 10	0.22 0.07 0.07	0.12 0.07 0.07	0.001 0.001 0.001	0.61 0.14 0.14
u_{method}	uncertainty due to different testing methodologies tested on reference material	$\sqrt{\frac{1}{n(n-1)}\sum_{i=1}^{n}(R_i-R_m)^2}$	all levels	40	0.004	0.004	no value	no value
$u_{repeatability}$	uncertainty considering the repeatability of test results under ideal conditions	$\sqrt{\frac{1}{n(n-1)}\sum_{i=1}^{n}(R_i-R_m)^2}$	all levels	30 × 10	0.006	0.005	0.001	0.02
$u_{reproducability}$	uncertainty considering the reproducability of test results under real conditions	$\frac{CV}{\sqrt{n}}$	all levels	30 × 10	0.05	0.05	0.01	0.18
u_c	combined standard uncertainty in accordance with Equation (3)		all levels level −1/−6 level −2/−6			0.14 0.10 0.10		
$U=k\cdot u_c$	expanded measurement uncertainty using $k = 2$		all levels level −1/−6 level −2/−6			0.29 0.21 0.20		

Table 3. Estimation of measurement uncertainty for determination of uniaxial compressive strength from point load test.

Uncertainty	Description	Calculation	Valid for	n	u_{mean}	u_{median}	Min.Value u	Max.Value u
			[-]	[-]	[-]	[-]	[-]	[-]
$u_{var,hor}$	uncertainty due to geological variability of the rock mass in horizontal direction	$\left(\frac{\lvert I_{s(50)\,ref,m}-I_{s(50)\,i,m}\rvert}{I_{s(50)\,ref,m}}\right)$	all levels	3	0.19	0.17	0.13	0.28
$u_{var,ver}$	uncertainty due to weathering processes progressing vertically downwards from the top level resulting from environmental influences	$\left(\frac{\lvert I_{s(50)\,level\ i}-I_{s(50)\,level\ j}\rvert}{I_{s(50)\,level\ i}}\right)$	all levels level −1/−6 level −2/−6	30 × 10 20 × 10 10 × 10	0.42 0.33 0.19	0.52 0.34 0.12	0.02 0.02 0.02	0.79 0.63 0.52
u_{method}	uncertainty due to different testing methodologies tested on reference material	$\sqrt{\frac{1}{n(n-1)}\sum_{i=1}^{n}\left(I_{s(50)_i}-I_{s(50)_m}\right)^2}$	all levels	20	0.025	0.025	no value	no value
$u_{repeatability}$	uncertainty considering the repeatability of test results under ideal conditions	$\sqrt{\frac{1}{n(n-1)}\sum_{i=1}^{n}\left(I_{s(50)_i}-I_{s(50)_m}\right)^2}$	all levels	2 × 10	0.003	0.003	no value	no value

Table 3. *Cont.*

Uncertainty	Description	Calculation	Valid for	n	u_{mean}	u_{median}	Min.Value u	Max.Value u
			[-]	[-]	[-]	[-]	[-]	[-]
$u_{reproducability}$	uncertainty considering the reproducability of test results under real conditions	$\frac{CV}{\sqrt{n}}$	all levels	15 × 10	0.05	0.05	0.03	0.08
u_c	combined standard uncertainty in accordance with Equation (3)		all levels level −1/−6 level −2/−6			0.58 0.39 0.20		
$U = k \cdot u_c$	expanded measurement uncertainty using k = 2		all levels level −1/−6 level −2/−6			1.16 0.78 0.41		

Table 4. Estimation of measurement uncertainty for determination of *LA*-value from fragmentation testing.

Uncertainty	Description	Calculation	Valid for	n	u_{mean}	u_{median}	Min.Value u	Max.Value u
			[-]	[-]	[-]	[-]	[-]	[-]
$u_{var,hor}$	uncertainty due to geological variability of the rock mass in horizontal direction	$\left(\frac{\lvert LA_{ref,m}-LA_{i,m}\rvert}{LA_{ref,m}}\right)$	all levels	3	0.11	0.15	0.02	0.17
$u_{var,ver}$	uncertainty due to weathering processes progressing vertically downwards from the top level resulting from environmental influences	$\left(\frac{\lvert LA_{level\ i}-LA_{level\ j}\rvert}{LA_{level\ i}}\right)$	all levels level −1/−6 level −2/−6	30 × 10 20 × 10 10 × 10	0.67 0.43 0.19	0.54 0.33 0.17	0.02 0.02 0.02	2.74 1.58 0.42
u_{method}	uncertainty due to different testing methodologies tested on reference material	$\sqrt{\frac{1}{n(n-1)}\sum_{i=1}^{n}(LA_i - LA_m)^2}$	all levels	no value [1]	0.06	0.06	no value	no value
$u_{repeatability}$	uncertainty considering the repeatability of test results under ideal conditions	$\sqrt{\frac{1}{n(n-1)}\sum_{i=1}^{n}(LA_i - LA_m)^2}$	no value	no value [1]	no value	no value	no value	no value
$u_{reproducability}$	uncertainty considering the reproducability of test results under real conditions	$\frac{CV}{\sqrt{n}}$	all levels	no value	0.16	0.16	no value	no value
u_c	combined standard uncertainty in accordance with Equation (3)		all levels level −1/−6 level −2/−6			0.58 0.40 0.29		
$U = k \cdot u_c$	expanded measurement uncertainty using k = 2		all levels level −1/−6 level −2/−6			1.17 0.80 0.57		

[1] in acc. to EN 1097-2.

The expanded uncertainty of measurement for the *LA* value from fragmentation testing lies between approximately 60% and 120% depending on whether the weathering effects in the two upper horizons are considered or not.

3.3.4. Overview of Measurement Uncertainty

Based on the evaluation of the test results in the sections above, the combined measurement uncertainty *u* for the different measurement methods is graphically shown in Figure 11 based on the values from Tables 2–4. The uncertainties differ significantly depending on the investigated rock level. There is a considerable difference within the combined measurement uncertainty between the in-situ methods, which are the rebound value and the point load test, resulting also because of the different number of conducted repeated test series. In the case of a larger number of measurements there is a higher certainty that the received value is true. The *LA* value is to be considered as the decisive parameter for the assessment of the suitability of the material. It is evident that the point load index $Is_{(50)}$ in Figure 11 reflects similar combined measurement uncertainties as the *LA* values as a result of the comparable number of test series.

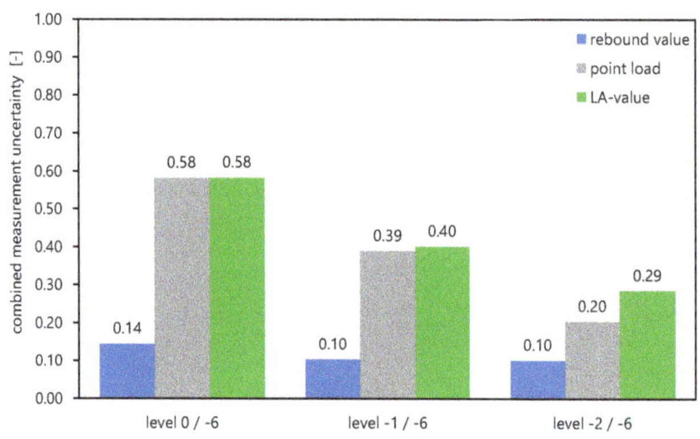

Figure 11. Comparison of combined measurement uncertainty for different testing methods.

The single contributions to the measurement uncertainty shown in Tables 2–4 are finally referenced to the determined combined uncertainty (all contributions subsequently represent 100%) deriving the percentage contribution of the different input variables as shown in Figure 12.

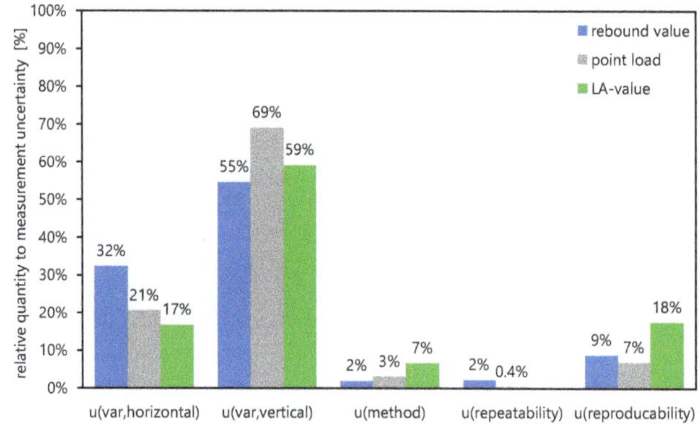

Figure 12. Percentage contribution of the different input variables on the total measurement uncertainty for different testing methods.

It is obvious that the percentages of the vertical variability represent the majority of the total measurement uncertainty. Therefore, it is obvious that is necessary and also technically sensible to distinguish areas with evidence of weathering from those with no evidence of weathering. This reduces the uncertainty of the measurement by up to 70%.

As already stated in Tables 2–4, the measurement uncertainty can be significantly reduced if weathered and unweathered rocks are considered separately.

For example, a reduction in the combined measurement uncertainty from 0.14 to 0.10 can be observed for the rebound value *R*, from 0.58 to 0.20 for the $Is_{(50)}$ from the point load test and from 0.58 to 0.29 for the *LA* value if the two upper weathered horizons are considered separately as shown in Table 5.

Table 5. Overview of the calculated combined uncertainty of measurement *u* for different conditions.

Expanded Measurement Uncertainty U for Testing Method	A Priori Consideration of Vertical Variation		A Priori Consideration of Horizontal Variation	
	All Levels Including Weathering	Level −2/−6 No Weathering	All Levels Including Weathering	Level −2/−6 No Weathering
RH	0.14	0.10	0.13	0.07
PLT	0.58	0.20	0.57	0.13
LAT	0.58	0.29	0.57	0.24

Additionally, as indicated in Table 5, if the horizontal geospatial variation can be a priori considered, the combined measurement uncertainty is reduced for the different testing methods to the following values: 0.13 to 0.07 for the re-bound value *R*, 0.57 to 0.13 for the $Is_{(50)}$ from the point load test, and 0.57 to 0.24 for the *LA* value.

4. Conclusions

In general, uncertainties in the measurement method are difficult to determine quantitatively, if no repeated measurements are feasible. Specifically, this is the case in the determination of parameters of natural materials that are subject to non-reproducible influences. For the validity of a measurement, the knowledge of the measurement uncertainty is essential. This is particularly the case when compliance with limit values or technical specifications is considered.

The measurement uncertainty itself therefore has a considerable influence on the significance and the reliability of measurements. This also applies to the methods for the determination of the rock characteristics in geology. In this publication, analyses are presented on the estimation of the measurement uncertainty in the determination of material parameters of natural rock mass.

The following conclusions can be made:

- the combined measurement uncertainty of the compared testing methods shows that, as is generally known, the measurement uncertainty decreases with an increasing number of tests, and the real value can be approximated with a higher accuracy;
- it is obvious that a consideration of a priori information as the different horizons/levels is a method to reduce the combined uncertainty *u*; this can be reduced to values of approximately 10–30% depending on the testing method and the number of test series also considering the weathering effects;
- as shown in Figure 10, the rebound value *R* is not considering the effects from weathering in the same way as the point load index $Is_{(50)}$ and the *LA* value; this approach therefore only has restricted significance.
- the relative influence on the combined uncertainty *u* of the different testing methods shows a range of $u_{var,hor}$ between 17% and 32% due to the geological variability in the horizontal direction; in comparison, the re-bound hammer method has the highest relative measurement uncertainty;

- the relative influence on the combined uncertainty reaches values for $u_{var,ver}$ from 55% to 69% as a result of the different vertical direction (levels) due to weathering phenomena at the surface;
- considering the vertical direction, the point load test shows the highest contribution to the relative quantity of measurement uncertainty with a value of 69%; however, such influence is quite comparable within all measurement methods;
- this leads to the conclusion that a higher weathering degree of the rock has an increasing effect on the measurement uncertainty; therefore, the use of a priori information for the realization of a testing task in rock is strongly recommended;
- weathering can affect both the discontinuity itself and the surrounding rock.

In terms of aggregate production, weathering of the discontinuity surface is almost unproblematic, but the progress of weathering into the surrounding rock results in a reduction in the rock (aggregate) strength. Starting from the surface level, weathering must always be considered as an influencing factor; this was also clearly observed in the present study. The observed variability generally implies a significant challenge in the assessment of the associated material parameters.

These influences are inadequately reflected by the RH. While the 'aleatoric' uncertainty is predetermined by the rock mass structure and cannot be statistically considered, the test method-related uncertainty increases the 'epistemic' part of the total uncertainty and thus the overall uncertainty. This means that if only the RH is used to assess the mechanical rock parameters during an on-site rock mass testing (this concerns both the addressing of rock in an existing quarry, as well as in investigations for the expansion of the quarry), it is possible that the quality of the aggregates produced from the rock mass will not be sufficient. Accordingly, it is strongly recommended by the authors to verify the rebound values by test results of the in-situ feasible PLT or the LAT in the laboratory. These more complex, but more meaningful test methods are associated with a significantly lower measurement uncertainty and are therefore recommended for the considered applications.

Author Contributions: Conceptualization, K.V., O.Z. and R.A.; Methodology, K.V. and K.W.; Software, K.W.; Validation, O.Z.; Formal analysis, O.Z. and P.G.; Investigation, K.V., P.G. and K.W.; Data curation, P.G. and K.W.; Writing—original draft, K.V., O.Z. and P.G.; Writing—review & editing, R.A.; Visualization, O.Z.; Supervision, R.A.; Project administration, K.V.; Funding acquisition, K.V. All authors have read and agreed to the published version of the manuscript.

Funding: This research received no external funding.

Institutional Review Board Statement: Not applicable.

Informed Consent Statement: Not applicable.

Data Availability Statement: The raw/processed data required to reproduce these findings cannot be shared at this time due to legal or ethical reasons.

Conflicts of Interest: The authors declare no conflict of interest.

References

1. Bieniawaski, Z. *Engineering Rock Mass Classifications: A Complete Manual for Engineers and Geolo-gists in Mining, Civil and Petroleum Engineering*; John Wiley & Sons: Hoboken, NJ, USA, 1989; p. 251.
2. Hoek, E.; Brown, E.T. Practical estimates of rock mass strength. *Int. J. Rock Mech. Min. Sci.* **1997**, *34*, 1165–1186. [CrossRef]
3. Brandecker, H. *Die Gestaltung von Böschungen in Lockermassen und in Fels*; Forschungsberichte. For-schungsgesellschaft Für Straßenwesen Im Österr.: Köln, Germany, 1971.
4. Price, D.G. *Engineering Geology: Principles and Practice*; Springer: Berlin, Germany, 2009; ISBN 978-3-540-29249-4.
5. Ugur, I.; Demirdag, S.; Yavuz, H. Effect of rock properties on the Los Angeles abrasion and impact test characteristics of the aggregates. *Mater. Charact.* **2009**, *61*, 90–96. [CrossRef]
6. Smart, K.; Ferrill, D.; McKeighan, C.; Chester, F. Estimating rock mechanical properties from microrebound measurements. *Eng. Geol.* **2023**, *312*, 106954. [CrossRef]
7. Luo, P.; Li, D.; Ma, J.; Zhou, A.; Zhang, C. Experimental investigation on me-chanical properties and deformation mechanism of soft-hard interbedded rock-like mate-rial based on digital image correlation. *J. Mater. Res. Technol.* **2023**, *24*, 1922–1938. [CrossRef]

8. Dong, L.; Zhang, L.; Liu, H.; Du, K.; Liu, X. Acoustic Emission b Value Charac-teristics of Granite under True Triaxial Stress. *Mathematics* **2022**, *10*, 451. [CrossRef]
9. Sun, X.; Cui, L.; Zhang, Y.; Wang, L.; Qi, Z. Mechanical properties of rock with pre-cracks anchored by constant resistance and large deformation cables based on particle flow codes. *Eng. Fail. Anal.* **2022**, *142*, 106781. [CrossRef]
10. Chen, Y.; Lin, H.; Xie, S. Effect of joint microcharacteristics on macroshear be-havior of single-bolted rock joints by the numerical modelling with PFC. *Environ. Earth Sci.* **2022**, *81*, 276. [CrossRef]
11. Ma, S.; Liu, K.; Guo, T.; Yang, J.; Li, X.; Yan, Z. Experimental and numerical in-vestigation on the mechanical characteristics and failure mechanism of cracked coal & rock-like combined sample under uniaxial compression. *Theor. Appl. Fract. Mech.* **2022**, *122*, 103583. [CrossRef]
12. Alshkane, Y.; Marshall, A.; Stace, L. Prediction of strength and deformability of an interlocked blocky rock mass using UDEC. *J. Rock Mech. Geotech. Eng.* **2017**, *9*, 531–542. [CrossRef]
13. Pan, C.; Li, X.; He, L.; Li, J. Study on the effect of micro-geometric heterogeneity on mechanical properties of brittle rock using a grain-based discrete element method coupling with the cohesive zone model. *Int. J. Rock Mech. Min. Sci.* **2021**, *140*, 104680. [CrossRef]
14. Zhang, B.; Mu, J.; Zheng, J.; Lv, Q.; Deng, J. A new estimation method and an an-isotropy index for the deformation modulus of jointed rock masses. *J. Rock Mech. Geotech. Eng.* **2022**, *14*, 153–168. [CrossRef]
15. Katz, O.; Reches, Z.; Roegiers, J.-C. Evaluation of mechanical rock properties using a Schmidt Hammer. *Int. J. Rock Mech. Min. Sci.* **2000**, *37*, 723–728. [CrossRef]
16. Aydin, A. ISRM Suggested Method for Determination of the Schmidt Hammer Rebound Hardness: Revised Version. In *The ISRM Suggested Methods for Rock Characterization. Testing and Monitoring: 2007–2014*; Elsevier: Amsterdam, The Netherlands, 2008.
17. Thuro, K. Recommendation No. 5 (revised) of the Commission on Rock Testing of the Deutsche Gesellschaft für Geotechnik e.V.—"point load tests on rock samples". *Bautechnik* **2010**, *87*, 322–331. (in German) [CrossRef]
18. ISRM International Society for Rock Mechanics. Suggested method for determining point load strength. *Int. J. Rock Mech. Min. Sci. Geomech. Abstr.* **1985**, *22*, 51–60. [CrossRef]
19. Tang, C.; Hudson, J.A. *Rock Failure Mechanisms: Illustrated and Explained*; CRC Press: Boca Raton, FL, USA, 2010.
20. *EN 1097-2 (2020-06)*; Tests for Mechanical and Physical Properties of Aggregates. Part 2: Methods for the Determination of Resistance to Fragmentation. Austrian Standards: Vienna, Austrian, 2020.
21. Voit, K.; Kuschel, E. Rock Material Recycling in Tunnel Engineering. *Appl. Sci.* **2020**, *10*, 2722. [CrossRef]
22. Twiss, R.; Moores, E. *Structural Geology*, 2nd ed.; Freeman and Company: New York, NY, USA, 2006; ISBN 978-0716749516.
23. Lamplmair, S.; Zeman, O.; Voit, K. Factors Influencing the Load-Bearing Capacity of Rock as Base Material for Post-Installed Anchors. *Materials* **2021**, *14*, 5130. [CrossRef]
24. Gupta, A.S.; Seshagiri Rao, K. Weathering effects on the strength and deformational behaviour of crys-talline rocks under uniaxial compression state. *Eng. Geol.* **2000**, *56*, 257–274. [CrossRef]
25. Ogunsola, N.O.; Olaleye, B.M.; Saliu, M.A. Effects of Weathering on some physical and mechanical properties of Ewekoro Limestone South-Western Nigeria. *Int. J. Eng. Appl. Sci.* **2017**, *4*, 72–81.
26. *JCGM 100*; Evaluation of Measurement Data–Guide to the Expression of Uncertainty in Measurement. Joint Committee for Guides in Metrology (JCGM): Sevres, France, 2008.
27. *JCGM 104*; Evaluation of Measurement Data—An Introduction to the "Guide to the Expression of Uncer-Tainty in Measurement" and Related Documents. Joint Committee for Guides in Metrology (JCGM): Sevres, France, 2009.
28. *JCGM 200*; International Vocabulary of Metrology—Basic and General Concepts and Associated Terms (VIM). Joint Committee for Guides in Metrology (JCGM): Sevres, France, 2012.
29. Der Kiureghian, A.; Ditlevsen, O. Aleatory or epistemic? Does it matter? *Struct. Saf.* **2009**, *31*, 105–112. [CrossRef]
30. Taffe, A. *DAfStb-Heft 574: Zur Validierung Quanti-Tativer Zerstörungsfreier Prüfverfahren im Stahlbe-Tonbau am Beispiel der Laufzeitmessung*; Deutscher Ausschuss für Stahlbeton: Berlin, Germany, 2008.
31. Schulmann, K.; Lexa, O.; Janousek, V.; Lardeaux, J.; Edel, J. Anatomy of a diffuse cryptic suture zone: An example from the Bohemian Massif. *Eur. Variscides. Geol.* **2014**, *42*, 275–278. [CrossRef]
32. Racek., M.; Lexa, O.; Schulmann, K.; Corsini, M.; Štípská, P.; Maierová, P. Re-evaluation of polyphase kinematic and 40Ar/39Ar cooling history of Moldanubian hot nappe at the eastern margin of the Bohemian Massif. *Int. J. Earth Sci.* **2016**, *106*, 394–420. [CrossRef]
33. Petrakakis, K. Evolution of Moldanubian rocks in Austria: Review and synthesis. *J. Metamor-Phic Geol.* **1997**, *15*, 203–222. [CrossRef]
34. ISRM International Society for Rock Mechanics. Rock characterisation, testing and monitoring. In *ISRM Suggested Methods*; Brown, E.T., Ed.; Pergamon: Oxford, UK, 1981; p. 211.
35. Broch, E.; Franklin, J.A. The point-load strength test. *Int. J. Rock Mech. Min. Sci.* **1972**, *9*, 669–697. [CrossRef]
36. Ballmann, P.; Collins, R.J.; Delalande, G.; Van den Elshout, J.P.; Mishellany, A.; Sym, R. Testing of industrial products—Aggregates for construction. *European Commission DG XII. Final Rep. Proj.* **1998**, 134.
37. Deree, D.; Miller, R. *Engineering Classification and Index Properties for Intact Rock (Technical Report No. AFWL-TR-65-116)*; Air force Weapons Laboratory: Wright-Patterson Air Force Base, OA, USA, 1996.

38. Singh, T.N.; Kainthola, A.; Venkatesh, A. Correlation between Point Load Index and UCS for different rock types. *Rock Mech. Rock Eng.* **2012**, *45*, 259–264. [CrossRef]
39. Singh, V.; Singh, D. Correlation between point load index and compressive strength for quartzite rocks. *Geotech. Geol. Eng.* **1993**, *11*, 269–272. [CrossRef]
40. Yagiz, S. Predicting uniaxial compressive strength, modulus of elasticity and index properties of rocks using the Schmidt hammer. *Bull. Eng. Geol. Environ.* **2009**, *68*, 55–63. [CrossRef]
41. Buyuksagis, I.S.; Goktan, R.M. The effect of Schmidt hammer type on uniaxial compressive strength prediction of rock. *Int. J. Rock Mech. Min. Sci.* **2007**, *44*, 299–307. [CrossRef]
42. Basu, A.; Aydin, A. A method for normalization of Schmidt hammer rebound values. *Int. J. Rock Mech. Min. Sci.* **2004**, *41*, 1211–1214. [CrossRef]
43. Bieniawski, Z. The point-load test in geotechnical practice. *Eng. Geol.* **1975**, *9*, 1. [CrossRef]
44. Brook, N. The equivalent core diameter method of size and shape correction in poin load testing. *Int. J. Rock Mech. Min. Sci. Geomech. Abstr.* **1985**, *22*, 61–70. [CrossRef]
45. Rusnak, J.; Mark, C. Using the point load test to determine the uniaxial compressive strength of coal measure rock. In Proceedings of the 19th International Conference on Ground Control in Mining, Morgantown, WV, USA, 8–10 August 2000; West Virginia University: Morgantown, WV, USA, 1999; pp. 362–371.
46. Kahraman, S.; Fener, M. Predicting the Los Angeles abrasion loss of rock aggregates from the uniaxial compressive strength. *Mater. Lett.* **2007**, *61*, 4861–4865. [CrossRef]
47. Hering, E.; Schönfelder, G. Messfehler, Messgenauigkeit und Messparameter. In *Sensoren in Wissenschaft und Technik*; Hering, E., Schönfelder, G., Eds.; Springer Vieweg: Wiesbaden, Germany, 2018. [CrossRef]

materials MDPI

Article

Fatigue Behavior of Sandstone Exposed to Cyclic Point-Loading: Implications for Improving Mechanized Rock Breakage Efficiency

Xin Cai [1,2], Jifeng Yuan [1], Zilong Zhou [1,2], Zhibo Wu [3], Jianmin Liu [4], Barkat Ullah [1] and Shaofeng Wang [1,2,*]

1 School of Resources and Safety Engineering, Central South University, Changsha 410010, China
2 Hunan Provincial Key Laboratory of Resources Exploitation and Hazard Control for Deep Metal Mines, Central South University, Changsha 410010, China
3 Norinmining Co., Ltd., Beijing 100055, China
4 CCFED Civil Engineering Co., Ltd., Changsha 410010, China
* Correspondence: sf.wang@csu.edu.cn

Abstract: During the process of mechanized excavation, rock is essentially subjected to cyclic point loading (CPL). To understand the CPL fatigue behavior of rock materials, a series of CPL tests are conducted on sandstone samples by using a self-developed vibration point-load apparatus. The effects of loading frequency and waveform on rock fatigue properties under CPL conditions are specifically investigated. The load and indentation depth histories of sandstone samples during testing are monitored and logged. The variation trends of fatigue life (failure time) under different loading conditions are obtained. Test results indicate that the fatigue life of the sandstone sample exposed to CPL is dependent on both loading frequency and waveform. As the loading frequency rises, the fatigue life of the sandstone first declines and then increases, and it becomes the lowest at 0.5 Hz. In terms of waveform, the fatigue life of the sandstone is largest under the trigonal wave and is least under the rectangular wave. These findings can provide valuable theoretical support for optimizing the rock cutting parameters to enhance the efficiency of mechanized excavation.

Keywords: cyclic point loading; rock fatigue; loading frequency; waveform; mechanized rock breakage; parameter optimization

Citation: Cai, X.; Yuan, J.; Zhou, Z.; Wu, Z.; Liu, J.; Ullah, B.; Wang, S. Fatigue Behavior of Sandstone Exposed to Cyclic Point-Loading: Implications for Improving Mechanized Rock Breakage Efficiency. *Materials* **2023**, *16*, 2918. https://doi.org/10.3390/ma16072918

Academic Editors: Enrique Casarejos and Andrei Victor Sandu

Received: 30 December 2022
Revised: 31 March 2023
Accepted: 4 April 2023
Published: 6 April 2023

1. Introduction

Rock breakage is imperative to a variety of rock engineering projects, such as mining and tunnelling [1–3]. With the accelerating progress of rock engineering, the drawbacks of the drilling-and-blasting method, such as great rock damage, violent ground vibration, significant operating risk, low energy utilization, and intermittent operation, have been increasingly exposed [4,5]. These disadvantages seriously restrict the safe and efficient construction of rock engineering projects. Thus, this addresses the urgent need to revolutionize the traditional approach of rock breakage by means of explosives. Due to its superiorities such as weak disturbance to surrounding rock and continuous operation, the mechanized excavation can provide more realistic solutions to the issues induced by blasting [6].

In fact, the manner of non-explosive mechanized excavation has been broadly used in soft rock strata, typically in coal mines [7,8], and has been preliminarily applied in hard rock mines [9,10] or tunnels [11]. Varied rock breakage machines have also been invented, two of which are most widely employed in the mining industry, i.e., roadheader with pick and high-frequency impact hammer with bucket teeth (Figure 1a,b). Whether the roadheader or the high-frequency impact hammer is used, the contact area between the cutter and the rock is very small. Such localized contact can be considered as a contact point. Moreover, the hard rock commonly experiences multiple cuttings before fracturing because the high cutting resistance of the hard rock makes the rock breakage difficult in one cutting cycle. The mechanism of rock cutting can be simplified as the fatigue failure of rock subjected to point loading (PL), as illustrated in Figure 1c. Hence, a comprehensive and in-depth understanding of the mechanical behavior of rock exposed to cyclic point loading (CPL) is significant for non-explosive mechanized excavation.

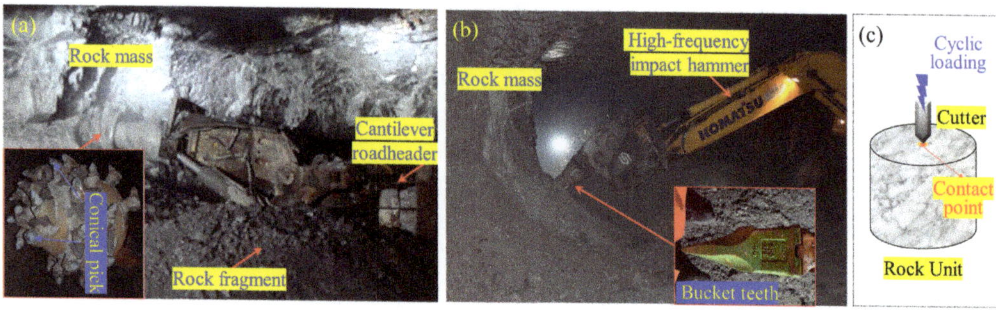

Figure 1. (**a**) Roadheader with pick; (**b**) high-frequency impact hammer with bucket teeth; (**c**) mechanical model of rock unit exposed to the disturbance from mechanized excavation.

In the past several decades, great efforts have been made to explore the rock fatigue behavior [12,13]. Numerous cyclic loading tests are performed on various rock or rock-like materials, such as salt [14,15], sandstone [16,17], dolomite [18], limestone [19,20], marble [21,22], granite [23–25], tuff [26,27], and concrete [28,29]. Based on these abundant publications, it is widely reported that many internal (including brittleness [19], strength [20]) and external factors (such as stress level [15,25], waveform [17], loading frequency [14,25], loading pattern [26,27], and confinement [14–16,22,23]) pronouncedly control the fatigue behavior of rock. Concretely speaking, Attewell and Farmer [18] pointed out that the fatigue life of limestone rises with the decreasing amplitude of cyclic load. Under a given condition of upper stress limit, loading frequency, and stress amplitude, the percentage of strain hardening is increased as the cyclic number rises. Fuenkajorn and Phueakphum [14] suggested that the number of cycles to failure is increased by the rising loading frequency but is reduced by the increasing upper stress limit. Bagde and Petroš [17] reported that under the same frequency and amplitude of cyclic loading, the square waveform results in the fastest damage accumulation in the rock sample, followed by sinusoidal and ramp waveforms. Liu and He [16] found that the confinement elevates both final deformation and the cyclic number of the accelerating phase of sandstone. These outcomes of the previous research greatly enhance the understanding of rock fatigue behavior. In the majority of previous investigations, the samples are, however, subjected to a uniform surface stress field rather than a concentrated point load [30]. The fatigue characteristics of rock materials under the CPL condition are still unknown.

This paper aims to study the fatigue behavior of sandstone exposed to cyclic point-loading. The laboratory vibration point-load apparatus that can apply CPL on rock samples is used to simulate the rock breakage process in mechanized excavation. The experimental procedures and measurement methods are also described. Several CPL tests are conducted on cylindrical sandstone samples to obtain the mechanical response of rock. The effects of the loading frequency and waveform of CPL on the fatigue behavior of the sandstone are revealed. The findings from the test results to the optimization of mechanized rock cutting are discussed.

2. Apparatus Design and Test Methodology

2.1. Apparatus Structure

The vibration point-load apparatus is placed in the advanced research center at Central South University, China. It mainly consists of a static uniaxial loading unit, a vibration excitation unit, and a point-load generating unit [30], as shown in Figure 2. The static loading unit is an electro-hydraulic servo testing machine (MTS Landmark), which has a stiffness of 467 mN/m, a maximum load of 100 kN, and a load measuring accuracy of 0.5%. The vibration excitation is controlled by an electro-hydraulic servo valve that can realize fatigue loading up to 80 Hz simultaneously with the static load. The waveforms can be adjusted by computer programs, such as triangle, rectangle, sinusoidal, and realistic complex waveforms. The loading patterns include constant-amplitude, variable-amplitude, and stochastic-amplitude classes. The point-load-generating unit is composed of the hydraulic clamp and the miniature cutter (Figure 2b). The shapes of cutters are substitutable according to the simulating rock breakage machines, such as the pick and bucket teeth. A cone-shaped cutter is designed in this study to simulate the conical pick installed in the roadheader, and its specific geometry is illustrated in Figure 2c. The cutter is made of hardened chromium alloy with 60 HRC so that they can considered as a rigid body to minimize their deformation and damage during testing. By this apparatus, we can conduct the axial CPL test on rock samples to investigate the rock breakage exposed to mechanical vibrating impact.

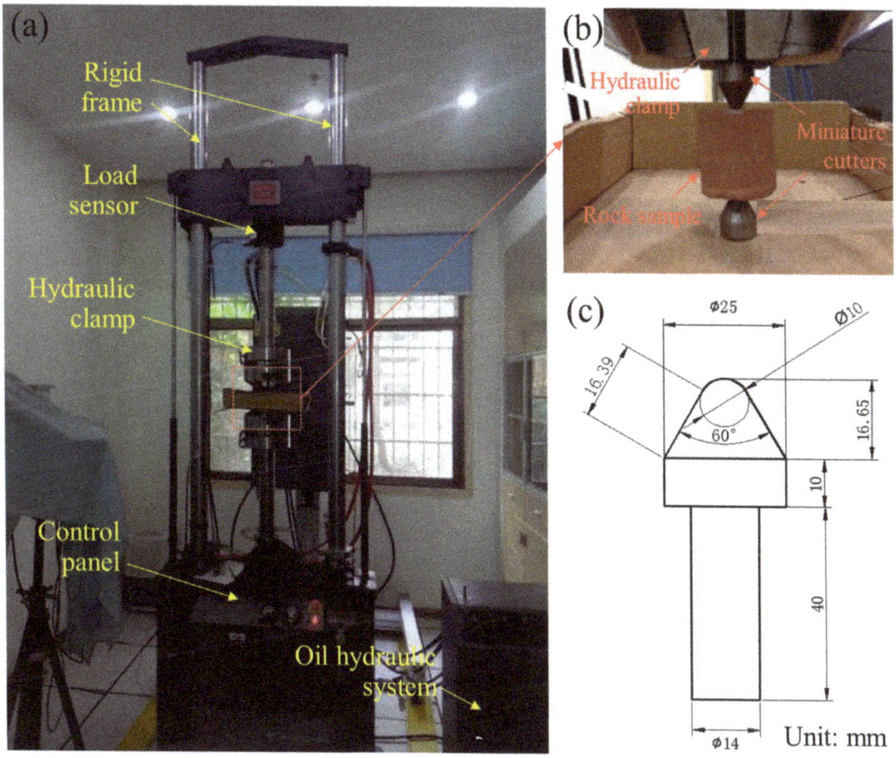

Figure 2. The vibration point-load apparatus: (**a**) photographic view; (**b**) detailed view of the rock sample and the miniature cutters; (**c**) geometry of the cone-shaped cutter.

2.2. Sample Requirements

The rock samples required for CPL tests are in accordance with the standard suggested by the International Society for Rock Mechanics and Rock Engineering (ISRM) [31,32]. Both cylindrical and block samples can be used for tests, as illustrated in Figure 3. Specific requirements for sample preparation are as follows:

(1) Rock samples should be retrieved from the rock slate without visible geological weakness to minimize the property dispersion across the samples.

(2) The size of the sample should be at least 10 times greater than the average grain size in the rock.
(3) Cylindrical samples with a length/diameter (L/D) ratio of 0.3–1.0 are preferable (Figure 3a). The ends of the sample should be polished to ensure that the ends are flat to 0.02 mm and depart from perpendicularity to the axis of the sample by less than 0.001 rad.
(4) A rock block is an alternative for the sample shape (Figure 3b). The ratio of thickness (T) to width (W) should be between 0.3 and 1.0. The main side length (L_m) should be at least 0.5 W.
(5) For routine testing, the sample should be dried (naturally dried or oven-dried) before testing to eliminate the moisture effect on the test results.

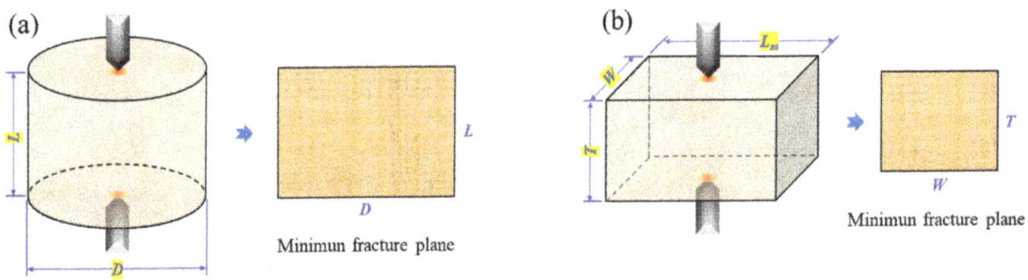

Figure 3. Schematics of the sample suggested in CPL testing: (**a**) cylindrical sample and (**b**) block sample (modified from [31]).

2.3. Testing Procedures

The CPL experiment should be executed according to the following procedures:

(1) The selected cutter is tightly installed at the hydraulic clamp.
(2) The sample is inserted between a pair of cutters that are closed to make contact along a line perpendicular to the end surfaces of the sample.
(3) The cyclic loading path is input into the computer program and then the desired load is applied on the sample until the failure of the sample.
(4) The applied load F and indentation depth δ (i.e., axial displacement) are monitored and recorded by a force sensor in the MTS landmark and a linear variable differential transformer (LVDT), respectively. The curve of F-δ of the rock sample is obtained, and the fatigue behavior of the sample can be determined accordingly. The moment when the force declines to zero is defined as the failure time of the sample.

2.4. Data Reduction

The point load strength (I_s) of the rock sample subjected to CPL can be determined as the form in the monotonous point loading test [31]:

$$I_s = \frac{P}{D_c^2} \tag{1}$$

where P is the peak load in CPL testing and D_c is the equivalent diameter of the sample [31]:

$$D_c^2 = \frac{4S}{\pi} \tag{2}$$

where S is the minimum cross-sectional area of a plane through the cutter contacting points. As shown in Figure 3, for the cylindrical sample, $S = DL$; for the block sample, $S = WT$ [31].

3. Experimental Schemes

3.1. Material Characterization and Sample Preparation

A sandstone collected from Yunnan province of China (Yunnan sandstone, YNS for short) is selected for testing in this study. The results of X-ray diffraction measurements show that the YNS sandstone is primarily composed of quartz (~54%) and feldspar (~29%) by weight. Other low-content minerals contain mica (~6%), calcite (~6%), and clay minerals (~6%). Several routine tests are conducted on the YNS samples to determine the essential physical and mechanical parameters. The YNS samples have a density of 2220 kg/m^3 and a porosity of 7.08% measured by the nuclear magnetic resonance technique. The average uniaxial compressive strength, obtained on four standard samples, is about 45.7 MPa, and the monotonous peak point load (PL, P_s) is 4.63 kN (i.e., 1.81 MPa PL strength).

All samples are drilled from a single rock slate to minimize the variation in engineering properties across the sample set. The cylindrical samples with a 50 mm diameter and 0.8 L/D ratio are manufactured. The maximum grain size of YNS is about 0.5 mm, which is far less than the sample size. The other requirements in Section 2.2. are strictly satisfied. All prepared samples are placed in a well-ventilated laboratory for air-drying over seven days and are then sealed with a plastic wrap to prevent moisture in the atmosphere before testing [33,34].

3.2. Loading Schemes

In the present study, two groups of tests are designed to investigate the influences of loading frequency and waveform on rock fatigue behavior under CPL conditions. The specific loading schemes are introduced in the following subsections.

3.2.1. CPL Testing with Different Loading Frequencies

In the first set of experiments, to study the effects of loading frequency on rock fatigue behavior subjected to CPL, the loading frequency is set as the single independent variable between the testing groups. As shown in Figure 4, the sinusoidal waveform is selected with the constant lower (R_{min})- and upper (R_{max})-load limits of 1.39 kN (~0.3 P_s) and 4.35 kN (~0.94 P_s) respectively, such that the amplitude of the wave is sustained at 0.64 P_s. The range of loading frequency is between 0.1 and 5 Hz.

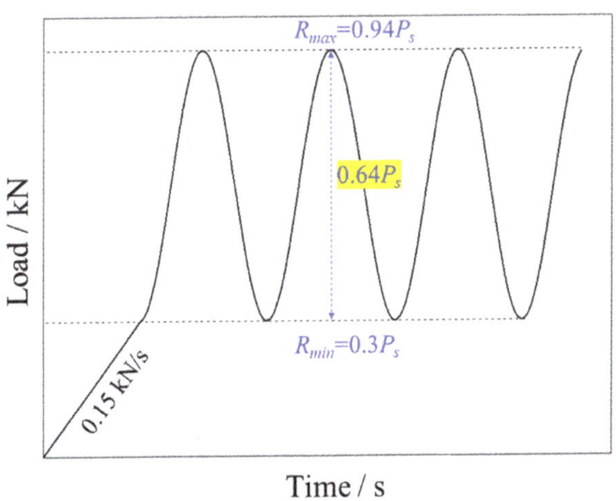

Figure 4. Schematic of force history in CPL tests.

The force control mode is chosen for loading in all tests. The static load applied on the rock sample is first increased from 0 to 1.39 kN at a constant loading rate of 0.15 kN/s. Subsequently, the static load is sustained, with which an axial CPL with 2.96 kN amplitude and various frequencies is superimposed until the failure of the sample. The loading schemes and test results are listed in Table 1.

Table 1. Test parameters and results in CPL testing with different loading frequencies.

Sample No.	Upper Limit (kN)	Lower Limit (kN)	f (Hz)	Number of Cycles	Fatigue Life [a] (s)
D-1-1 D-1-2 D-1-3	0.94 P_s	0.3 P_s	0.1	68 69 69	690.4 700.5 705.1
D-2-1 D-2-2 D-2-3	0.94 P_s	0.3 P_s	0.2	60 55 59	310.1 288.2 305.3
D-3-1 D-3-2 D-3-3 D-3-4 D-3-5	0.94 P_s	0.3 P_s	0.5	8 9 6 7 9	25.6 28.3 22.5 25.6 28.1
D-4-1 D-4-2 D-4-3 D-4-4 D-4-5	0.94 P_s	0.3 P_s	1.0	244 309 290 339 308	255.1 316.7 300.4 350.1 319.1
D-5-1 D-5-2 D-5-3 D-5-4 D-5-5	0.94 P_s	0.3 P_s	5	2846 1961 2741 2856 2796	580.3 403.1 559.1 582.1 570.1

[a] The time duration from the loading initiation to rock failure is defined as the fatigue life of the sample.

3.2.2. CPL Testing with Different Waveforms

In the second set of experiments, the waveform is the single independent variable. Three waveforms, i.e., trigonal, sinusoidal, and rectangular waves, are used for tests with a constant loading frequency of 0.5 Hz. In addition, the lower and upper load limits are 1.39 kN and 4.35 kN, respectively, and are consistent in all tests.

Similar to the first set in Section 3.2, the load is exerted through the force control mode. First, the static load is applied to 1.39 kN by a rate of 0.15 kN/s. The CPLs with different waveforms are superimposed with the static load until the rock failure. The loading parameters and experimental results are summarized in Table 2.

Table 2. Test parameters and results in CPL testing with different waveforms.

Sample No.	Waveform	Upper Limit (kN)	Lower Limit (kN)	f (Hz)	Number of Cycles	Fatigue Life [b] (s)
T-1 T-2 T-3	Trigonal	0.94 P_s	0.3 P_s	0.5	84 80 90	690.4 700.5 705.1
S-1 S-2 S-3	Sinusoidal	0.94 P_s	0.3 P_s	0.5	5 12 8	19.8 31.8 25.7
R-1 R-2 R-3	Rectangular	0.94 P_s	0.3 P_s	0.5	<1 [a] <1 [a] <1 [a]	9.4 9.4 9.4

[a] The samples fail in the first cycle of point loading. [b] The time duration from the loading initiation to rock failure is defined as the fatigue life of the sample.

4. Experimental Results and Discussion

4.1. Effects of Loading Frequency on Rock Behavior under Cyclic Point Loading

4.1.1. Failure Pattern

Figure 5 illustrates the failure patterns of YNS samples in monotonous and cyclic point loading tests. Regardless of the loading condition, the failure patterns are very similar. A couple of spherical indentations is clearly observed on the center of the sample surfaces. A diametrical fracture surface splits the sample into halves, which agrees with the failure pattern of Gosford Sandstone subjected to monotonous point loading [35]. This diametrical fracturing should be attributed to the tensile failure induced by point indentation.

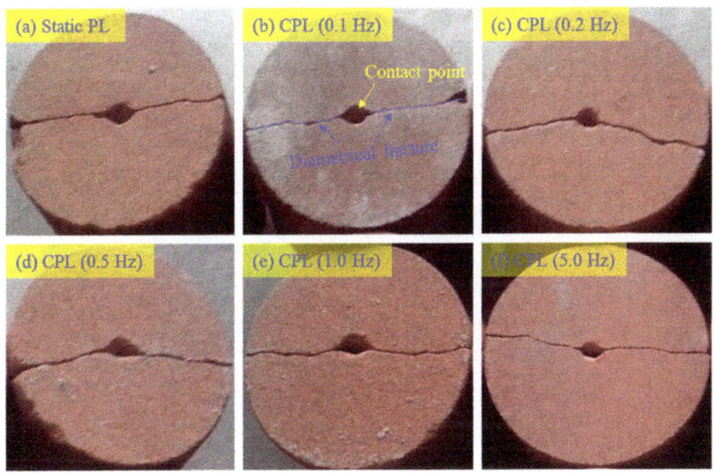

Figure 5. Failure pattern of YNS subjected to (**a**) monotonous point loading and (**b–f**) CPLs with different frequencies.

4.1.2. Load-Indentation Depth Curves and Fatigue Life of Sandstone

Figure 6 presents the typical histories of force (F) and indentation depth (δ) of samples in CPL tests with the loading frequencies of 0.2, 0.5, and 1.0 Hz. The lower and upper load limits, as well as loading frequencies, are precisely imposed on the rock sample according to our settings. The wave amplitude is constant before the rock failure. In each loading cycle, the indentation depth is changed synchronously with the load. From the view of the whole loading process, the lower and upper values of the indentation depth gradually rise with increasing cyclic number. The hysteresis loops are clearly observed in the F-δ curve. The density of the hysteresis loop apparently increases as the cyclic number rises until the failure of the rock sample. Once the rock sample fails, the load drops abruptly while the indentation depth surges sharply. All observations indicate that the proposed apparatus can implement the CPL test to obtain the point-loading fatigue behavior of the rock sample.

The fatigue life is an important indicator for evaluating the efficiency of rock breakage. In this study, we regard the time duration from the loading initiation to the total failure of rock as the fatigue life of the sample. From Figure 6, it can be seen that under various loading frequencies, the experienced cycle numbers of the rock sample are much different. This means that the loading frequency plays the role of a controlling factor in the fatigue life of the rock sample. Figure 7a further gives the variation in the fatigue life of the YNS samples against the loading frequency. Interestingly, with the increase in loading frequency, the fatigue life of rock declines first and then rises. The value of the fatigue life becomes the least at the loading frequency of 0.5 Hz. The number of cycles changes against the loading frequency in a similar manner of fatigue life (Figure 7b).

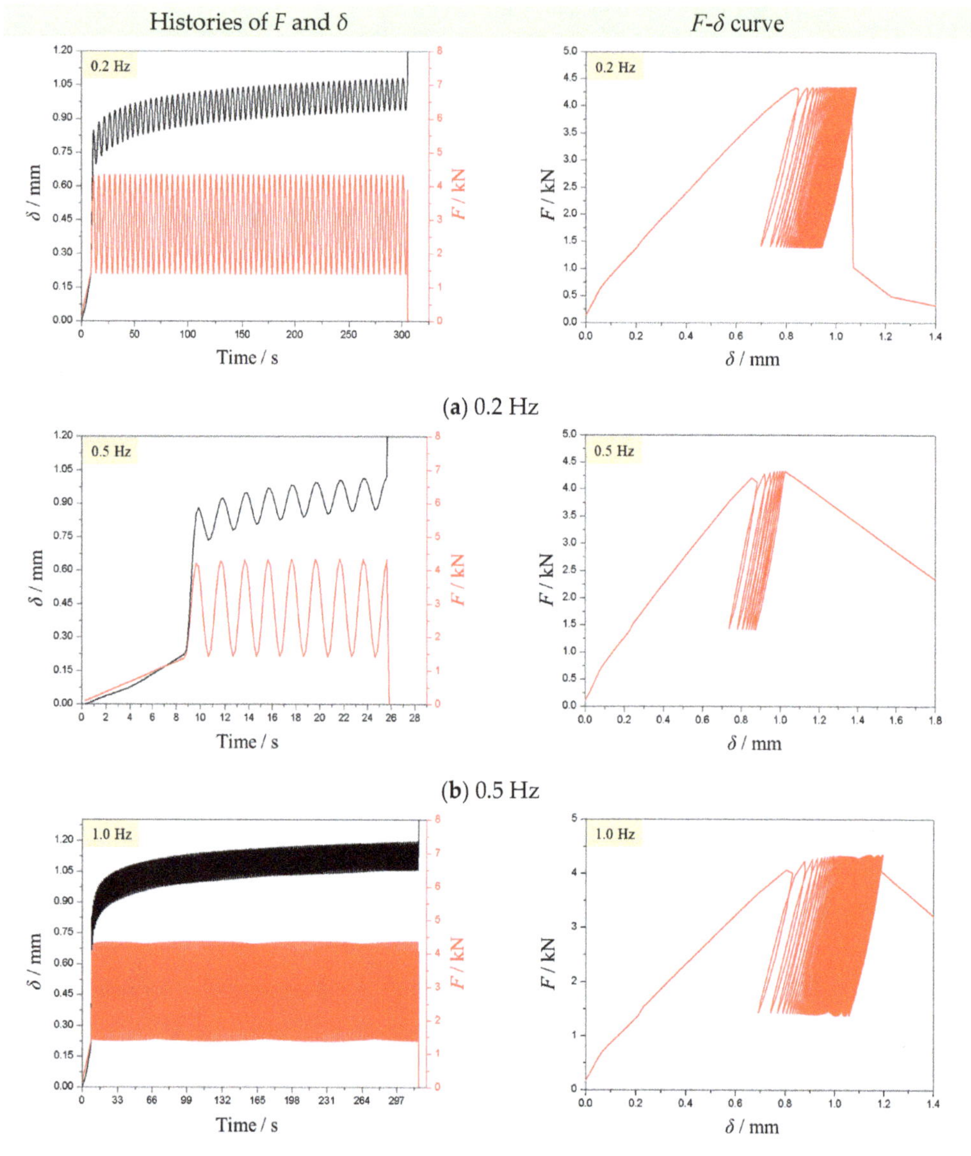

Figure 6. Typical curves of load and indentation depth in CPL tests with the frequencies of (**a**) 0.2 Hz, (**b**) 0.5 Hz, and (**c**) 1.0 Hz (the left column shows the histories of F and δ, and the right column shows the F-δ curves).

It is shared that the fatigue life of rock mainly depends on the correlation between the loading rate and the speed of crack growth [12]. If the loading rate is high enough (high loading frequency), the rapid increase in load limits the development and nucleation of cracks or flaws, and further raises the fatigue life of rock. On the contrary, when the loading rate is very low, the majority of cracks developed in the cycle are closed again before the beginning of the next loading cycle. Hence, we can reasonably speculate that rock material has a minimum fatigue life at a specific loading frequency. In other words, there is an optimal loading frequency for rock breakage. The optimal rock breaking frequency for the tested YNS sample is nearly 0.5 Hz under the fixed upper and lower load limits.

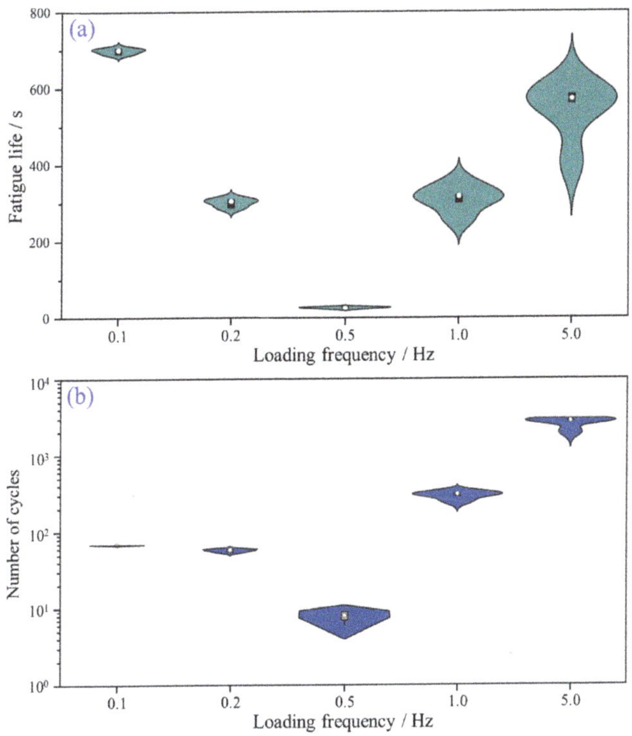

Figure 7. Variations in (**a**) fatigue life and (**b**) number of cycles of YNS sample against loading frequency under CPL condition.

It is worth noting that the observations in Figure 7 disagree with the prior study [25], in which the number of cycles of Alvand monzogranite monotonously rises at higher loading frequencies (from 0.1 to 5 Hz) under cyclic uniaxial compression. One possible reason is that the optimal loading frequencies for rock failure are different for various rock types. The other interpretation is that the alteration in loading type (point loading or surface loading) changes the optimal loading frequency.

4.2. Effects of Waveform on Rock Behavior under Cyclic Point Loading

4.2.1. Failure Pattern on Rock Behavior under Cyclic Point Loading

The failure patterns of the YNS samples subjected to CPL with trigonal and sinusoidal waves are similar to those in monotonous PL tests, as shown in Figure 8. An interesting observation is that under the CPL with rectangular waves, the fractured zones occur near the contact point in the majority of samples. It is hypothesized that the steep front of the rectangular wave leads to a larger damage area on the end surface of the sample. Contrarily, both the trigonal and sinusoidal waves have a slower rising front such that the force is more concentrated on the contact point. The damage is totally generated on the contact point rather than on the surface near the contact point.

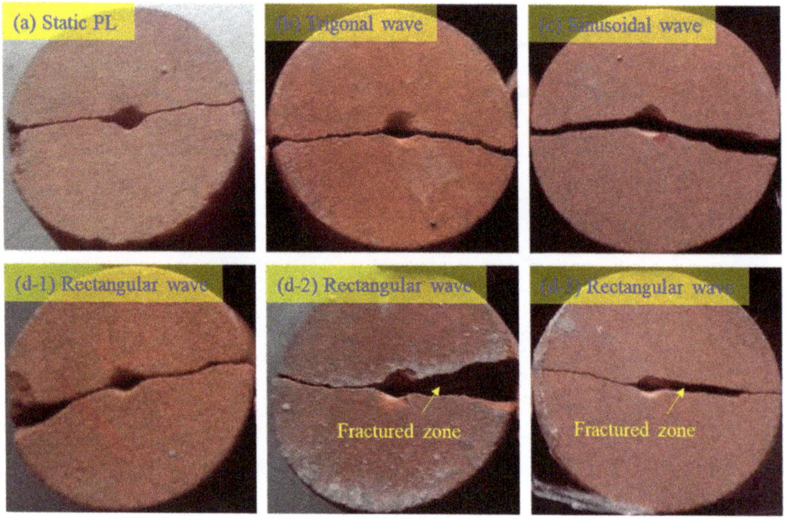

Figure 8. Failure patterns of the YNS sample subjected to (**a**) static point loading and (**b–d**) CPLs with different waveforms.

4.2.2. Load-Indentation Depth Curves and Fatigue Life of Sandstone

Figure 9 displays the typical histories of F and δ of samples in CPL tests with three different waveforms. The waveform significantly affects the CPL fatigue behavior of the rock sample. For the trigonal wave (Figure 9a), the YNS sample fails after more than 80 point-loading cycles. When the sinusoidal wave is applied, the sample only experiences much fewer cycles (5–12 cycles). The point load with the rectangular wave destroys the sample in the first cycle. This means that the rectangular wave can notably promote the rock breakage.

To understand the effect of waveform on the CPL fatigue life of rock, the histogram of the CPL fatigue life of each YNS sample exposed to trigonal, sinusoidal, and rectangular waves is plotted in Figure 10. It can be clearly observed that under the trigonal wave, the YNS sample has the average CPL fatigue life of 698.7 s, which is the maximum in the three tested waves. The CPL fatigue life of the sample subjected to the sinusoidal wave is 77.3 s, which is reduced by 88.9% compared to the life under the trigonal wave. The fatigue life of the sample under the rectangular wave is the least, which is less than 10 s. According to previous studies [17,36], the shape of the waveform plays a predominant role in the fatigue life of rock. At a given loading frequency and amplitude, the rectangular waveform is the most severe testing condition and leads to the shortest fatigue life of rock due to its high loading rate, a great change in loading rate, and long residence period. The trigonal waveform has the slowest damage accumulation probably because of its constant and lowest loading rate. From the view of mechanized excavation, this phenomenon indicates that the rectangular wave has the highest efficiency of rock breakage. This provides a valuable hint that the rectangular wave should be prioritized for the rock breakage machine.

Histories of F and δ | F-δ curve

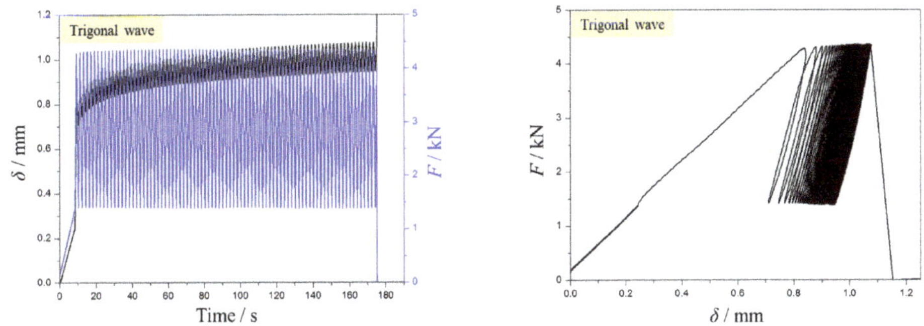

(**a**) Trigonal wave

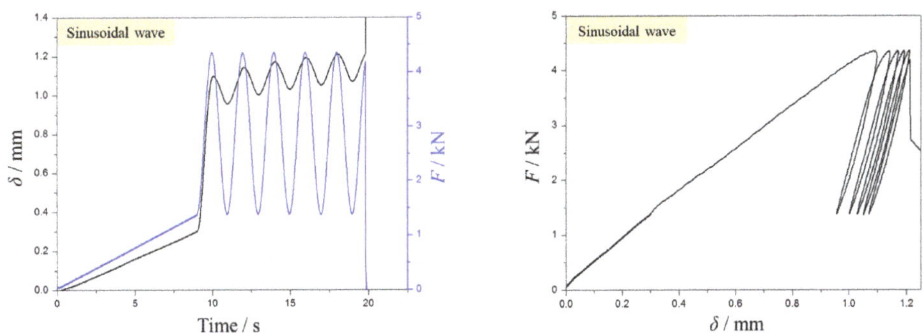

(**b**) Sinusoidal wave

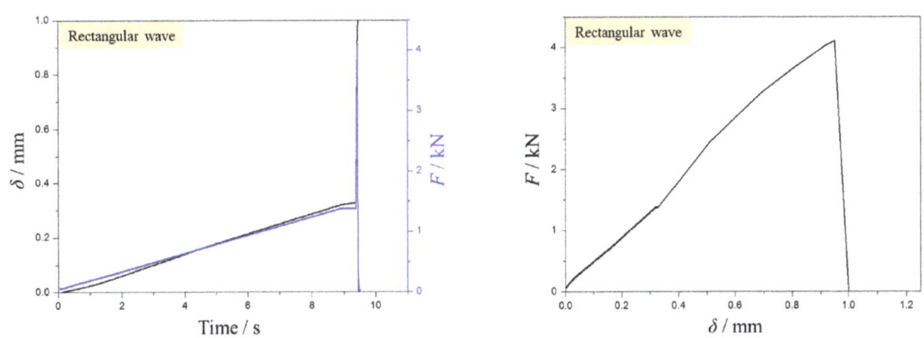

(**c**) Rectangular wave

Figure 9. Typical curves of load and indentation depth in CPL tests with (**a**) trigonal, (**b**) sinusoidal, and (**c**) rectangular waves (the left column shows the histories of F and δ, and the right column shows the F-δ curves).

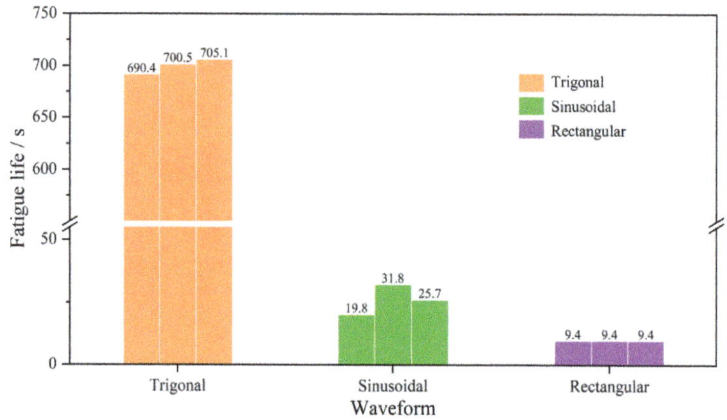

Figure 10. CPL fatigue lives of the YNS samples subjected different waveforms.

5. Conclusions and Prospects

In this study, a self-developed vibration point-load apparatus determining the fatigue behavior of rock materials under cyclic point loading (CPL) is introduced. By this device, two sets of CPL tests are performed on cylindrical YNS samples. The influences of loading frequency and waveform on the fatigue responses of the YNS samples are preliminarily investigated. The main conclusion of this study can be drawn as follows:

(1) The fatigue behavior of rock under CPL conditions is greatly dependent on loading frequency. The fatigue life of the YNS sample shows a trend of "decline followed by rise" with the increase in loading frequency. The minimum value of the YNS sample is 0.5 Hz.

(2) The waveform also plays a controlling role in the fatigue behavior. The order of the fatigue life from largest to least is as follows: trigonal wave > sinusoidal wave > rectangular wave. In addition, when subjected to rectangular waveforms, the fractured zone can be observed on the rock surface. The rectangular waveform has the most severe damage on rock among the three tested waves.

(3) The enlightenment of this work to mechanized excavation is that the efficiency of rock breakage (i.e., fatigue life) is significantly controlled by the parameters of the rock breaking machine, such as loading frequency and waveform. For a given rock type, there is an optimal combination of rock cutting parameters probably including loading frequency, waveform, amplitude, and upper and lower load limits. In rock engineering practice, similar CPL tests should be first conducted on the rock sample gathered from the site to predetermine the optimal combination of rock cutting parameters. Then, the optimal parameters should be applied to the mechanized machine, such that the efficiency of rock breakage can be markedly improved.

This study is still limited to one rock type (Yunnan sandstone) and two loading parameters (loading frequency and waveform). In future work, the impacts of other loading parameters such as cutter shape, amplitude, and upper and lower load limits should be explored intensively. Moreover, the CPL tests will be further performed on other types of hard rocks, such as limestone, marble, and granite, to testify whether our findings are more generally applicable.

Author Contributions: X.C.: Conceptualization, Methodology, Validation, Formal Analysis, Resources, Data Curation, Writing—Review and Editing, Supervision, Project Administration, Funding Acquisition. J.Y.: Writing—Original Draft, Investigation, Data Curation. Z.Z.: Supervision, Funding Acquisition. Z.W.: Investigation, Data Curation, Resources. J.L. and B.U.: Writing—Review and Editing, Resources, Data Curation. S.W.: Writing—Review and Editing, Project Administration. All authors have read and agreed to the published version of the manuscript.

Funding: The work is supported by financial grants from the National Natural Science Foundation of China (52104111, 52174099, 52274249), and the Natural Science Foundation of Hunan (2021JJ30819). The authors are very grateful to the financial contribution and convey their appreciation for supporting this basic research.

Conflicts of Interest: The authors declare that they have no conflict of interest.

References

1. Al-Bakri, A.; Hefni, M. A review of some nonexplosive alternative methods to conventional rock blasting. *Open Geosci.* **2021**, *13*, 431–442. [CrossRef]
2. Li, X.F.; Li, H.B.; Zhang, Q.B.; Jiang, J.L.; Zhao, J. Dynamic fragmentation of rock material: Characteristic size, fragment distribution and pulverization law. *Eng. Fract. Mech.* **2018**, *199*, 739–759. [CrossRef]
3. Li, X.; Zhao, Q.; Li, H. Grain-based discrete element method (GB-DEM) modelling of multi-scale fracturing in rocks under dynamic loading. *Rock Mech. Rock Eng.* **2018**, *51*, 3785–3817. [CrossRef]
4. Cheng, R.; Zhou, Z.; Chen, W.; Hao, H. Effects of Axial Air Deck on Blast-Induced Ground Vibration. *Rock Mech. Rock Eng.* **2022**, *55*, 1037–1053. [CrossRef]
5. Zhou, Z.; Cai, X.; Li, X.; Cao, W.; Du, X. Dynamic Response and Energy Evolution of Sandstone Under Coupled Static–Dynamic Compression: Insights from Experimental Study into Deep Rock Engineering Applications. *Rock Mech. Rock Eng.* **2020**, *53*, 1305–1331. [CrossRef]
6. Wang, S.; Li, X.; Yao, J.; Gong, F.; Li, X.; Du, K.; Tao, M.; Huang, L.; Du, S. Experimental investigation of rock breakage by a conical pick and its application to non-explosive mechanized mining in deep hard rock. *Int. J. Rock Mech. Min. Sci.* **2019**, *122*, 104063. [CrossRef]
7. Miao, X.; Zhang, J.; Feng, M. Waste-filling in fully-mechanized coal mining and its application. *J. China Univ. Min. Technol.* **2008**, *18*, 479–482. [CrossRef]
8. He, S.; Song, D.; Li, Z.; He, X.; Chen, J.; Zhong, T.; Lou, Q. Mechanism and Prevention of Rockburst in Steeply Inclined and Extremely Thick Coal Seams for Fully Mechanized Top-Coal Caving Mining and Under Gob Filling Conditions. *Energies* **2020**, *13*, 1362. [CrossRef]
9. Wang, S.; Sun, L.; Tang, Y.; Jing, Y.; Li, X.; Yao, J. Field application of non-blasting mechanized mining using high-frequency impact hammer in deep hard rock mine. *Trans. Nonferrous Met. Soc. China* **2022**, *32*, 3051–3064. [CrossRef]
10. Li, X.; Yao, J.; Du, K. Preliminary study for induced fracture and non-explosive continuous mining in high-geostress hard rock mine-a case study of Kaiyang phosphate mine. *Yanshilixue Yu Gongcheng Xuebao/Chin. J. Rock Mech. Eng.* **2013**, *32*, 1001–1111.
11. Liu, Z.; Li, L.; Fang, X.; Qi, W.; Shen, J.; Zhou, H.; Zhang, Y. Hard-rock tunnel lithology prediction with TBM construction big data using a global-attention-mechanism-based LSTM network. *Autom. Constr.* **2021**, *125*, 103647. [CrossRef]
12. Cerfontaine, B.; Collin, F. Cyclic and Fatigue Behaviour of Rock Materials: Review, Interpretation and Research Perspectives. *Rock Mech. Rock Eng.* **2018**, *51*, 391–414. [CrossRef]
13. Liu, Y.; Dai, F. A review of experimental and theoretical research on the deformation and failure behavior of rocks subjected to cyclic loading. *J. Rock Mech. Geotech. Eng.* **2021**, *13*, 1203–1230. [CrossRef]
14. Fuenkajorn, K.; Phueakphum, D. Effects of cyclic loading on mechanical properties of Maha Sarakham salt. *Eng. Geol.* **2010**, *112*, 43–52. [CrossRef]
15. Song, R.; Yue-Ming, B.; Jing-Peng, Z.; De-Yi, J.; Chun-He, Y. Experimental investigation of the fatigue properties of salt rock. *Int. J. Rock Mech. Min. Sci.* **2013**, *64*, 68–72. [CrossRef]
16. Liu, E.; He, S. Effects of cyclic dynamic loading on the mechanical properties of intact rock samples under confining pressure conditions. *Eng. Geol.* **2012**, *125*, 81–91. [CrossRef]
17. Bagde, M.N.; Petroš, V. Waveform Effect on Fatigue Properties of Intact Sandstone in Uniaxial Cyclical Loading. *Rock Mech. Rock Eng.* **2005**, *38*, 169–196. [CrossRef]
18. Attewell, P.B.; Farmer, I.W. Fatigue behaviour of rock. *Int. J. Rock Mech. Min. Sci. Geomech. Abstr.* **1973**, *10*, 1–9. [CrossRef]
19. Nejati, H.R.; Ghazvinian, A. Brittleness Effect on Rock Fatigue Damage Evolution. *Rock Mech. Rock Eng.* **2014**, *47*, 1839–1848. [CrossRef]
20. Voznesenskii, A.S.; Krasilov, M.N.; Kutkin, Y.O.; Tavostin, M.N.; Osipov, Y.V. Features of interrelations between acoustic quality factor and strength of rock salt during fatigue cyclic loadings. *Int. J. Fatigue* **2017**, *97*, 70–78. [CrossRef]
21. Song, Z.; Wang, Y.; Konietzky, H.; Cai, X. Mechanical behavior of marble exposed to freeze-thaw-fatigue loading. *Int. J. Rock Mech. Min. Sci.* **2021**, *138*, 104648. [CrossRef]
22. Yang, S.-Q.; Tian, W.-L.; Ranjith, P.G. Experimental Investigation on Deformation Failure Characteristics of Crystalline Marble Under Triaxial Cyclic Loading. *Rock Mech. Rock Eng.* **2017**, *50*, 2871–2889. [CrossRef]
23. Song, Z.; Wu, Y.; Yang, Z.; Cai, X.; Jia, Y.; Zhang, M. Mechanical Responses of a Deeply Buried Granite Exposed to Multilevel Uniaxial and Triaxial Cyclic Stresses: Insights into Deformation Behavior, Energy Dissipation, and Hysteresis. *Adv. Mater. Sci. Eng.* **2021**, *2021*, 3160968. [CrossRef]
24. Akesson, U.; Hansson, J.; Stigh, J. Characterisation of microcracks in the Bohus granite, western Sweden, caused by uniaxial cyclic loading. *Eng. Geol.* **2004**, *72*, 131–142. [CrossRef]

25. Momeni, A.; Karakus, M.; Khanlari, G.R.; Heidari, M. Effects of cyclic loading on the mechanical properties of a granite. *Int. J. Rock Mech. Min. Sci.* **2015**, *77*, 89–96. [CrossRef]
26. Erarslan, N.; Alehossein, H.; Williams, D.J. Tensile Fracture Strength of Brisbane Tuff by Static and Cyclic Loading Tests. *Rock Mech. Rock Eng.* **2014**, *47*, 1135–1151. [CrossRef]
27. Ghamgosar, M.; Erarslan, N. Experimental and Numerical Studies on Development of Fracture Process Zone (FPZ) in Rocks under Cyclic and Static Loadings. *Rock Mech. Rock Eng.* **2016**, *49*, 893–908. [CrossRef]
28. Song, Z.; Konietzky, H.; Cai, X. Modulus degradation of concrete exposed to compressive fatigue loading: Insights from lab testing. *Struct. Eng. Mech.* **2021**, *78*, 281–296. [CrossRef]
29. Song, Z.; Frühwirt, T.; Konietzky, H. Inhomogeneous mechanical behaviour of concrete subjected to monotonic and cyclic loading. *Int. J. Fatigue* **2020**, *132*, 105383. [CrossRef]
30. Zhou, Z.; Wu, Z.; Li, X.-B.; Li, X.; Ma, C. Mechanical behavior of red sandstone under cyclic point loading. *Trans. Nonferrous Met. Soc. China* **2015**, *25*, 2708–2717. [CrossRef]
31. Franklin, J.A. Suggested method for determining point load strength. *Int. J. Rock Mech. Min. Sci. Geomech. Abstr.* **1985**, *22*, 51–60. [CrossRef]
32. Ulusay, R. (Ed.) *The ISRM Suggested Methods for Rock Characterization, Testing and Monitoring: 2007–2014*; Springer International Publishing: Cham, Switzerland, 2015. [CrossRef]
33. Cai, X.; Zhou, Z.; Zang, H.; Song, Z. Water saturation effects on dynamic behavior and microstructure damage of sandstone: Phenomena and mechanisms. *Eng. Geol.* **2020**, *276*, 105760. [CrossRef]
34. Cai, X.; Zhou, Z.; Du, X. Water-induced variations in dynamic behavior and failure characteristics of sandstone subjected to simulated geo-stress. *Int. J. Rock Mech. Min. Sci.* **2020**, *130*, 104339. [CrossRef]
35. Masoumi, H.; Horne, J.; Timms, W. Establishing Empirical Relationships for the Effects of Water Content on the Mechanical Behavior of Gosford Sandstone. *Rock Mech. Rock Eng.* **2017**, *50*, 2235–2242. [CrossRef]
36. Gong, M.; Smith, I. Effect of Waveform and Loading Sequence on Low-Cycle Compressive Fatigue Life of Spruce. *J. Mater. Civ. Eng.* **2003**, *15*, 93–99. [CrossRef]

materials

MDPI

Article

Experimental Investigation of Mechanical and Fracture Behavior of Parallel Double Flawed Granite Material under Impact with Digital Image Correlation

Lei Zhang [1,2], Zhijun Zhang [1,2], Ying Chen [1,2,*], Yong Liu [3], Xinyao Luo [4] and Bing Dai [1,2,*]

1 School of Resource Environment and Safety Engineering, University of South China, Hengyang 421001, China
2 Hunan Province Engineering Technology Research Center for Disaster Prediction and Control on Mining Geotechnical Engineering, Hengyang 421001, China
3 School of Physics and Optoelectronic Engineering, Shenzhen University, Shenzhen 518060, China
4 School of Resource and Safety Engineering, Central South University, Changsha 410083, China
* Correspondence: usc_chen@usc.edu.cn (Y.C.); daibingusc@usc.edu.cn (B.D.); Tel.: +86-18569069231 (Y.C.); +86-17873415528 (B.D.)

Abstract: During the excavation of underground projects, the rock masses left as the bearing support system are also subjected to dynamic loads from the excavation activities ahead. These rock masses have been damaged and fractured during the initial exposure (dynamic loads) and are subjected to static loads in the subsequent process as the support system. In this study, granite rock samples and specimens with different angles were produced, preloaded with different confining pressure, and under a combination of dynamic and static loading tests using a modified dynamic and static loading system: split Hopkinson pressure bar (SHPB). The peak strain and dynamic modulus of elasticity are weakened by the inclination angle in a similar way to the strength, with the specimens showing an evolutionary pattern from tensile strain to shear damage. The change in the inclination angle of flaws would weaken the dynamic and combined strengths, and a larger inclination flaw results in a significant decrease in its strength. Fractal analysis revealed that the fractural dimension was closely related to the fissure angle and showed a good linear correlation with the strain rate. This study will provide an important security assurance for deep mining.

Keywords: DIC; double fissures; SHPB; damage characteristics; fractal

Citation: Zhang, L.; Zhang, Z.; Chen, Y.; Liu, Y.; Luo, X.; Dai, B. Experimental Investigation of Mechanical and Fracture Behavior of Parallel Double Flawed Granite Material under Impact with Digital Image Correlation. *Materials* **2023**, *16*, 2263. https://doi.org/10.3390/ma16062263

Academic Editor: René de Borst

Received: 14 October 2022
Revised: 16 February 2023
Accepted: 3 March 2023
Published: 11 March 2023

1. Introduction

Rocks are solid materials with natural defects, including joints, fractures, laminations, and even pores, which reduce the dynamic and static strength of the rock to varying degrees [1,2]. With the development of urbanization, more and more underground works (shopping malls, passages, pipelines, and other underground works) have been developed to supplement the urban land [3,4]. The construction of underground works does not excavate all the rocks, but always leaves some of them as part of the support system of the underground works [5,6]. In addition, both mechanical excavation and blasting excavation are processes that use dynamic loads to destroy the original state of the rock [7–9]. As a result, the part of the rocks with natural defects is left as the support system, bearing the dynamic loads of the subsequent excavation and static loads (support system). The stability of this part of the rock mass has a deep relationship with the reliability of the engineering structure (support mechanism) [10–12].

With the aim of ensuring safety in deep mining and other underground projects, many studies have been carried out on fractured rock masses [13–18], and some researchers have studied the factual damage of preflawed rocks in different ways [19–21]. As known to all, the loading condition could profoundly affect rock distress, as well as the rock type [22]. Rock specimens exhibit mainly mixed-mode fracture under quasi-static loads dominated by shear [23,24], where they generally fail via tension mode under dynamic loads. It is believed that both static prestress and dynamic load can enlarge the dynamic zone of a rock, but this knowledge is limited for coupled tension tests [25–27]. Additionally, in the study of the rock failure problem, the method of numerical simulation is welcomed by many scholars. Compared with others, the method of numerical simulation can not only save time and funds, but also overcome the shortcomings of the test operations, and can fully consider various complex conditions. Many scholars have studied the dynamic load of rock by using particle flow code (PFC), such as AUTODYN software and FLAC3D, to establish a numerical split Hopkinson pressure bar (SHPB) test system. Numerical dynamic tests under different impact velocities were conducted. It showed that there was a clear relationship between the variation of the reflected wave and the stress equilibrium state in the specimens.

Kaiser et al. [28] found that recent analyses support the existence of failure modes that change with increasing confinement—a transition from the axial splitting at unconfined or low confinement to the shear failure at high confinement. The results showed that angle, length, number, and lamination impact the mechanical properties of the fractured rock's mass under static loading conditions. With the aim of enhancing the support effect of the fissure development area, Deng et al. [29] conducted uniaxial compression tests on a single flaw sandstone of various angles and lengths using acoustic emission devices. The results showed that the maximum axial load-bearing capacity of the specimens decreased with increasing fissure length. In addition, the inclination angle of the prefabricated fissure had some influence on the maximum axial load-bearing capacity of the sandstone, but it was less than the influence of the fissure length on the load-bearing capacity of the specimens. Dou et al. [30,31] performed uniaxial compression tests on sandstones with precast fractures of different inclination angles. The results revealed that the fissure inclination angle had a large influence on the mechanical strength of the fractured rock. With the increase in the fissure inclination angle, the peak strength and elastic modulus of the prefractured rock first decreased and then increased. When the fissure inclination angle was 45° with the direction of the applied axial stress parallel to the loading direction, the initial damage and damage rate of the rock during the fissure evolution were the largest, and the resistance to deformation and damage was the weakest. In addition to the properties of the fissures, the mechanical properties of the rock mass are also influenced by the fissures of different laminae. Lin et al. [32] conducted uniaxial compression tests on a jointed rock mass with two different laminae to investigate the influence of the joint angle and rock bridge angle on the mechanical behavior and damage process of the laminated rock mass. The results indicated that the peak strength of the rock mass was not only related to the joint and bridge pinch angles but was also influenced by the laminae, and the rock mass damage was mainly caused by the extension of cracks in the low-strength laminae. In addition, the number of fissures also affects the mechanical properties of the rock mass under static loading.

The above studies are all about the mechanical properties and damage characteristics of fractured rock under static loads. However, during the excavation of underground projects, fractured rock is often subjected to coupled dynamic and static loads. Regarding this problem, some scholars have also carried out relevant studies [4,18,33,34].

Wang et al. [35] carried out SHPB impact tests with boundary conditions on fractured rocks and found that the boundary conditions play an important role in the rock fracture process. In addition, the fracture angle also affects the dynamic strength of the fractured rock mass. Dai et al. [36] carried out an experimental study of dynamic and static coupling loading using a fractured rock mass. The result revealed that the coupling strength initially decreased with increasing fracture inclination, reached a minimum at approximately 45°, and then increased with increasing fracture inclination. Li et al. [37] used an improved SHPB device to study the energy consumption of sandstone samples with prefabricated fissures and holes under combined dynamic and static loading. Weng et al. [38] used rock-like specimens with single fractures and an MTS-793 test system to study the energy damage characteristics of rock loaded with a low strain rate under a one-dimensional dynamic and static combined loading. Tao et al. [39] studied the stress concentration and energy dissipation around the cavity of the rock sample under the combined action of static load and stress wave, and the results demonstrated that a higher prestatic load is more likely to cause cavity damage. To a certain extent, the dissipated energy of the rock reflected the damage inside the rock.

In addition to the mechanical properties, studies of fracture development under dynamic and static loadings also yielded interesting phenomena. Liu et al. and Yan et al. [40,41] found that the length of the FPZ (fracture development zone) was significantly longer under dynamic loading than under static loading, and the length of the FPZ generally increased with an increase in strain rate, showing a clear rate dependence. Furthermore, the fracture inclination angle had a significant effect on the fracture development and damage mode of the rock mass under dynamic loading. Ai et al. [42] conducted impact tests on coal bodies containing vertical and horizontal joints at different impact velocities. The results showed that the joint orientation had a significant effect on the crack expansion path. Furthermore, numerical simulations demonstrated that the central crack development accelerates with the increase in impact velocity.

The existing studies have been carried out on the dynamic strength and fracture development of single fractured rock masses, but most rock masses contain many fractures in the natural environment. In addition, the rocks excavated under dynamic loads continue to bear the dynamic loads that were generated from the excavation blast at the front and affected by both dynamic and static loads (Figure 1). The realization of advanced optical instruments is instrumental in exploring many of the details in the break process, and is particularly promising for the research of damage and stability in deep mining. In these studies, the technique was used to collect the data during the test. In this paper, the bi-parallel fractured rock mass will be used for dynamic and static loading tests. The damage process and the variation of the strain field of a bi-parallel fractured rock under different dip angles and envelope pressure conditions are discussed.

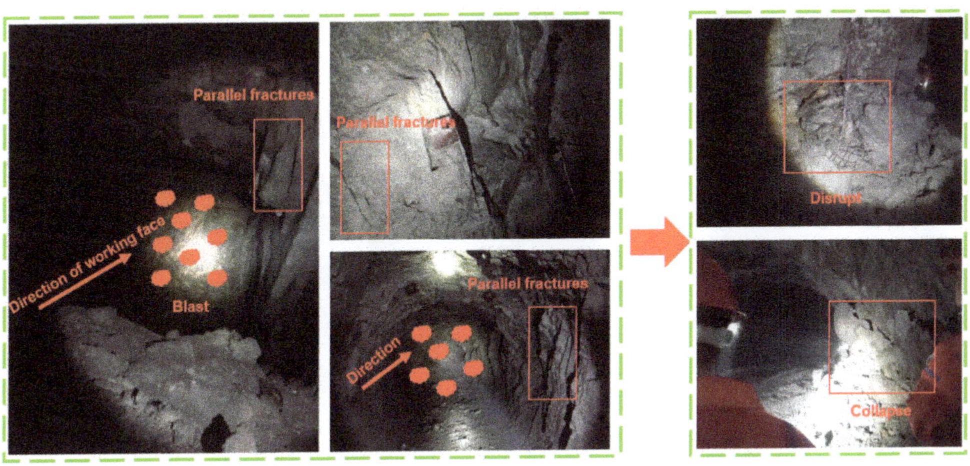

Figure 1. Hazards caused by dynamic and static coupling loads.

The structure of this paper is as follows: Section 2 presents the experimental methodology, such as the materials, equipment and the data processing methods. Section 3 demonstrates the experimental results, including the dynamic stress-strain curves of specimens with different angles, the influences of flaw angle and axial pressure on the dynamic and combined strengths, peak strain and modulus of elasticity. Section 4 discusses the influences of flaw angle and axial pressure on the damage process, especially on the evolution characteristics of the strain field at the crack end. Section 5 illustrates the strain field evolution laws of specimens under different axial pressures. Section 6 discusses the final damage pattern of granite specimens at the same fracture inclination, influenced by the magnitude of the applied static load. Section 7 concludes the whole study.

2. Experimental Methodology

2.1. Specimen Preparation

First, the rock samples were cut from the rock mass in the undisturbed part. In order to obtain a clearer observation of crack evolution, a single granite with good geometric integrity and uniformity was cut to obtain prismatic samples with a height, width and thickness of 45, 45 and 20 mm, respectively, which proved the feasibility and effectiveness of prismatic rock samples in an SHPB test. Then, the high-pressure water jet cutting machine was used to cut the rock specimens into length, width and thickness of 10 mm, 1 mm and 1 mm parallel cracks at different angles (0°, 45° and 90°). The geometry of the precut specimens is shown in Figure 2. After cutting the specimen rocks, the loading ends of all specimens were polished according to the standards of the International Society of Rock Mechanics (ISRM). The samples with different flaws were divided into 6 groups. The detailed geometric dimensions and mechanical parameters of each group are shown in Table 1. Among them, Int represents the complete sample, and S_0-flaw0°-1 represents the first sample of a 0° fracture sample at 0% UCS.

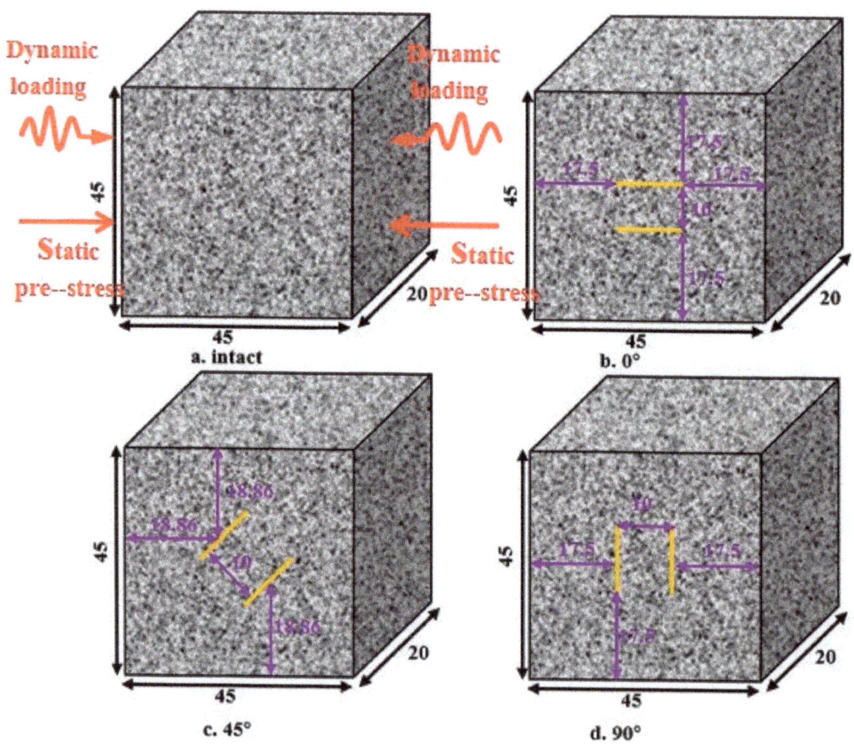

Figure 2. Granite specimen.

Table 1. Physical parameters and experimental results of granite specimens under combined static and dynamic loads.

Specimen Number	Length /mm	Height /mm	Thickness /mm	Fissure Angle/°	Axial Pressure/MPa	Dynamic Strength/MPa	Combined Strength/MPa	Peak Strain/10^{-3}	Strain Rate/s^{-1}
S_0-flaw0°-1	45.37	45.1	20.2	0	0% UCS	155.60	155.60	8.83	109.16
S_A-flaw0°-2	45.24	45.14	20.53	0	10% UCS	196.50	210.50	6.62	98.37
S_B-flaw0°-3	45.08	45.4	19.88	0	20% UCS	181.63	209.53	5.17	73.93
S_C-flaw0°-1	45.04	45.37	19.86	0	30% UCS	189.45	231.35	5.53	114.61
S_D-flaw0°-1	45.26	45.29	20.13	0	50% UCS	166.90	236.70	4.28	140.65
S_E-flaw0°-1	45.09	45.33	20.20	0	60% UCS	149.32	233.12	4.04	131.82
S_0-flaw45°-1	45.27	45.22	20.28	45	0% UCS	134.79	134.79	5.30	153.50
S_A-flaw45°-2	45.30	45.15	20.27	45	10% UCS	161.25	175.25	6.54	139.22
S_B-flaw45°-2	45.16	45.26	20.24	45	20% UCS	132.27	160.17	5.46	144.60
S_C-flaw45°-2	45.04	45.43	20.31	45	30% UCS	132.68	174.58	5.37	154.26
S_D-flaw45°-1	45.16	45.25	20.18	45	50% UCS	102.26	172.06	3.37	164.08
S_E-flaw45°-1	45.37	45.34	20.29	45	60% UCS	106.25	190.05	3.57	162.34
S_0-flaw90°-1	45.21	45.29	20.31	90	0% UCS	141.32	141.32	6.07	136.89
S_A-flaw90°-2	45.06	45.37	20.36	90	10% UCS	179.34	193.34	7.13	105.62
S_B-flaw90°-2	45.05	45.3	20.26	90	20% UCS	159.76	187.66	5.11	119.33
S_C-flaw90°-3	45.22	45.13	20.30	90	30% UCS	170.19	212.09	5.73	142.65
S_D-flaw90°-1	45.25	45.14	20.21	90	50% UCS	145.30	215.10	4.38	121.70
S_E-flaw90°-2	45.15	45.27	20.42	90	60% UCS	111.07	194.87	3.54	147.74

2.2. Experimental Apparatus and Testing Procedure

In this paper, the one-dimensional dynamic and static combined loading test is based on the SHPB device system of Central South University, Changsha, China, as shown in Figure 3. The test system includes a stress wave generation device, a stress transfer mechanism, axial static pressure loading parts, confining pressure (static pressure) loading parts and a data acquisition and processing device. During the test, the axial static pressure device is first used to apply the axial pressure required for the test to the sample, and then the stress wave generating device is started. The alloy bullet impacts the elastic rod to generate a sinusoidal-shaped loading stress wave, and then the stress wave propagates along the incident rod. Transmission and reflection waves occur on the contact surfaces of the rock sample and the elastic rod. The transmitted stress wave continues to pass forward to the transmission rod, and the reflected wave returns to the incident rod. The transient stress wave signal in the rod is collected by the strain gauge attached to the incident and transmission rods, then the signal is transmitted to the micro-computer system for processing, and then the various parameters of the sample are tested.

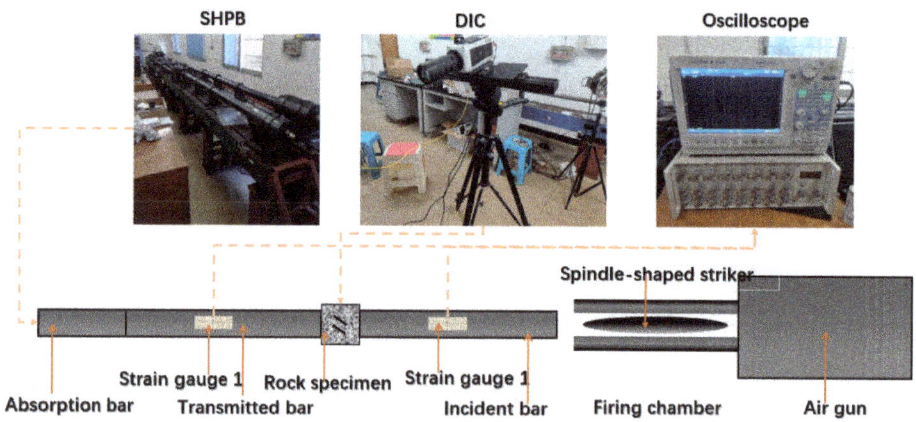

Figure 3. SHPB test device.

The digital image correlation (DIC) technique is a noncontact measurement that can obtain the full field displacement and strain distribution of the sample surface by matching the positions of the speckle before and after deformation in the region of interest (region of interest, ROI). With the help of a high-speed camera, this study aims to obtain the crack growth process and deformation field distribution at a higher resolution. The ROI region is composed of many subsets, which contain speckles of different quantities and sizes, making each subset unique. When the specimen surface was deformed, the subsets in the deformed and reference images were matched by the analysis of DIC, and finally a grid containing displacement and strain information relative to the reference image were obtained.

The displacement field is calculated by the following formula [43]:

$$\begin{cases} x_i' = x_i + u + \frac{\partial u}{\partial x}\Delta x + \frac{\partial u}{\partial y}\Delta y \\ y_i' = y_i + v + \frac{\partial v}{\partial x}\Delta x + \frac{\partial v}{\partial y}\Delta y \end{cases} \tag{1}$$

where u and v are displacement components of the center point along the x- and y- direction, respectively; x_i, y_i, x_i' and y_i' are the coordinate components of each point; $\frac{\partial u}{\partial x}$, $\frac{\partial u}{\partial y}$, $\frac{\partial v}{\partial x}$ and $\frac{\partial v}{\partial y}$ are the first-order displacement gradients of the reference subset.

In this study, as DIC technology was used to obtain the strain field on the rock sample surface, artificial speckle should be created on the sample observation surface in advance. First, a layer of white matte paint was uniformly sprayed on the test surface to form the background color, and the random speckle was created with black matte paint. A high-speed camera (V711, Phantom, Vision Research, Wayne, NJ, USA) was used to photograph the sample surface to analyze the changes in the strain field. In this experiment, the shooting resolution of the high-speed camera was 256 × 256, the shooting speed was 79,166 fps (i.e., a frequency of 153,000 frame/s) and the TTL signal was synchronously triggered by connecting with the hyperdynamic strain gauge. The shooting angle of the two cameras was not more than 60°, and the distance between the camera and the sample was approximately 1.5 m.

2.3. Data Processing Method

For fractured rock samples, the cracks in the sample will inevitably have a certain influence on the propagation of stress waves. Therefore, before the analysis of the results, it is necessary to clarify the effectiveness of the SHPB test in studying the dynamic mechanical properties of fractured rock samples. In the SHPB test, only when the stress state in the sample reaches the stress balance before the failure is the dynamic strength data valid. The SHPB test system needs to meet two basic assumptions: (1) One-dimensional stress wave hypothesis, the stress wave on the bar in the SHPB system is assumed to be a one-dimensional stress wave and the specimen is also in a one-dimensional loading state, ignoring the wave form dispersion effect. (2) In the SHPB test, the stress field and strain field inside the specimen should be ensured to be uniform, so as to avoid premature failure before the specimen reaches the peak stress, especially for the specimen with cracks.

Based on the one-dimensional stress wave theory, the stress (σ_s), strain (ε) and strain rate ($\dot{\varepsilon}_s$) histories of specimens can be derived using three waves [44]. According to the references [45] and based on the 1D wave theory, the average strain $\varepsilon_s(t)$, strain rate $\dot{\varepsilon}_s(t)$ and stress $\sigma_s(t)$ in the specimen can be derived as follows [46]:

$$\begin{cases} \varepsilon_s(t) = -\frac{2C_e}{L_s}\int_0^t \varepsilon_R(t)dt \\ \dot{\varepsilon}_s(t) = -\frac{2C_e}{L_s}\varepsilon_R(t) \\ \sigma_s(t) = \frac{E_e A_e}{A_s}\varepsilon_T(t) \end{cases} \quad (2)$$

where $\varepsilon_R(t)$, $\varepsilon_T(t)$, A_e, C_e, and E_e are the reflected strain, transmitted strain, cross-sectional area, P-wave velocity and Young's modulus of the elastic bar, respectively; A_s and L_s are the cross-sectional area and length of the specimen, respectively.

3. Results

3.1. Stress Equilibrium

For the accuracy of the test, the dynamic stress balance of the test must be resolved. It can be seen from Figure 4 that the dynamic loading force at both ends of the specimen is basically equal during the SHPB dynamic loading process. The curve variation of the superposition of the incident stress and the reflected stress (In + Re) is basically the same as that of the transmission stress (Tr), indicating that each test specimen has reached the dynamic stress balance during the loading process. Under the stress balance, because the difference in loading force between the two ends of the specimen is small, the inertial force caused by the inertial effect can be ignored, indicating that the specimen can achieve and maintain the dynamic stress balance condition during the dynamic loading process, thus verifying the validity of the test results.

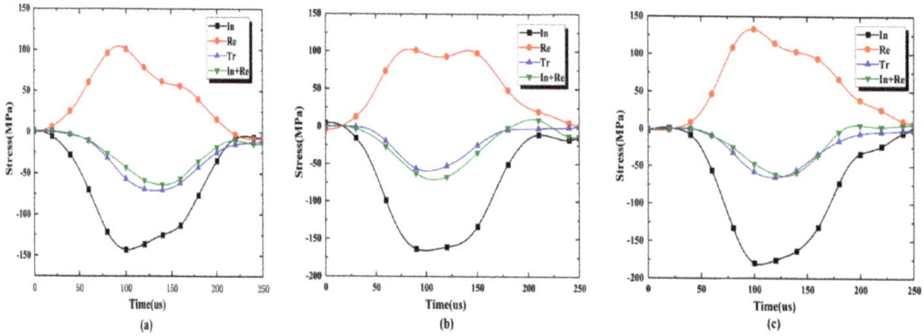

Figure 4. Dynamic stress equilibrium in specimens: (**a**) S_A-flaw0°-2, (**b**) S_B-flaw45°-1 and (**c**) S_C-flaw90°-2.

Figure 5 shows that the dynamic stress-strain curve changes from linear to nonlinear before the peak stress, which can be roughly divided into a micro-crack compaction stage, an elastic deformation stage and a post-peak failure stage. The unloading stage of impact stress after peak stress is the stage where dynamic strain increases to the end of impact stress unloading. The reason is that the rock is in the elastic deformation process when the impact disturbance occurs after the internal micro-cracks of the rock are compacted, resulting in the linear growth of the dynamic stress-strain curve. With the increase in impact stress, the micro-cracks in the rock germinate, expand and penetrate and plastic deformation occurs. Especially when the stress reaches the yield stress of the rock, the dynamic stress-strain curve enters the nonlinear stage. The reason why dynamic strain increases at the end of unloading is that in the process of disturbance impact, local damage occurs inside the rock, and the elastic energy stored inside is released more, which is not enough to resist the compressive strength caused by impact stress.

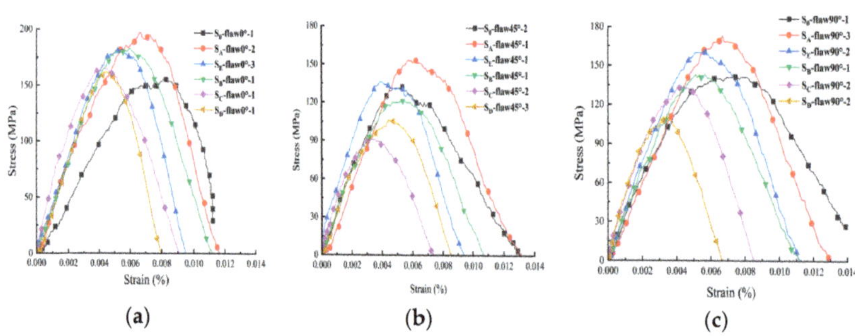

Figure 5. Dynamic stress-strain relationship under different axial static pressures for: (**a**) specimens with 0° flaw; (**b**) specimens with 45° flaw; (**c**) specimens with 90° flaw.

3.2. Strength and Deformation Properties

As shown in Figures 6 and 7, with the increase in the applied static axial pressure, the dynamic strength showed an increasing trend and then a decreasing trend, reaching its maximum at an axial pressure of 14 MPa (10% UCS). The combined strength shows an overall increase at 0–14 MPa (0–10% UCS) axial pressure. When the axial pressure is from 14 MPa to 69.8 MPa (10–50% UCS), the strength exhibits a slight increase at 27.9 MPa (20% UCS) axial pressure. Afterward, the combined strength increases slowly in steps and decreases when the axial pressure is from 69.8 MPa to 83.8 MPa (50–60% UCS). Under a certain axial compression, the dynamic strength and combined strength of rock samples first decrease and then increase with the increase in fracture angle, which indicates that the changes in the parallel double fissure inclination angle can weaken the dynamic strength and combined strength, consistent with the results of numerous studies.

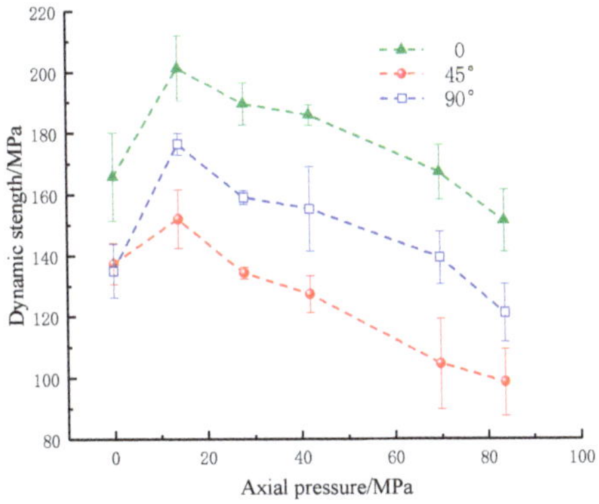

Figure 6. Dynamic strength of specimens with an artificial flaw under different axial prestresses.

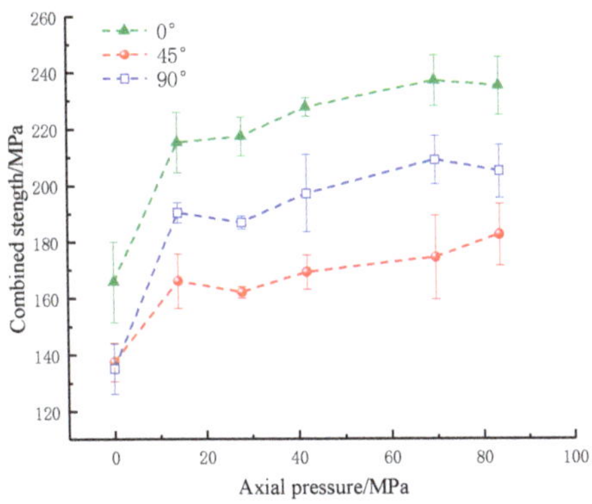

Figure 7. Combined strength of specimens with an artificial flaw under different axial prestresses.

The analysis showed that the change in the inclination angle influenced the distribution of the internal stress field of the rock during the impact of the specimen. Macro-scopically, this change manifested as a weakening of dynamic and combined strengths. For larger fracture inclination angles, the strength reduction was significant. According to the theory of fracture mechanics, the stress level required for a prefabricated fracture inclination angle of $(90° - \varphi)/2$ is minimal for the same effective shear force, thus explaining the phenomenon of the minimum strength of the 45° fracture specimens. To a certain extent, the axial pressure induces the closure of micro-fractures within the rock, increasing the dynamic and combined strengths. When the axial pressure gradually increases, the excessive static pressure causes the sprouting and expansion of the internal micro-fractures, decreasing the residual bearing capacity under dynamic loading. However, the increase in axial pressure causes the combined loading of the rock to show intensified characteristics. When the applied static pressure is too large and intensifies the microfracture expansion within the rock, the decrease rate of dynamic strength increases, showing a decrease in the combined strength.

The strain and modulus of elasticity are important indicators of the deformation characteristics of the rocks. The relationship between peak strain, dynamic modulus of elasticity and axial pressure is shown in Figures 8 and 9. It can be seen that the weakening of the peak strain and dynamic elastic modulus by the inclination angle is similar to that of the strength, both of them first decrease and then increase with the increases in the crack angle. The peak strain shows an overall increasing trend and then a decreasing trend as the applied axial static pressure increases. The reason for its stepwise decrease is that the increase in the axial pressure reduces the remaining deformation of the rock. At the axial pressure of 27.9 MPa (20% UCS), the dynamic modulus of elasticity at different fracture inclinations of the precast specimens shows an increase with axial static pressure. At axial pressures of 0~69.8 MPa (0~50% UCS), the overall trend increases step by step, with inflection points at 27.9 MPa (20% UCS) and 41.9 MPa (30% UCS) and a decreasing trend at 69.8–83.8 MPa (50–60% UCS). The reason for the rise in the dynamic modulus of elasticity is similar to that of the combined strength and is related to the compression-density effect of the axial pressure. The decrease at high axial pressures (50–60% UCS), on the other hand, is due to the larger loads making the rock enter an earlier state of damage, weakening its effective load-bearing capacity in the dynamic impact loading and leading to a decrease in the dynamic modulus of elasticity. In this process, the above-mentioned information indices produced a significant turnaround in the trend at axial pressures of 27.9 MPa (20% UCS) and 41.9 MPa (30% UCS). Therefore, the measurement indices of the parallel double fracture specimens at 20–30% static axial pressures should be noted.

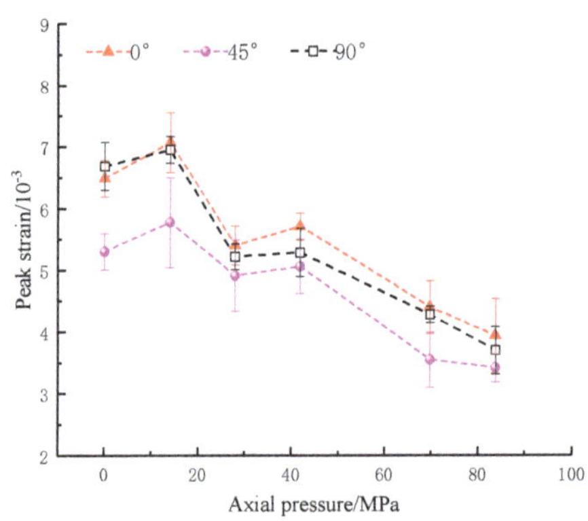

Figure 8. Peak strain of specimens with an artificial flaw under different axial prestresses.

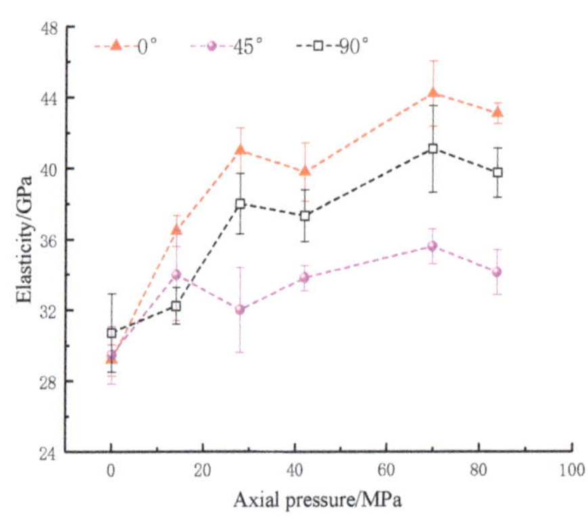

Figure 9. Elasticity of specimens with an artificial flaw under different axial prestresses.

4. Damage Process

Due to the different types and sizes of subsurface rockwork disturbances, it is necessary to find the influence of dynamic impact loading on the mechanical properties of rocks. In damage characteristics, there is also a lack of findings on dynamic monitoring of crack extension and full-field strain measurements, especially on the evolution characteristics of the strain field at the crack end. Therefore, the damage process was analyzed in detail in this study. Due to space limitations, only the test results for 10%, 30% and 60% axial pressure were analyzed.

4.1. The Damage Process of the Specimen with Different Parallel Bi-Flaws at the 10% Axial Pressure

Figures 10–15 show the stress-time curve and damage process of the specimens with 0°, 45°, 90° parallel bi-flaws at the 10% axial pressure, respectively.

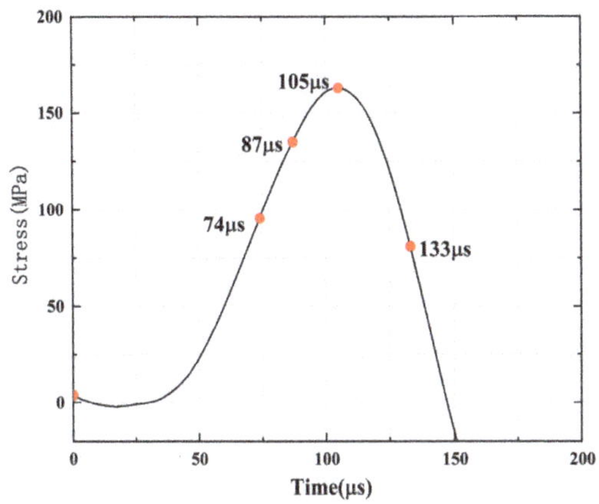

Figure 10. Stress-time curve of a typical specimen with a 0° flaw (S_A-flaw0°-1).

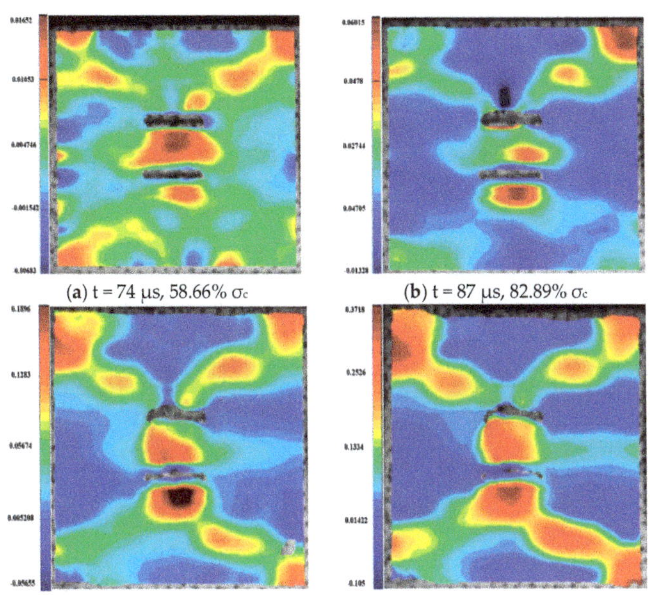

(**a**) t = 74 μs, 58.66% σ_c (**b**) t = 87 μs, 82.89% σ_c

(**c**) t = 105 μs, 100% σ_c (**d**) t = 133 μs, 49.65% σ_c

Figure 11. Typical damage process of a rock specimen with a 0° flaw (S_A-flaw0°-1).

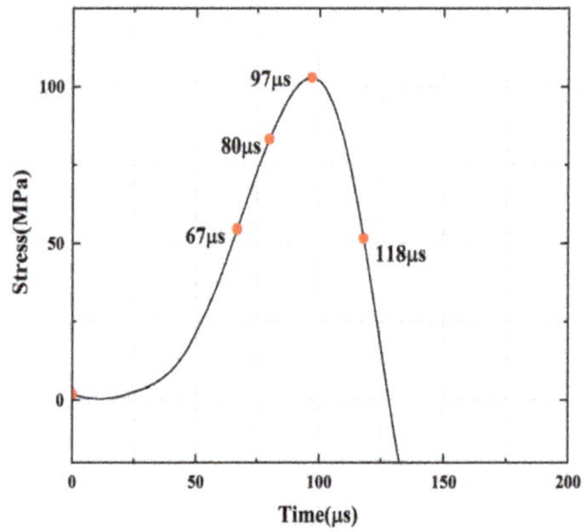

Figure 12. Stress-time curve of a typical specimen with a 45° flaw (S_A-flaw45°-3).

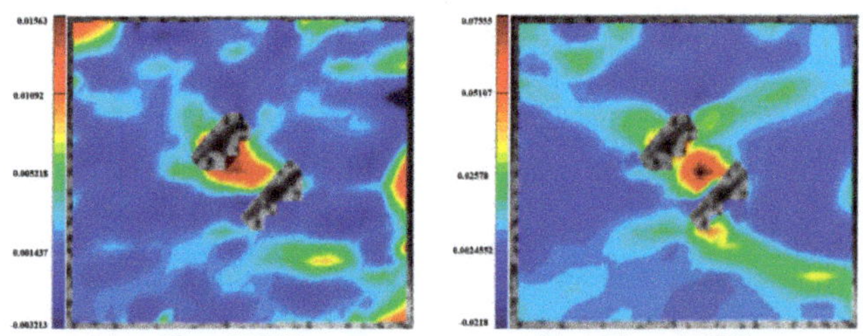

(**a**) t = 67 μs, 53.01% σ_c (**b**) t = 80 μs, 80.79% σ_c

(**c**) t = 97 μs, 100% σ_c (**d**) t = 118 μs, 50.10% σ_c

Figure 13. Typical damage process of a rock specimen with a 45° flaw (S_A-flaw45°-3).

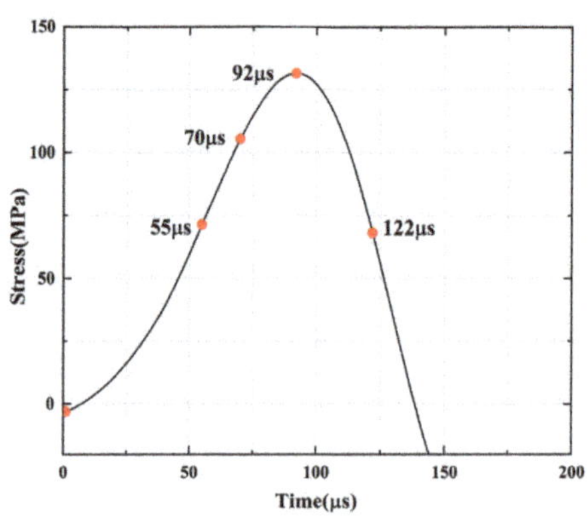

Figure 14. Stress-time curve of a typical specimen with 90° flaw (S_A-flaw90°-2).

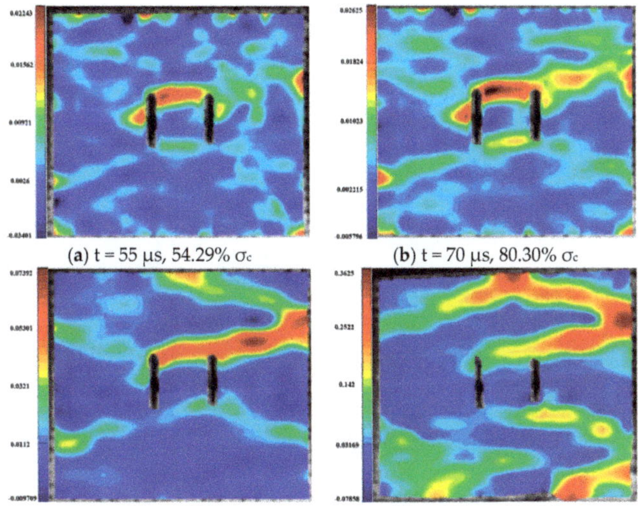

(**a**) t = 55 μs, 54.29% σc (**b**) t = 70 μs, 80.30% σc (**c**) t = 92 μs, 100% σc (**d**) t = 122 μs, 51.71% σc

Figure 15. Typical damage process of a rock specimen with a 90° flaw (S_A-flaw90°-2).

For the rock specimens with 0° bi-flaws, the stress field and crack propagation mode gradually change under the loading of a static pressure of 10% UCS. We selected four typical stress fields in the loading process and analyzed their changes, as shown in Figure 10. Figure 11a shows that large TSZs (tensile strain zones) have been developed in the middle, upper and lower parts of the parallel bi-flaws, in the upper left and upper right corners, and in the lower left corner. In Figure 11b, the upper part penetrated the upper flaw, forming two oblique SSZ (shear strain zones). Figure 11c reveals that the upper and lower parts of the flaws have been partially destroyed, and the TSZ above and below the flaws have developed into an "X"-shaped shear strain zone centered on the flaws. In addition, the SSZ in Figure 11d penetrated each other, and the central TSZ formed a TS (tensile strain) crack.

For the rock specimens with 45° bi-flaws, the stress field and crack propagation mode gradually change under the loading of a static pressure of 10% UCS. We selected four typical stress fields in the loading process and analyzed their changes, as shown in Figure 12. During loading, a large tensile strain deformation occurred in the bridge region of the parallel bi-flaws (Figure 13a). At the same time, TSZ also appeared in the upper left and right ends and also in the lower part of the specimen. At the end of the flaws in Figure 13b, both inverse and along the SSZ fissure direction appear, and the TS in the center continues to increase. In Figure 13c, the SSZ extended to the upper diagonal of the specimen, while the strain zone extending to the upper diagonal developed horizontally in Figure 13d, forming a TSZ.

For the rock specimens with 90° bi-flaws, the stress field and crack propagation mode gradually changed under the loading of a static pressure of 10% UCS. We selected four typical stress fields in the loading process and analyzed their changes, as shown in Figure 14. In Figure 15a, TSZ is generated at the upper end of the parallel bi-flaws during loading. In Figure 15b, the TS (tensile strain) continues to increase in the TSZ. In Figure 15c, the TSZ has extended to the right-hand boundary, and Figure 15d shows that the strain zone extends to the middle of the top of the specimen and develops horizontally, forming a TS crack and causing damage to the specimen.

4.2. The Damage Process of the Specimen with Different Parallel Bi-Flaws at the 30% Axial Pressure

Figures 16–22 show the stress-time curve and damage process of the specimens with 0°, 45° and 90° parallel bi-flaws at the 30% axial pressure, respectively.

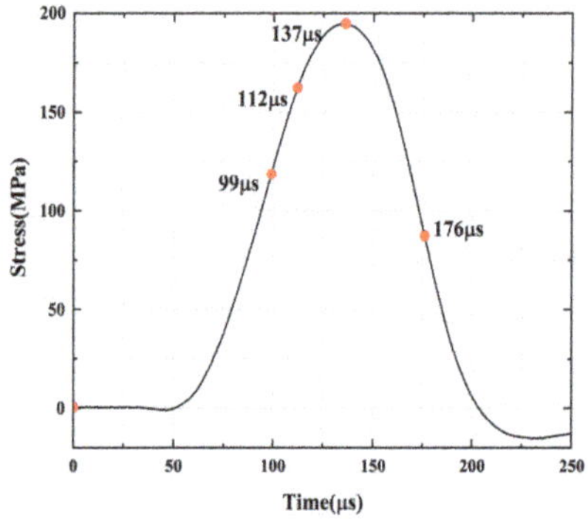

Figure 16. Stress-time curve of a typical specimen with a 0° flaw (S_B-flaw0°-2).

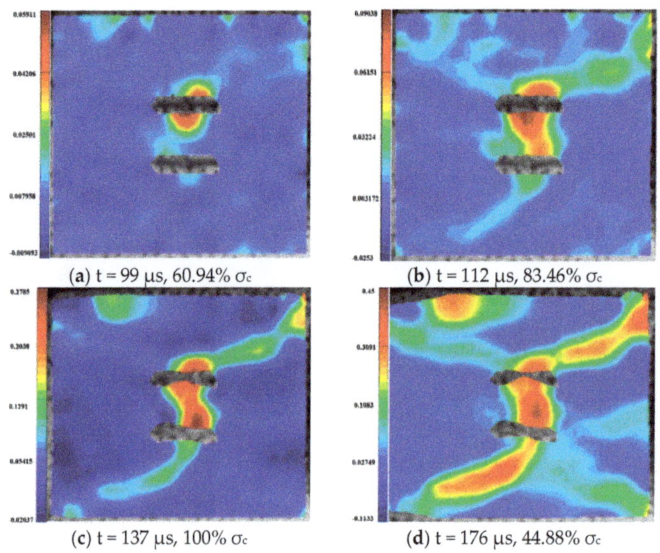

Figure 17. Typical damage process of a rock specimen with a 0° flaw (S_B-flaw0°-2).

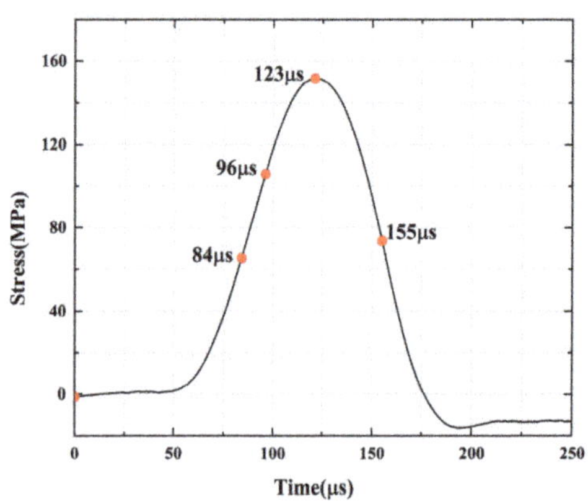

Figure 18. Stress-time curve of a typical specimen with a 45° flaw (S_B-flaw45°-3).

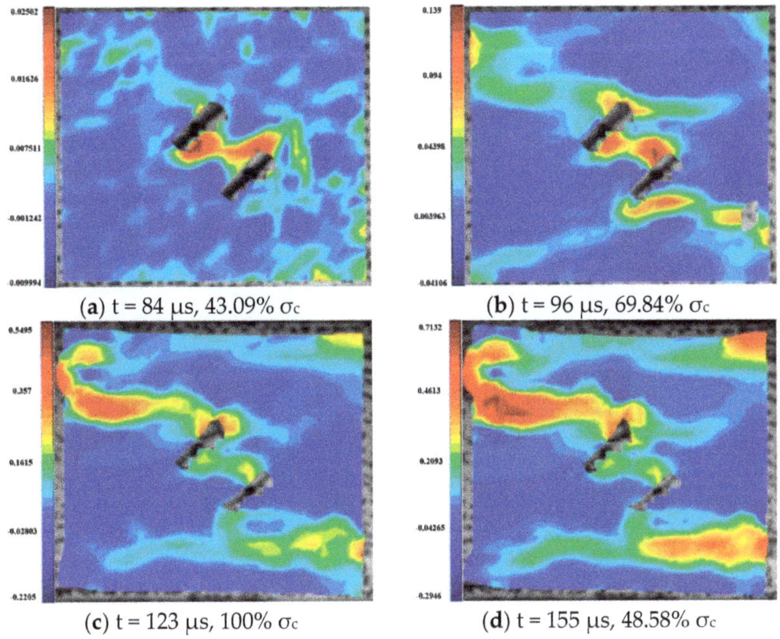

Figure 19. Typical damage process of a rock specimen with a 45° flaw (S_B-flaw45°-2).

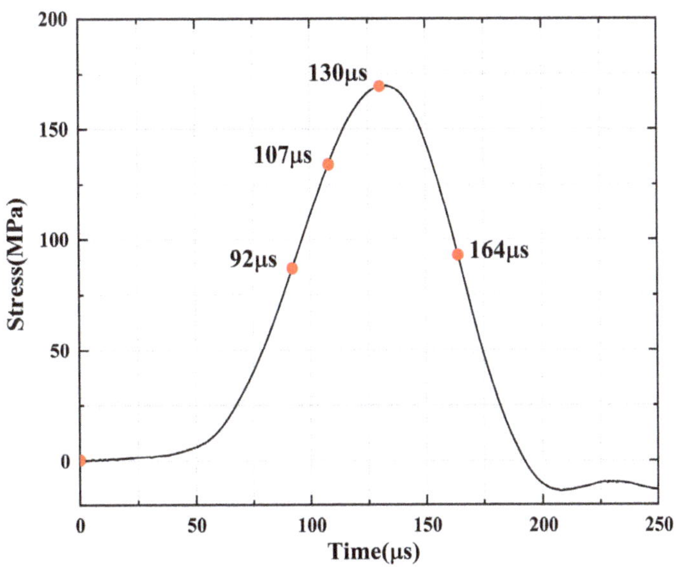

Figure 20. Stress-time curve of a typical specimen with a 90° flaw (S_B-flaw90°-3).

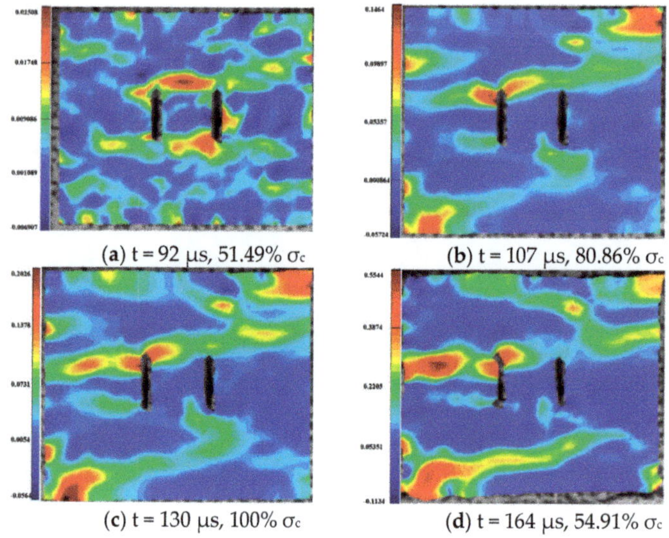

Figure 21. Typical damage process of a rock specimen with a 90° flaw (S_B-flaw90°-2).

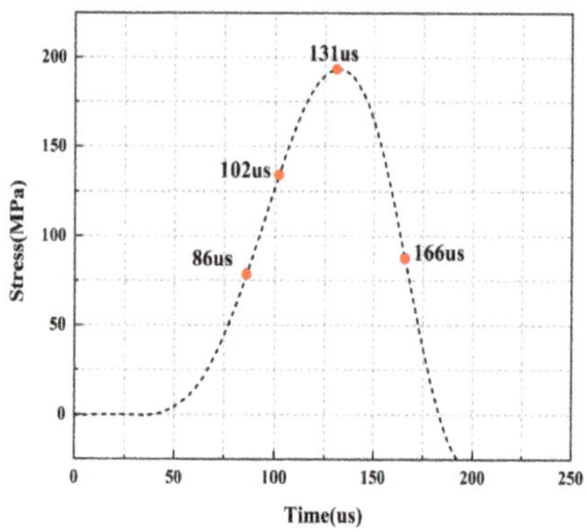

Figure 22. Stress-time curve of a typical specimen with a 0° flaw (S_E-flaw0°-2).

For the rock specimens with 0° bi-flaws, the stress field and crack propagation mode gradually change under the loading of static pressure of 30% UCS. We selected four typical stress fields in the loading process and analyzed their changes, as shown in Figure 16. Figure 17b reveals that the upper flaw penetrates the upper right corner to form a shear strain crack, generating a larger TSZ in the middle. The lower part of the lower flaw begins to develop a lower strain, and partial shear damage develops, as shown in Figure 17c. In Figure 17d, two shear cracks extend to the diagonal, connecting the ends of the flaws, which eventually cause damage to the rock.

For the rock specimens with 45° bi-flaws, the stress field and crack propagation mode gradually change under the loading of static pressure of 30% UCS. We selected four typical stress fields in the loading process and analyzed their changes, as shown in Figure 18. As shown in Figure 19a, a larger TSZ appears on the rock bridge with a smaller inverse TSZ at the outer end of the flaw. In Figure 19b, the inverse TSZ develops and extends diagonally towards the upper left and lower right, along with subshear strain cracks towards the flaw direction. Figure 19c shows that the diagonal shear damage pattern developed through the parallel bi-flaws. Eventually, the specimen was damaged by a parallel TSZ developing at the upper end, as shown in Figure 19d.

For the rock specimens with 90° bi-flaws, the stress field and crack propagation mode gradually change under the loading of static pressure of 30% UCS. We selected four typical stress fields in the loading process and analyzed their changes, as shown in Figure 20. In Figure 21a, two TSZ were formed at the upper and lower ends of the two flaws and developed toward the gauge. Figure 21b shows that the TSZ in the upper part expands diagonally to form the SSZ. Furthermore, the strain zone at the lower end developed in the lower left corner to form the SSZ. In Figure 21c, a primary shear strain formed in the lower part, and a shear crack developed in the upper SSZ next to the upper left corner. Finally, the damage was caused by two shear cracks and a tensile crack, as shown in Figure 21d.

4.3. *The Damage Process of the Specimen with Different Parallel Bi-Flaws at the 60% Axial Pressure*

Figures 23–26 show the stress-time curve and damage process of the specimens with 0°, 45° and 90° parallel bi-flaws at the 60% axial pressure, respectively.

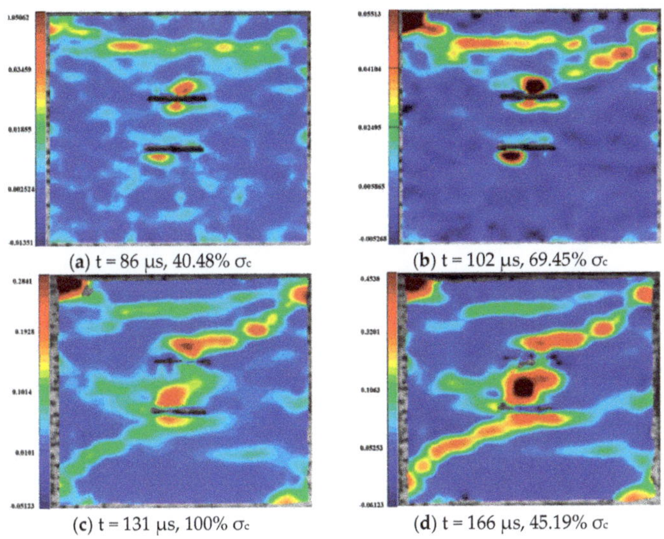

Figure 23. Typical damage process of a rock specimen with a 0° flaw (S_E-flaw0°-2).

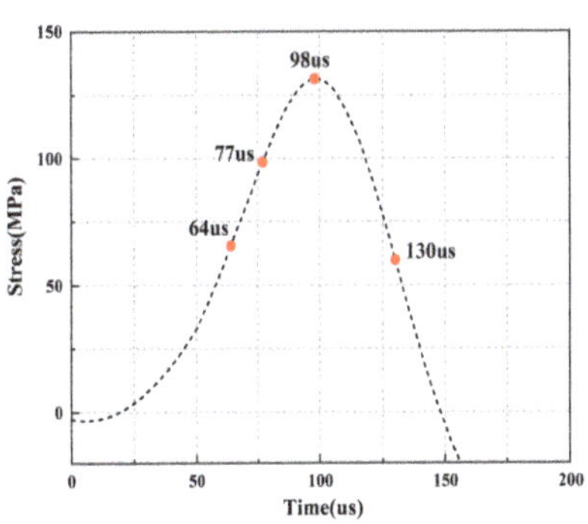

Figure 24. Stress-time curve of a typical specimen with a 45° flaw (S_E-flaw45°-3).

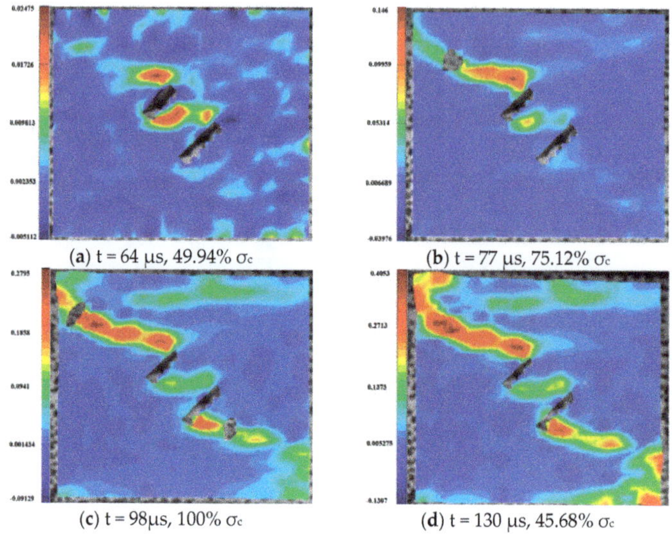

(**a**) t = 64 μs, 49.94% σ_c (**b**) t = 77 μs, 75.12% σ_c

(**c**) t = 98μs, 100% σ_c (**d**) t = 130 μs, 45.68% σ_c

Figure 25. Typical damage process of a rock specimen with a 45° flaw (S_E-flaw45°-3).

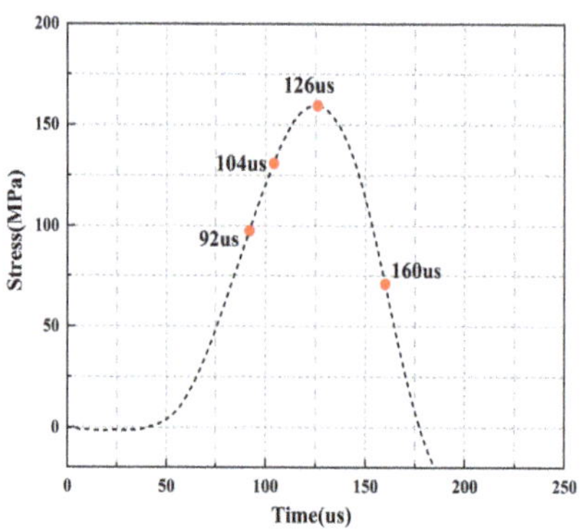

Figure 26. Stress-time curve of a typical specimen with a 90° flaw (S_E-flaw90°-1).

For the rock specimens with 0° bi-flaws, the stress field and crack propagation mode gradually change under the loading of a static pressure of 60% UCS. We selected four typical stress fields in the loading process and analyzed their changes, as shown in Figure 22. Figure 23a shows that a TSZ first develops in the upper part, then develops in both the upper and lower parts. As shown in Figure 23b, the upper TSZ further develops into a SSZ towards the upper left corner. In Figure 23c, the TSZ in the lower flaw and center of the bi-flaws also starts to develop, and an SSZ develops toward the lower left corner. Afterward, the SSZ in Figure 23d penetrates both ends and is accompanied by the final damage caused by the newly developed shear cracks at the upper and lower ends.

For the rock specimens with 45° bi-flaws, the stress field and crack propagation mode gradually change under the loading of a static pressure of 60% UCS. We selected four typical stress fields in the loading process and analyzed their changes, as shown in Figure 24. From Figure 25a, it can be seen that a large TSZ is formed in the bridge area of the parallel bi-flaws. At the upper end of the upper flaw, a TSZ is formed in the anticlinal direction. Figure 25b shows that the anticlinal strain zone continues to develop along the diagonal, forming a smaller strain zone in the upper flaw along the fracture direction and in the lower part of the lower flaw. In Figure 25c, the end of the flaw is destroyed, and the shear strain propagates through the specimen along the loading direction. The main shear crack and a horizontal TSZ in the upper part are formed, which eventually lead to the shear damage, as shown in Figure 25d.

For the rock specimens with 90° bi-flaws, the stress field and crack propagation mode gradually change under the loading of a static pressure of 60% UCS. We selected four typical stress fields in the loading process and analyzed their changes, as shown in Figure 26. From Figure 27a, it can be seen that TSZ is developed at the upper left and lower ends of the bi-flaws, and the upper left corner of the rock. Figure 27b shows that the TSZ at the upper left fissure end expands diagonally to form an SSZ, while the TSZ at the lower end of the bi-flaws continues to develop on both sides. In Figure 27c, the TSZ at the lower part of the specimen is largely formed. In Figure 27d, the TSZ in the upper left corner also extends horizontally into a tensile crack.

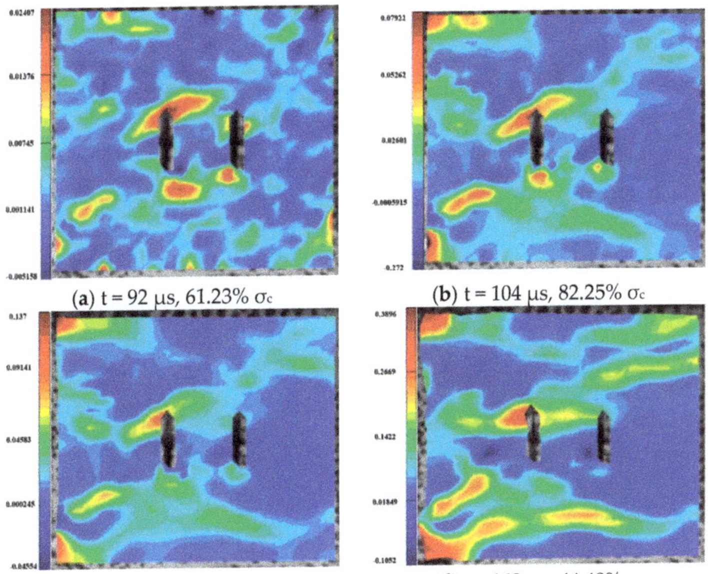

(**a**) t = 92 μs, 61.23% σ_c (**b**) t = 104 μs, 82.25% σ_c (**c**) t = 126 μs, 100% σ_c (**d**) t = 160 μs, 44.48% σ_c

Figure 27. Typical damage process of a rock specimen with a 90° flaw (S_E-flaw90°-1).

5. Strain Field Evolution Laws

5.1. Effects of Axial Pressure on the Strain Field Evolution Laws of Specimens with 0° Flaw

The fracture of rock contains the process of crack initiation, propagation and coalescence. The damage evolution process is very complex and has very important engineering significance. Through DIC technology, the stress field variation characteristics of fractured rock after failure under dynamic and combined loadings were analyzed [46,47]. The maximum principal strains of 0° specimens under different axial pressures are given in Figure 28, where representative strains from each group were selected for analysis. As can be seen from the figures, the damage patterns of the 0° inclined specimens under different axial pressures are still very different. Under a combined, dynamic and static loading of 0% UCS, the granite with parallel double fissures breaks with three TSZ running through the two flaws of the specimen. The uppermost TSZ shows a significant increase in tensile strain values, eventually presenting significant tensile damage.

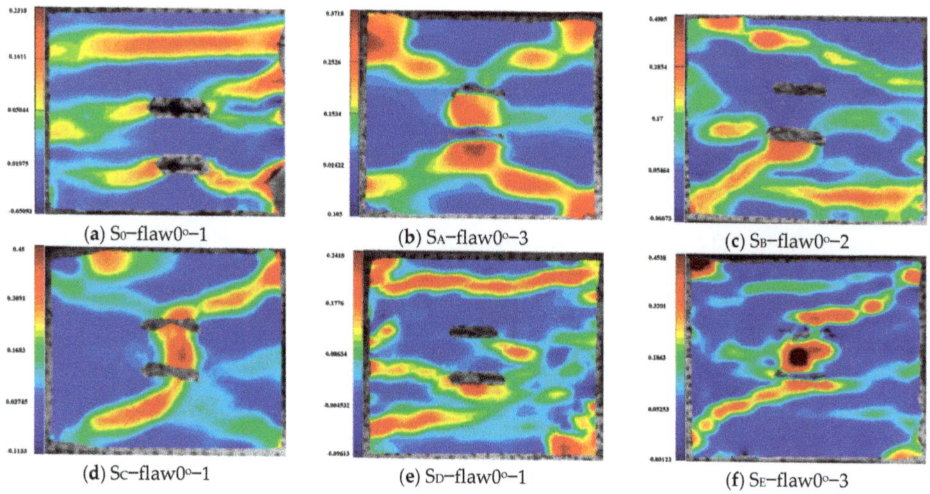

Figure 28. Maximum principal strains during the damage of granite specimens containing 0° parallel double fissures under different combinations of dynamic and static loading.

Under a combined dynamic and static loading of 10% UCS, the damage pattern of the granite with parallel double fissures changes to two diagonal TSZ through the upper and lower flaws. A main tensile crack through one side is produced between the parallel bi-flaws, and a larger TSZ is produced in the middle of the bi-flaws. Under a combined dynamic and static loading of 20% UCS, a diagonal SSZ through the specimen appears in the upper part of the granite containing parallel double fissures. The lower part of the fissure shows a composite strain zone with shear and tensile strains interspersed.

The granite with parallel bi-flaws shows an obvious "X"-shaped SSZ developing in the center of the fissure under a combined dynamic and static loading of 30% UCS. Under a combined dynamic and static loading of 50% UCS, the upper part of the granite with parallel double fissures has a TSZ through the fissures in the parallel loading direction. In the lower part, there is a composite tensile-shear strain zone through the flaws and a large local TSZ at the lower right end. Under a combined, dynamic and static loading of 60% UCS, the damage to granite with parallel double fissures is caused by a primary SSZ through the lower fissure and a secondary SSZ connecting the upper fissure to the upper right end. A large TSZ is located in the middle of the parallel fissure and at the upper left end of the specimen, with the middle TSZ penetrating the left end and two TSZ at each of the top and bottom of the specimen.

5.2. *Effects of Axial Pressure on the Strain Field Evolution Laws of Specimens with a 45° Flaw*

Figure 29 shows the maximum principal strains of the 45° inclined angle specimens with parallel double fissures damaged. The specimens with a 45° all develop an inverse SSZ during the damage of the dynamic and static loading. At the end of the parallel double fissure, the penetration forms a TSZ parallel to the direction of stress loading. Under the static and dynamic combined load of 0% UCS, the granite containing parallel double fissures produces two parallel TSZs with significant tensile strain values along the upper left and lower right corners. These zones penetrate both ends of the specimen and intersect with the diagonal inverse SSZ, causing damage to the specimen. Under the combined, dynamic and static loading of 10% UCS, the granite containing parallel double fissures produces two parallel TSZ from the upper left to the lower right. These zones penetrate both ends of the specimen and intersect with the diagonal inverse SSZ, causing damage to the specimen.

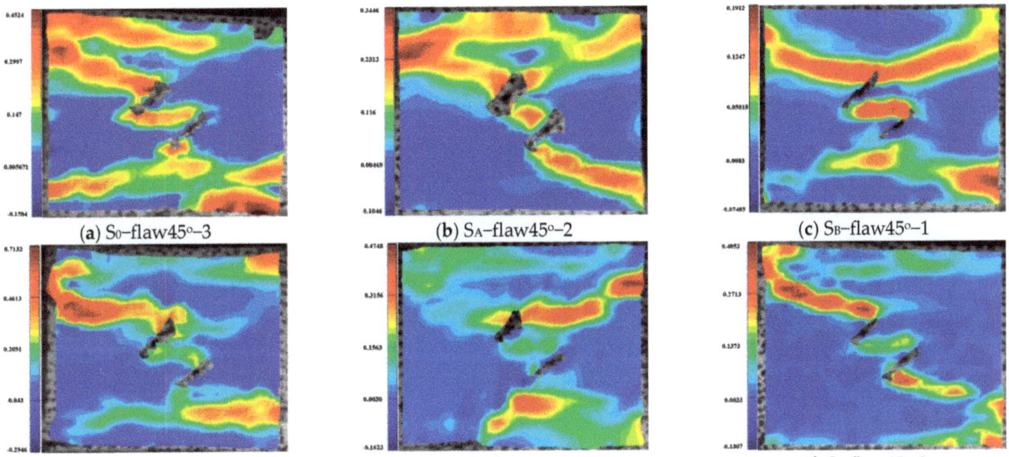

(**a**) S_0–flaw45°–3 (**b**) S_A–flaw45°–2 (**c**) S_B–flaw45°–1

(**d**) S_C-flaw45°-2 (**e**) S_D-flaw45°-3 (**f**) S_E-flaw45°-3

Figure 29. Maximum principal strain clouds during the damage of granite specimens containing 45° parallel double fissures under different combinations of dynamic and static loading.

In addition to the diagonal SSZ, a horizontal TSZ is generated at the upper end of the granite, which passes through the specimen in the parallel loading direction, causing large deformation damage at the intersection. Under a combined, dynamic and static loading of 20% UCS, a horizontal TSZ is generated in the upper and lower parts. The TSZ is formed at the inner end towards the horizontal bi-flaws due to the penetration, in which an inverse SSZ may also exist. In addition to the inverse SSZ through the diagonal of the specimen, a smaller SSZ along the fissure direction intersects the inverse SSZ, which extends to the end of the specimen and causes damage.

In the case of 30% UCS, the SSZ is also smaller in the direction of the fissure and extends to the end of the specimen. In the upper part of the flaw end, a smaller SSZ in the flaw direction is formed, while a TSZ appears in the lower part of the flaw end. At the same time, there is also a TSZ in the upper part of the granite, causing damage. In addition to them, under the combined dynamic-static loading of 50% UCS, one TSZ appeared above and below the crack tip and gradually expanded in the loading direction. Under a combined, dynamic and static loading of 60% UCS, granite damage in parallel double specimens is produced by the typical SSZ in the diagonal direction. At the end of the contact between the specimen and the bar, there is an enhanced strain signal and a horizontal TSZ in the upper part of the granite.

5.3. Effects of Axial Pressure on the Strain Field Evolution Laws of Specimens with a 90° Flaw

Figure 30 shows the maximum principal strains at the damage site of the 90° inclined angle specimen with parallel double fissures under different axial pressures. The two primary TSZ and one secondary TSZ were formed in the granite containing parallel bi-flaws under a combined, static and dynamic loading of 0% UCS. One of the primary TSZs appears through the upper end of the parallel bi-flaws, and the extension of the three primary strain zones damages the granite. Under a combined static and dynamic loading of 10% UCS, the upper part generates a tensile crack through the upper end of the flaws, which is related to a tensile crack in the upper part of the granite. The tensile crack at the upper end of the parallel double fissure penetrates the TSZ at the right end, while a penetrating SSZ is generated at the lower end of the parallel double fissure.

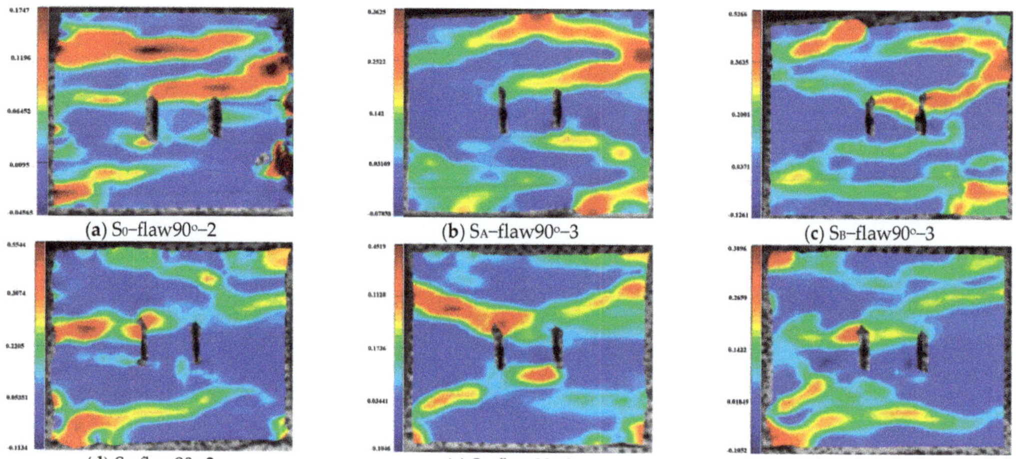

(**a**) S_O–flaw90°–2 (**b**) S_A–flaw90°–3 (**c**) S_B–flaw90°–3

(**d**) S_C-flaw90°-2 (**e**) S_D-flaw90°-2 (**f**) S_E-flaw90°-2

Figure 30. Maximum principal strain clouds during the damage of granite specimens containing 90° parallel double fissures under different combinations of dynamic and static loading.

Two TSZ were generated at the lower part of the granite under a combined static and dynamic loading of 20% UCS. One is approximately parallel to the loading direction through both ends of the granite, and the other develops through the end of the parallel double fissure. The upper part of the specimen shows a significant diagonal development of SSZ, which also develops through the end of the parallel double fissure. Under the combined loading of 30% UCS, two SSZ were generated in the upper left and lower left parts of the specimen. Both zones develop similarly to the "X"-type SSZ, which finally causes damage to the specimen. Additionally, under the combined static and dynamic loading of 50% UCS, two SSZ are generated through the upper and lower ends of the parallel double fissures. In the lower part of the specimen, a TSZ is generated under the combined static and dynamic loading of 60% UCS. At the same time, an SSZ develops diagonally from the ends of the granite towards the end of the flaw in the upper part.

5.4. Fractal Analysis

According to the mass fractal model of rock fragment distribution of rock fragments established by Mandelbrot and other scholars, the fractal dimension D can be obtained according to the mass-frequency relationship of the screening test. The distribution equation of rock fragments under impact load is:

$$M(x)/M_T = (x/x_m)^{3-D} \tag{3}$$

In the formula, $M(x)$ and M_T represent the total mass of the fragments and the cumulative mass under the sieve, respectively; x and x_m represents the particle size and maximum particle size of the fragment; D is the fractal dimension of fragment distribution.

By taking the logarithm of both sides of the above formula at the same time, we can obtain:

$$\lg[M(x)/M_T] = (3-D)\lg(x/x_m) \tag{4}$$

The fractural dimension of rock specimens could directly and quantitatively reflect the degree of rock fracture [48]. The larger the dimension of the specimen, the more fragments; the smaller the volume, and the higher the degree of fragmentation. In order to analyze the fractural characteristics, the pieces were collected for sieving after the impact was completed (Figure 31). In order to quantitatively analyze the failure state under different conditions, the fractural dimensions were calculated, and the relationship between the fractural dimension and the strain rate, the axial pressure and the fissure angle were discussed. It can be seen from Figure 32a that the fractural dimension increased with the increase in the strain rate, which showed a good linear correlation. In the test, the granite strain rate increased from 73.93 s^{-1} to 164.08 s^{-1}, and the fractural dimension increased from 1.5963 to 2.8014. As shown in Figure 32b, the fractural dimension is closely related to the fissure angle. The fractural dimension showed the change law of decreasing and then increasing. When the angle is 90°, the dimension is the smallest, and at 45°, it reaches the maximum. Figure 32c showed that the fractural dimension first decreased and then increased with the increase in axial pressure, but it is slightly higher when the axial pressure is 0% UCS, which is opposite to the dynamic strength, which illustrates that in engineering practices, bridge angle and fractural dimension affect the rock fragmentation degree. Therefore, appropriately increasing the angle of the excavation fissure and reasonably utilizing the natural in situ stress will help to improve the efficiency of rock excavation and rock breakage to some extent.

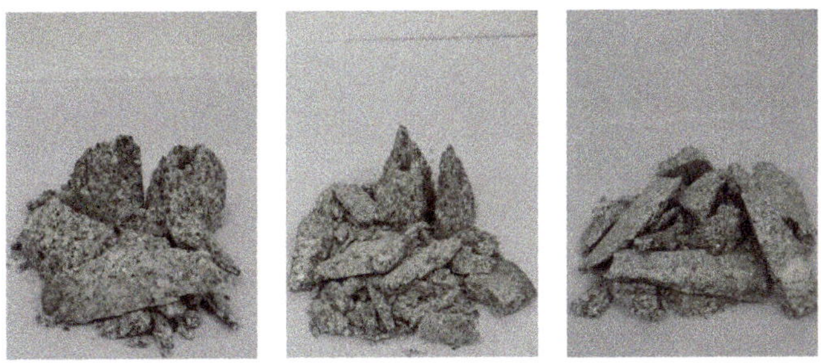

Figure 31. Fracture morphology of disturbed samples after impact failure.

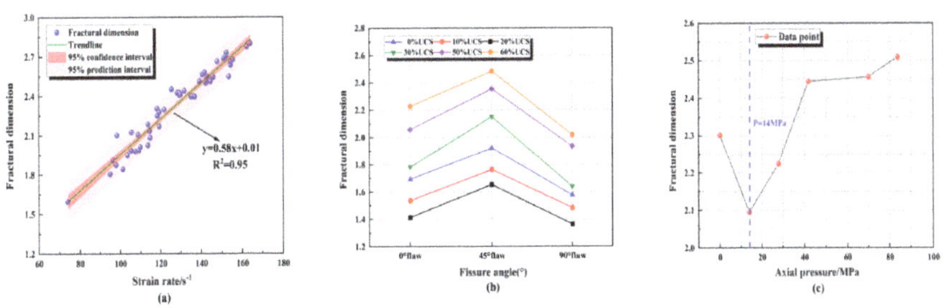

Figure 32. The correlation between fractural dimension and (**a**) strain rate, (**b**) fissure angle and (**c**) axial pressure.

6. Discussion

The final damage pattern of granite specimens at the same fracture inclination is influenced by the magnitude of the applied static load: (1) zero degree parallel bi-flaws. The granite with parallel double fissures at the combined dynamic and static loading of 0° produces three tensile strain cracks parallel to the loading direction, which eventually leads to tensile damage of specimens. In contrast, a distinct SSZ appears with combined dynamic and static loading, with typical tensile-shear damage at the combined dynamic and static loadings of 20% and 50% under uniaxial compression, and shear damage at the combined dynamic and static loadings of 10%, 30% and 60%. Significantly, the shear damage pattern is dominant at the combined dynamic and static loadings of 10%, 30% and 60% under uniaxial compression. (2) forty-five degree parallel bi-flaws. At a loading of 0–60%, the change in axial pressure has little effect on the damage mode of the specimens with 45° cracks, which basically form an inverse SSZ along the end of the LAW.

With the increase in axial pressure, this shear mode becomes increasingly significant, and there is a tendency to transition from shear-tension damage mode to shear damage mode. (3) ninety-degree parallel bi-flaws. The shear strain zone and the length and number of shear cracks increase with increasing axial pressure, and there is a tendency to transition from tensile to shear-tensile damage modes.

7. Conclusions

Analysis of the damage process of fractured granite under combined dynamic and static loading by the DIC technique has obtained many details and processes of damage that cannot be observed by the naked eye, providing powerful approaches to understanding the damage of rock under dynamic and static loading. This is of great significance for further study of the mechanism of rockburst and the prevention of rockburst. However, although disturbance excavation is the most common way of deep rock mass engineering, it provides a new idea for the efficient crushing of rock mass if the high storage energy in rock mass can be effectively utilized to induce rock breaking. This study can be summarized as follows:

1. As the angle of inclination increases, the peak stress increases, and it can be concluded that the external impact resistance of the rock is minimized when the inclination angle of the fissure within the rock is approximately 45°. Changes in the angle produce a certain degree of weakening of the dynamic and combined strengths. Changes in the inclination angle affect the distribution of the internal stress field of the rock during the impact of the specimen, which is macroscopically manifested by the weakening of the dynamic and combined strengths.
2. The peak strain and dynamic modulus of elasticity, affected by the weakening of the inclination angle, are similar to the strength. The peak strain shows an overall trend that increases and then decreases with the increase in axial pressure. The inflection point of the trend occurs at an axial pressure of 27.9 MPa (20% UCS). The overall axial pressure at 0~50% UCS increases step by step.
3. The analysis of the maximum principal strains of the specimens during the damage shows that the specimens all eventually evolved from tensile strain to shear damage. Significantly, the parallel fracture specimens with an inclination angle of 45° show an inverse shear strain zone at different axial pressure conditions, which does not appear in the 0° and 90° specimens. The damage patterns of the 0° and 90° inclined specimens at different axial pressures differed significantly. The 45° inclined specimens show an inverse flank shear strain zone at combined dynamic and static loading damage, while the 0° and 90° inclined specimens show an "X"-shaped shear strain zone at different axial pressures.
4. Both the axial pressure and flaw inclination angle affect the fractal dimension of double-flawed rocks. With the increase in fracture angle, the fractal dimension first increases and then decreases, reaching its maximum at 45°. The fractural dimension increases with the increase in the strain rate, which showed a good linear correlation.

Author Contributions: Conceptualization, L.Z. and Y.L.; methodology, Z.Z. and L.Z.; writing—original draft preparation: Y.C. and B.D.; Data Curation, Z.Z. and Y.L.; writing—review and editing, Y.L. and X.L.; funding acquisition, Y.C. and B.D. All authors have read and agreed to the published version of the manuscript.

Funding: This research is supported by the National Natural Science Foundation of China (51804163, 52004130), the China Postdoctoral Science Foundation (2021M693799) and the Provincial Natural Science Foundation of Hunan (2022JJ40373). The authors are very grateful for the financial contribution and convey their appreciation to the organization for supporting this basic research.

Data Availability Statement: Data sharing not applicable.

Conflicts of Interest: The authors declare no conflict of interest.

References

1. Wang, S.; Tang, Y.; Li, X.; Du, K. Analyses and Predictions of Rock Cuttabilities under Different Confining Stresses and Rock Properties Based on Rock Indentation Tests by Conical Pick. *Trans. Nonferrous Met. Soc. China* **2021**, *31*, 1766–1783. [CrossRef]
2. Dang, W.; Chen, J.; Huang, L. Experimental Study on the Velocity-Dependent Frictional Resistance of a Rough Rock Fracture Exposed to Normal Load Vibrations. *Acta Geotech.* **2021**, *16*, 2189–2202. [CrossRef]
3. Du, K.; Li, X.; Tao, M.; Wang, S. Experimental Study on Acoustic Emission (AE) Characteristics and Crack Classification during Rock Fracture in Several Basic Lab Tests. *Int. J. Rock Mech. Min. Sci.* **2020**, *133*, 104411. [CrossRef]
4. Wang, S.; Li, X.; Yao, J.; Gong, F.; Li, X.; Du, K.; Tao, M.; Huang, L.; Du, S. Experimental Investigation of Rock Breakage by a Conical Pick and Its Application to Non-Explosive Mechanized Mining in Deep Hard Rock. *Int. J. Rock Mech. Min. Sci.* **2019**, *122*, 104063. [CrossRef]
5. Zhao, Z.; Sun, W.; Chen, S.; Feng, Y.; Wang, W. Displacement of Surrounding Rock in a Deep Circular Hole Considering Double Moduli and Strength-Stiffness Degradation. *Appl. Math. Mech.-Engl. Ed.* **2020**, *41*, 1847–1860. [CrossRef]
6. Dai, B.; Shan, Q.; Chen, Y.; Luo, X. Mechanical and Energy Dissipation Characteristics of Granite under Cyclic Impact Loading. *J. Cent. South Univ.* **2022**, *29*, 116–128. [CrossRef]
7. Qiu, J.; Li, D.; Li, X.; Zhu, Q. Numerical Investigation on the Stress Evolution and Failure Behavior for Deep Roadway under Blasting Disturbance. *Soil Dyn. Earthq. Eng.* **2020**, *137*, 106278. [CrossRef]
8. Dang, W.; Chen, J.; Huang, L.; Ma, J.; Li, X. Frictional Behavior of Granular Materials Exposed to Dynamic Normal Load. *Eng. Geol.* **2021**, *295*, 106414. [CrossRef]
9. Wang, S.; Huang, L.; Li, X. Analysis of Rockburst Triggered by Hard Rock Fragmentation Using a Conical Pick under High Uniaxial Stress. *Tunn. Undergr. Space Technol.* **2020**, *96*, 103195. [CrossRef]
10. Zhou, Z.; Cai, X.; Li, X.; Cao, W.; Du, X. Dynamic Response and Energy Evolution of Sandstone Under Coupled Static–Dynamic Compression: Insights from Experimental Study into Deep Rock Engineering Applications. *Rock Mech. Rock Eng.* **2020**, *53*, 1305–1331. [CrossRef]
11. Dang, W.; Tao, K.; Chen, X. Frictional Behavior of Planar and Rough Granite Fractures Subjected to Normal Load Oscillations of Different Amplitudes. *J. Rock Mech. Geotech. Eng.* **2021**, *14*, 746–756. [CrossRef]
12. Zhang, L.; Zhang, Z.; Chen, Y.; Dai, B.; Wang, B. Crack Development and Damage Patterns under Combined Dynamic-Static Loading of Parallel Double Fractured Rocks Based on DIC Technique. *Acta Geotech.* **2022**, *18*, 877–901. [CrossRef]
13. Wang, S.; Tang, Y.; Wang, S. Influence of Brittleness and Confining Stress on Rock Cuttability Based on Rock Indentation Tests. *J. Cent. South Univ.* **2021**, *28*, 2786–2800. [CrossRef]
14. Du, K.; Sun, Y.; Zhou, J.; Wang, S.; Tao, M.; Yang, C.; Khandelwal, M. Low Amplitude Fatigue Performance of Sandstone, Marble, and Granite under High Static Stress. *Geomech. Geophys. Geo-Energy Geo-Resour.* **2021**, *7*, 68. [CrossRef]
15. Qiu, J.; Li, X.; Li, D.; Zhao, Y.; Hu, C.; Liang, L. Physical Model Test on the Deformation Behavior of an Underground Tunnel Under Blasting Disturbance. *Rock Mech. Rock Eng.* **2021**, *54*, 91–108. [CrossRef]
16. Dang, W.; Tao, K.; Huang, L.; Li, X.; Ma, J.; Zhao, T. A New Multi-Function Servo Control Dynamic Shear Apparatus for Geomechanics. *Measurement* **2022**, *187*, 110345. [CrossRef]
17. Dang, W.; Wu, W.; Konietzky, H.; Qian, J. Effect of Shear-Induced Aperture Evolution on Fluid Flow in Rock Fractures. *Comput. Geotech.* **2019**, *114*, 103152. [CrossRef]
18. Du, K.; Yang, C.; Su, R.; Tao, M.; Wang, S. Failure Properties of Cubic Granite, Marble, and Sandstone Specimens under True Triaxial Stress. *Int. J. Rock Mech. Min. Sci.* **2020**, *130*, 104309. [CrossRef]
19. Brugo, T.M.; Campione, I.; Minak, G. Investigation by Digital Image Correlation of Mixed-Mode I and II Fracture Behavior of Polymeric IASCB Specimens with Additive Manufactured Crack-Like Notch. *Materials* **2021**, *14*, 1084. [CrossRef]
20. Khan, S.; Khulief, Y.; Al-Shuhail, A.; Bashmal, S.; Iqbal, N. The Geomechanical and Fault Activation Modeling during CO_2 Injection into Deep Minjur Reservoir, Eastern SaudiArabia. *Sustainability* **2020**, *12*, 9800. [CrossRef]
21. Li, D.; Zhang, C.; Zhu, Q.; Ma, J.; Gao, F. Deformation and fracture behavior of granite by the short core in compression method with 3D digital image correlation. *Fatigue Fract. Eng. Mater. Struct.* **2022**, *45*, 425–440. [CrossRef]
22. Basu, A.; Mishra, D.A.; Roychowdhury, K. Rock Failure Modes under Uniaxial Compression, Brazilian, and Point Load Tests. *Bull. Eng. Geol. Environ.* **2013**, *72*, 457–475. [CrossRef]
23. Amann, F.; Button, E.A.; Evans, K.F.; Gischig, V.S.; Blümel, M. Experimental Study of the Brittle Behavior of Clay Shale in Rapid Unconfined Compression. *Rock Mech. Rock Eng.* **2011**, *44*, 415–430. [CrossRef]
24. Li, X.; Gong, F.; Tao, M.; Dong, L.; Du, K.; Ma, C.; Zhou, Z.; Yin, T. Failure mechanism and coupled static−dynamic loading theory in deep hard rock mining: A review. *J. Rock Mech. Geotech. Eng.* **2017**, *9*, 767–782. [CrossRef]
25. Wang, S.; Sun, L.; Li, X. Experimental investigation and theoretical analysis of indentations on cuboid hard rock using a conical pick under uniaxial lateral stress. *Geomech. Geophys. Geo-Energy Geo-Resour.* **2022**, *8*, 34. [CrossRef]
26. Wang, S.; Liu, K.; Wang, S. Three-dimensional stochastic distribution characteristics of void fraction in longwall mining-disturbed overburden. *Bull. Eng. Geol. Environ.* **2022**, *81*, 414. [CrossRef]
27. Zhou, Z.; Li, X.; Zou, Y.; Jiang, Y.; Li, G. Dynamic Brazilian Tests of Granite Under Coupled Static and Dynamic Loads. *Rock Mech. Rock Eng.* **2014**, *47*, 495–505. [CrossRef]
28. Kaiser, P.K.; Kim, B.-H. Characterization of Strength of Intact Brittle Rock Considering Confinement-Dependent Failure Processes. *Rock Mech. Rock Eng.* **2015**, *48*, 107–119. [CrossRef]

29. Deng, M.; Zhang, Z.; Xin, J.; Yu, W.; Yu, X. Acoustic Emission Fractal Characteristics of Different Prefabricated Fractured Sandstone under Uniaxial Compression Tests. *Geotech. Geol. Eng.* **2021**, *39*, 3161–3170. [CrossRef]
30. Dou, L.; Yang, K.; Chi, X. Fracture Behavior and Acoustic Emission Characteristics of Sandstone Samples with Inclined Precracks. *Int. J. Coal Sci. Technol.* **2021**, *8*, 77–87. [CrossRef]
31. Wang, Y.; Deng, H.; Deng, Y.; Chen, K.; He, J. Study on Crack Dynamic Evolution and Damage-Fracture Mechanism of Rock with Pre-Existing Cracks Based on Acoustic Emission Location. *J. Pet. Sci. Eng.* **2021**, *201*, 108420. [CrossRef]
32. Lin, Q.; Cao, P.; Wen, G.; Meng, J.; Cao, R.; Zhao, Z. Crack Coalescence in Rock-like Specimens with Two Dissimilar Layers and Pre-Existing Double Parallel Joints under Uniaxial Compression. *Int. J. Rock Mech. Min. Sci.* **2021**, *139*, 104621. [CrossRef]
33. Wang, S.; Sun, L.; Li, X.; Wang, S.; Du, K.; Li, X.; Feng, F. Experimental Investigation of Cuttability Improvement for Hard Rock Fragmentation Using Conical Cutter. *Int. J. Geomech.* **2021**, *21*, 06020039. [CrossRef]
34. Li, X.; Zhou, T.; Li, D. Dynamic Strength and Fracturing Behavior of Single-Flawed Prismatic Marble Specimens Under Impact Loading with a Split-Hopkinson Pressure Bar. *Rock Mech. Rock Eng.* **2017**, *50*, 29–44. [CrossRef]
35. Wang, M.; Wang, F.; Zhu, Z.; Dong, Y.; Mousavi Nezhad, M.; Zhou, L. Modelling of Crack Propagation in Rocks under SHPB Impacts Using a Damage Method. *Fatigue Fract. Eng. Mater. Struct.* **2019**, *42*, 1699–1710. [CrossRef]
36. Yan, Z.; Dai, F.; Zhu, J.; Xu, Y. Dynamic Cracking Behaviors and Energy Evolution of Multi-Flawed Rocks Under Static Pre-Compression. *Rock Mech. Rock Eng.* **2021**, *54*, 5117–5139. [CrossRef]
37. Li, D.; Xiao, P.; Han, Z.; Zhu, Q. Mechanical and Failure Properties of Rocks with a Cavity under Coupled Static and Dynamic Loads. *Eng. Fract. Mech.* **2020**, *225*, 106195. [CrossRef]
38. Weng, L.; Li, X.; Taheri, A.; Wu, Q.; Xie, X. Fracture Evolution Around a Cavity in Brittle Rock Under Uniaxial Compression and Coupled Static–Dynamic Loads. *Rock Mech. Rock Eng.* **2018**, *51*, 531–545. [CrossRef]
39. Tao, M.; Ma, A.; Cao, W.; Li, X.; Gong, F. Dynamic Response of Pre-Stressed Rock with a Circular Cavity Subject to Transient Loading. *Int. J. Rock Mech. Min. Sci.* **2017**, *99*, 1–8. [CrossRef]
40. Yan, Z.; Dai, F.; Liu, Y.; Li, A.; Du, H. Numerical Assessment of the Rate-Dependent Cracking Behaviours of Single-Flawed Rocks in Split Hopkinson Pressure Bar Tests. *Eng. Fract. Mech.* **2021**, *247*, 107656. [CrossRef]
41. Qian, X.K.; Liang, Z.Z.; Liao, Z.Y.; Wang, K. Numerical Investigation of Dynamic Fracture in Rock Specimens Containing a Pre-Existing Surface Flaw with Different Dip Angles. *Eng. Fract. Mech.* **2020**, *223*, 106675. [CrossRef]
42. Ai, D.; Zhao, Y.; Wang, Q.; Li, C. Crack Propagation and Dynamic Properties of Coal under SHPB Impact Loading: Experimental Investigation and Numerical Simulation. *Theor. Appl. Fract. Mech.* **2020**, *105*, 102393. [CrossRef]
43. Feng, P.; Zhao, J.; Dai, F. Mechanical behaviors of conjugate-flawed rocks subjected to coupled static–dynamic compression. *Acta Geotech.* **2022**, *17*, 1765–1784. [CrossRef]
44. Ai, D.; Zhao, Y.; Wang, Q.; Li, C. Experimental and Numerical Investigation of Crack Propagation and Dynamic Properties of Rock in SHPB Indirect Tension Test. *Int. J. Impact Eng.* **2019**, *126*, 135–146. [CrossRef]
45. Han, Z.; Li, D.; Zhou, T.; Zhu, Q.; Ranjith, P.G. Experimental Study of Stress Wave Propagation and Energy Characteristics across Rock Specimens Containing Cemented Mortar Joint with Various Thicknesses. *Int. J. Rock Mech. Min. Sci.* **2020**, *131*, 104352. [CrossRef]
46. Zhou, Z.; Li, X.; Ye, Z.; Liu, K. Obtaining Constitutive Relationship for Rate-Dependent Rock in SHPB Tests. *Rock Mech. Rock Eng.* **2010**, *43*, 697–706. [CrossRef]
47. Zhu, Q.; Li, D.; Han, Z.; Li, X.; Zhou, Z. Mechanical properties and fracture evolution of sandstone specimens containing different inclusions under uniaxial compression. *Int. J. Rock Mech. Min. Sci.* **2019**, *115*, 33–47. [CrossRef]
48. Zhou, X.; Zhang, J.; Wong, L. Experimental study on the growth, coalescence and wrapping behaviors of 3D cross-embedded flaws under uniaxial compression. *Rock Mech. Rock Eng.* **2018**, *51*, 1379–1400. [CrossRef]

materials

MDPI

Article

Deformation Mechanism and Control of In-Situ Assembling Caisson Technology in Soft Soil Area under Field Measurement and Numerical Simulation

Jie Zhou [1,*], Chengjun Liu [1], Jie Xu [2], Zhenguang Zhang [2] and Zeyao Li [1]

[1] Department of Geotechnical Engineering, Tongji University, Shanghai 200092, China
[2] Shanghai Road and Bridge Group Co. Ltd., Shanghai 200433, China
* Correspondence: zhoujie1001@tongji.edu.cn; Tel.: +86-136-3665-1710; Fax: +86-021-6598-7079

Abstract: With urban space becoming much more crowded, the construction of underground spaces continues to expand to deeper, and the requirements for the large depth and minor deformation in urban engineering construction are getting more urgent. A new kind of in-situ assembling caisson technology (called VSM) is a vertical shaft method (VSM), which excavates the stratum under water with a mechanical arm and assembles the prefabricated caisson segments at the same time. This paper takes the Shanghai Zhuyuan Bailonggang Sewage Connecting Pipe Project as an example, which is the first construction project in the soft soil area, such as Shanghai, and makes a technical analysis of the VSM by comparing the field measurement and numerical simulation. Ground settlements and layered deep displacements were monitored in the field measurement during the VSM construction. It shows that the maximum ground settlement caused by the VSM is 15.2 mm and the maximum horizontal displacement is 3.74 mm. The influence range of the shaft excavation on the ground settlement is about 30 m away from the shaft center. The results demonstrate that the VSM construction has great applicability in the soft soil area. A finite element simulation model of the VSM shaft is established and verified by field measurement. There is a certain error between the traditional theoretical calculation by analogy to the common retaining walls of the deep foundation pit and the measured results, while the simulation results are relatively consistent with field measurements. The reasons for the difference are well-analyzed. Finally, the effects of the VSM construction method on the engineering environment are analyzed, and the suggestions for deformation control in the future are put forward.

Keywords: in-situ assembling caisson; VSM construction method; ground settlement; deep layered deformation; field measurement; stability

Citation: Zhou, J.; Liu, C.; Xu, J.; Zhang, Z.; Li, Z. Deformation Mechanism and Control of In-Situ Assembling Caisson Technology in Soft Soil Area under Field Measurement and Numerical Simulation. *Materials* **2023**, *16*, 1125. https://doi.org/10.3390/ma16031125

Academic Editors: Krzysztof Schabowicz and Francesco Fabbrocino

Received: 27 November 2022
Revised: 29 December 2022
Accepted: 21 January 2023
Published: 28 January 2023

1. Introduction

A shaft is an important structural form of underground space construction. In recent years, with the continuous development of urban construction, underground space excavation has had a complex and diversified development. Underground engineering construction puts forward higher and higher requirements for shaft excavation.

The VSM applies to the excavation of small diameter shafts in urban areas and does not require dewatering. It has the advantages of deep excavation depth, small construction disturbance, fast construction speed, small site use, high economic efficiency, and strong stratum applicability. This kind of method has been used in shaft engineering in Europe, the United States [1], Singapore [2], and other places. The VSM construction method is mainly used in subway ventilation shaft construction and presently has a maximum excavation depth of 115.2 m [3,4].

There are relatively few engineering applications of the VSM in China [5]; Zhang et al. [6], Huang et al. [7], Jiang et al. [8] took the ultra-deep prefabricated shaft project in Jianye District, Nanjing as an example and mainly introduced the key technology of ultra-deep prefabricated shaft undrained construction in water-rich sand stratum; but the requirements for the surrounding engineering environment is not very high since it is away from the downtown. In general, there are relatively few studies of the VSM construction method in soft clay area, such as Shanghai, as well as the impact of this new technology on the engineering environment.

Even though few specific numerical simulations of in-situ assembling caisson technology are conducted at present, previous research on the excavation of deep foundation and traditional caisson can provide good significance. Ma [9] studied ground settlement and the stability of the retaining wall deformation in the process of excavation by COMSOL and also made a comparative analysis with the monitoring data during the excavation of a deep foundation pit in Shanghai. Lin et al. [10] established a numerical simulation model by FLAC3D to study the pile displacement and ground settlement with the exaction time under different computation boundaries for considering creep and seepage. Shi et al. [11] focused on the mechanical characteristics and cracking control of a large diameter caisson in the initial sinking stage in the FEM numerical simulation by ANSYS. Zhao et al. [12] studied the stress and deformation performance of an anchor caisson foundation in sands by model tests and the numerical simulation in PLAXIS 3D; and the mechanism on the interaction of the soil-structure (anchor caisson wall) was also analyzed. Different from deep foundation construction with underground diaphragm wall, or traditional caisson, the construction speed is much faster and the construction condition is underwater; generally, the construction depth can also be much deeper. Most importantly, nowadays deep shafts are rarely constructed by traditional caisson in urban cities due to the severe engineering environmental impact, instead deep foundation pits by underground diaphragm wall construction are always used. However, the leakage problem greatly impedes the construction depth induced by the diaphragm wall quality; simultaneously, dewatering is really not environmentally-friendly in the soft soil area. Therefore, the deformation mechanism and control of this new in-situ assembling caisson technology should be specifically analyzed for potential broad utilization, especially in Shanghai, where thick, soft mucky clays are widely distributed.

Taking the first engineering project in Shanghai as an example, the 17# shaft in the Shanghai Zhuyuan Bailonggang Sewage Connecting Pipe Project, where a specifically designed field measurement of the ground settlement and the deep layered deformation of surrounding soils are analyzed in this paper. Simultaneously, the traditional theoretical calculations for design and numerical simulation are both conducted to compare with field measurements. In addition, the effects of the VSM construction method on the engineering environment are analyzed, and the suggestions for the deformation control in the future are put forward.

2. VSM Construction Technology

The VSM construction method is a submerged in-situ assembling shaft excavation construction method in which the mechanical arm is used to excavate the stratum underwater and assemble the prefabricated caisson segments at the same time. The VSM construction equipment (as shown in Figure 1) is mainly composed of the excavating main engine, slurry inlet pipe, slurry outlet pipe, prefabricated caisson segment, and sinking unit.

While the mechanical arm attached to the excavating main engine is excavating the soil in the shaft, prefabricated caisson segments are being assembled at the ground surface, and the sinking unit is being operated to sink the caisson segments. Bentonite is grouted between the prefabricated segments and the soil during the sinking to mostly reduce friction resistance. During the process of excavation and sinking, water should be injected into the shaft in time to ensure that the water level in the shaft is always higher than the groundwater level to keep excavation stability. Each prefabricated caisson segment is connected with each other by radial longitudinal bolts. During the sinking of the prefabricated caisson segments, the sinking unit is always connected to the bottom edge of the first prefabricated caisson segment through steel strands to precisely control the sinking speed. Bentonite grouting holes are designed at the lower part of the shaft, which is connected to the bentonite grouting pipe. The process of excavation, assembling, and sinking is cycled until the shaft excavation reaches the design depth.

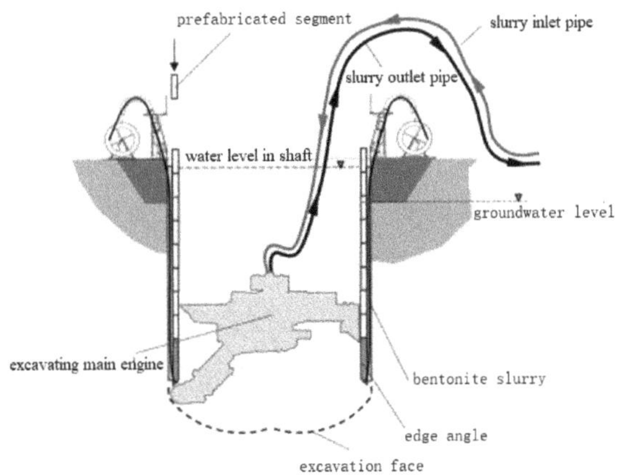

Figure 1. Schematic diagram of VSM construction method.

Once the excavation reaches the design depth, the bottom sealing shall be operated by pouring concrete underwater to form concrete slab. Then, the cement slurry is used to replace the bentonite filled outside the prefabricated caisson segments. After that, the water in the shaft is pumped, and the inner wall of shaft is flushed by a high-pressure water gun. Finally, shaft floor slab construction works, such as binding steel bars and pouring concrete, are completed.

Compared with other shaft construction methods, the VSM has the following advantages:

(1) The diameter range of the shaft is 4.5~18 m.
(2) The construction depth can reach over 120 m below the groundwater level.
(3) Prefabricated caisson segments greatly increase the stability of the shaft.
(4) VSM can be operated remotely.
(5) Surrounding environmentally-friendly by underwater sinking without lowering the groundwater level.
(6) Fast construction speed and controllable structure sinking process.
(7) The construction site is small.

3. Project Overviews

Zhuyuan Bailonggang Sewage Connecting Pipe Project is located in the Pudong New Area, Shanghai. The total length of connecting pipeline is about 19.8 km, which is constructed by the shield method and pipe jacking method.

In this project, the 17# shaft on connecting pipeline line is constructed by the VSM. Each ring of prefabricated caisson segments is made up of 6 identical segments, which are 1500 mm wide. It is connected by 12 radial and 18 longitudinal bolts. The segment concrete strength grade is C60 (the uniaxial compressive strength reaches 60 MPa of a national standard specimen after a curing duration of 28 d), and the impermeability grade is P12 (which can resist hydrostatic pressure of 1.2 MPa (equivalent to 120 m underwater) in a national standard specimen after a curing duration of 28 d). The design depth of the 17# shaft is 39 m, and the final sinking elevation of the blade foot is about −35.400 m. The shaft top is cemented with a 2.4 m height cast-in-place connection wall, which would be connected to the inside of the foundation ring beam after the shaft completely sinks in place. The foundation ring beam is used as the load-bearing structure of the VSM ground equipment during the shaft sinking process, as well as an anti-floating structure during the service stage. The section size of the foundation ring beam is about 2600 mm × 2500 mm with the inner and outer diameters of 13 m and 18.2 m, respectively. In total, 12 Φ800 mm diameter bored piles (cast-in-situ) are set under the foundation ring beam to bear the pressure load of the VSM equipment and the buoyancy of the shaft during the service stage. The section of the shaft structure is shown in Figure 2, technical parameters are shown in Table 1, and the formation parameters are shown in Table 2.

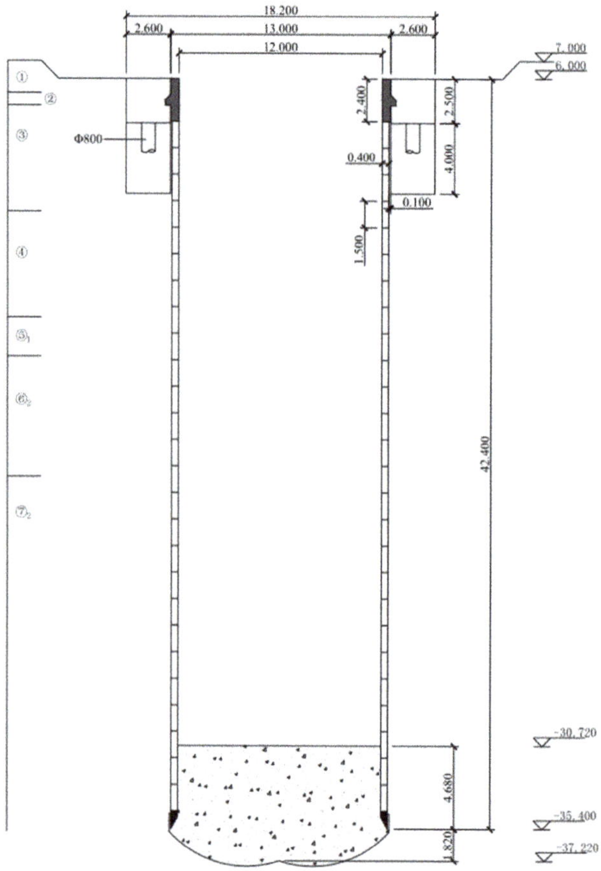

Figure 2. Shaft section layout (m).

Table 1. Technical parameters in VSM construction.

Shaft Structural Parameters	Mechanical Arm Parameters	Excavation and Assembly Speed	Bentonite Parameters
Excavation depth: 43.22 m Excavation radius: 6.5 m Inside diameter of shaft: 6 m Over excavation depth: 15~20 cm Prefabricated segments: C60 concrete Maximum depth: 120 m	Elongation: 0~1000 mm Swing scope: −10°~+47° Rotation scope: ±190°	Up to 4.5 m/d	Marsh funnel viscosity: 90 Ms/L Static yield point: 40 N/mm^2 API filtration: up to 20 mL

Table 2. Soil layer characteristics.

Number	Soil Layer	Unit Weight/kg/m^3	Void Ratio	Compression Modulus /MPa	Cohesion /kPa	Internal Friction Angle/°	Thickness/m
①	plain fill/flush fill	1800	1 *	2 *	10 *	10 *	1.8
②	sandy silt	1840	0.85	5.5	5	30.5	0.7
③	clay	1780	1.11	2.2	12	17.5	6.0
④	silty clay	1690	1.38	2.1	12	12	6.0
⑤$_1$	clay	1800	0.97	4.5	18	17	2.2
⑥$_2$	clay mixed with silt	1820	0.9	6	19	19	6.8
⑦$_2$	sandy silt	1890	0.76	13.5	3	34	35.5

Note: * is empirical value.

4. Construction Process

The construction started at the end of 2021. After the preliminary work, such as equipment debugging and site leveling, the VSM officially initiated excavation on 13 January 2022. Due to the traditional Chinese Spring Festival holiday, the entire excavation process was divided into two stages. In total, 32.2 m was excavated for 17 days in the first stage from 13 January to 29 January, and 5.1 m was excavated for 3 days in the second stage from 8 February to 10 February. The excavation rate was 1.9 m per day on average, which was roughly the same throughout the excavation process. The bottom of the shaft was sealed on 21 February, the bentonite replacement was completed on 2 March, and the construction of the shaft bottom was completed on 22 March. The construction process is shown in Figure 3, and the construction excavation rate is shown in Figure 4.

Figure 3. Construction node.

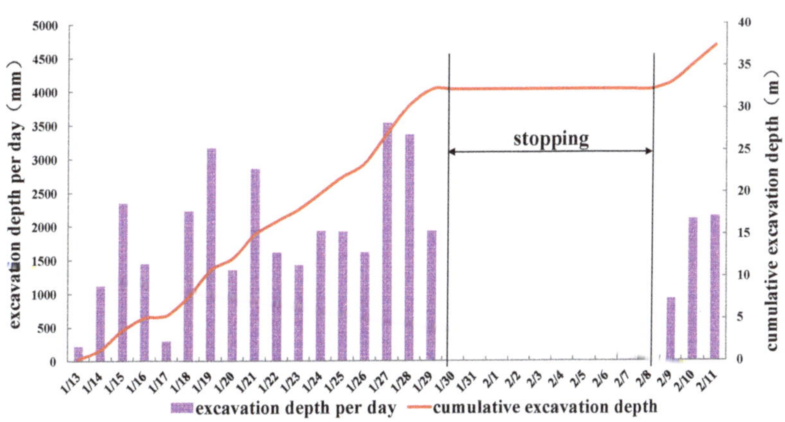

Figure 4. Construction progress.

5. Earth Pressure Design Theory in Shaft Lining

5.1. Theoretical Calculation of Earth Pressure

Lots of researchers discussed the earth pressure theories in shaft engineering in cohesionless soils and the saturated undrained soft clay [13,14]. They all agree that the active Rankine values at the greater depth are advisable in the design purpose, which is also consistent with the technical code for excavation engineering (DG/TJ 08-61-2018) [15] in Shanghai. In Prater's [13] discussion, the earth pressure according to Berezantzev, reaches a limiting value asymptotically, which is much smaller than the Rankine value at greater depths, with a reduction in pressure at the greater depth, due to the arching action. Similar is known to exist for retaining walls not fulfilling the plasticity deformation conditions. While in the soft soil area, especially in Shanghai, as mentioned above, the large thickness (in this case, subsurface 20~40 m) soft mucky clay has poor permeability and high plasticity. Both with the plain assumptions of the Rankine earth pressure theory (the same as the Coulomb method for cohesionless soils), the active Rankine values are most conservative, and it would be advisable for design purposes. The Rankine earth pressure theory could be used to calculate the earth pressure in shaft design.

$$p_{ak} = \sigma_{ak}K_{ai} - 2c\sqrt{K_{ai}}$$

$$K_{ai} = tan^2(45^\circ - \varphi/2)$$

$$\sigma_o = K_0\gamma h$$

Here, p_{ak} is the active earth pressure on the outside soil of support structure (kPa); σ_{ak} is the vertical stress in the soil layer at the calculated position outside the support structure (kPa); φ is the internal friction angle (°); σ_o is the static earth pressure (kPa); γ is the weight of soil (kN/m^3); h is the thickness of the soil (m).

5.2. Theoretical Calculation of Surface Deformation

Bowles [16] proposed a method to predict the ground surface settlement of the cohesive soil layer without considering the consolidation settlement. The calculation is as follows [17]:

(1) Calculate the lateral deformation of the support structure;
(2) Calculate the horizontal volume Vs of soil outside foundation pit;
(3) Estimate the influence area D of soil settlement outside foundation pit;

$$Vs = (H_e + H_d)tan(45^\circ - \varphi/2)$$

where H_e is the excavation depth (m); H_d is the excavation width or diameter (m).

(4) Calculate the maximum surface subsidence δ

$$\delta = (\frac{V_S}{D})(\frac{x}{D})^2$$

where x is the distance from the calculation point to the support structure.

6. Shaft Monitoring

6.1. Monitoring Purpose

In this project, on the one hand, monitoring work was implemented in order to ensure the construction quality and safety of shaft; the monitoring data were used to timely adjust the shaft excavation and the sinking speed to prevent shaft instability. On the other hand, the surrounding environment impact was also monitored by ground settlement and deep layered deformations at different locations.

6.2. Monitoring Content

In this project, the deformation and stress conditions of the soil surrounding shaft were monitored. Total stations were used to monitor the ground settlement around the shaft; fixed inclinometers were used to monitor the horizontal displacement of stratums around the shaft along depth; and the buried earth pressure gauges were used to monitor the soil stress state.

In the construction site, 5 ground surface settlement monitoring points and 5 deep soil inclination observation points are arranged at 15 m, 30 m, 55 m, 80 m, and 110 m from the center of shaft, with depths of 5 m, 15 m, 25 m, 35 m, and 45 m; the positions of the observation points are shown in Figure 5. In total, 5 earth pressure gauges were buried at the depths of 5 m, 15 m, 25 m, 35 m, and 45 m, at the distance of 15 m from the shaft center.

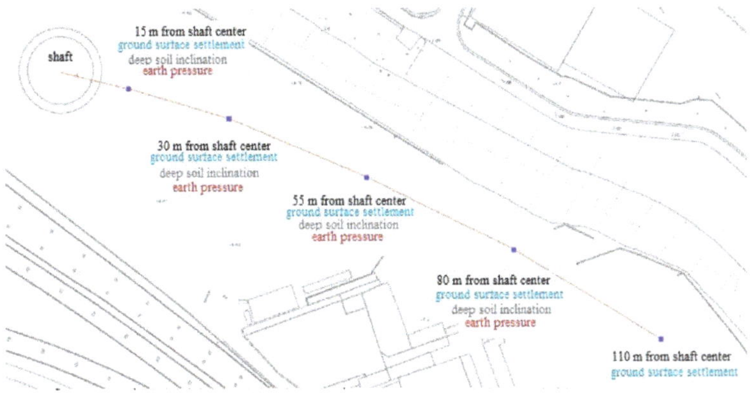

Figure 5. Layout of monitoring points.

6.3. Analysis of Measured Deformation during Shaft Construction

6.3.1. Ground Settlement Analysis

The ground settlement results of 15 m, 30 m, 55 m, 80 m, and 110 m from the shaft center are shown in Figure 6. The ground settlement was greater near the shaft. It was not obvious in the first 7 days of the excavation and increased rapidly after that. When the shaft excavation was suspended, the ground surface continued to settle for a period of time (7 days). This reveals that, in the soft clay layer with high plasticity and poor permeability, the ground settlement continued even without the disturbance of excavation. It will increase for a period of time after the excavation ends. This phenomenon is well-known to exist in the traditional slow excavation in the deep foundation pit of retaining walls (diaphragm walls), in which large deformation always happened if the support could be adjusted in time. Monitoring, as well as engineering control, during this period are still vital. After the bottom sealing was completed, the ground surface settlement increased slightly during the bentonite replacement stage; there was no significant change in the ground settlement during the shaft floor construction stage.

It can be discovered from Figure 6 that the excavation of the VSM has a significant impact on the ground settlement within 30 m from the shaft center, which forms a settlement tank. By the end of the shaft construction, the maximum ground settlement was about 14.56 mm at 15 m away from the shaft center and the minimum ground settlement was about 6.62 mm at 110 m away from the shaft center. In total, the effect of the shaft construction on the ground settlement was relatively small under the VSM construction speed, especially in the shallow part, even including the additional settlement due to temporally stopping during the holidays.

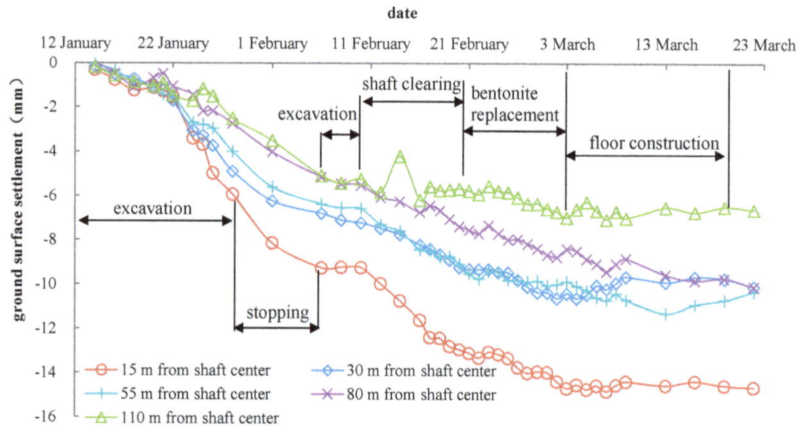

Figure 6. Monitoring results of ground settlement.

6.3.2. Deep Layered Displacements Analysis

It can be noticed from Figure 7 that the maximum deep layered displacement of the stratum was about 3.74 mm at 15 m away from the shaft center. During the construction of the shaft, the deep layered displacements increased continuously, and the growth rate was getting faster and faster. Except for the stratum 5 m deep underground, the deeper the depth, the smaller the horizontal displacement. (The small horizontal displacement of the stratum 5 m depth underground may be due to the surface hardening around the shaft and the construction equipment placed on the ground surface, which constrained the horizontal deformation near the ground surface to a certain extent.)

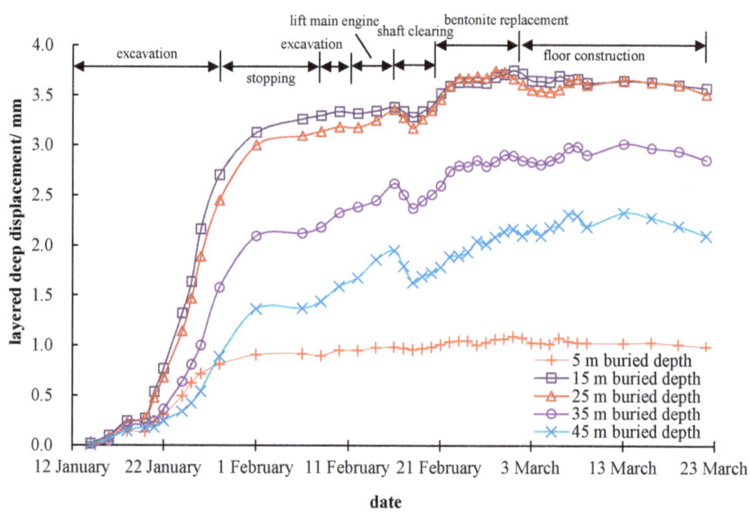

Figure 7. Monitoring results of the deep layered displacement at 15 m away from the shaft center.

When the excavation ends, the deep layered displacement of the monitoring point at 15 m away from the shaft center immediately stopped increasing. It can be inferred that the horizontal displacement has a strong time correlation with the excavation operation. From Figures 8 and 9, the maximum horizontal displacements of the stratum, measured by the inclinometer pipes 30 m and 85 m away from the shaft center, were about 2.20 mm and 0.63 mm. The horizontal displacement started to increase after 3–5 days after the construction began and did not stop rising when the excavation was completed. The horizontal displacement growth rate was roughly the same, and there was almost no accelerated growth trend. The reason for this phenomenon is that the deep viscoplastic soft soil layer in the construction site prolonged the deformation loading transfer time and weakened the deformation strength. Comparing the measurement results of the horizontal displacement at the same depth and different distances from the shaft, it could be figured out that the closer the distance to the shaft, the greater the horizontal displacement caused by the excavation. The maximum horizontal displacement is about 4.71 mm at 15 m from the shaft center and around 15 m in depth. In general, the displacement and deformation during the whole construction process were small, which meets the requirements of the technical code for excavation engineering (DG/TJ 08-61-2018) [15] for Shanghai.

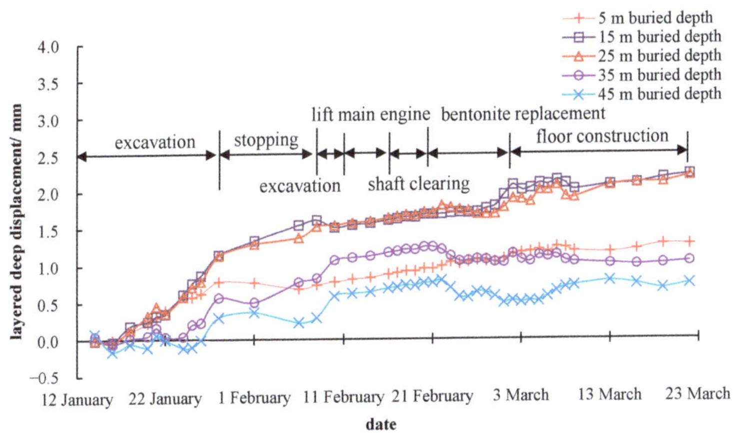

Figure 8. Monitoring results of the deep layered displacement at 30 m away from the shaft center.

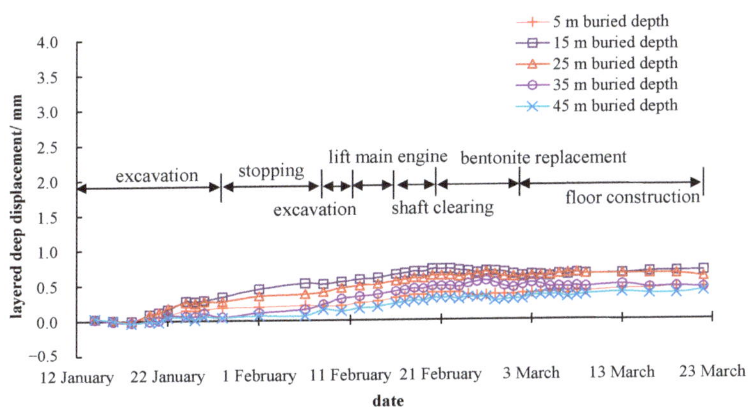

Figure 9. Monitoring results of the deep layered displacement at 85 m away from the shaft center.

6.3.3. Lateral Earth Pressure Monitoring and Analysis

The lateral earth pressure monitoring results are shown in Figure 10. It can be figured out from Figure 10 that the lateral earth pressure decreased slightly during the excavation stage, about 5%. The overall lateral earth pressure changes little during the whole construction process, indicating that the shaft construction had a small disturbance on the stress state of the surrounding soil.

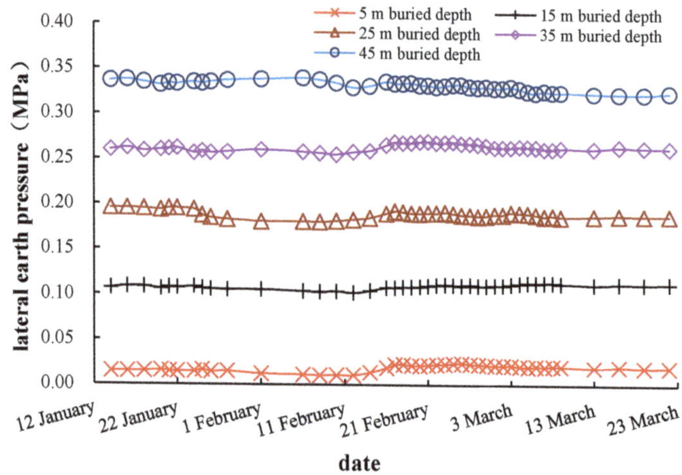

Figure 10. Monitoring results of lateral earth pressure at 15 m away from the shaft.

7. Numerical Simulation and Analysis

An axisymmetric steady state numerical simulation model is established to analyze the construction process of the VSM. Due to the fast construction speed and mostly poor permeability of Shanghai soft soils, the undrained condition is considered, and the consolidation of soil is neglected during the VSM excavation. However, before the excavation, a Biot consolidation model was used for the initial stress state computation under self-weight. This process is called self-weight balance, in which it calculates and equilibrates the initial ground stress and makes the soil model fully consolidated, and the deformation gets stable under the self-weight, to form the initial stress field for the subsequent VSM construction model. In VSM excavation, fluid-solid coupling is considered, in which an elastic-plastic Mohr-Coulomb model governs stress-strain field; and Darcy's law governs the seepage field. It is carried out by effective stress. The change of water head by the seepage in the shaft causes the change of the pore water pressure in the soil layers and then changes the effective stress and causes deformation. The soil layers in the construction site are: ① plain fill/flush fill, ② sandy silt, ③ clay, ④ silty clay, ⑤$_1$ clay, ⑥$_2$ clay mixed with silt, and ⑦$_2$ sandy silt. Considering that the engineering properties of the ② sandy silt, ④ silty clay and ⑤$_1$ clay layer are roughly the same, and the effects of the thinner ① plain fill/flush fill and ② sandy silt are ignored; this model generalizes the stratum as a combination of ④ silty clay and ⑦$_2$ sandy silt. The simulation model is shown as Figure 11, soil parameters are shown in Table 3, and structure parameters are shown in Table 4.

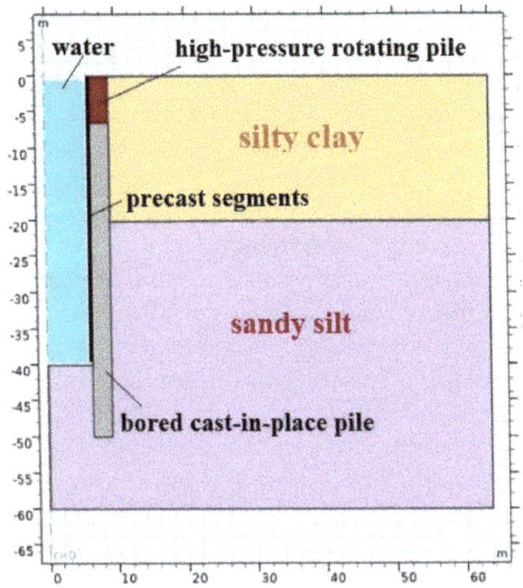

Figure 11. Numerical simulation model.

Table 3. Soil parameters in numerical simulation model.

Soil Layer	Unit Weight /kg/m^3	Void Ratio	Young's Modulus/Mpa	Cohesion/kPa	Internal Friction Angle/°	Permeability /cm/s
silty clay	1690	1.38	6.57	12	12	1.43×10^{-7}
sandy silt	1890	0.76	19.2	3	34	4.69×10^{-5}

Table 4. Structure parameters in numerical simulation model.

Material	Unit Weight/kg/m^3	Young's Modulus/Mpa	Poisson's Ratio
precast segments	2300	2.5×10^4	0.2
high-pressure rotating pile	1800	20	0.2
bored cast-in-place pile	2300	3.15×10^4	0.2

The consolidation of soil was completed before excavation in the numerical simulation model. Resultantly, the whole VSM model has an initial stress and strain field under self-weight. Furthermore, the shaft excavation and segment sinking are calculated at the same time every 5 m. As explained above, consolidation is not considered during the excavation process because of the rapid construction speed, short construction time, and low permeability of soils. The normal displacement of the side boundary and the lower boundary of the model are constrained. The hydraulic pressure is applied to the permeable excavation face and the inner wall of the prefabricated concrete segment. On the ground surface, the gravity of the prefabricated segment in the shaft is loaded on the concrete ring beam, and the pavement load of 5 kPa is loaded within 50 m from the outer wall of the shaft.

7.1. Simulation Results of Ground Settlement

Ground settlements of simulation results during 15 m, 25 m, and 35 m excavation, along different distances from the shaft center, are presented in Figure 12, combined together with field measurements and theoretical calculations. They all show that the ground settlement is mainly concentrated in the range of 30 m from the shaft center. The ground settlements all behave as a shape of a spoon type. As the excavation increases, the maximum ground settlement gets slightly larger, but the location is almost the same, all-around 15–20 m distance from the shaft center. The numerical simulation results in the range of 30 m from the shaft center are roughly the same as the field measurements. When the excavation of the shaft was basically completed (35 m excavation depth), the ground surface settlement 15 m from the shaft center obtained by numerical simulation is about 13.6 mm, and the measured value is about 13.4 mm. The simulation value is relatively consistent with the measured value; on another aspect, the simulation results also demonstrate the validity and effectiveness of the field measurement points design. The theoretical calculation result is about 19.7 mm, which is larger than the simulation and measured values. These results may be due to the small diameter of the VSM shaft, and the annular lining structure has great bending resistance in the radial direction. However, the stress and deformation theory (explained above) simplified segments into a two-dimensional structure (radial and axial directions), which reduces the radial bending resistance, thus making the theoretical calculation settlement larger. The field measurement shows that there is still a lot of ground settlement beyond 30 m away from shaft center, which is obviously larger than the numerical simulation and theoretical calculation results. The reason for this phenomenon may be that the construction site has a lot of construction facilities and building materials stacking in this range. These temporary overloadings greatly increase the ground settlement at this construction site. The numerical simulation and theoretical calculation are insufficient to estimate the construction overload, resulting in the phenomenon that the measured value is obviously larger than the simulated and calculated values. Both the simulation and measurement results show that the ground settlement caused by the VSM excavation is not greater than 15 mm, which is relatively small compared with other underground structure construction, inferring that the VSM construction has little impact on the ground settlement.

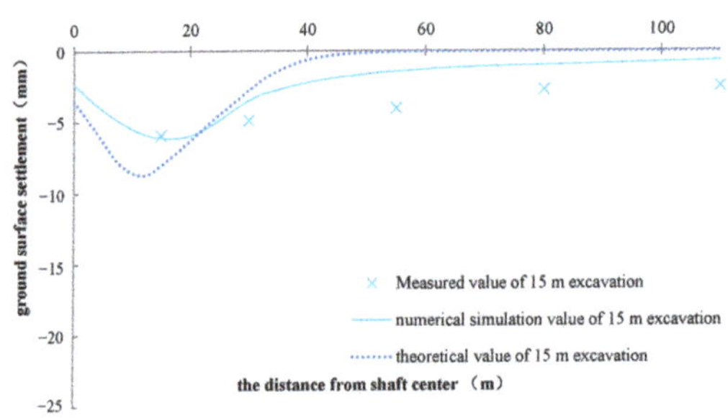

(a)

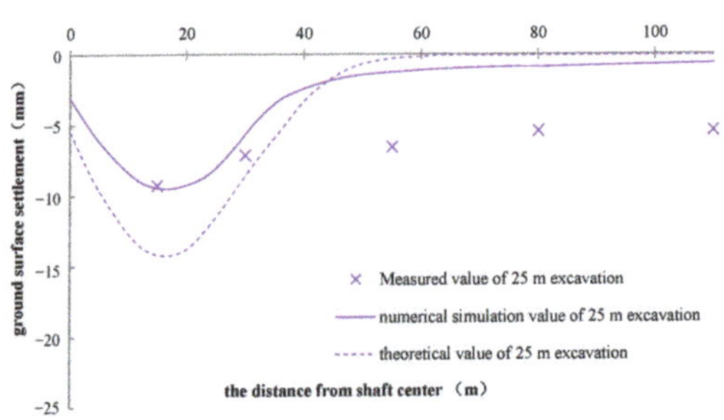

(b)

Figure 12. *Cont.*

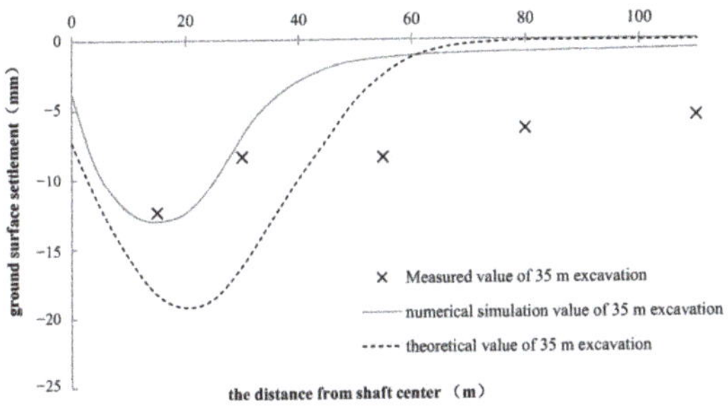

(c)

Figure 12. Simulation results of surface settlement ((**a**), 15 m excavation; (**b**), 25 m excavation; (**c**), 35 m excavation).

7.2. Analysis of Uplift Deformation on Shaft Excavation Face

The vertical uplift results of the soil near the excavation surface by simulation are shown as Figure 13, compared with field measurements. Both the numerical simulation and field measurement show that, during the excavation, a certain uplift would occur near the excavation surface; the deeper the excavation depth, the more obvious the uplift. The numerical simulation results of the vertical deformation near the excavation face are generally slightly larger than the field measurement results. The numerical simulation results of the excavation face uplift after shaft excavation are about 10.7 cm, while the measured result is about 8.6 cm. The reason for this phenomenon is that the caisson bottom soil has a certain elastoplasticity, causing a slow rebound rate. Under the rapid construction rate of the VSM, the foundation soil is not fully rebounded until the construction of the bottom sealing is completed. In general, the numerical simulation model is able to simulate the real excavation face uplift, and the results tend to be a good reference before the construction.

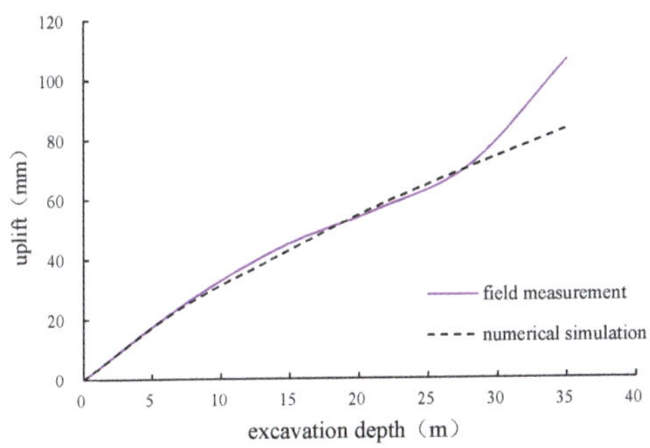

Figure 13. Uplift simulation results near the excavation face.

7.3. Analysis of Horizontal Displacement along Depth

In the construction site, the horizontal displacement simulation and the measured results along the depth of 15 m away from the shaft center are shown in Figure 14. Comparing the numerical simulation with the field measurements, the simulation values of the horizontal displacement under different excavation depths are slightly smaller than the measurement values. When the construction is finished (35 m excavation), at 15 m away from shaft center, the simulation result for the maximum horizontal displacement is about 3.26 mm, while the field measurement value is about 3.45 mm, and the error is only 6%. However, the main difference here is that the numerical simulation shows that the maximum horizontal displacement is located at the depth of 30 m to 40 m, while the measured results show that the maximum deformation occurs at the depth of 15 m to 20 m. This is mainly because this depth of 15–20 m is at the interface of the soft clay and silt. When excavating in the upper soft clay layer, due to the thixotropy of the soft clay, it is necessary to grout the bentonite between the prefabricated segments and the soil to ensure the stability of the shaft. The parameters of bentonite need to be strictly controlled in this process. The soil parameter changes abruptly at the interface of the stratum, and the bentonite parameter fails to adjust in time, leading to a large deformation. This process is not simulated by the model, leading to differences in results.

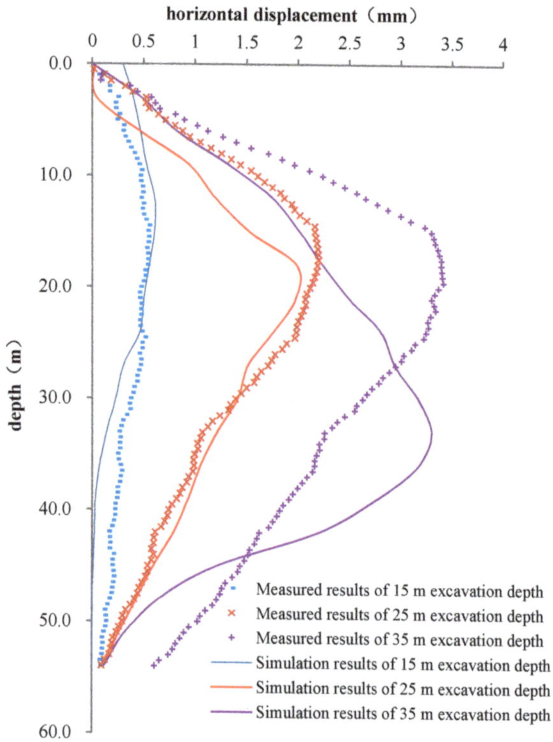

Figure 14. Simulation and measured results of the horizontal displacement of the shaft along depth of 15 m away from shaft center.

7.4. Analysis of Lateral Earth Pressure

The lateral earth pressure results along the excavation depth at 15 m away from the shaft center by simulation are shown in Figure 15, compared with the field measurements. The numerical simulation results show that the lateral earth pressure decreases slightly during the excavation process; the lateral earth pressure at the construction ending is about 7% lower than that at the construction beginning. In general, there is little difference between the simulated and measured values of the lateral earth pressure along the excavation depth, and the numerical simulation could roughly simulate the earth pressure during the excavation process.

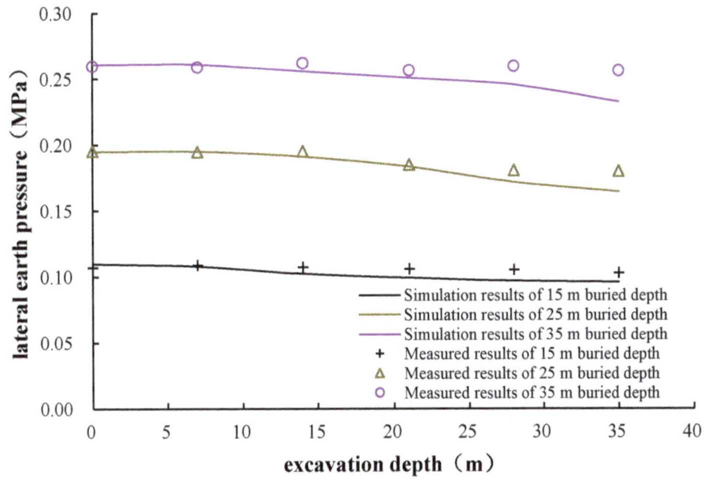

Figure 15. Simulation and measured results of lateral earth pressure at 15 m away from shaft center.

When the excavation finished, the numerical simulation, field measurement, and theoretical calculation of the lateral earth pressure at different depths of 15 m away from shaft center are shown in Figure 16. Both the numerical simulation results and the measured results show that the lateral earth pressure is slightly smaller than the static earth pressure and much larger than the active earth pressure. This shows that the horizontal deformation of shaft segments is relatively small during the excavation process, and the caisson soil is far away from the stress state that actively and completely squeezes the segments. The shaft excavation has little effect on the lateral earth pressure of the soil.

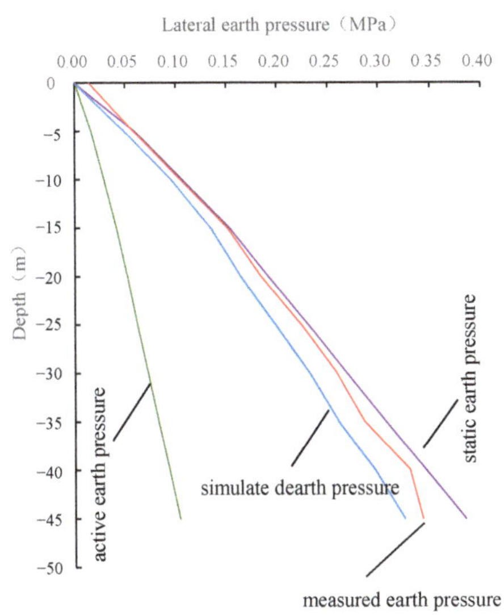

Figure 16. Simulation and measured results of lateral earth pressure at 15 m away from shaft center.

7.5. *Stability Analysis during Shaft Excavation*

During the traditional excavation of the shaft, the soil is probably at risk of instability [18]. The monolithic stability safety factor is an important evaluation index for the stability of the shaft excavation and the inrush risk on the excavation surface.

$$F_s = \frac{\sum R_n}{\sum T_n}$$

where F_s is the stability safety factor; R_n is the normal shear stress at a point on the sliding surface (kN); T_n is the normal shear strength at a certain point on the sliding surface (kN).

The numerical simulation model can automatically search for the most likely failure surface and calculate the stability safety factor of the shaft during the excavation process. The simulation results of the stability safety factor are shown in Figure 17. It can be seen from Figure 17 that during the excavation of the VSM shaft, the stability safety factors are all greater than 1.8, the overall risk of soil instability and water inrush is very small, and the construction of the VSM shaft is relatively safe due to the injection of water always during the construction. In the range of the excavation depth less than 15 m, the stability safety factor decreases continuously with the excavation of the shaft. The stability of the shaft is the worst when the excavation depth is in the range of 15–20 m. When the excavation depth exceeds 20 m, the stability increases slightly. The reason for this phenomenon is that, with the increase in the excavation depth, the sliding force provided by the overlying soil is far greater than the anti-sliding force provided by it. The stability of the structure is in the most dangerous state when the excavation reaches the interface between the soft clay and silt. However, the scope of the most dangerous failure surface expands when the depth exceeds a certain depth, which inhibits the tendency of the soil to become unstable. Then, the deeper the excavation depth, the smaller the hydraulic gradient of the pore water in the soil infiltrating into the shaft, and lower the seepage force, which also increases the stability in the deeper stage of the excavation. In general, the risk of soil instability and water inrush during the excavation process of the shaft is very small. The VSM is a relatively reliable construction method for the shaft when full of water inside the caisson. The only attention that should be paid is when the water discharged; but at that moment the caisson bottom has already been sealed.

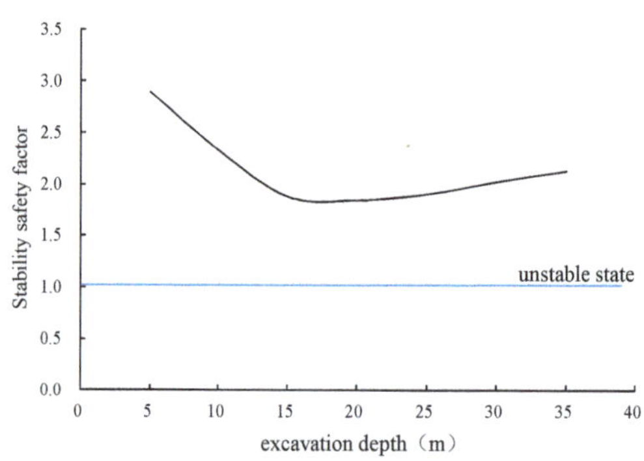

Figure 17. Stability safety factor of shaft excavation.

8. Discussion

The soft clay is widely distributed in the Shanghai area. Because soft clay has a strong viscoplasticity and also poor permeability, the construction speed of the underground space structure would greatly affect the soil deformation. The longer the construction time, the greater the soil deformation. The construction speed of in-situ assembling caisson technology is significantly faster than that of the cast-in-place concrete structure, which reduces the disturbance of the construction on the surrounding soil.

The process of the underground space excavation inevitably encounters temporary stopping. Engineering practice shows that, in the Shanghai area where soft clay is widely distributed, the construction stopping of underground buildings would still bring certain deformation and settlement to the environment because of creep [19,20]. The VSM is no exception. However, because the VSM construction is an undrained operation, the stress disturbance to the surrounding soil is very small, and the safety and stability of the shaft during the stopping period are relatively high. The construction of the entire VSM is preferred to the continuous operation to minimize the creep deformation. During the temporary stopping, the water level in the shaft should be raised to reduce settlement. The relevant parameters of bentonite grouting should be adjusted in time to reduce the lateral deformation of the soil layer when excavating to the interface of stratum, which ensures the construction safety and controls the deformation.

9. Conclusions

This paper takes the Shanghai Zhuyuan Bailonggang Sewage Connecting Pipe Project as the engineering background and briefly describes the technical aspects of in-situ assembling caisson technology and construction procedures of VSM. Ground settlements and deep layered displacements were monitored during the VSM construction, and corresponding theoretical calculations and numerical simulations were compared. Several important conclusions are drawn as follows:

(1) The VSM has the advantages of deep excavation depth, fast construction speed, small site usage, high economic efficiency, and strong stratum adaptability. It is an excellent construction method for urban vertical excavation using prefabricated in-situ assembly to build caisson, also with surrounding environmentally-friendly merit. The ground settlement and deep layered deformation during construction, both from measured results and numerical simulation, show that the VSM could be constructed underwater with little effect on the surrounding environment and high construction safety. It has obvious advantages in soft soil areas, which generally require engineering dewatering.

(2) Ground settlement is mainly concentrated within 30 m away from the shaft center. The maximum ground settlement caused by VSM excavation is about 15.2 mm and the maximum horizontal displacement is about 3.74 mm during the excavation depth of 40 m. When the shaft is excavated to the soil interface, the horizontal displacement of the stratum is the largest. The effect of shaft excavation on the lateral earth pressure is very small. The lateral earth pressure is slightly reduced by about 7% during the construction process. Compared with other construction techniques, its disturbance to the surrounding soil is really small.

(3) During the VSM construction excavation of shaft, the stability safety factors are all greater than 1.8. The risk of soil instability and water inrush during the shaft construction excavation is very small. The VSM is a relatively safe and reliable shaft construction method underwater, without the requirement of dewater in the soft soil area. The seepage stability of the VSM should be paid attention when water discharging.

(4) The numerical simulation results of the ground settlement, deep layered displacements, the uplift of the excavation bottom, and the lateral earth pressure are relatively accurate, verified by field measurement. It can effectively simulate the construction site conditions. The traditional foundation pit stress and deformation theory is relatively accurate for the calculation of the earth pressure, but the calculation results of the ground deformation have larger errors to field measurements, so the estimation accuracy is not high.

(5) The shaft should be constructed continuously to avoid the creep deformation of the soft clay during temporary stopping. If one has to, the water level in the shaft should be raised. The relevant parameters of bentonite grouting should be elaborately controlled in time to reduce the lateral deformation of the soil layer when excavating to the interface of stratum for stability control.

Author Contributions: Conceptualization, J.Z.; methodology, C.L.; software, C.L. and Z.L.; validation, J.Z., C.L., J.X. and Z.Z.; formal analysis, C.L., J.X. and Z.Z.; investigation, J.Z., C.L., J.X.; resources,

J.Z., J.X. and Z.Z.; data curation, C.L. and Z.Z.; writing—original draft preparation, C.L. and Z.Z.; writing—review and editing, J.Z. and C.L.; visualization, J.Z.; supervision, J.Z.; project administration, J.Z., J.X. and Z.Z.; funding acquisition, J.Z., J.X. and Z.Z. All authors have read and agreed to the published version of the manuscript.

Funding: The research work herein was sponsored by Shanghai Sailing Program, Science and Technology Commission of Shanghai Municipality (20YF1416200) and Tongji consulting service program for Shanghai Road and Bridge (Group) Co., Ltd (kh0023020212484).

Institutional Review Board Statement: Not applicable.

Informed Consent Statement: Not applicable.

Data Availability Statement: Not applicable.

Conflicts of Interest: The authors declare no conflict of interest.

References

1. First application of vertical shaft method (VSM) in North America. *Undergr. Eng. Tunn.* **2012**, *2*, 66.
2. Ma, Z. The engineering application of vertical sinking method (VSM). *Urban Constr. Theory Res.* **2021**, *25*, 67–68.
3. Peter, S. Vertical shaft machines: State of the art and vision. *Acta Montan. Slovaca* **2007**, *12*, 208–216.
4. Schmäh, P.; Berblinger, S. VSM shaft sinking technology: Mechanized shaft sinking with the VSM in different projects for subway ventilation shafts. *Proc. World Tunn. Congr.* **2014**, *III*, 123.
5. Vertical shaft sinking machine (VSM). *Equip. Geotech. Eng.* **2017**, *18*, 47–48.
6. Zhang, Z.; Xu, J.; Wang, S.; Jiang, H.; Hao, L.; Fu, W. Underwater mechanical tunneling construction technology of ultra-deep assembled shaft in water-rich strata-Taking a caisson parking facility construction project in Nanjing as an example. *Tunn. Constr.* **2022**, *42*, 492–500.
7. Huang, M.; Zhang, Z.; Xu, J.; Jiang, H.; Bao, H.; Liu, X. Field measurement of shaft wall stress based on VSM caisson construction process-a case study of Nanjing caisson underground intelligent parking garage project. *Tunn. Constr.* **2022**, *42*, 1033–1043.
8. Jiang, H.; Bao, H.; Lin, Y. Application of fabricated shaft design and construction technology-a case study of a caisson underground garage project in Nanjing. *Tunn. Constr.* **2022**, *42*, 463–470.
9. Ma, J. Numerical simulation of retaining wall in deep foundation based on COMSOL Multiphysics. *Highw. Eng.* **2018**, *1*, 256–260.
10. Lin, Z.; Zhang, B.; Yang, D. Simulation Study on Foundation Pit Excavation in Soft Soil Considering Effect of Creep and Seepage. *Mod. Tunn. Technol.* **2020**, *57*, 91–98.
11. Shi, Z.; Li, J.; Qin, B.; Li, B.; Li, S. Mechanical Characteristics and Cracking Control of Large Caisson Foundation in Initial Sinking Stage. *China Railw. Sci.* **2021**, *42*, 9–18.
12. Zhao, X.; Zhan, W.; Yan, X.; Wang, J.; Yang, Z.; Gong, X. Experimental study and simulation on deformation characteristics of caissons in sand under horizontal loads. *Chin. J. Geotech. Eng.* **2021**, *43*, 80–83.
13. Prater, E.G. An examination of some theories of earth pressure on shaft lining. *Can. Geotech. J.* **1977**, *14*, 91–106. [CrossRef]
14. Zhu, Z.; An, C.; Zhu, Y.; Liu, Z. Research on soil pressure theory of deep subway shaft. *Chin. J. Rock Mech. Eng.* **2013**, *2*, 3776–3783.
15. Committee of Housing and Urban-Rural Development of Shanghai. *Technical Code for Excavation Engineering (DG/TJ 08-61-2018)*; Tongji University Press: Shanghai, China, 2018.
16. Bowles, J.E. *Foundation Analysis and Design*; McGraw Hill: New York, NY, USA, 1988.
17. Hsieh, P.G.; Ou, C.Y. Shape of ground surface settlement profiles caused by excavation. *Can. Geotech. J.* **1998**, *35*, 1004–1017. [CrossRef]
18. Clough, G.W.; Duncan, J.M. Finite element analyses of retaining wall behavior. *J. Soil Mech. Found. Div.* **1971**, *97*, 1657–1673. [CrossRef]
19. Chen, Y.; Chen, J.; Zhu, W.; Lin, Y.; Xia, H. Measurement of mechanical performance of circular caisson in deep soft soil layer. *Build. Technol. Dev.* **2021**, *48*, 84–86.
20. Deng, Y.; Mao, W.; Yu, L.; Zhu, Y.; Xie, K. Consolidation and creep coupling model for soft soil considering temperature effect. *Eng. Mech.* **2022**, *39*, 103–113.

materials

MDPI

Article

A Model of Stress-Damage-Permeability Relationship of Weakly Cemented Rocks under Triaxial Compressive Conditions

Shizhong Zhang [1,2], Gangwei Fan [1,*], Dongsheng Zhang [1], Wenping Li [2], Tao Luo [1], Shuaishuai Liang [3] and Zhanglei Fan [1]

1 School of Mines, China University of Mining & Technology, Xuzhou 221116, China
2 School of Resources and Earth Science, China University of Mining & Technology, Xuzhou 221116, China
3 School of Mines, Saint Petersburg Mining University, 199106 Saint Petersburg, Russia
* Correspondence: fangw@cumt.edu.cn; Tel.: +86-136-5520-3693

Abstract: To unravel the permeability variation mechanism of weakly cemented rocks (WCR), the paper conducted triaxial permeability tests on weakly cemented sandstones (WCS) collected from the Jurassic formation in northwest China. The paper identified the correlation of WCS permeability versus porosity, cementation structure, and mineral composition, further developing a model to characterize the WCS stress–damage–permeability relationship. The research indicated that the WCS permeability was initially high due to the naturally high porosity, large pore diameter, and loose particle cementation, thus favoring a significant decline as pore convergence in the compaction stage. In the residual stage, kaolinite and montmorillonite minerals disintegrated into water and narrowed fractures, causing a slight permeability increase from the initial to the maximum and residual stages. The WCS matrix fracturing was phenomenologically accompanied by clay mineral disintegration. By assuming that the matrix can be compressed, jointed, and fractured, the paper defined a damage variable *D* and accordingly developed a stress–damage–permeability relationship model that incorporated matrix compression, jointing, and fracturing. The model can describe the WCS permeability regime regarding the high initial permeability and slight difference of the maximum and residual permeabilities versus the initial.

Keywords: weakly cemented rocks; triaxial compression; permeability; matrix fracturing; stress–damage–permeability

Citation: Zhang, S.; Fan, G.; Zhang, D.; Li, W.; Luo, T.; Liang, S.; Fan, Z. A Model of Stress-Damage-Permeability Relationship of Weakly Cemented Rocks under Triaxial Compressive Conditions. *Materials* **2023**, *16*, 210. https://doi.org/10.3390/ma16010210

Academic Editor: Carmelo Majorana

Received: 26 November 2022
Revised: 23 December 2022
Accepted: 23 December 2022
Published: 26 December 2022

1. Introduction

Weakly cemented rocks (WCR) widely exist in nature and contribute to the impact of geotechnical projects on the localized water system and ecological environment. Understanding WCR permeability responses to different stress conditions helps provide guidance for operating hydropower, underground tunneling, and mine projects [1–4]. In northwest China, the rock units in Jurassic and Cretaceous formations are often weakly cemented, generally featuring young sedimentation, low strength, and loose cementation. Such properties cause weakly cemented rocks to be sensitive to engineering perturbations [5,6]. Moreover, the arid and semi-arid climate, water resource scarcity, and eco-environmental fragility in northwest China further aggravate the impact of WCR permeability increase on groundwater depressurization and thus environmental deterioration [7,8]. Therefore, understanding the WCR permeability variation mechanism is of great significance for both geotechnical practice and eco-environmental protection.

The current progress of weakly cemented rocks is mainly in rock properties and permeability [9–11]. In rock mechanics, Sharma and Wang et al. conducted laboratorial tests and quantified the correlation of bulk density and P-wave velocity versus axial compressive strength and elastic modulus of weakly cemented mudstone (WCM) and sandstone

(WCS) [12,13]. Konstantinou et al. analyzed the relationship of WCS tensile strength with fracture patterns, surface features, and cementing degree [14]. Li and Zhao et al. studied the stress–strain, acoustic emission, and energy evolution characteristics of weakly cemented rocks under Brazilian test conditions, by which their relationship was discussed [15,16]. Liu et al. conducted triaxial compression tests and analyzed the impact of water content and confining pressure on fracture pattern and stress–strain curves of WCM and WCS samples [17]. Tommasi and Ye et al. proposed that WCR mechanical behavior is correlated to physical and structural properties, further analyzing the impact of pore abundance, mineral composition, and matrix-pore continuity on rock strength [18,19]. Regarding WCR hydraulics, Nguyen and Ruistuen et al. analyzed the impact of the stress path on WCS permeability and identified that the permeability had a significant decline when the shear strength was exceeded, which was attributed to particle friction and pore tortuosity increase [20,21]. You et al. and Liu et al. analyzed the impact of temperature, confining pressure, and water content on WCR permeability and developed a fractional quaternion model to describe WCR creep behavior in three continuous stages [22,23]. Liu et al. analyzed WCS permeability variation with porosity and mercury injection rate, by which the pore–throat ratio was considered a major indicator of permeability mutation [24]. Lyu et al. conducted laboratorial tests and analyzed WCS permeability features with varying particle diameters, revealing that the permeability coefficient was positively correlated to particle diameter in the same confining pressure and hydraulic pressure condition, between which the pore water pressure had greater impacts [25].

Previous research has investigated WCR strength properties and tried qualitative analysis on the permeability variation from physical perspectives, consolidating the basis for this paper. According to knowledge from geotechnical practices, the microstructural damage induced by stress environment change should be the critical contribution to WCR permeability variation. In this context, the paper collected WCR samples from the Jurassic formation in Ili mine fields, China and conducted complete stress–strain permeability tests using Rock Top multi-physics platform. The paper further studied the impact of porosity feature, cementation structure, and mineral composition on permeability variation. A WCR representative element volume (REV) model was constructed by considering physical mechanics and permeability property, by which, coupled with defining a damage variable D, a WCR stress–damage–permeability relationship model was developed. This research can provide theoretical reference for geotechnical operation and eco-environmental protection in a weakly cemented rock environment.

2. Materials

2.1. Sampling

The research first collected WCS samples from the Middle Jurassic Xishanyao Formation in Ili No. 4 Mine in Ili, Xinjiang Province, China. The downhole sampling adopted pump-suction reverse circulation drilling technology to maintain sample intactness to the maximum extent. The sampling position was 138.55 to 147.34 m below the topography, as shown in Figure 1.

2.2. Rock Physics

As indicated by some studies, rock mass permeability is closely correlated to the porosity, cementing form, and sediment percentage [26–29]. Before laboratorial permeability tests, the paper conducted mercury intrusion tests, scanning electron microscope (SEM, Tescan, Brno, Southern Moravia, Czech Republic) imaging, and X-ray diffraction (XRD, Bruker, Billerica, MA, USA) measurements to obtain WCS porosity, cementation structure, and mineral composition data.

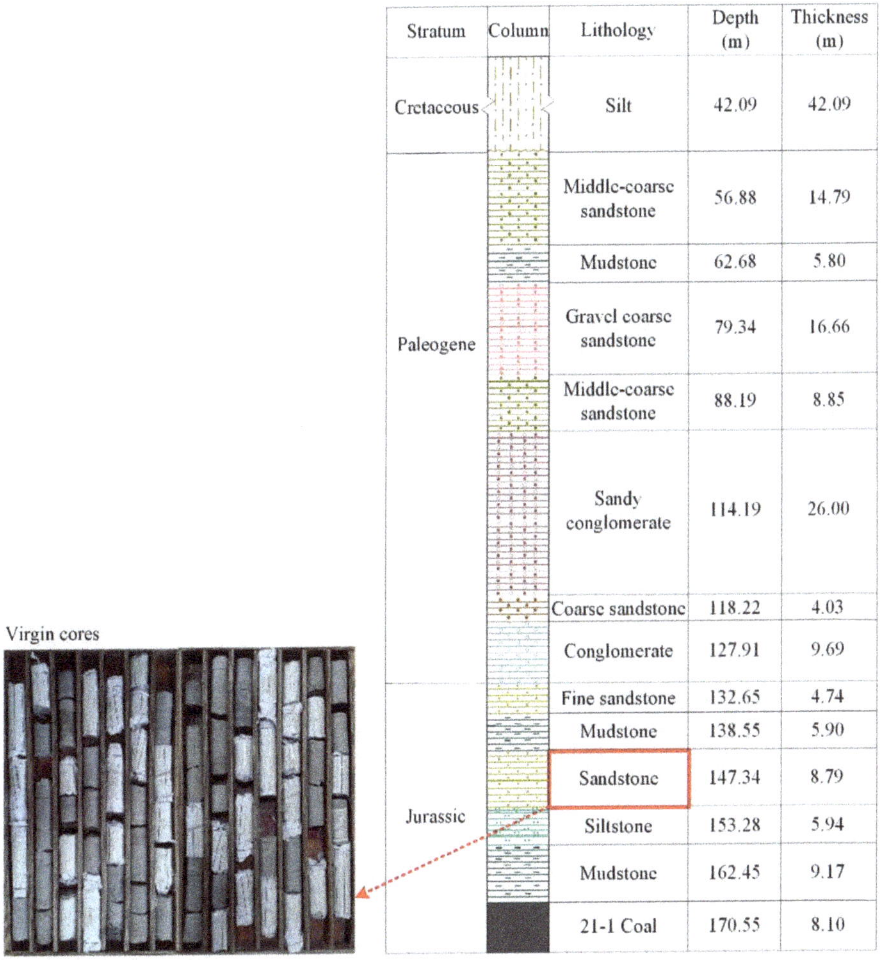

Figure 1. Borehole profile and WCS sampling position.

Mercury intrusion tests measured WCR porosity of 14.78%. The mercury injection rate of WCR was fast, indicating that the proportion of large-diameter pores in rocks was high. At the same time, the mercury ejection rate of WCR was slow, indicating that the pore connectivity of WCR was poor and the pores were mainly closed or partially closed. Rock pores characterize medium diameters: 1.8–10 µm diameter dominated, taking 33.53% of the total amount, followed by 10–50 µm diameter by 30.43%. WCR was mainly composed of intergranular middle pores [30].The mercury injection–ejection curves and porosity distribution are shown in Figure 2a,b, respectively. Figure 2c shows SEM imaging results, indicating that the pores were extensive and partially filled with fragments, compared to which the mineral particles had greater diameters about 25–60 µm and loosely distributed with explicit boundaries. The XRD measurement obtained WCS mineral compositions, as shown in Figure 2d, revealing that quartz, kaolinite, and montmorillonite took 90% of the total mineral content, among which the quartz and kaolinite minerals were about 34.7% and 33.2%, almost equivalent. As indicated by the XRD results, WCS samples were rich in clay minerals including kaolinite, montmorillonite, and illite, with a percentage of about 55.9%.

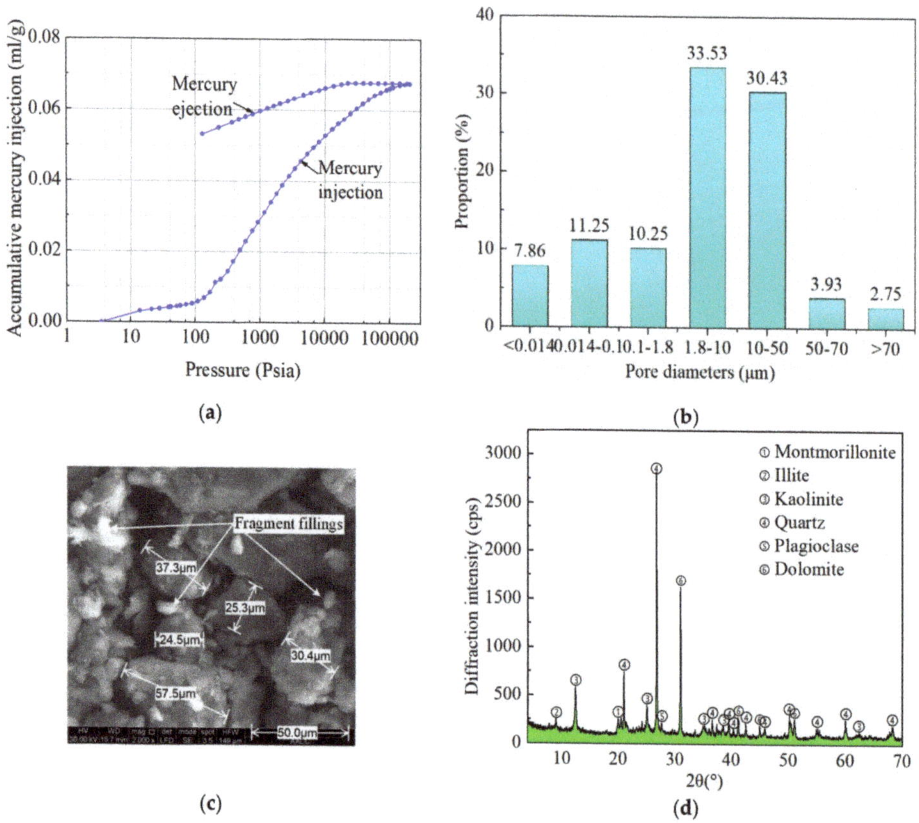

Figure 2. Physical properties of weakly cemented rocks: (**a**) mercury injection–ejection curves; (**b**) pore diameter distribution; (**c**) microscopic structures of pores and minerals via SEM imaging; (**d**) mineral composition via XRD measurement.

3. Complete Stress–Strain Permeability Tests

3.1. Methods

Permeability tests were conducted using the Rock Top multi-physics platform. The platform included a controller system for parameter and servo-control configuration, a hydraulic loading system for executing loading and unloading operations along axial and radius directions, and a monitoring system for sample encapsulation and data (displacement, pressure, and flux) monitoring. The maximum axial stress, maximum confining pressure, and maximum water pressure of the loading system were 500 MPa, 60 MPa, and 50 MPa, respectively. In order to accurately measure the axial deformation and radial strain, two Linear Variable Displacement Transducer (LVDT) displacement sensors were placed in parallel on both sides of the sample with a range of 12 mm and an accuracy of 0.001 mm [31]. Figure 3 exhibits the platform and its three components. The virgin cores were standardized into cylinders with a 100 mm height and 50 mm diameter; the parallelism tolerance of both ends was controlled within 0.1 mm and the diameter difference of both ends were within 0.2 mm. The tests adopted the displacement control method. The samples were saturated using pure water and then tested by keeping the hydraulic pressure difference and loading rate constant. Parameters of the permeability tests as shown in Table 1.

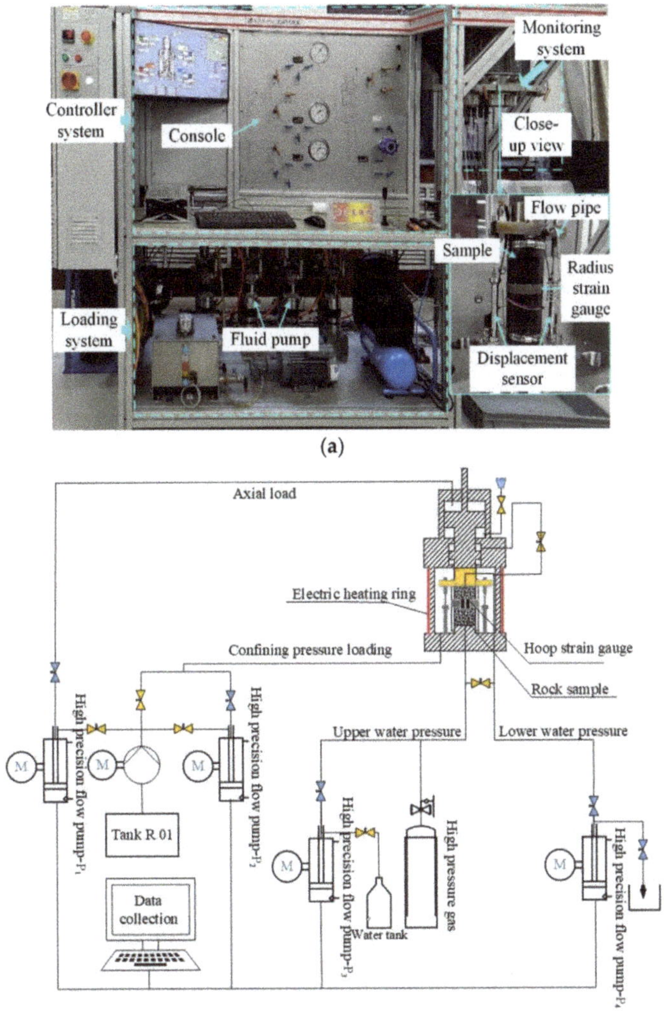

(a)

(b)

Figure 3. Rock Top multi-physics platform: (**a**) physical picture; (**b**) schematic diagram.

Table 1. Parameters of the permeability tests.

Projects	Confining Pressure (MPa)	Osmotic Pressure by Both Ends (MPa)	Hydraulic Difference (MPa)	Loading Controlled by	Loading Rate (mm/min)
1	2.0	Top: 0.5 Bottom: 1.5	1.0	Displacement	0.03
2	3.5				
3	5.0				

3.2. Results

Figures 4 and 5 show WCS stress–strain–permeability curves and sample failure scenarios, respectively.

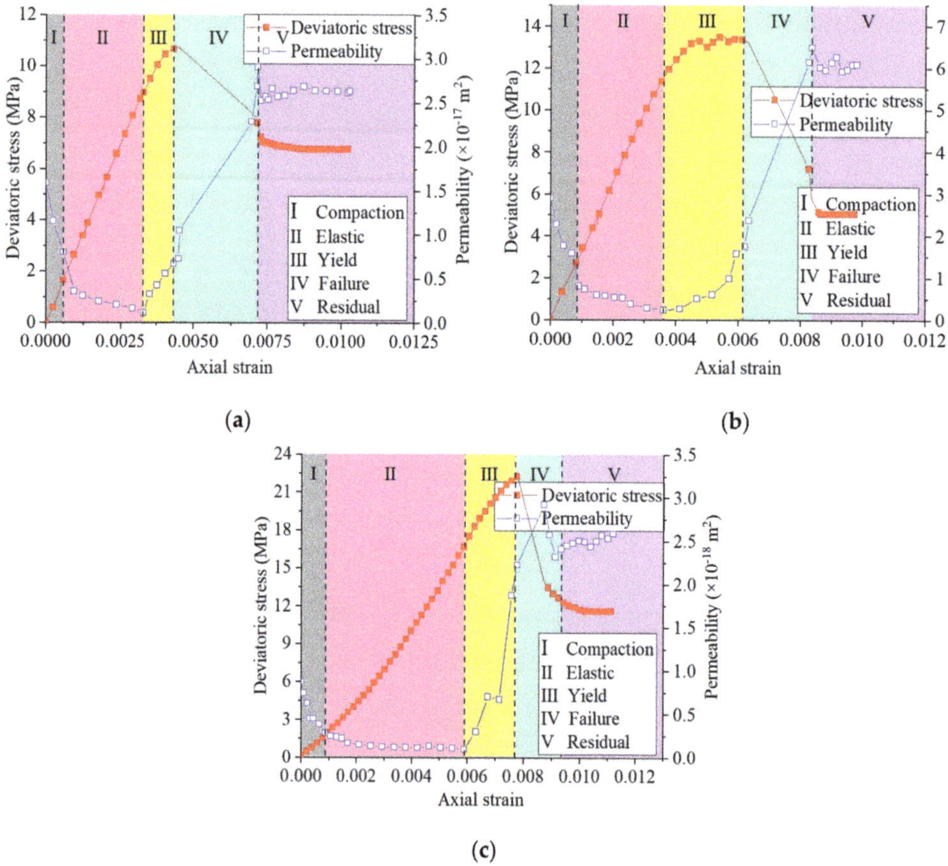

Figure 4. WCS stress–strain–permeability curves under different confining pressures: (**a**) confining pressure 2.0 MPa; (**b**) confining pressure 3.5 MPa; (**c**) confining pressure 5.0 MPa.

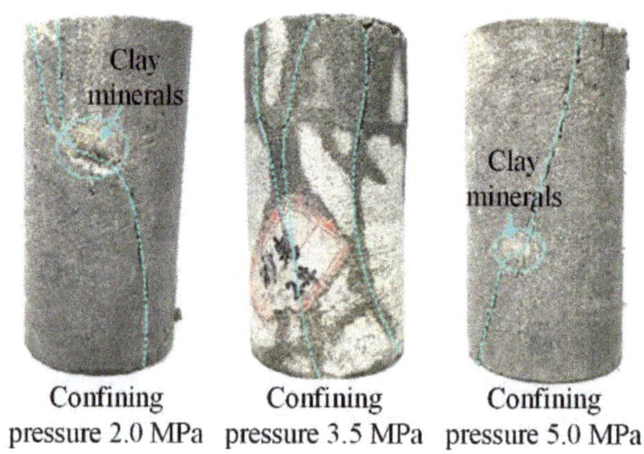

Figure 5. WCS sample failure.

Figure 4 shows that WCS permeability fluctuated throughout the complete stress–strain stages. The permeability first declined due to rock matrix compression, followed by a further but slight decline to the minimum in the elastic stage. Since rock yielding, the permeability increased and reached the maximum in the middle failure stage. There was again a permeability decline in the later failure stage, after which the permeability was almost unchanged in the residual stage. Of note is that the peak permeability was behind the peak deviatoric stress. An explicit feature of weakly cemented rocks was that the initial permeability (k_0) was relatively high and had a slight difference against the maximum (k_{max}) and residual permeability (k_{res}); for example, under the 5 MPa confining pressure condition, the ratio k_{max}/k_0 and k_{res}/k_0 was only 3.34 and 2.85, respectively. Table 2 lists the three permeability thresholds.

Table 2. WCS permeability thresholds (initial, maximum, and residual) in the complete stress–strain stages.

Lithology	Confining Pressure (MPa)	k_0 (m^2)	k_{max} (m^2)	k_{res} (m^2)	Ratio		
					k_{max}/k_0	k_{res}/k_0	k_{max}/k_{res}
Sandstone	2.0	1.59×10^{-17}	2.91×10^{-17}	2.61×10^{-17}	1.83	1.64	1.11
	3.5	2.91×10^{-18}	6.50×10^{-18}	6.06×10^{-18}	2.23	2.08	1.07
	5.0	8.75×10^{-19}	2.92×10^{-18}	2.49×10^{-18}	3.34	2.85	1.17

From Figure 5, it can be seen that the WCS sample failure was mainly characterized by shear fracturing; along fractures there was explicit disintegration of clay minerals.

3.3. Relationship between WCR Permeability and Physical Properties

The permeability variation indicated that WCR permeability was closely correlated to the porosity, cementation structure, and mineral composition. The high initial permeability was directly dependent on the high virgin porosity, large pore diameter, and loose connection of mineral particles. The permeability significantly decreased in the compaction stage, practically resulting from the loading-induced compression on extensive pores and thus flow pathways. In the failure stage, the maximum permeability was slightly different than the initial permeability since the abundant clay minerals such as kaolinite and montmorillonite expanded due to encountering water and partially narrowed the fluid flow pathways. As a result, the maximum permeability kept a low level. The residual stage experienced fracture propagation and closure and clay mineral disintegration, thus favoring an almost constant permeability due to a dynamic balance between fresh and existing fractures. In addition, clay minerals took up to 50%, being a major part of the WCR matrix. The clay mineral disintegration suggested that the WCR matrix fractured under the triaxial compression condition.

4. Stress–Damage–Permeability Relationship Model

The above analysis indicated that the high porosity, weak cementation structure, and abundant clay minerals caused WCRs to characterize distinctive permeability regimes. Considering that permeability variation is a reflection of rock material damage induced by pressurization, the paper introduced a damage variable D to help quantify WCS permeability variation. On this basis, a stress–damage–permeability relationship model was then developed.

4.1. Model Construction

Based on WCS physics and permeability features identified above and existing studies [32], the paper placed the following hypotheses [33–36].

(1) Weakly cemented rocks were composed of matrix and fractures; the matrix contained natural pores that can be squeezed and closed in the compaction stage and allow matrix materials to be jointed.

(2) The matrix can fracture in response to triaxial compression, further affecting the effective aperture of fractures and thus permeabilities.

The weakly cemented rocks were described using representative element volume (REV) units as shown in Figure 6.

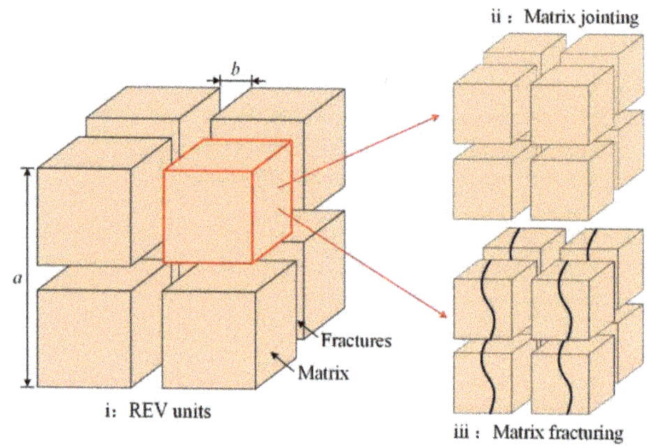

Figure 6. REV description of weakly cemented rocks.

The REV was the smallest micro element structure that represents the deformation of loaded coal rock mass. Accordingly, the porosity of REV units can be expressed via fracture aperture and matrix geometry [37]:

$$\frac{\phi_f}{\phi_{f0}} = \frac{a_0}{b_0} \bullet \frac{b}{a} = (1 + \frac{\Delta b}{b_0})\frac{a_0}{a} \tag{1}$$

where ϕ_f and ϕ_{f_0} are respectively the current and initial porosity of fractures, a_0 and b_0 are the initial geometry of matrix and pores, a and b are the current geometry of matrix and pores, and Δb represents the variation of pore geometry.

Existing studies on WCR porosity variation generally placed less focus on the impact of matrix deformation on permeability. In this research, the laboratorial tests revealed that weakly cemented rocks were of high porosity and large pore diameter; the large pores can be geometrically compressed in the compaction stage, and WCR samples well degraded in water. By implication, the paper assumed that the fractured matrix can joint in compressive and fluid flow conditions.

According to this hypothesis, the WCR pore geometry variation Δb was considered composed of matrix geometry change b_M and fracture aperture change b_F, with proportion of γ_M and γ_F respectively. Therefore,

$$\begin{cases} \Delta b = b_M + b_F \\ \gamma_M = b_M / \Delta b \\ \gamma_F = b_F / \Delta b \end{cases} \tag{2}$$

where b_M and b_F are respectively the contribution of matrix and fracture to pore geometry variation, and γ_M and γ_F are ratios of the respective contribution.

The volumes of fracture and matrix are:

$$\begin{cases} V_F = a^2 \bullet b_F; \quad V_{F0} = {a_0}^2 \bullet b_{F0} \\ V_M = a^2 \bullet b_M; \quad V_{F0} = {a_0}^2 \bullet b_{M0} \end{cases} \tag{3}$$

where V_F and V_M are volumes of fracture and matrix, with initial values expressed as V_{F0} and V_{M0}, respectively.

The fracture volume and matrix volume have another expression in natural strain and engineering strain [38]:

$$\begin{cases} V_F = V_{F0}\exp(-\frac{\Delta\sigma_e}{K_F}) \\ V_M = V_{M0}(1-\frac{\Delta\sigma_e}{K_M}) \end{cases} \tag{4}$$

where $\Delta\sigma_e$ is effective stress increment, K_F is fracture bulk modulus, and K_M is matrix bulk modulus. Considering that the matrix was difficult to compress, this work focused on the influence of K_F on the WCR permeability. Meanwhile, the determination of K_F was mainly obtained by fitting the stress–strain and permeability curves of WCR.

Combining Equations (2)–(4) obtained the pore geometry variation Δb:

$$\Delta b = (\frac{a_0}{a})^2\left[b_{F0}\exp(-\frac{\Delta\sigma_e}{K_F})+b_{M0}(1-\frac{\Delta\sigma_e}{K_M})\right] \tag{5}$$

The matrix itself was rather rigid but can deform with the contribution of matrix jointing. Consider that the matrix geometry variation obeyed a function of strain ε_e, and

$$b_M = b_{M0}\frac{a_0}{a}(\frac{1}{f(\varepsilon_e)})^2 \tag{6}$$

where ε_e is matrix strain and $f(\varepsilon_e)$ the matrix geometry change, following $f(\varepsilon_e) = \exp(-\varepsilon_e)$.

Substituting Equation (6) into Equation (5) obtained:

$$\Delta b = (\frac{a_0}{a})^2\left[b_{F0}\exp(-\frac{\Delta\sigma_e}{K_F})+b_{M0}(\frac{1}{f(\varepsilon_e)})^2\right] \tag{7}$$

Incorporating Equations (1), (2), and (7), with cubic low, obtained WCR permeability expression:

$$k = k_0(\frac{a_0}{a})^3\left\{1+(\frac{a_0}{a})^2\left[\gamma_F\exp(-\frac{\Delta\sigma_e}{K_F})+(1-\gamma_F)(\frac{1}{f(\varepsilon_e)})^2\right]\right\}^3 \tag{8}$$

where k_0 and k are respectively the initial and current permeability.

With continuous pressurization, WCR units damaged and then fractured as the damage gradually accumulated to a certain extent. The fracture initiation and development were accompanied by the matrix geometry variation. By assuming the number of natural fractures and fractures corresponding to the WCR unit damage was respectively N_0 and N, the matrix geometry variation can be written as:

$$\frac{a_0}{a} = \frac{N}{N_0} = \frac{N_0+\Delta N}{N_0} = 1+\frac{\Delta N}{N_0} = 1+D\varsigma \tag{9}$$

where ΔN refers to the fracture number increment, D is damage variable, and ς is the coefficient of matrix geometry variation.

Substituting Equation (9) into Equation (8) obtained a stress–damage–permeability relationship:

$$k = k_0\underbrace{(1+D\varsigma)^3}_{①}\left\{1+\underbrace{(1+D\varsigma)^2}_{①}\left[\underbrace{\gamma_F\exp(-\frac{\Delta\sigma_e}{K_F})}_{②}+\underbrace{\left(1-\gamma_F\right)(\frac{1}{f(\varepsilon_e)})^2}_{③}\right]\right\}^3 \tag{10}$$

where items ①, ②, and ③ represent the impact of fracture number, aperture convergence, and matrix joint on permeability, respectively.

4.2. Damage Variable Determination

Assume the damage variable D obeying Weibull distribution [39]:

$$P(F^*) = \frac{m_0}{F_0}\left(\frac{F^*}{F_0}\right)^{m_0-1}\exp\left[-\left(\frac{F^*}{F_0}\right)^{m_0}\right] \tag{11}$$

where F^* represents microstructure strength, such as deformation modulus, tensile strength, and cohesion, m_0 is coefficient of average degree, and F_0 is mean value of the strength parameter F^*.

By Drucker–Prager failure criterion, the microstructure strength F^* can be expressed as:

$$\begin{cases} F^* = f(\sigma^*) = \alpha_0 I_1 + J_2^{1/2} \\ \alpha_0 = \frac{\sin\varphi}{\sqrt{9+3\sin^2\varphi}} \\ I_1 = \sigma_1 + \sigma_2 + \sigma_3 \\ J_2 = \frac{1}{6}\left[(\sigma_1-\sigma_2)^2 + (\sigma_2-\sigma_3)^2 + (\sigma_1-\sigma_3)^2\right] \end{cases} \tag{12}$$

where φ is angle of friction, and σ_1, σ_2, and σ_3 refer to three effective principal stress. In this test, the second and third effective principal stresses (σ_2 and σ_3) were consisted equivalent. I_1 and J_2 are respectively the first and second invariants of effective stress tensor.

Considering that the damage degree varied with positions, it was obtained via elastic damage theory [33,40]:

$$\begin{cases} \sigma = E\varepsilon(1 - C_n D) \\ D = 1 - \exp\left[-\left(\frac{F^*}{F_0}\right)^{m_0}\right] \end{cases} \tag{13}$$

where E is deformation modulus, and C_n is proportionality coefficient of the damage, reflecting the residual strength of WCR material.

To better reflect WCR mechanics and hydraulics regarding high initial permeability, rapid decline in the compaction stage, and slight change in the residual stage, the overall permeability variation was segmented into compaction elasticity, yielding damage, failure, and residual stages, corresponded to the complete stress–strain profile. Combining Equations (11)–(13) obtained a WCR statistical damage model:

$$\sigma_1 = \begin{cases} E\varepsilon_1 + 2\mu\sigma_3 & \varepsilon_1 \le \varepsilon_{ela} \\ E(\varepsilon_1 - \varepsilon_{ela})\left(1 - C_n + C_n \exp\left(-\frac{\frac{E(\varepsilon_1-\varepsilon_{ela})}{\sqrt{3}(\sigma_1-2\mu\sigma_3)}\left(\frac{\sin\varphi(\sigma_1-\sigma_{ela}+2\sigma_3)}{\sqrt{3+\sin^2\varphi+\sigma_1-\sigma_3}}\right)+\sigma_1-\sigma_{ela}-\sigma_3}{F_0}\right)^{m_0}\right) & \\ +\sigma_{ela} + 2\mu\sigma_3 & \varepsilon_{ela} \le \varepsilon_1 \le \varepsilon_{max} \\ E\left(\frac{\varepsilon_{max}}{\varepsilon_1}\right)^2 + 2\mu\sigma_3 & \varepsilon_{max} < \varepsilon_1 \le \varepsilon_{res} \\ \sigma_{res} & \varepsilon_1 > \varepsilon_{res} \end{cases} \tag{14}$$

where ε_{ela} is the strain by the end of the elastic stage, ε_{max} is the peak strain, ε_{res} is the strain at the start of the residual stage, and σ_{ela} measures the axial stress when axial strain is ε_{ela}.

Fitting the stress–strain data of each damage stage and residual coefficient determined from residual strength obtained m_0, F_0, ζ, and further the damage variable D of the WCR stress–strain curves under different confining pressures. Table 3 lists the curve fitting results.

Table 3. WCR damage parameters.

Lithology	Confining Pressure (MPa)	Statistical Parameters		
		m_0	F_0	ζ
Sandstone	2.0	3.59	7.02	1.74
	3.5	3.74	9.09	1.52
	5.0	1.74	14.47	1.02

4.3. Model Verification

To verify the WCR stress–damage–permeability relationship model, the statistical damage parameters summarized in Table 3 were further used to determine the damage variable D. Then, the WCR permeability variation under different confinement conditions can be calculated by substituting D into Equation (10). The comparison between calculation results and measurement is shown in Figure 7.

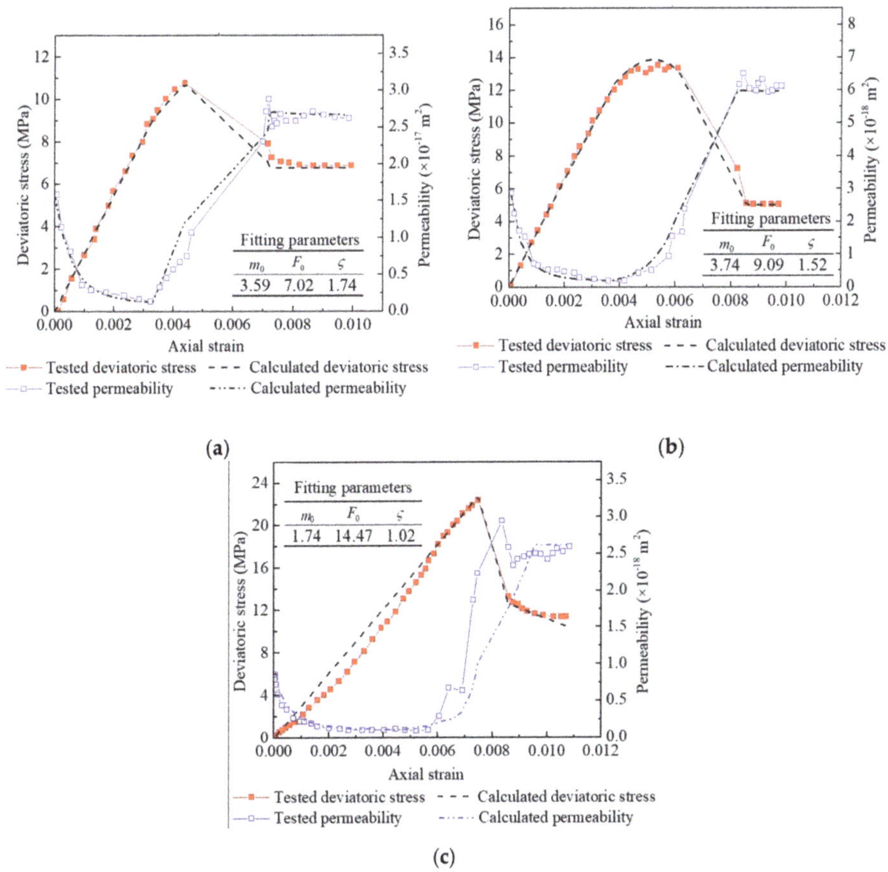

Figure 7. WCR permeability variation curves under different confinement conditions: (**a**) confining pressure 2.0 MPa; (**b**) confining pressure 3.5 MPa; (**c**) confining pressure 5.0 MPa.

Figure 7 suggests a good agreement between theoretical and laboratorial test results, with an error of 12% in deviatoric stress and 15% in permeability, Table 4. This result indicates that the stress–damage–permeability relationship model (Equation (10)) can well describe WCR permeability regime in complete stress–strain stages.

Table 4. Error of theoretical WCR deviatoric stress and permeability versus laboratorial measurements.

Lithology	Confining Pressure (MPa)	Mean Relative Error (%)	
		Deviatoric Stress	Permeability
Sandstone	2.0	7.24	7.13
	3.5	5.59	6.32
	5.0	11.39	14.79

5. Discussion

The WCRs in Jurassic and Cretaceous generally characterize young sedimentation, low strength, and loose cementation. In this work, the Jurassic weakly cemented sandstone is selected from Ili No. 4 Coal Mine in Xinjiang Province, China. The sample selection is very typical and representative. Therefore, the authors believe that the results of permeability test and the established stress–damage–seepage relationship model of WCR based on the typical samples can express its general permeability characteristics. Of course, in the future, the WCRs in more different areas should be tested and fitted to further verify the validity and universal applicability of the established stress–damage–seepage relationship model.

It can be seen from Figure 7c that when the confining pressure is 5.0 MPa, the fitting error of the deviatoric stress and permeability of weakly cemented sandstone is more than 10%. This may be due to the fact that when the confining pressure is 5.0 MPa, the weakly cemented sandstone has obvious elastic brittle failure characteristics (as shown in Figure 7c), and its stress–strain and permeability curves have mutation points (or distortion points). The existence of mutation points destroys the continuity of the curves, which makes its fitting error larger. In future studies, more weakly cemented samples under high confining pressure should be tested and fitted to further explain the above phenomenon.

6. Conclusions

Weakly cemented rocks showed permeability fluctuation in the complete stress–strain stages, including a decrease in compaction and elastic segments, an increase in yield and failure segments, and an again decrease to a leveling-off value in the ultimate residual segment. Such a permeability regime was physically correlated to WCR porosity, cementation structure, and mineral composition properties. WCR material had a high initial permeability due to the naturally high porosity, large pore diameter, and loose cementation structure, thus favoring a rapid permeability decline due to pore compression. Since the yield stage, WCR clay minerals such as kaolinite and montmorillonite decomposed into water and narrowed the fractures, causing that the maximum and residual permeabilities were slightly greater than the initial. This minor difference indicated that WCR materials had a limited permeability responses in triaxial confinement conditions, which helps WCR strata maintain a good water-resisting functionality.

XRD measurements revealed that WCR materials were rich in clay minerals. Kaolinite, montmorillonite, and illite minerals took more than 50%, constituting a major part of the WCR matrix. Evident clay mineral disintegration was observed with sample fracturing, implying matrix fracturing in response to triaxial compression.

By assuming that WCR matrix can be compressed, jointed, and fractured, the paper defined a damage variable D and further developed a stress–damage–permeability relationship model. Then, a D determination method was established according to Weibull distribution and the Drucker–Prager principle. The relationship model was verified with errors in deviatoric stress and permeability within 15% with respect to laboratorial measurements, suggesting that the model can represent the permeability regime of weakly cemented rocks in the complete stress–strain stages.

Author Contributions: Conceptualization, D.Z. and W.L.; data curation, S.L. and Z.F.; investigation, T.L. and S.L.; methodology, G.F.; validation, S.Z., T.L. and Z.F.; writing—original draft; S.Z.; writing—review and editing; G.F., D.Z. and W.L. All authors have read and agreed to the published version of the manuscript.

Funding: This research was funded by the National Natural Science Foundation of China (Grant No. 52204161; 51974291), the Fundamental Research Funds for the Central Universities (Grant No. 2022QN1008; 2021ZDPY0226), the Jiangsu Funding Program for Excellent Postdoctoral Talent (Grant No. 2022ZB511), the Shanxi Province Unveils Bidding Project (Grant No. 20201101009).

Institutional Review Board Statement: Not applicable.

Informed Consent Statement: Not applicable.

Data Availability Statement: Not applicable.

Acknowledgments: The authors are grateful to the anonymous reviewers for their helpful comments.

Conflicts of Interest: The authors declare no conflict of interest.

References

1. Grgic, D.; Al Sahyouni, F.; Golfier, F.; Moumni, M.; Schoumacker, L. Evolution of gas permeability of rock salt under different loading conditions and implications on the underground hydrogen storage in salt caverns. *Rock Mech. Rock Eng.* **2021**, *55*, 691–714. [CrossRef]
2. Rutqvist, J.; Borgesson, L.; Chijimatsu, M.; Hernelind, J.; Jing, L.; Kobayashi, A.; Nguyen, S. Modeling of damage, permeability changes and pressure responses during excavation of the TSX tunnel in granitic rock at URL, Canada. *Environ. Geol.* **2009**, *57*, 1263–1274. [CrossRef]
3. Akbarimehr, D.; Aflaki, E. Site investigation and use of artificial neural networks to predict rock permeability at the Siazakh Dam, Iran. *Q. J. Eng. Geol. Hydrogeol.* **2019**, *52*, 230–239. [CrossRef]
4. Khanal, M.; Guo, H.; Adhikary, D. 3D Numerical Study of Underground Coal Mining Induced Strata Deformation and Subsequent Permeability Change. *Geotech. Geol. Eng.* **2019**, *37*, 235–249. [CrossRef]
5. Zhao, Z.; Lv, X.; Wang, W.; Tan, Y. Damage evolution of bi-body model composed of weakly cemented soft rock and coal considering different interface effect. *Springerplus* **2016**, *5*, 292. [CrossRef]
6. Zhang, S.; Fan, G.; Zhang, D.; Li, Q. Physical simulation research on evolution laws of clay aquifuge stability during slice mining. *Environ. Earth Sci.* **2018**, *77*, 278. [CrossRef]
7. Fan, G.; Chen, M.; Zhang, D.; Wang, Z.; Zhang, S.; Zhang, C.; Li, Q.; Cao, B. Experimental study on the permeability of weakly cemented rock under different stress states in triaxial compression tests. *Geofluids* **2018**, *2018*, 9035654. [CrossRef]
8. Zhang, S.; Fan, G.; Zhang, D.; Li, S.; Fan, Y.; Luo, T. Impacts of longwall mining speeds on permeability of weakly cemented strata and subsurface watertable: A case study. *Geomat. Nat. Hazards Risk* **2021**, *12*, 3063–3088. [CrossRef]
9. Dong, J.; Tu, C.; Lee, W.; Jheng, Y. Effects of hydraulic conductivity/strength anisotropy on the stability of stratified, poorly cemented rock slopes. *Comput. Geotech.* **2012**, *40*, 147–159. [CrossRef]
10. Holtzman, R. Micromechanical model of weakly-cemented sediments. *Int. J. Numer. Anal. Methods Geomech.* **2012**, *36*, 944–958. [CrossRef]
11. Yeh, P.; Lee, K.Z.; Chang, K. 3D Effects of permeability and strength anisotropy on the stability of weakly cemented rock slopes subjected to rainfall infiltration. *Eng. Geol.* **2020**, *266*, 105459. [CrossRef]
12. Wang, Z.; Li, W.; Wang, Q.; Liu, S.; Hu, Y.; Fan, K. Relationships between the petrographic, physical and mechanical characteristics of sedimentary rocks in Jurassic weakly cemented strata. *Environ. Earth Sci.* **2019**, *78*, 131. [CrossRef]
13. Sharma, M.S.R.; O'Regan, M.; Baxter, C.D.P.; Moran, K.; Vaziri, H.; Narayanasamy, R. Empirical relationship between strength and geophysical properties for weakly cemented formations. *J. Pet. Sci. Eng.* **2010**, *72*, 134–142. [CrossRef]
14. Konstantinou, C.; Biscontin, G.; Logothetis, F. Tensile strength of artificially cemented sandstone generated via microbially induced carbonate precipitation. *Materials* **2021**, *14*, 4735. [CrossRef]
15. Zhao, Y.; Liu, B. Deformation field and acoustic emission characteristics of weakly cemented rock under brazilian splitting test. *Nat. Resour. Res.* **2021**, *30*, 1925–1939. [CrossRef]
16. Li, P.; Li, Z.; Liu, B.; Teng, T.; Guo, J. Experimental investigation on the tensile properties of weakly cemented sandstone in China Shendong mining area. *Therm. Sci.* **2020**, *24*, 3987–3994. [CrossRef]
17. Liu, H.; Zhang, D.; Zhao, H.; Chi, M.; Yu, W. Behavior of weakly cemented rock with different moisture contents under various tri-axial loading stats. *Energies* **2019**, *12*, 1563. [CrossRef]
18. Yu, W.; Li, K.; Liu, Z.; An, B.; Wang, P.; Wu, H. Mechanical characteristics and deformation control of surrounding rock in weakly cemented siltstone. *Environ. Earth Sci.* **2021**, *80*, 337. [CrossRef]
19. Tommasi, P.; Verrucci, L.; Rotonda, T. Mechanical properties of a weak pyroclastic rock and their relationship with microstructure. *Can. Geotech. J.* **2015**, *52*, 211–223. [CrossRef]

20. Ruistuen, H.; Teufel, L.W.; Rhett, D. Influence of reservoir stress path on deformation and permeability of weakly cemented sandstone reservoirs. *SPE Reserv. Eval. Eng.* **1999**, *2*, 266–272. [CrossRef]
21. Nguyen, V.H.; Gland, N.; Dautriat, J.; David, C.; Wassermann, J.; Guélard, J. Compaction, permeability evolution and stress path effects in unconsolidated sand and weakly consolidated sandstone. *Int. J. Rock Mech. Min. Sci.* **2014**, *67*, 226–239. [CrossRef]
22. You, S.; Ji, H.; Wang, T.; Song, Z. Thermal and mechanical coupling effects on permeability of weakly cemented sandstone. *Emerg. Mater. Res.* **2018**, *7*, 100–108. [CrossRef]
23. Liu, J.; Jiang, H.; Meng, B.; Wang, L.; Yang, J.; Zhang, X. A four-element fractional creep model of weakly cemented soft rock. *Bull. Eng. Geol. Environ.* **2020**, *79*, 5569–5584. [CrossRef]
24. Liu, Q.; Sun, Y.; Li, J. Experimental study on seepage characteristics of Jurassic weakly cemented sandstone under water-rock interaction. *Geofluids* **2020**, *2020*, 854368. [CrossRef]
25. Lyu, X.; Zhao, Z.; Wang, X.; Wang, W. Study on the Permeability of Weakly Cemented Sandstones. *Geofluids* **2019**, *2019*, 8310128. [CrossRef]
26. Han, T.; Best, A.I.; Sothcott, J.; North, L.J.; Macgregor, L.M. Relationships among low frequency (2 Hz) electrical resistivity, porosity, clay content and permeability in reservoir sandstones. *J. Appl. Geophys.* **2015**, *112*, 279–289. [CrossRef]
27. Chao, Z.; Ma, G.; He, K.; Wang, M. Investigating low-permeability sandstone based on physical experiments and predictive modeling. *Undergr. Space* **2021**, *6*, 364–378. [CrossRef]
28. Singh, V.K.; Kumar, D.; Kashyap, P.S.; Singh, P.K.; Kumar, A.; Singh, S.K. Modelling of soil permeability using different data driven algorithms based on physical properties of soil. *J. Hydrol.* **2020**, *580*, 124223. [CrossRef]
29. Liu, X.; Xu, M.; Wang, K. Mechanism of permeability evolution for reservoir sandstone with different physical properties. *Geofluids* **2018**, *2018*, 5327895. [CrossRef]
30. Saki, M.; Siahpoush, S.; Khaz'ali, A.R. A new generalized equation for estimation of sandstone and carbonate permeability from mercury intrusion porosimetry data. *J. Pet. Explor. Prod. Technol.* **2020**, *10*, 2637–2644. [CrossRef]
31. Chen, L.; Zhang, D.; Zhang, W.; Guo, J.; Yao, N.; Fan, G.; Zhang, S.; Wang, X.; Wu, P. Experimental Investigation on Post-Peak Permeability Evolution Law of Saturated Sandstone under Various Cyclic Loading-Unloading and Confining Pressure. *Water* **2022**, *14*, 1773. [CrossRef]
32. Greco, O.D.; Ferrero, A.M.; Oggeri, C. Experimental and analytical interpretation of the behaviour of laboratory tests on composite specimens. *Int. J. Rock Mech. Min. Sci. Geomech. Abstr.* **1993**, *30*, 1539–1543. [CrossRef]
33. Fan, Z.; Fan, G.; Zhang, D. Permeability and energy evolution characteristics of heterogeneous coal and rock mass. *Nat. Resour. Res.* **2021**, *30*, 4493–4514. [CrossRef]
34. Li, L.; Tu, W.; Zhou, Z.; Shi, S.; Zhang, M.; Chen, Y. Dynamic unloading instability mechanism of underground cavern based on seepage-damage coupling. *KSCE J. Civ. Eng.* **2020**, *24*, 1620–1631. [CrossRef]
35. Jiang, A.; Yang, X.; Xu, M.; Jiang, T. Coupled hydrologic-mechanical-damage analysis and its application to diversion tunnels of hydropower station. *Adv. Civ. Eng.* **2021**, *2021*, 8341528. [CrossRef]
36. Wang, S.; Yang, T.; Zhang, Z.; Sun, Z. Unsaturated seepage-stress-damage coupling and dynamic analysis of stability on discrete fractured rock slope. *Environ. Earth Sci.* **2021**, *80*, 660. [CrossRef]
37. Fan, Z.; Fan, G.; Zhang, D.; Zhang, L.; Zhang, S.; Liang, S.; Yu, W. Optimal injection timing and gas mixture proportion for enhancing coalbed methane recovery. *Energy* **2021**, *222*, 119880. [CrossRef]
38. Qiao, T.; Zhang, G.; Hu, S.; Zhao, S. Research on constitutive model of microcrack compaction section in brittle rock. *J. Saf. Sci. Technol.* **2017**, *13*, 128–135. [CrossRef]
39. Pan, X.; Guo, W.; Wu, S.; Chu, J. An experimental approach for determination of the Weibull homogeneity index of rock or rock-like materials. *Acta Geotech.* **2020**, *15*, 375–391. [CrossRef]
40. Xiao, W.; Zhang, D.; Wang, X.; Yang, H.; Wang, X.; Wang, C. Research on Microscopic Fracture Morphology and Damage Constitutive Model of Red Sandstone Under Seepage Pressure. *Nat. Resour. Res.* **2020**, *29*, 3335–3350. [CrossRef]

materials

MDPI

Article

Experimental Investigation of Pre-Flawed Rocks under Dynamic Loading: Insights from Fracturing Characteristics and Energy Evolution

Guifeng Zhao [1,*], Lei Zhang [1], Bing Dai [1,*], Yong Liu [1,2], Zhijun Zhang [1] and Xinyao Luo [1,3]

1 School of Resource Environment and Safety Engineering, University of South China, Hengyang 421001, China
2 School of Physics and Optoelectronic Engineering, Shenzhen University, Shenzhen 518060, China
3 School of Resources and Safety Engineering, Central South University, Changsha 410000, China
* Correspondence: zgf_184@163.com (G.Z.); daibingusc@usc.edu.cn (B.D.); Tel.: +86-184-8360-3537 (G.Z.); +86-178-7341-5528 (B.D.)

Abstract: Different fractures exist widely in rock mass and play a significant role in their deformation and strength properties. Crack rocks are often subjected to dynamic disturbances, which exist in many fields of geotechnical engineering practices. In this study, dynamic compression tests were carried out on rock specimens with parallel cracks using a split hopkinson pressure bar apparatus. Tests determined the effects of strain rate and crack intensity on dynamic responses, including progressive failure behavior, rock fragmentation characteristics, and energy dissipation. Based on the crack classification method, tensile–shear mixed cracking dominates the failure of rock specimens under the action of impact loading. Increasing the flaw inclination angle from 0°–90° changes the predominant cracking mechanism from tensile cracking to mixed tensile–shear cracking. The larger the loading rate, the more obvious the cracking mechanism, which indicates that the loading rate can promote the cracking failure of rock specimens. The fragmentation analysis shows that rock samples are significantly broken at higher loading rates, and higher loading rates lead to smaller average fragment sizes; therefore, the larger the fractal dimension is, the more uniform the broken fragments of smaller sizes are. Energy utilization efficiency decreases while energy dissipation density increases with increasing strain rate. For a given loading rate, the energy absorption density and energy utilization efficiency first decrease and then increase with increasing flaw inclination, while the rockburst tendency of rock decreases initially and then increases. We also find that the elastic–plastic strain energy density increases linearly with the total input energy density, confirming that the linear energy property of granite has not been altered by the loading rate. According to this inherent property, the peak elastic strain energy of the crack specimen can be calculated accurately. On this basis, the rockburst proneness of granite can be determined quantitatively using the residual elastic energy index, and the result is consistent with the intensity of actual rockburst for the specimens.

Keywords: SHPB; flawed rocks; cracking mechanism; energy dissipation; rock-burst

Citation: Zhao, G.; Zhang, L.; Dai, B.; Liu, Y.; Zhang, Z.; Luo, X. Experimental Investigation of Pre-Flawed Rocks under Dynamic Loading: Insights from Fracturing Characteristics and Energy Evolution. *Materials* **2023**, *15*, 8920. https://doi.org/10.3390/ma15248920

Academic Editor: René de Borst

Received: 16 October 2022
Accepted: 6 December 2022
Published: 13 December 2022

1. Introduction

An increasing demand of energy and resources means rock engineering plays an increasingly important role in the construction of deep underground spaces in which the defective rock is disturbed by the dynamic loads of blasting and impact points [1,2]. Static compressive loads are derived from gravitational stress and tectonic stress [3–5], and dynamic loads can be inferred from measurements, such as drilling, quarrying, and earthquakes, as shown in Figure 1. The most typical rock engineering involves the excavation of underground caverns in which the dynamic loads from blasting act on the rock column. Natural rock inherently contains many pre-existing natural defects which significantly affect the stability of underground engineering structures, so accurately obtaining the fracture behavior of stable rock containing pre-existing defects under the impact load is of vital importance for construction safety and rock engineering disaster prevention [6,7].

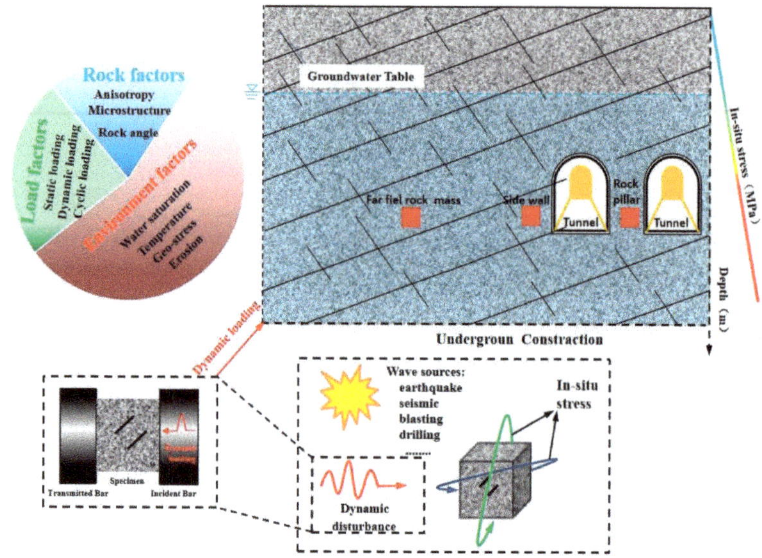

Figure 1. An overview of the influencing factors in rock engineering structures suffering from different impact loads.

In the past few decades, people have devoted themselves to the study of dynamic strength and fracture behavior under shock loading. The dynamic strength of rock materials has been shown to have a pronounced rate-dependent resistance. So far, studies on defective stable rock have mainly focused on purely static loading conditions. A number of static uniaxial compression tests were performed on rocks containing single or multiple defects. It is very difficult to use classical fracture mechanics to study the failure mechanism of rocks due to the heterogeneity of rock materials. Therefore, laboratory tests, as a basic method, have been widely used to study the mechanics and fracture of rocks under various loading conditions [8–11].

For instance, a group at MIT led by Einstein systematically investigated the strength and fracturing of rock and rock-like (i.e., gypsum) materials containing pre-existing flaws under static compressions [12–14] and found great impacts on the strength, deformation, crack propagation, and coalescence of the rock under static loading. Wu et al. [15] studied the crack strength of sandstone material with two parallel inclined cracks under uniaxial compression and concluded that it mainly contains three modes of shear, mixed shear tension, and wing crack tension. Wong and Chau et al. [16] studied the effect of pre-existing elliptical defects on the deformation and fracture behavior of natural rocks under static uniaxial compression and concluded that the geometry of elliptical defects determines the principal stress and ultimate failure mode of the rock specimen and elliptical form defects weaken its compressive strength and elastic modulus. Dai et al. [17–19] studied the mechanical and energy dissipation characteristics of granite under cyclic impact loading and the results showed different axial pressures. The damage value criterion established based on energy dissipation could characterize the relationship between damage and the number of impacts, which showed a slow increase, a steady increase, and a high-speed increase, and the damage value depended mainly on the last impact. Zhang et al. [20,21] studied crack development and damage patterns under combined dynamic–static loading of parallel double fractured rocks and concluded that under a constant strain rate in the triaxial compression tests, the confining pressure effect is very obvious for the dynamic triaxial compressive strength and secant modulus and the 45° fractures represent high synergy between the slip rate of the cracks in the bridge zone and the dissipation energy generated by the specimens. Bobet et al. [22] and Li et al. [23] studied the mechanical behavior and dynamic cracking of single and multi-defect rocks under static uniaxial compressive loading tests and concluded that they are both significantly affected by the shape stability of geometric defects, including defect length, angle, spacing between the two defects, and the length and angle of the rock bridge. These results show that the mechanical behaviors of the defective rock under static and dynamic loading are different, but they are all significantly affected by the geometry of the defect, including its length, spacing, and angle. For static experiments on specimens with defects in 3D conditions, Li et al. [24] tested rock-like specimens with two and three parallel pre-existing fissures under uniaxial loading. In their study, seven types of crack coalescence were identified from the experimental results of specimens containing various fissure angles. However, although rock-like materials can simulate some behaviors of real rock, it is very difficult to simulate all properties of real rock materials, such as heterogeneity, mineral grains, boundary effect, and cementation. Li et al. [25] investigated the real-time cracking behavior of sandstone specimens containing two coplanar flaws with different flaw angles under uniaxial compression. The relationship between the coplanar flaw angle and crack coalescence stress was constructed in accordance with the photographic monitoring results. Furthermore, Yang et al. [26] tested the strength and failure behavior of sandstone specimens with a pair of unparallel flaws under uniaxial compression. They adopted photographic monitoring and the AE technique and analyzed crack coalescence and AE evolution behavior depending on various flaw angles.

In addition to static load conditions, dynamic loads, such as earthquakes, explosions, and rockbursts are also widely present in mining and geotechnical engineering. Under dynamic loading, it is recognized that the strength of the rock (i.e., compressive, tensile, and shear strength) depends on the loading rate; that is, rock strength increases with the loading rate [27,28]. Under dynamic loading, the propagation and crack growth rates of stress waves are much higher than those of static loading, resulting in different mechanical responses of rock specimens containing defects. Based on split Hopkinson bar SHPB equipment, Yan et al. [29,30] carried out a series of dynamic load tests on rock samples with a single defect and used a high-speed camera to analyze the fracture process of the samples. They found that the mechanical parameters of the defective specimens show obvious rate dependence under dynamic disturbance and the observed shear cracks are the main crack types under dynamic loading in an "X"-shaped shear failure. Zhang et al. [31] systematically studied the effects of strain rate and crack density on the mechanical properties and energy characteristics of rock specimens with parallel fractures and concluded that more mixed cracks would appear on the surface of the specimen with an increase in crack strength and that crack networks have become more complex. Both dynamic strength and deformation modulus showed a decreasing trend with the increasing defect strength. Yan et al. [32] and Fan et al. [33] studied the modeling problem of dynamic fractures in rocks and concluded that the Numerical Population Method (NMM) can be used to study the propagation of stress waves in fractured rocks. To investigate the effect of dynamic loading on crack propagation, Wu. et al. [34] used Split Hopkinson Pressure Bars (SHPB) to test rhombohedral marbles containing single defects with different dip angles, resulting in the existence of artificial defects that change the failure mode of marble under dynamic loading from the cracking of intact specimens to the shearing of defective specimens. Li et al. [35] studied the effect of a single pre-existing cavity (a circular or elliptical cavity) on the fracture behavior of marble specimens in SHPB experiments and concluded that shear cracks were more efficient than tensile cracks under dynamic loading conditions, which appeared earlier and dominated the fracture process. Zhou et al. [36] conducted dynamic tests on the behavior of cracking at the edge and proved that crack velocity is not constant and its propagation toughness is related to crack propagation velocity. In addition, Zhang and zhao [37] used a digital imaging technique (DIC) to measure the full-field strain on the surface of rock specimens under dynamic loading and showed the microscopic mechanism of brittle failure before the appearance of a macroscopic crack.

Previous studies have shown that subterranean rock masses with prefabricated fractures are often subjected to static compression and dynamic loads. However, although single defect or multiple defects are widely distributed in natural rock mass, no scholars have ever studied the mechanical behavior and energy consumption characteristics of parallel crack rocks with different fracture angles under different loading rates. Therefore, in this study, granite with prefabricated parallel double defects was introduced in the dynamic load test through an improved SHPB device to simulate the engineering rock with prefabricated defects under the action of a certain loading rate. The SHPB system is the first to study the effects of loading rate and fracture angle on the dynamic strength, deformation characteristics, failure mode, rockburst, and energy dissipation of fractured rocks. The fracture distribution of rock samples with parallel fractures under the loading rate and the failure modes of rocks at different fracture inclination angles are comprehensively analyzed through high-speed camera images. Section 2 first introduces the sample preparation and test setup and then evaluates the energy calculation method for the SHPB test system. Section 3 systematically expounds the experimental results, including dynamic stress balance, dynamic strength, rock progressive cracking behavior, rockburst law, and energy evolution. Section 4 comprehensively discusses some issues of crack classification and potential guidance for practical engineering. Section 5 discusses and interprets the discussion. Section 6 summarizes the entire study.

2. Experimental Setup

2.1. Specimen Preparation

We used granite for the experimental research due to its mechanical isotropy and structural homogeneity, and we made rock samples with parallel double fractures with different fracture angles. In order to obtain a clearer observation of crack evolution, a single granite with good geometric integrity and uniformity was cut to obtain prismatic samples with a height, width, and thickness of 45, 45, and 20 mm, respectively, which proved the feasibility and effectiveness of the prismatic rock samples in the SHPB test. Then, the high pressure water jet cutting machine (Changsheng stone firm, Ningbo, China) was used to cut the rock specimens with a length, width, and thickness of 10 mm, 1 mm, and 1 mm parallel cracks at different angles (0°, 45°, and 90°). The upper and lower surfaces of the sample were polished with a grinding machine (Haideli Machinery Company, Yueyang, China) during the processing of the sample preparation, and the processing accuracy strictly conformed to the ISRM (International Rock Mechanics) standard. The flatness of the surface was controlled at 0.05 mm, and the vertical deviation of the upper and lower surfaces was controlled at 0.25°. The samples were divided into complete samples and samples with parallel double cracks with different inclination angles. The average density of the samples was 2580 kg/m^3, and the elastic modulus was about 9.86 GPa. A total of 60 samples were produced, as shown in Figure 2.

2.2. Experimental Apparatus and Techniques

SHPB devices in Central South University (Difeisi Dynamic High Voltage Technology Company, Nanjing, China) have been widely used to study the dynamic mechanical properties of rocks under different loading rates according to the recommendation of the International Society of Rock Mechanics. In this study, a SHPB device with a diameter of 50 mm was used, which consisted of an impact rod, an incident rod, and a transmission rod with lengths of 300, 3000, and 2000 mm, respectively. The stress transfer components included impact rods, incident rods, and transmission rods with lengths of 300, 3000, and 2000 mm, respectively. The entire system was composed of switches and safety valves and could generate a slowly rising half-sine wave to eliminate the wave oscillation effect and achieve a constant strain rate on brittle materials. All rods were made of ultra-high-strength titanium alloy with a density of 7800 kg/m^3 and an elastic modulus of 211 GPa. A thin copper sheet connected to the free end of the incident rod acted as a pulse shaper, and two strain gauges were installed separately. The two strain gauges were used to record the strain signal during dynamic loading at the center of the incident rod and the transmission rod.

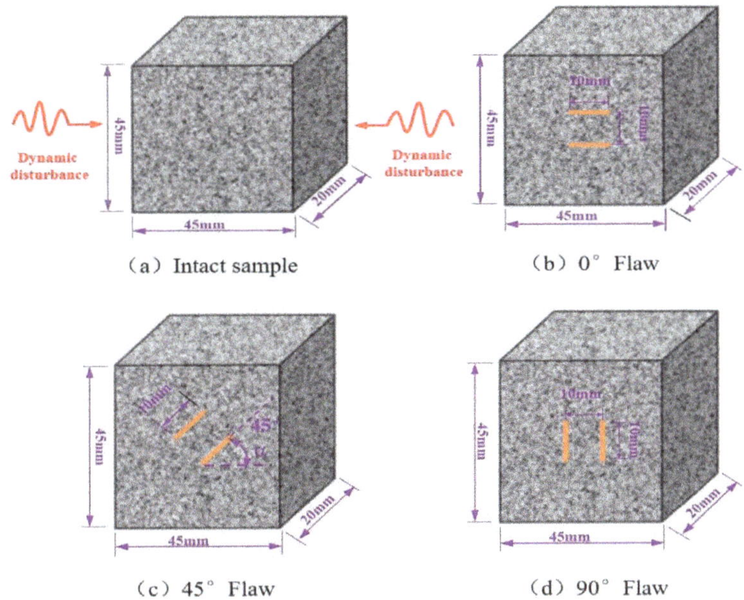

Figure 2. Geometrical images of the specimens.

During the experiment, the sample was sandwiched between the incident rod and the transmission rod, and the interface between the rod and the sample was evenly coated with lubricant to reduce the friction effect. Once ready, the conical impact bullet was fired from the air gun at high velocity and struck the front end of the injection rod. The resulting incident wave propagated along the strip, and a part was reflected onto the incident rod when the wave reached the incident rod-specimen interface (reflected wave), while the rest (transmitted wave) passed through the specimen and was further transmitted to the transmitted rod superior. The magnitude of the transmitted wave depended on the difference in wave impedance between the rod and the specimen. The magnitudes of the incident, reflected, and transmitted waves were collected using strain gauges attached to the surfaces of the incident and the transmitted rods, and the collected wave forms were displayed on an oscilloscope. The strain rate, peak stress, and stress–strain curve of the specimen was obtained after processing the data based on the three-wave method.

The high-speed (HS) photography system was composed of an HS camera (Ketianjian Photoelectric Technology Company, Hunan, China), a pair of high-intensity flashes (Jiangsu Xuanlang Lighting Company, Yangzhou, China), and a high-performance laptop (Dell, TX, USA). The HS camera recorded images with a resolution of 128 × 256 fps pixels and an area of 50 × 100 mm^2 at an inter-frame time of 6.5 μs. The pulse signal released by the oscilloscope triggered the HS camera when the strain gauge signal on the incident rod arrived so that the HS image could be synchronized with the dynamic loading stress. The HS camera was located at a safe distance of 1.5 m from the specimen to prevent damage caused by breaking rock fragments punching out, as shown in Figure 3.

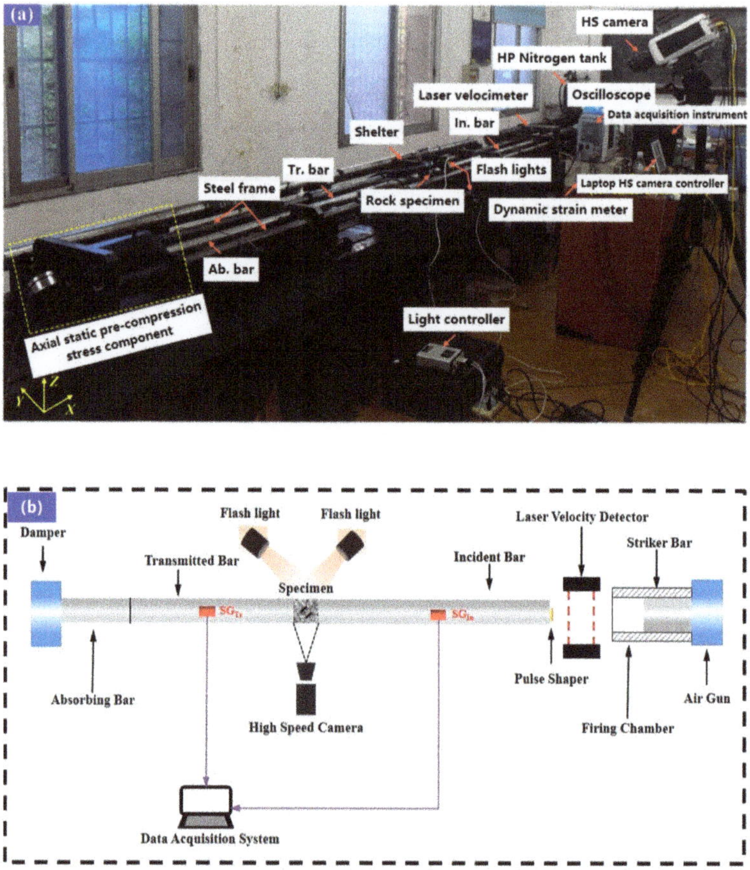

Figure 3. The photographic view of the experimental setup: (**a**) Overview, (**b**) SHPB test system.

2.3. Data Processing

One-dimensional stress propagation is still guaranteed for dynamic SHPB compression tests. According to previous studies, the baselines for incident, reflected, and transmitted waves are all zero. A strain gauge can record the entire strain signal, including the strain. The dynamic load at the end of the sample was determined using the Formulas (1)–(3). Based on the one-dimensional stress wave theory, mechanical parameters, such as the average dynamic stress, strain, and strain rate of the sample, can be calculated using the following formulas [38]:

$$\sigma_{(t)} = \frac{A_e}{2A_s}[\sigma_I(t) - \sigma_R(t) - \sigma_T(t)] \tag{1}$$

$$\varepsilon(t) = \frac{1}{\rho_e C_e L_s}\int_0^t [\sigma_I(t) + \sigma_R(t) - \sigma_T(t)]d_t \tag{2}$$

$$\dot{\varepsilon(t)} = \frac{1}{\rho_e C_e L_s}[\sigma_I(t) + \sigma_R(t) - \sigma_T(t)] \tag{3}$$

where $\sigma_I(t)$, $\sigma_R(t)$, $\sigma_T(t)$ are the incident stress, reflected stress, and transmission stress of the rod, respectively; A_e, ρ_e, and C_e are the cross-sectional area, density, and longitudinal wave velocity of the rod, respectively; and A_s and L_s are the cross-sectional area and length of the sample, respectively. In the SHPB impact test, the incident wave energy caused by the impact load is mainly converted into the energy of the reflected wave and transmitted wave as well as the energy absorbed by the rock sample. Incident energy (E_I), emitted energy (E_R), and transmitted energy (E_T) can be indirectly calculated from the corresponding three stress wave signals using Equations (4)–(6) [39]:

$$E_I = \frac{A_e}{C_e \rho_s} \int_0^t \sigma_I^2(t) d_t \quad (4)$$

$$E_R = \frac{A_e}{C_e \rho_s} \int_0^t \sigma_R^2(t) d_t \quad (5)$$

$$E_T = \frac{A_e}{C_e \rho_s} \int_0^t \sigma_T^2(t) d_t \quad (6)$$

The energy absorbed by the rock specimen (E_A), the energy utilization efficiency N_d defined in this study, and the energy absorption density of the rock specimen can be determined as follows using Equations (7)–(9):

$$E_A = E_I - E_R - E_T \quad (7)$$

$$N_d = \frac{E_A}{E_I} \quad (8)$$

$$e_d = \frac{E_A}{V_0} \quad (9)$$

where V_0 is the volume of the rock specimen and the absorbed energy can be divided into three main groups: crack propagation and fracture and damage energy of micro-cracks in the sample, the kinetic energy of flying fragments, and other forms of consumption which can be neglected energy, such as heat and sound. The energy absorption rate can reflect the amount of energy absorbed by the rock sample during the dynamic loading process, while the energy absorption density represents the energy absorption of flawed rocks per unit volume.

2.4. Sieving Tests

Debris distribution is a key indicator in rockburst and rock excavation. In this study, rock fragments were collected in a plastic collection box after the shock loading experiment was completed. Analyses were performed using standard sieves with mesh sizes of 0.075, 0.25, 0.5, 1, 2, 5, 10, 20, and 40 mm,. After sieving, the rock was divided into fragments of different sizes. The cumulative mass distribution can be obtained by weighting each group of rock fragments. To quantify the rock fragmentation at different loading rates, the average size of the rock fragments can be determined using Equation (10):

$$d_m = \left(\sum w_i \overline{d_i}\right) / \sum w_i \quad (10)$$

where d_i is the mean size of the fragments situated between two levels of mesh size and w_i is the interval mass percent of the rock fragments corresponding to $\overline{d_i}$.

It is generally believed that rock fragments are similar, and fractal geometry theory is widely used in rock fragmentation analysis. In this study, the fractal dimension D is used to quantify the fragment size distribution, as shown in Figure 4, using Equation (11):

$$y_i = \frac{M(< d_i)}{M} = \left(\frac{d_i}{d_{max}}\right)^{3-D} \quad (11)$$

where $M (<d_i)$ is the cumulative mass of fragment size less than d_i; M is the total mass of all fragments; And d_{max} represents the maximum size of the rock fragments.

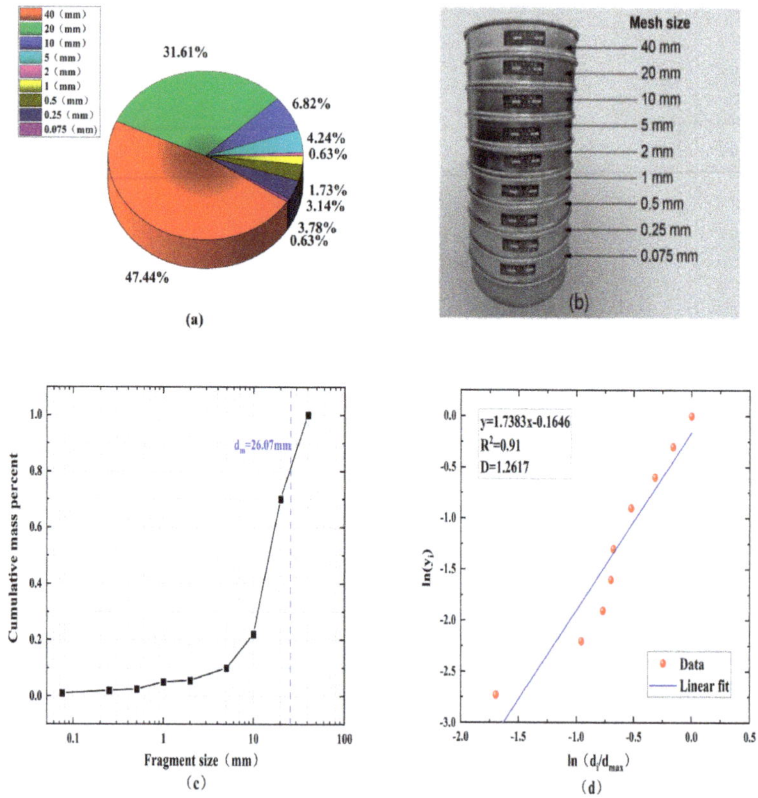

Figure 4. (**a**) A standard sieve used in the sieving tests. (**b**) Sieved fragments with different size intervals. (**c**) A cumulative mass curve against the fragment sizes. (**d**) The fractal dimensions of the rock fragments.

For the sieving test, the values of d_i and d_{max} were determined using the mesh of the sieve device and M and M ($<d_i$) were calculated after sieving, so the fractal dimension D of the rock can be obtained using Equation (11).

The best fit can be derived using least squares linear regression through the data points in Figure 4d based on using fractal theory to quantify the blockiness of the rock after fragmentation and taking the logarithm of both sides of Equation (11). The slope of the linear fit line is 3–D, and its fractal dimension D can be further obtained.

3. Experimental Results

3.1. Dynamic Stress Equilibrium Check

Dynamic stress balance before failure is a prerequisite for a valid dynamic impact specimen and can be checked by comparing the dynamic stress across the specimen during the entire test period. Figure 5 shows the stress balance at both ends of the three test specimens under different loading rate conditions in the dynamic test, and the loading rate is represented by the slope of the straight line segment before the peak of the stress time history curve. The dynamic stress is the sum of the incident stress and the reflected stress at the incident end of the specimen, expressed as incident stress+ reflected stress. The dynamic stress is caused by the transmitted wave at the transmission end of the specimen, marked as transmitted stress. It is obvious that the dynamic stress on both sides is almost the same throughout the loading process which indicates that the loading experiment has achieved the stress balance on the specimen and the axial inertia effect can be neglected since there is no force difference in the specimen that causes inertial force.

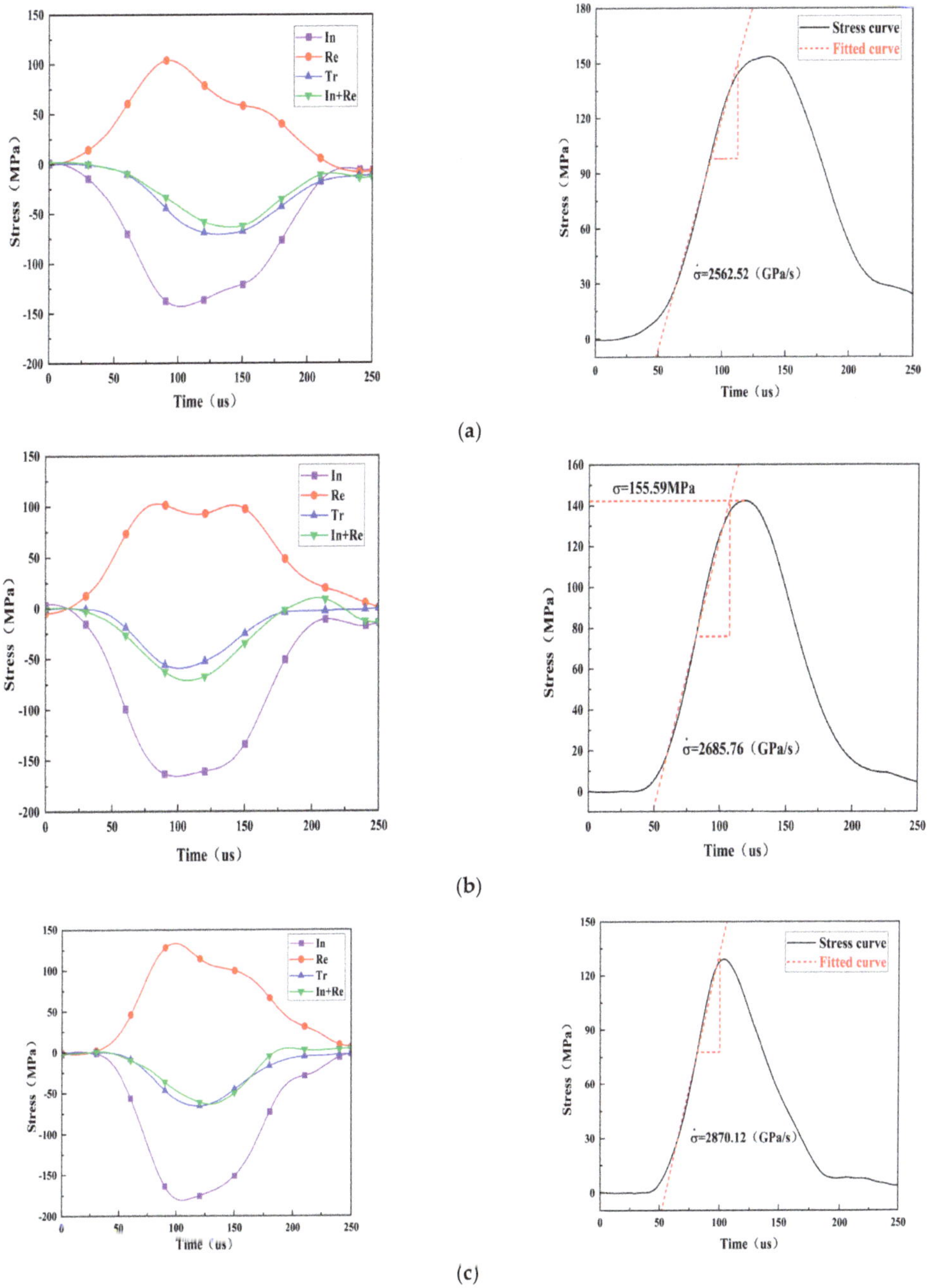

Figure 5. Dynamic stress equilibrium check for the rock specimens with different crack angles: (**a**) rock specimen with a 0° angle, (**b**) rock specimen with a 45° angle, and (**c**) rock specimen with a 90° angle.

3.2. Dynamic Deformation Characteristics

Figure 6 shows the dynamic stress–strain curves of rock specimens with different fracture angles. The results show that the strain rate significantly affects the shape of the dynamic stress–strain curve. The rock sample has no strain recovery under the action of each loading rate, and it increases monotonically during the entire loading process. The stress–strain curve of the specimen does not appear in the compaction stage under a certain loading, indicating that the granite has good compactness. The stress–strain curve includes three stages, including the elastic stage, the yield stage, and the post-peak failure stage. The the elastic stage and yield stage of the specimens develop faster as the loading rate continues to increase, but the yield stage shortens, mainly because the increasing loading rate leads to increasing linear elastic deformation which shortens the development process of linear elasticity and the yield stage in the curve.

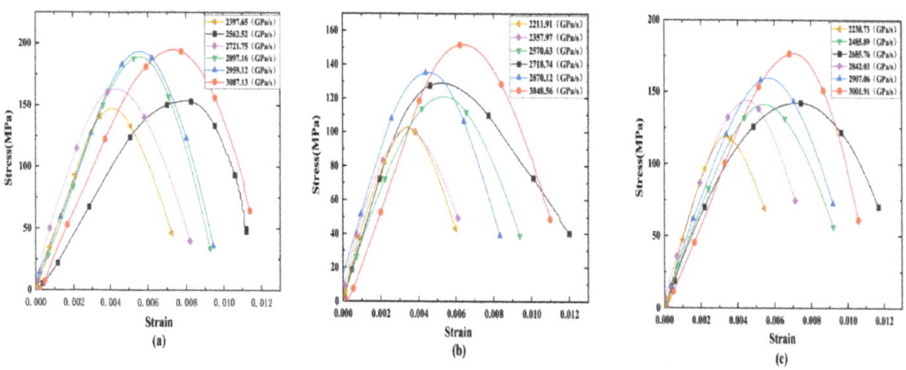

Figure 6. Typical stress–strain relationships of crack specimens: (**a**) rock specimen with a 0° angle, (**b**) rock specimen with a 90° angle, and (**c**) rock specimen with a 90° angle.

4. Progressive Cracking Behaviors

4.1. Dynamic Fracture Process of Cracked Rock in Impact Test

The failure mode of the specimen is a key indicator to reveal the failure mechanism of the rock, and Figure 7 shows the different dynamic fracture processes captured by the high-speed camera. In this section, three typical fractured rock samples with different inclination angles were selected to study progressive cracking behavior under impact load. Only the dynamic fracture process of representative samples was selected due to a space limitation. For each specimen, five snapshots were selected covering the initial stage, the fracture initiation stage, the stable crack propagation stage, the peak stress stage, and the post-peak failure stage. Three typical crack types were highlighted in each diagram to illustrate the fracture mechanism of some local cracks more clearly. For each photograph, the left and right sides of the tested specimen were the transmitted bar and the incident bar, respectively.

Cracks are mainly divided into three types: tensile cracks, shear cracks, and mixed tensile–shear cracks. Tensile cracks are consistent with the loading direction of the specimen, while shear cracks expand in the oblique direction, and tensile–shear mixed cracks are common cracks in rock samples with parallel loading directions and vertical loading directions. The failure of the specimen is mainly caused by the initiation, expansion, and penetration of the crack at the tip of the parallel crack to both ends of the specimen, and it is connected to the tensile shear crack formed at both ends of the specimen. New cracks are formed under the action of a higher loading rate. It is generally accepted that the wing crack (primary crack) initiates at an angle with the pre-existing fissure and extends along the loading direction during the loading process of the pre-cracked specimens. However, secondary crack initiate after wing crack in two initiation directions, including coplanar with the fissure (coplanar secondary crack) or at an angle (oblique secondary crack), and one is similar to the wing crack but in the opposite direction, as shown in Figure 7.

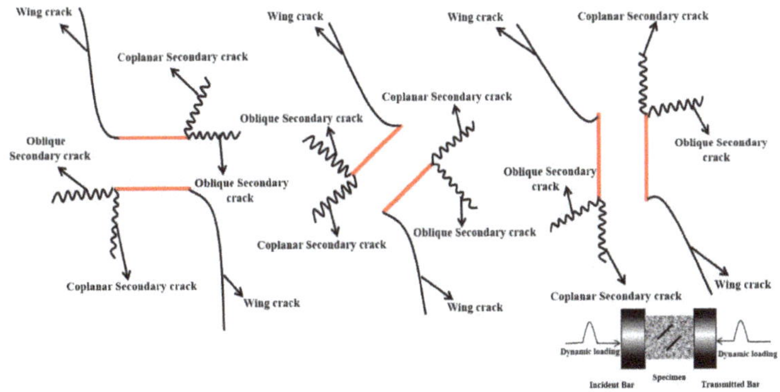

Figure 7. Simplified crack pattern observed in a pre-cracked specimen in dynamic loading.

Figure 8a shows the typical cracking behavior of the 0° crack under the loading rate of 2721.75 GPa/s. The first picture corresponds with the situation where the rock begins to load (0 μs), and the sample is not deformed at this time. Two tensile cracks (FT crack) begin to form in the middle and lower part of the parallel cracks when the rock is loaded for 99 μs, and no obvious cracks form at the crack tip. As the loading continues, mixed tensile shear cracks (MTS crack) occurr in the upper left and lower left corners of the specimen at 112 μs, and the two cracks expand diagonally and gradually connect with the middle tensile crack. When the rock loads to the peak stress (137 μs), two tensile–shear mixed cracks form in the upper and lower right corner of the sample, respectively, and the propagation speed is faster than the previous two. At the beginning, the tensile crack in the middle of the parallel crack continues to expand and connect with the crack at the left and right ends of the specimen. The MTS crack and the FT crack are connected and penetrated as the stress gradually decreases in the post-peak stage, and new tensile shear cracks (FS crack) are generated at the left and right ends of the sample. Finally, the specimen forms an "X"-shaped shear tensile failure.

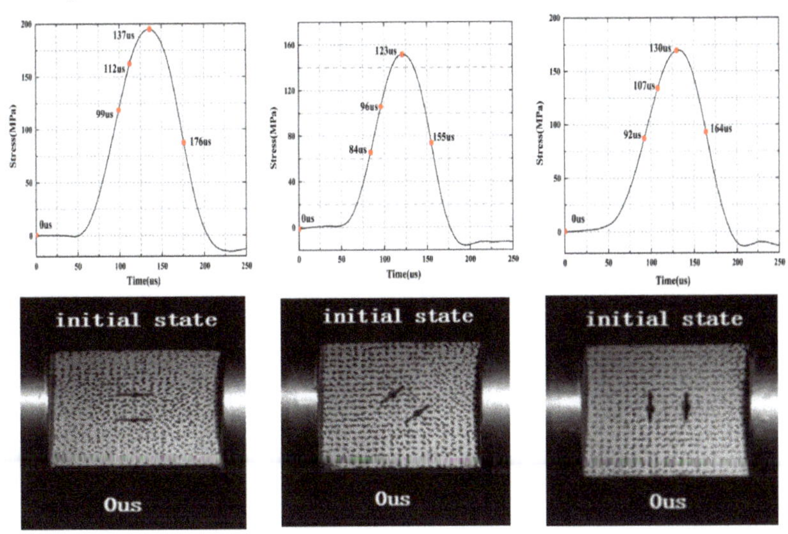

Figure 8. *Cont.*

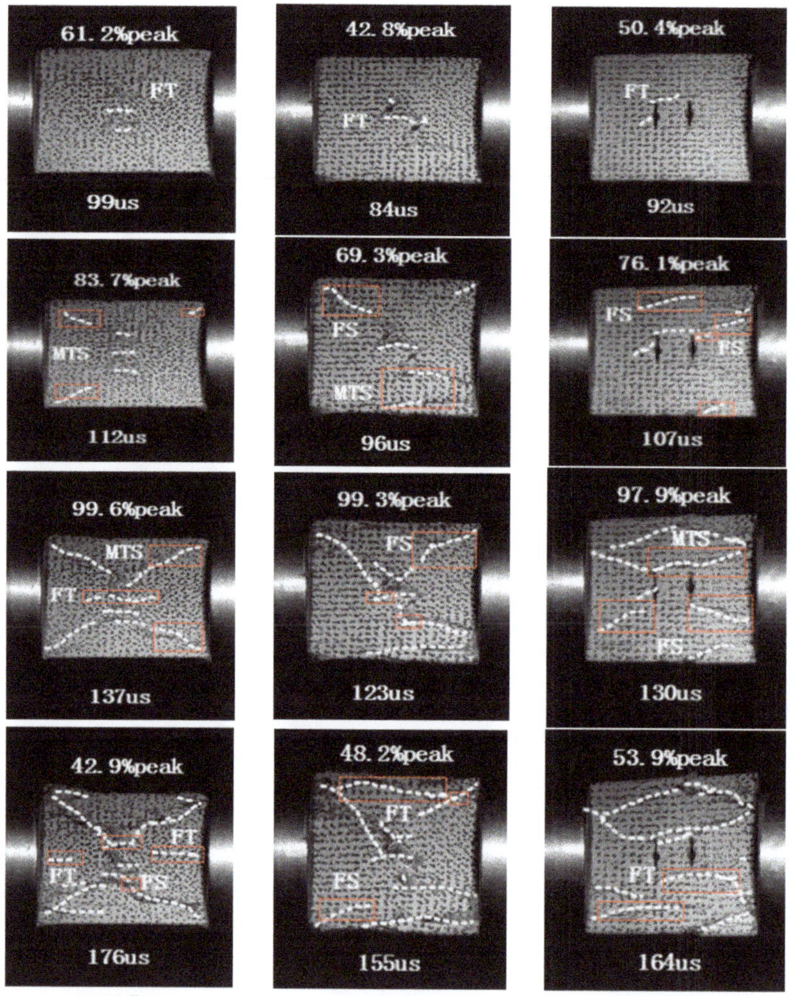

(**a**) 0° flaw (**b**) 45° flaw (**c**) 90° flaw

Figure 8. Progressive failure processes of the three typical tested rock specimens with different crack angles: (**a**) rock specimen with a 0° angle, (**b**) rock specimen with a 45° angle, and (**c**) rock specimen with a 90° angle. The right side is the incident rod while the left side is the transmission rod.

Figure 8b shows the typical cracking behavior of the 45° cracked rock under the loading rate of 2397.57 GPa/s. The first picture corresponds to a situation where the rock starts to load (0 μs), and the sample is not deformed at this time. At 84 μs, a short tensile crack is formed in the central rock bridge region of the parallel crack, and a new crack begins to sprout at the tip of the crack. A shear crack (FS crack) occurs at the upper left corner of the sample as the loading continues, while mixed tensile cracks (MTS crack) appear at the upper and lower right ends of the specimen, and a shear back-wing crack appears at the tip of the crack, which gradually connects with the lower mixed tensile crack. The shear cracks at the upper and left ends of the sample extend to the parallel cracks along the diagonal of the specimen when the rock is loaded to peak stress (123 μs). At this time, a new shear crack appears at the upper right part and gradually expands along the crack tip. As the stress gradually decreases in the post-peak stage, a horizontally developed tensile crack appears in the upper part of the sample and connects with the shear cracks at both ends, resulting in the final tensile shear failure.

Figure 8c shows the typical cracking behavior of the 90° crack specimen under the loading rate of 2485.89 GPa/s. The first picture corresponds to the situation where the rock starts to load (0 μs), and the sample is not deformed at this time. The tensile cracks (FT crack) parallel to the loading direction are generated in the tip and middle of the crack when the rock is loaded at 92 μs. As the loading continues, new shear cracks (FS crack) occur at the upper and right ends of the specimen at 107 μs, and they connect wit the tensile crack generated at the tip of the crack, and the shear crack at the upper end continues to expand along the loading direction. The cracks around the crack tip connect to each other to form an "X"-shaped crack when the sample is loaded to the peak stress (130 μs), and the shear crack in the lower part of the sample gradually expands. New tensile cracks occur at the lower end of the crack and the lower right corner of the specimen as the stress gradually decreases to 53.9% of the peak stress, and this eventually results in an "X" shear tensile failure.

4.2. Final Failure Modes of Rocks with Different Fractures under Impact Loading

The final failure mode of the specimen under impact loading has important scientific significance and engineering application value for studying the crack propagation characteristics and the overall failure mechanism of rock mass. Three main failure modes are classified according to the final dynamic fracture modes and the positional relationship between the new and pre-cracked cracks, as depicted in Table 1.

Table 1. Comparison of the failure modes of flawed rock specimens with three different inclination angles under distinct loading rates.

Loading rate (GPa/s) / Angle(°)	2397.65	2562.52	2721.75	2897.16	3029.12	3087.13
0	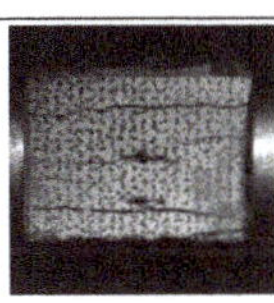			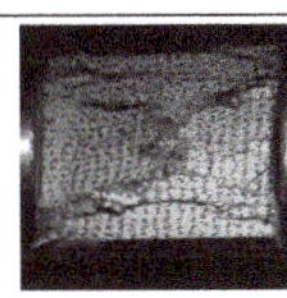	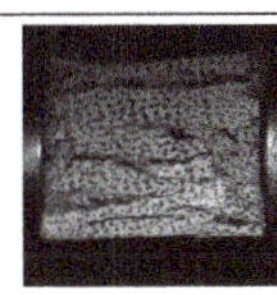	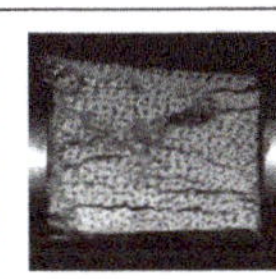
Loading rate (GPa/s) / Angle(°)	2211.91	2357.97	2570.63	2718.74	2870.12	3048.56
45				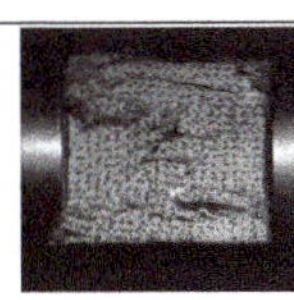	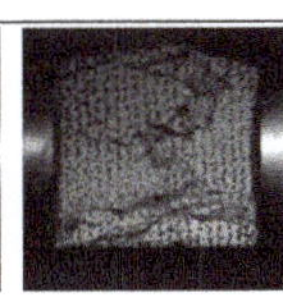	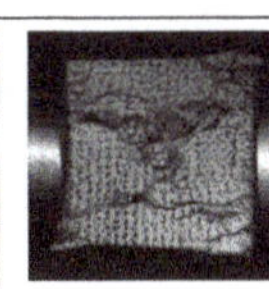
Loading rate (GPa/s) / Angle(°)	2238.73	2485.89	2685.76	2842.03	2907.06	3001.91
90					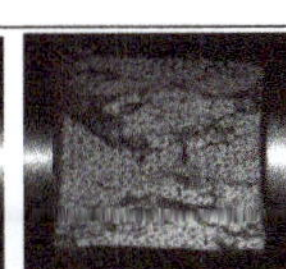	

Faliur type I is axial splitting tensile failure along the loading direction of the specimen, mainly for 0° parallel crack specimens. The macroscopic failure crack of the sample generally expands along the loading direction parallel to the direction of the crack inclination angle due to the increase in the impact load, and a composite tensile shear crack occurs at the crack tip. The initial macroscopic cracks in the main diagonal directions (two and four quadrants) and the far-field cracks at the top and bottom of the specimen surface usually propagate through the surface along the loading direction to the complete failure of the specimen.

Faliur type II is "X" type" shear failure, mainly for specimens with 45° parallel cracks and crack specimens under the 2357.97 GPa/s loading rate. The failure modes are rock bridge penetration and crack initiation at the crack tip X-shaped shear failure extending in the diagonal direction. The macroscopic tensile crack is generated at the top and bottom of the crack, and the macroscopic initial crack formed at the loading section generally extends along the parallel end face. The secondary and far-field cracks in the main diagonal direction propagate through the entire surface of the specimen along the loading direction until it fails.

Faliur type III is tensile–shear composite failure mode, mainly samples with 90° parallel cracks. Shear cracks first occur at the tip of the crack of the sample, and with the increasing loading rate, the shear crack gradually develops from a local shear crack distributed at and near the tip of the crack to a tensile extension extending from the tip of the crack along the top and bottom of the sample. The crack then propagates along the loading direction and penetrates toward the incident and transmission end, respectively, resulting in the tensile–shear composite failure of the specimen.

In addition to newly created cracks, some small rock debris may be exfoliated from the behavior of the failed specimen. Figure 9 shows the final failure modes of all fractured specimens, where the solid symbols, half-filled symbols, and hollow symbols indicate Type I (axial splitting tensile failure), Type II ("X" type shear failure), and Type III (tensile–shear hybrid failure) rock failure mode. In our tests, the Type I failure mode is the most common, and the percentage of each failure mode is about 38%, 33%, and 13%. The Type I failure mode becomes more obvious as the loading rate increases, which means the loading rate plays an important role in the failure mode, and each crack angle contains three failure modes, which indicates that the crack angle has no obvious effect on its failure mode.

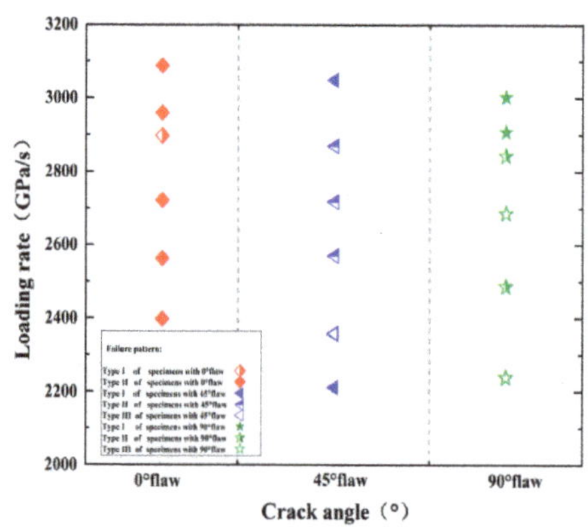

Figure 9. Failure patterns of the flawed specimens under different crack angles and loading rates.

The cumulative mass classification curve of the sample can be obtained via the sieving test. The results show that the rock loading rate has a great influence on the failure mode of the sample. Regardless of the size of the crack inclination, the rock fracture increases with the increase in the loading rate, but the specimens remained intact or slightly split at low loading rates. However, the rock is broken into large strips or fragments under the action of a medium loading rate. The rock sample is crushed into small fragments or even powder under the action of a high loading rate. The cumulative mass percentage gradually increases to 100% as the fragment size increases from 0.075 mm to 40 mm, as shown Figure 10.

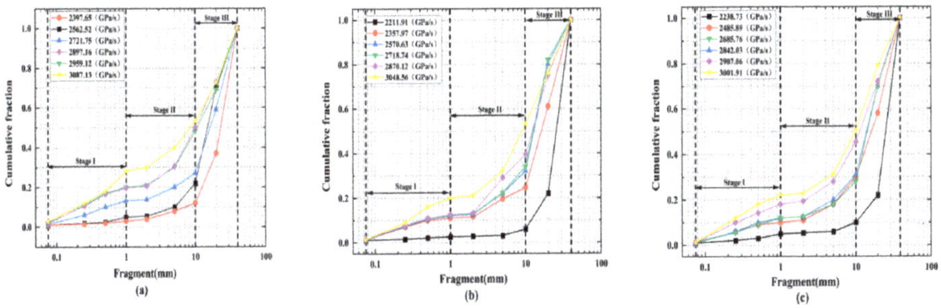

Figure 10. Cumulative mass percent of the pre-cracked rock specimens under different loading rates: (**a**) rock specimen with a 0° angle, (**b**) rock specimen with a 45° angle, (**c**) rock specimen with a 90° angle.

Table 2 and Figure 11 show the effects of loading rates and fracture angles on average rock fragment sizes and fractal dimensions. The fractal dimension is smaller at a lower loading rate, while the factual dimension increases as the loading rate increases to a higher level, and its increasing speed gradually slows down, and its maximum value can reach 2.1. For a given loading rate, the fractal dimension decreases with an increasing fracture angle and reaches a maximum when the fracture angle is 0°. The average fragment size for most specimens remains in the range of 19–24 mm. The average fragment size generally decreases as the loading rate increases, and the higher the mass fraction of small-sized fragments, the faster the decline rate; therefore, it can be inferred that a higher loading rate can cause a more significant effect within the rock specimen and produce a smaller average size. In general, both the loading rate and the fracture angle have certain effects on the rock crushing characteristics.

Table 2. Summary of mean fragment sizes and fractal dimensions of cracked rock specimens under different loading rates.

Specimen	Crack Angles (°C)	Loading Rate (GPa/s)	Mean Fragment Size (mm)	Fractal Dimension
Flaw0°-D1-1	0	2397.65	30.15	1.906
Flaw0°-D2-2	0	2562.52	27.07	1.938
Flaw0°-D3-2	0	2721.75	25.40	2.001
Flaw0°-D4-3	0	2897.16	24.35	2.013
Flaw0°-D5-2	0	2929.12	21.13	2.037
Flaw0°-D6-3	0	3087.13	19.61	2.101
Flaw45°-D1-1	45	2311.91	26.15	1.744
Flaw45°-D2-2	45	2457.97	24.80	1.773
Flaw45°-D3-2	45	2570.63	21.85	1.793
Flaw45°-D4-3	45	2718.74	21.69	1.850
Flaw45°-D5-2	45	2870.12	21.04	1.898
Flaw45°-D6-3	45	3048.56	19.13	1.937

Table 2. *Cont.*

Specimen	Crack Angles (°C)	Loading Rate (GPa/s)	Mean Fragment Size (mm)	Fractal Dimension
Flaw90°-D1-1	90	2238.73	29.69	1.738
Flaw90°-D2-2	90	2385.89	23.97	1.777
Flaw90°-D3-2	90	2585.76	23.68	1.968
Flaw90°-D4-3	90	2642.03	22.29	1.842
Flaw90°-D5-2	90	2707.06	20.79	1.689
Flaw90°-D6-3	90	2801.91	19.76	1.591

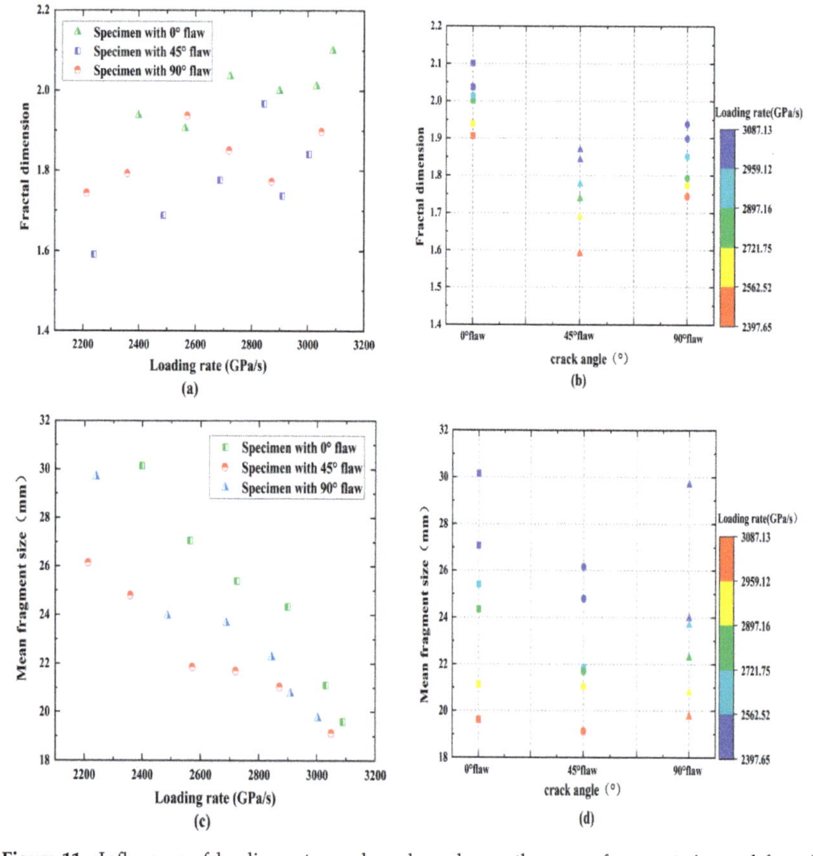

Figure 11. Influences of loading rates and crack angles on the mean fragment size and fractal dimension: (**a**) fractal dimension versus loading rate, (**b**) fractal dimension versus crack angle, (**c**) mean fragment size versus loading rate, and (**d**) mean fragment size versus crack angle.

5. Energy Dissipation

5.1. Energy Utilization Efficiency and Energy Absorption Density

Figure 12 shows the effect of loading rates and fracture angles on the energy utilization efficiency and energy absorption density of the rock samples, and the table summarizes the energy distribution of the fractured samples. Under the same fracture inclination angle, the energy utilization efficiency of the fractured rock samples is between 20% and 40%, and the energy absorption rate decreases with the increasing loading rate, regardless of the size of the fracture angle, as shown in Table 3 and Figure 10. It can be speculated that most of the energy will not be absorbed by the rock when the incident energy is not enough to break the rock and that the energy utilization efficiency is very low. The applied dynamic load is not so large for rock breaking engineering. Instead, there is a reasonable threshold, and using a certain method to control the loading level effectively improves the energy utilization rate. On the contrary, the figure shows that the loading rate promotes the energy absorption of the rock sample, and the absorbed energy per unit volume increases gradually. Specifically, the energy absorption density increases from 0.6 J/m^3 to 1.4 J/m^3 as the loading rate increases from 2200 GPa/s to 3100 GPa/s. In addition, for a given loading rate, the absorbed energy per unit volume and energy absorption rate first decreases and then increases with the increasing crack angle. For example, when the loading rate is 2400 GPa/s, the energy absorption density corresponding to each crack specimen is 0.92, 0.63, and 0.88 J/cm^3, respectively. This decrease–increase trend is somewhat like that of dynamic strength due to the fact that higher strength means more energy will be dissipated for rock failure.

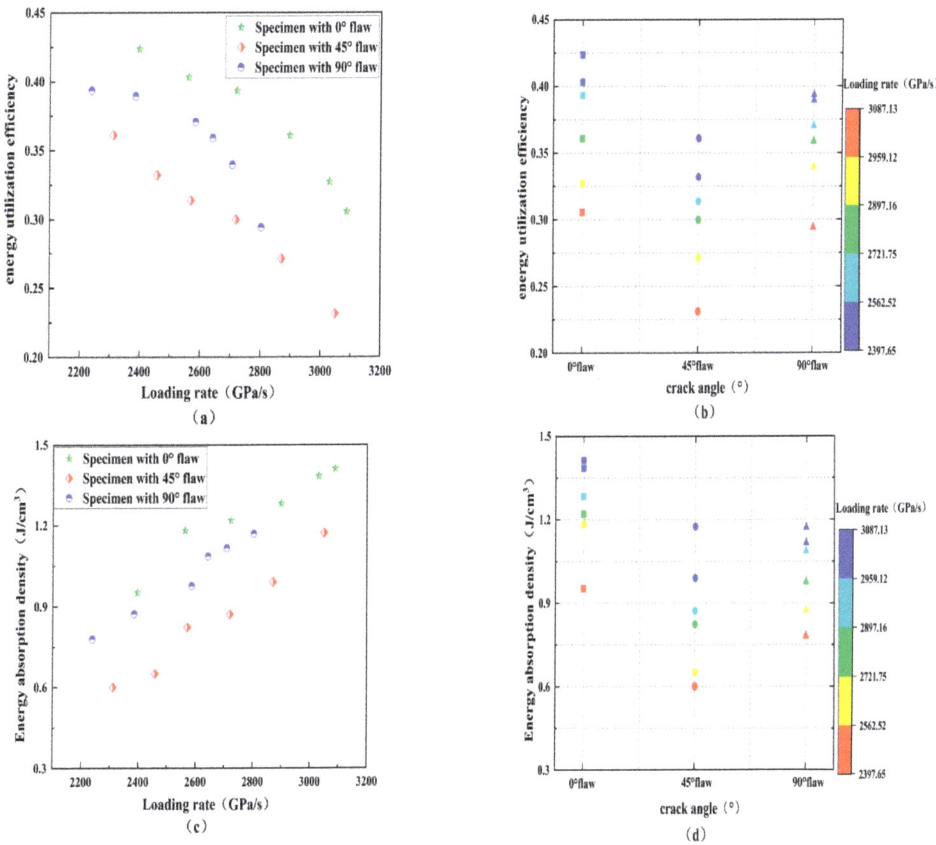

Figure 12. Influences of the loading rate and crack angle on the energy utilization efficiency and energy dissipation density: (**a**) energy utilization efficiency versus loading rate; (**b**) energy utilization efficiency versus crack angle; (**c**) energy dissipation density versus loading rate; and (**d**) energy dissipation density versus crack angle.

Table 3. Summary of energy partitions of pre-cracked rock specimens under different loading rates.

Specimen	Crack Angle (°)	Loading Rate (GPa/s)	Energy Utilization	Energy Absorption Density (J/m^3)
Flaw0°-D1-1	0	2397.65	0.42	0.95
Flaw0°-D2-2	0	2562.52	0.40	1.18
Flaw0°-D3-2	0	2721.75	0.39	1.22
Flaw0°-D4-3	0	2897.16	0.36	1.28
Flaw0°-D5-2	0	2929.12	0.32	1.38
Flaw0°-D6-3	0	3087.13	0.30	1.41
Flaw45°-D1-1	45	2311.91	0.36	0.60
Flaw45°-D2-2	45	2457.97	0.33	0.65
Flaw45°-D3-2	45	2570.63	0.31	0.82
Flaw45°-D4-3	45	2718.74	0.29	0.87
Flaw45°-D5-2	45	2870.12	0.20	0.99
Flaw45°-D6-3	45	2870.12	0.23	1.17
Flaw90°-D1-1	90	2238.73	0.39	0.78
Flaw90°-D2-2	90	2385.89	0.38	0.87
Flaw90°-D3-2	90	2585.76	0.37	0.97
Flaw90°-D4-3	90	2642.03	0.35	1.08
Flaw90°-D5-2	90	2707.06	0.33	1.11
Flaw90°-D6-3	90	2801.91	0.29	1.17

5.2. Rockburst Proneness of Pre-Crack Specimens

5.2.1. Rockburst Characteristics of Pre-Crack Specimens at Different Loading Rates

Rockburst is an extremely violent and destructive phenomenon, usually accompanied by a large number of rock fragments ejected in a short period of time; therefore, high-speed camera systems are used to record the "rockburst phenomenon". The rockburst phenomena of fractured rocks with different angles at different loading rates are similar. First, the rock particles are ejected or chipped, and then the cracks gradually expand with the increasing loading road, and the fragments on the rock are separated from the sample, and finally, a large amount of rock fragments and powders are produced, and a loud cracking sound is emitted. There is a strong rock fragment ejection phenomenon during the failure process for all fractured rocks with small loading rates. The rockburst tendency is lower for samples with higher loading rates, while the ejection velocity of rock blocks is lower, and the spray distance is short. It is worth noting that the duration of rock macroscopic failure increases significantly at higher loading rates. This indicates that the loading rate promotes the initiation of micro-cracks in the rock, thereby delaying the occurrence of rockburst. However, the micro-defects of the rock sample increase significantly due to the higher loading rate, which reduces its deformation ability and prolongs the stable failure stage of the sample.

In addition, quantitative analysis of rockburst phenomenon of fractured rock under impact load is still the most critical issue. In deep rock engineering, comprehensive judgment is often made based on the distance, distribution range, and acoustic properties of rock fragments ejected during rockburst. The ejection mass ratio outside the lever, M_E (mass ratio of rock fragment outside the lever to total exfoliated rock fragment, defined by Gong et al., 2018b), was used as a quantitative index to evaluate the degree of rockburst occurrence. M_E reflects the distance and mass of the ejected rock fragments and also reflects their kinetic energy (to a certain degree). The classification of M_E is [40]

$$M_E = 0, \ (No \ rockburst \ proneness) \tag{12}$$

$$< M_E \leq 0.4, \ (Low \ rockburst \ proneness) \tag{13}$$

$$0.4 < M_E \leq 0.6, \ (Medium \ rockburst \ proneness) \tag{14}$$

$$M_E > 0.6, \ (High \ rockburst \ proneness) \tag{15}$$

The statistical results of M_E for the specimens with cracks are presented in Figure 13. The M_E tends to increase as the loading rate increases, and the high rockburst area is concentrated in the loading rate of 2600 GPa/s–3100 GPa/s. The results show that the loading rate has a significant effect on the rockburst under the impact load. The inter-particle distances within the rock decrease as the loading rate increases and the interaction between particles enhances, which creates additional stress. The M_E of the rock sample is lower than 0.6 when the loading rate is lower than 2600 GPa/s, while the occurrence of rockburst is not obvious, and medium rockburst occurs. The reason is that the mineral composition and structure inside the rock are gradually destroyed under a certain impact load, resulting in a significant reduction in the releasable elastic energy of the rock and weakening of the rockburst strength. The degree of rockburst decreases first and then increases as the fissure inclination angle increases, and the magnitude relationship is 90° > 0° > 45°; however, the M_E of 90° fractured rock suddenly increases under the action of a large impact. It can be seen that the fracture angle is not obvious for judging the rockburst tendency, but M_E could accurately reflect the rockburst tendency of rock under different loading rates.

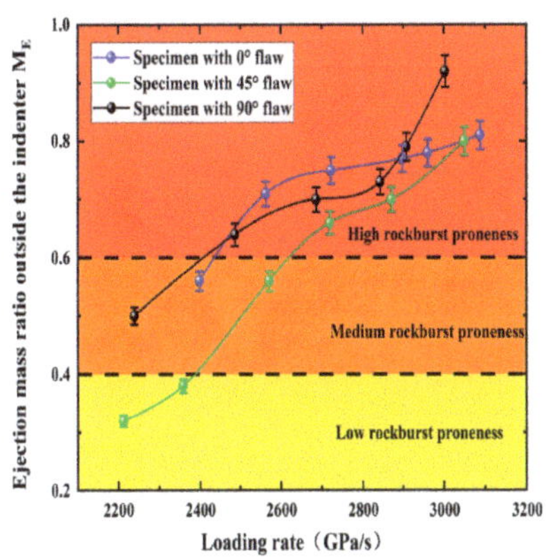

Figure 13. M_E of the pre-crack specimens at different loading rates.

5.2.2. Quantitative Evaluation of Rockburst Proneness of the Pre-Crack Specimens under Loading Impact

It is well-known that the occurrence of rockburst is an outward expression of residual elastic energy release inside rocks. Higher rock fragment ejection velocity and rockburst propensity corresponds to larger elastic residual elastic energy. Therefore, it is extremely important to calculate the residual energy accurately during the impact process of cracked specimens with respect to energy storage and dissipation.

It is assumed that the granite sample follows the first law of thermodynamics during the impact process; in other words, the sample does not interchange energy with the outside, and the energy of the rock only comes from the impact of the compression level on the specimen. Energy density is used as a characterization of strain energy results to eliminate the effect of the sample volume. Therefore, the strain energy variations of the granite specimen can be described as:

$$u_t = u_d + u_k \tag{16}$$

where u_t is the total input strain energy density; u_k is the releasable strain energy stored in the rock, i.e., elastic strain energy density; and u_d is the energy used for compaction of pores, crack propagation, and plastic deformation inside the rock, i.e., the dissipated strain energy density.

Figure 14 shows the u_k and u_d of the fractured granite specimen under six loading levels are accurately calculated during the impact loading process, and the calculation formula of u_t, u_k, and u_d are as follows:

$$u_t = \int_0^{\varepsilon_1} f(\varepsilon)d\varepsilon \tag{17}$$

$$u_e = \int_{\varepsilon_0}^{\varepsilon_1} f_u(\varepsilon)d\varepsilon \tag{18}$$

$$u_d = u_t - u_e = \int_0^{\varepsilon_1} f(\varepsilon)d\varepsilon - \int_{\varepsilon_0}^{\varepsilon_1} f_u(\varepsilon)d\varepsilon \tag{19}$$

$$u_a = \int_0^{\varepsilon_u} f(\varepsilon)d\varepsilon \tag{20}$$

$$u_t^p = u_t/v_0, u_d^p = u_d/v_0, u_k^p = u_k/v_0, u_a^p = u_a/v_0 \tag{21}$$

where $f(\varepsilon)$ and $f_{\mathrm{u}}(\varepsilon)$ represent the loading curves; ε_0 represents the elastic strain point; and ε_1 represents the plastic deformation of specimen after loading.

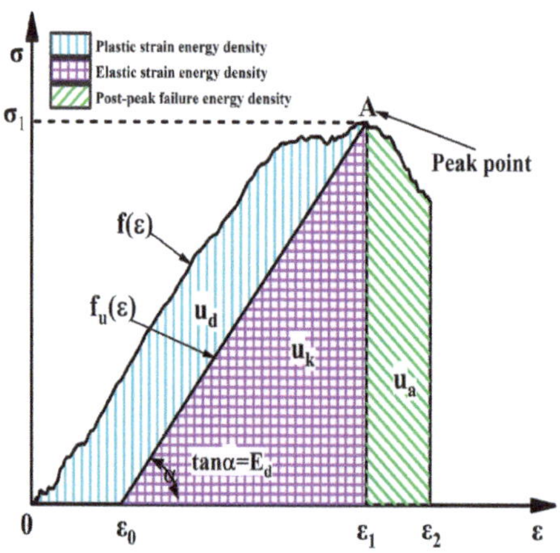

Figure 14. Schematic diagram of calculation for A_{EF}.

According to Equations (17)–(21), we could obtain the u_d^p, u_k^p, and u_a^p of fractured rock. The highly linear relations between u_k^p and u_t^p, u_d^p, and u_t^p are found during the shock loading process of fractured rocks by analyzing the results of these strain energies, as illustrated in Figure 15. These results show that the linear energy storage and dissipation laws can be used for dynamic loading experiments of fractured rock samples. Theoretically, external loads are a prerequisite for rock energy accumulation, and both elastic strain energy density and plastic strain energy density increase with increasing input energy density. Additionally, the elastic strain difficulty density first increases and then decreases with the increasing crack angle, while the energy density first decreases and then increases.

The coefficient of determination of linear fitting for u_d^p and u_t^p are 0.99, 0.93, and 0.90, respectively, and the coefficient of linear fitting for u_k^p and u_t^p are 0.90, 0.98, and 0.97.

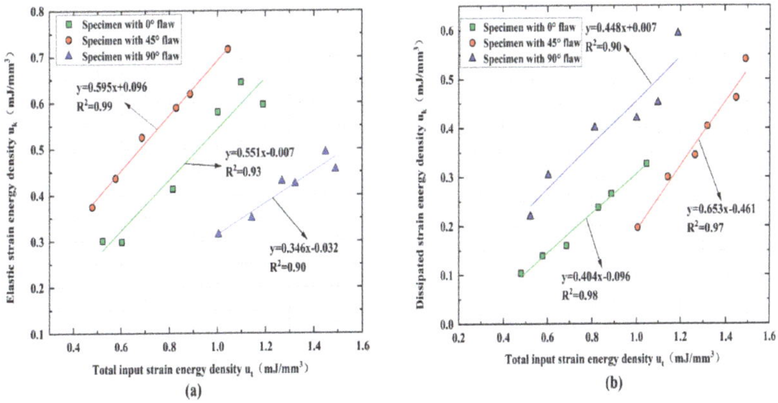

Figure 15. Linear relationships between the three energy densities of pre-crack specimens: (**a**) u_k versus u_t and (**b**) u_d versus u_t.

Based on these findings, we can accurately calculate the relationship between the strain energy densities at the peak strength (i.e., peak elastic strain energy density u_d^p and peak plastic strain energy density u_k^p) of the specimens and the loading rate by using the linear storage and dissipation laws. Both the loading rate and the crack angle have a certain influence on the change law of the strain energy accompanying the impact. As shown in Figure 16, the input strain energy density first increases and then decreases with the increasing loading rate for the rock specimens with a 0° and 45° flaw, while the rock specimens with a 90° flaw are on the rise. The elastic strain energy density and plastic strain energy density of rock specimens with 0°, 45°, and 90° flaws increase continuously with the increasing of the loading rate during the whole loading process, and the elastic strain energy density is always greater than the plastic strain energy density, which indicates that the energy absorbed and stored inside the rock is mainly converted into the energy required for rock failure and fracture expansion.

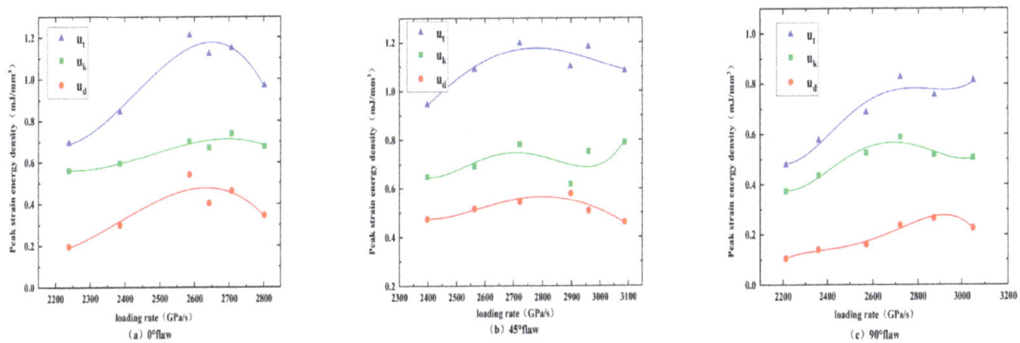

Figure 16. u_t, u_k, and u_d of pre-crack specimens: (**a**) rock specimen with a 0° angle; (**b**) rock specimen with a 45° angle; and (**c**) rock specimen with a 90° angle.

$K_s = u_d/u_k$ is defined based on the elastic strain energy storage law, and K_s is used to represent the stable state of the sandstone system under storage and dissipation. Figure 17 shows the changes in K_s with the loading rate, where K_s first increases and then decreases with the increasing loading rate, and the value of K_s is always less than 1, which indicates that most of the strain energy of the rock is transformed into elasticity under impact loading. The specimen generally absorbs the energy during the impact loading, which also verifies the conclusion that the rock sample absorbs energy when rock-burst occurs. With the increasing of K_s, the initiation and propagation of cracks are accompanied by the increasing of elastic strain energy. At this time, less plastic strain energy is generated and gradually released outward which reduces the ability of the rock to resist external loads. The system of the crack sample gradually changes from an unsteady state to a steady state, and the energy storage capacity also gradually weakens. The K_s of the 45° and 90° fracture specimens begin to decrease with the increasing loading rate when the loading rate increases to 2500 GPa/s, and a large amount of strain energy is used for the sliding friction of rock fracture, which indicates that the closure of pores and micro-cracks is gradually completed. However, the elastic strain energy continues to increase, the entire rock system still maintains a stable state, the reversible elastic strain energy in the rock system gradually accumulates, the energy behavior continues to be energy storage, and its ability to store energy is further enhanced. As K_s gradually decreases, the degree of steady state increases gradually while the stability increases gradually. K_s reaches the minimum value under the action of large impact loading, and u_d/u_k is also the minimum value at this time, while the energy storage capacity of the sample gradually increases. The turning point occurs when the loading rate is 2800 GPa/s for the specimens with a 0° flaw, which indicates that the 0° crack rock has stronger impact resistance and needs more external energy to break the rock with the increasing loading rate, but its K_s has the same variation law of specimens with the 45° and 90° rocks.

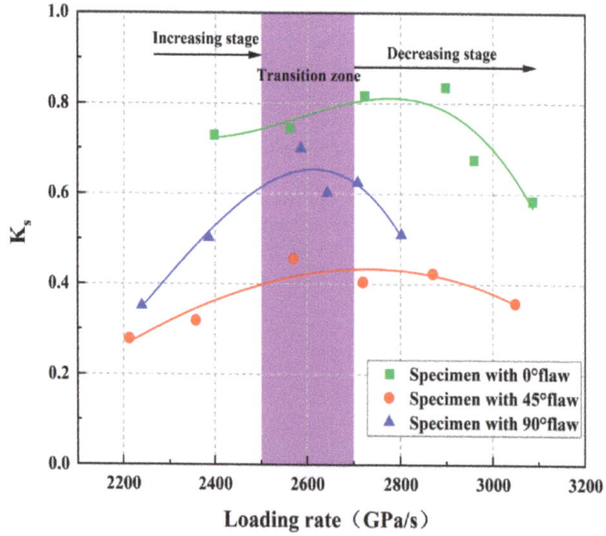

Figure 17. The evolution process of K_s under impact loading.

We can calculate the absolute residual energy of the rock at failure and instability (i.e., the residual elastic energy density, A_{EF}, or the difference between u_d^p and post-peak failure energy density u_a^p) when the rock is at failure and instability in order to verify the accuracy of the proposed rockburst classification method, as shown in Figure 17. We obtain the A_{EF} (a new rockburst classification method) to evaluate the rockburst proneness of fractured rocks based on this approach, and the calculation formula and classification criterion of A_{EF} are expressed as follows [41]:

$$A_{EF} = u_e^p - u_a \tag{22}$$

$$A_{EF} < 50\ \text{KJ/m}^3 \tag{23}$$

$$50\ \text{KJ/m}^3 \le A_{EF} \le 150\ \text{KJ/m}^3 \tag{24}$$

$$150\ \text{KJ/m}^3 \le A_{EF} \le 200\ \text{KJ/m}^3 \tag{25}$$

$$A_{EF} > 200\ \text{KJ/m}^3 \tag{26}$$

As shown in Figure 18, the rockburst tendency of the rock samples has an upward trend with the increasing loading rate. Specifically, the maximum A_{EF} of the fractured granite is 500 kJ/m^3, and the minimum A_{EF} is 100 kJ/m^3. The rock A_{EF} is lower than 200 kJ/m^3 when the loading rate is lower than 2600 GPa/s, while the occurrence of rockburst is not obvious, and moderate rockburst occurs. The inclination angle of the fracture is also related to the rockburst tendency of the rock sample during the impact process. Only the 45° fractured rock occurs in the low rockburst area, while the 0°, 45°, and 90° fractured rocks almost all have medium and high rockburst, with the rockburst tendency first decreasing and then increasing with the increasing fissure angle, and its magnitude relationship is 0° > 90° > 45°, which also indicates that there is less correlation between rockburst tendency and fracture angle.

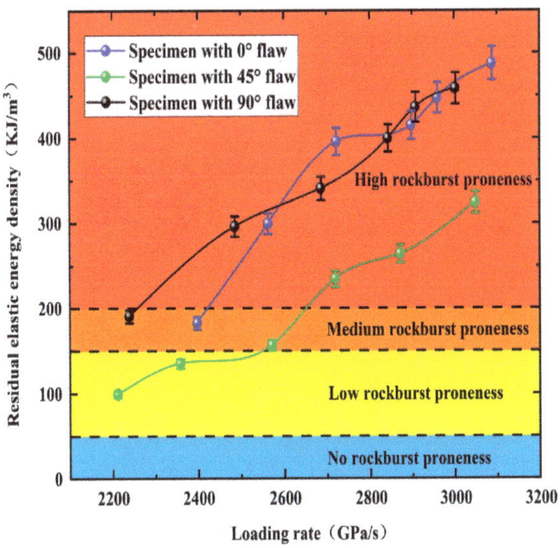

Figure 18. A_{EF} of pre-crack specimens at different loading rates.

5.2.3. Comprehensive Analysis of Rockburst Proneness of Pre-Crack Specimens

Both M_E and A_E reflect the failure degree and residual energy of the fractured rock when the rockburst occurs, respectively. Figure 19 shows the relationship between M_E and A_{EF} of the rock samples at different fracture angles. M_E gradually increases as A_{EF} increases. This indicates that A_{EF} is consistent with the actual rockburst intensity, and it also means that A_{EF} can accurately reflect the rockburst tendency of rock samples during dynamic loading. Specifically, the rockburst tendency gradually decreases when the fracture angle increases from 0° to 45°. The rockburst tendency increases gradually when the fracture angle increases from 45° to 90°. Therefore, 45° can be regarded as the threshold angle for rockburst proneness. This can be explained because the fracture angle is more than 45° and the effect of the prefabricated fractures reduces the interaction force between the particles and thus generates additional stress, leading to stress concentration inside the rock mass and reducing the intensity of rock-burst. However, releasable elastic energy within the rock during the loading process increases significantly when the fracture angle is greater than 45° and thus enhances the intensity of rockburst.

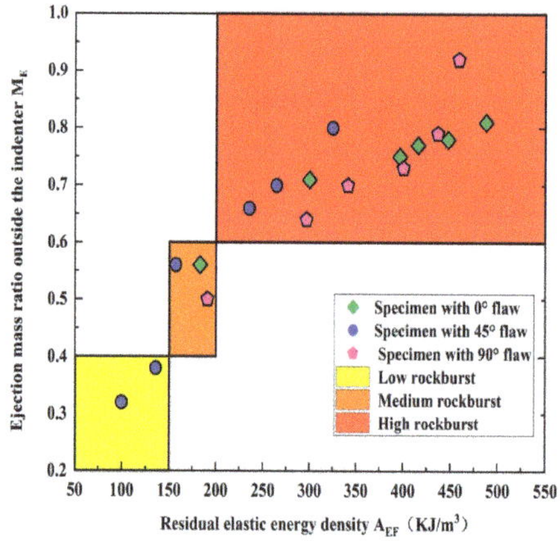

Figure 19. Relationship between the M_E and A_{EF}.

6. Discussion

The sudden release of internal elastic energy during the excavation of deep rock projects may lead to the occurrence of rockburst where the damage process is mainly the release of energy. In essence, rock compression damage is an energy-driven process as the mechanical properties of the rock are changed by the crack fractures, while energy storage is a prerequisite for internal energy. Therefore, studying the law change of the internal elastic energy of rock specimens with loading rates during impact disturbance can reveal the elastic energy generated under the rockburst. However, only analyzing the change law of elastic energy cannot reflect the absolute amount of kinetic energy released from the rock, and further improvements are required. According to the law of conservation of energy, when the rockburst occurs, the kinetic energy of the ejected rock comes from the residual elastic energy after the overall failure of the rock, and the greater the remaining elastic energy, the greater the kinetic energy ejected from the rock fragments and the higher the rockburst tendency is. Therefore, the measure of the intensity of the rock explosion should be judged based on the absolute energy value remaining when the rock is destroyed as an indicator, which provides a theoretical reference for the prevention of rockburst.

The similarities and differences between the kinetic characteristics and energy evolution laws of granites containing prefabricated fractures and general homogeneous rocks are first explored in order to apply the research results to engineering practice. Compared to the intact rocks, there is no significant difference in the final failure mode and energy dissipation law of the fractured rocks, nor are there significant differences in the patterns of elastic and plastic energies that accompany the tests as loading rates increase. The difference lies in their brittleness and resistance to external energy, while the brittleness of fractured rocks is weaker than that of general homogeneous rocks, but the maximum elastic energy stored under the same condition is higher than that of general intact rocks.

The change law of plastic energy, reflected energy, and transmitted energy with loading rate can reveal the degree of damage inside the rock specimen and reflect the degree of development of fissures and cracks in the rock mass. Monitoring the energy of impact stress waves generated during the blasting disturbance and combining this with the variation law of the energy generated during the blasting disturbance can infer the damage degree of the rock mass, which provides a theoretical reference to analyze the integrity of the rock mass and facilitate service blasting mining, rock enclosure support, and stability control of the extraction area. A certain amount of elastic energy is stored inside the rock under the condition of high static stress. Rockburst will occur instantly when the energy exceeds the storage limit of the rock specimen. However, the energy requires impact perturbation to induce a rockburst if the energy is below the energy storage limit of the rock specimen.

Although some energy is released during the rockburst, the rock mainly absorbs energy during the whole loading process, which means that the energy absorbed by the rock from the dynamic disturbance is greater than the energy released.

7. Conclusions

Many underground engineering projects show that rockburst can occur in flaw rocks under dynamic loading. In this work, dynamic compression tests were conducted on pre-flawed rock specimens using a constrained SHPB system, and the influences of loading rates and flaw angles on deformation properties, fragment distributions, failure patterns, energy storages, dissipations, and surpluses of rocks were analyzed. Moreover, high-speed and crack classification methods were used to study the progressive cracking behavior and final failure form of fractured specimens. The loading rate and fracture angle comprehensively affect the progressive cracking behavior of the rock. The following major conclusions can be drawn:

1. Loading rates can promote the cracking of the rock for a given fracture angle, and only tensile cracks appear when the loading rate is small. Shear cracks become more prominent at higher fracture angles as the loading rate increases, while tensile cracks become more prominent at lower fracture angles. The dominant cracking mechanism changes from tensile to hybrid tensile–shear cracking as the fracture angle increases from 0° to 90°.
2. The three main failure types I-III are tensile failure, X-type shear failure, and tensile–shear mixed failure. The failure type of the specimen is Type I when the crack angle and loading rate are low. Increasing loading rates and crack angles mean the failure modes of other specimens belong to type II and III.
3. The fractal dimension increases with the loading rate, while the average fragment size does the opposite. The fractal dimension decreases as the fracture angle increases, and the average fragment size first increases and then decreases. The increasing loading rate reduces the energy utilization efficiency while it promotes the energy dissipation density for a fixed crack angle. The energy dissipation density first decreases and then increases for a given loading rate, while the energy utilization efficiency first increases and then decreases with the increasing loading ratio. Additionally, all rock specimens feature positive absorbed energy values in the dynamic tests.
4. A a strong ejection phenomenon occurs in the granite specimen under the action of a large loading rate, which means a large number of rock fragments and powders are ejected when it is broken. The rockburst tendency first decreases and then increases with the increasing fracture angle. The M_E values of granite with different fracture angles under impact loading increase with the increasing loading rate, which indicates that the rockburst tendency of rock samples increases with the increasing loading rate.
5. Based on accurate calculations of u_d^p and u_a^p for the granite specimens at different crack angles, A_{EF} is employed to evaluate the rockburst proneness, and the results are in good agreement with the statistical results of M_E. Strong linear relationships exist between u_e and u_t, u_d, and u_t during the shock compression process of granite, and the crack angle does not alter the linear energy storage and dissipation laws of granite.

Author Contributions: Conceptualization, L.Z.; methodology, Z.Z.; writing—original draft preparation, G.Z.; data curation, X.L.; writing—review and editing, Y.L. and B.D.; funding acquisition, B.D. All authors have read and agreed to the published version of the manuscript.

Funding: This research is supported by the National Natural Science Foundation of China (51804163, 52004130), the China Postdoctoral Science Foundation (2021M693799), the Provincial Natural Science Foundation of Hunan (2022JJ40373), and the 2021 Hunan Provincial Technology Key Research Projects (2021SK1050), the Hunan Provincial Department of Education Key Projects (22A0293), the Hunan Provincial Department of Education General Projects (22C0235). The authors are very grateful for the financial contributions and convey their appreciation of the organizations for supporting this research.

Institutional Review Board Statement: Not applicable.

Informed Consent Statement: Not applicable.

Data Availability Statement: The data presented in this study are available on request from the corresponding author.

Conflicts of Interest: The authors declare no conflict of interest.

References

1. Xu, Y.; Dai, F.; Du, H. Experimental and numerical studies on compression-shear behaviors of brittle rocks subjected to combined static-dynamic loading. *Int. J. Mech. Sci.* **2020**, *175*, 105520. [CrossRef]
2. Xia, K.W.; Yao, W. Dynamic rock tests using split Hopkinson (Kolsky) bar system-A review. *J. Rock Mech. Geotech.* **2015**, *7*, 27–59. [CrossRef]
3. Du, H.B.; Dai, F.; Xu, Y.; Yan, Z.L.; Wei, M.D. Mechanical responses and failure mechanism of hydrostatically pressurized rocks under combined compression-shear impacting. *Int. J. Mech. Sci.* **2020**, *165*, 105219. [CrossRef]
4. Braunagel, M.J.; Griffith, W.A. The effect of dynamic stress cycling on the compressive strength of rocks. *Geophys. Res. Lett.* **2019**, *46*, 79–86. [CrossRef]
5. You, W.; Dai, F.; Liu, Y.; Yan, Z. Effect of Confining Pressure and Strain Rate on Mechanical Behaviors and Failure Characteristics of Sandstone Containing a Pre-existing Flaw. *Rock Mech. Rock Eng.* **2022**, *55*, 2091–2109. [CrossRef]
6. Liu, Y.; Dai, F. A review of experimental and theoretical research on the deformation and failure behavior of rocks subjected to cyclic loading. *J. Rock Mech. Geotech.* **2021**, *13*, 1203–1230. [CrossRef]
7. Liu, Y.; Dai, F. A damage constitutive model for intermittent jointed rocks under cyclic uniaxial compression. *Int. J. Rock Mech. Min.* **2018**, *103*, 289–301. [CrossRef]
8. Huang, C.; Yang, W.; Duan, K.; Fang, L.; Wang, L.; Bo, C. Mechani-cal behaviors of the brittle rock-like specimens with multi-non-persistent joints under uniaxial compression. *Constr Build Mater.* **2019**, *220*, 426–443. [CrossRef]
9. Li, X.F.; Li, H.B.; Zhang, Q.B.; Jiang, J.L.; Zhao, J. Dynamic fragmenta-tion of rock material: Characteristic size, fragment distribution and pulverization law. *Eng Fract Mech.* **2018**, *199*, 739–759. [CrossRef]
10. Zou, C.; Wong, L.N.Y.; Loo, J.J.; Gan, B.S. Different mechanical and cracking behaviors of single-flawed brittle gypsum specimens under dynamic and quasi-static loadings. *Eng. Geol.* **2016**, *201*, 71–84. [CrossRef]
11. Liu, Y.; Dai, F.; Pei, P. A wing-crack extension model for tensile response of saturated rocks under coupled static-dynamic load-ing. *Int. J. Rock Mech. Min.* **2021**, *146*, 1048922.
12. Wong, L.N.Y.; Einstein, H.H. Crack coalescence in molded gyp-sum and carrara marble: Part 2—Microscopic observations and interpretation. *Rock Mech. Rock Eng.* **2009**, *42*, 513–545. [CrossRef]
13. Wong, L.N.Y.; Einstein, H.H. Systematic evaluation of cracking behavior in specimens containing single flaws under uniaxial com-pression. *Int. J. Rock Mech. Min. Sci.* **2009**, *46*, 239–249. [CrossRef]
14. Zhu, Q.Q.; Li, D.Y.; Han, Z.Y.; Li, X.B.; Zhou, Z.L. Mechanical prop-erties and fracture evolution of sandstone specimens containing different inclusions under uniaxial compression. *Int. J. Rock Mech. Min. Sci.* **2019**, *115*, 33–47. [CrossRef]
15. Wu, Z.J.; Wong, L.N.Y.; Fan, L.F. Study of dynamic fracture problems in viscoelastic sedimentary rocks using the numerical manifold method. *Rock Mech. Rock Eng.* **2013**, *46*, 1415–1457. [CrossRef]
16. Wong, R.H.C.; Chau, K.T. Crack coalescence in a rock-like material containing two cracks. *Int. J. Rock Mech. Min. Sci.* **1998**, *35*, 147–164. [CrossRef]
17. Huang, Y.H.; Yang, S.Q.; Zhao, J. Three-dimensional numerical simulation on triaxial failure mechanical behavior of rock-like specimen containing two unparallel fissures. *Rock Mech. Rock Eng.* **2016**, *49*, 4711–4729. [CrossRef]
18. Bing, D.; Xinyao, L.U.; Qiwei, S.H.; Ying, C.H.; Yong, L.I. Analysis on damage characteristics and energy dissipation of rock with a single hole under cyclic impact loads. *China Saf. Sci. J.* **2020**, *30*, 69–77.
19. Dai, B.; Chen, Y. A novel approach for predicting theheight of the water-flow fracture zone in undersea safety min-ing. *Remote Sens.* **2020**, *12*, 358. [CrossRef]
20. Dai, B.; Chen, Y.; Zhao, G.; Liang, W.; Wu, H. A numer-ical study on the crack development behavior of rock-like material containing two intersecting flaws. *Mathematics* **2019**, *7*, 1223. [CrossRef]
21. Zhang, L.; Zhang, Z.; Chen, Y.; Dai, B.; Wang, B. Crack development and damage patterns under combined dynamic-static loading of parallel double fractured rocks based on DIC technique. *Acta Geotech.* **2022**, *17*, 3099–3115. [CrossRef]
22. Dai, B.; Shan, Q.; Chen, Y.; Luo, X. Mechanical and Energy Dissipation Characteristics of Granite under Cyclic Impact Loading. *J. Cent. South Univ.* **2022**, *29*, 116–128. [CrossRef]
23. Bobet, A.; Einstein, H.H. Fracture coalescence in rock-type materi-als under uniaxial and biaxial compression. *Int. J. Rock Mech. Min.* **2020**, *35*, 863–888. [CrossRef]
24. Li, Y.P.; Chen, L.Z.; Wang, Y.H. Experimental research on pre-cracked marble under compression. *Int. J. Solids Struct.* **2005**, *42*, 2505–2516. [CrossRef]

25. Li, D.Y.; Xiao, P.; Han, Z.Y.; Zhu, Q.Q. Mechanical and failure proper-ties of rocks with a cavity under coupled static and dynamic loads. *Eng Fract Mech.* **2020**, *225*, 106195. [CrossRef]
26. Yang, S.Q.; Jing, H.W. Strength failure and crack coalescence behavior of brittle sandstone samples containing a single fissure under uniaxial compression. *Int. J. Fract.* **2011**, *168*, 227–250. [CrossRef]
27. Yang, S.Q.; Liu, X.R.; Jing, H.W. Experimental investigation on fracture coalescence behavior of red sandstone containing two unparallel fissures under uniaxial compression. *Int. J. Rock Mech. Min.* **2013**, *63*, 82–92. [CrossRef]
28. Li, X.B.; Zhou, T.; Li, D.Y. Dynamic strength and fracturing behav-ior of single-flawed prismatic marble specimens under impact loading with a split-Hopkinson pressure bar. *Rock Mech. Rock Eng.* **2017**, *50*, 29–44. [CrossRef]
29. Yan, Z.; Dai, F.; Zhu, J.; Xu, Y. Dynamic cracking behaviors and energy evolution of multi-flawed rocks under static pre-compression. *Rock Mech. Rock Eng.* **2021**, *54*, 5117–5139. [CrossRef]
30. Zou, C.; Wong, L.N.Y. Experimental studies on cracking processes and failure in marble under dynamic loading. *Eng. Geol.* **2014**, *173*, 19–31. [CrossRef]
31. Zhang, Q.B.; Zhao, J. A review of dynamic experimental tech-niques and mechanical behaviour of rock materials. *Rock Mech. Rock Eng.* **2014**, *47*, 1411–1478. [CrossRef]
32. Yan, Z.; Dai, F.; Liu, Y.; Du, H. Experimental investigations ofthe dynamic mechanical properties and fracturing behavior of cracked rocks under dynamic loading. *Eng. Geol. Environ.* **2020**, *18*, 209–224.
33. Fan, L.F.; Yi, X.W.; Ma, G.W. Numerical manifold method (NMM) simulation of stress wave propagation through fractured rock mass. *Int. J. Appl. Mech.* **2013**, *5*, 249–268. [CrossRef]
34. Wu, W.; Li, H.; Zhao, J. Dynamic responses of non-welded andwelded rock fractures and implications for p-wave attenuation in a rock mass. *Int. J. Rock Mech. Min.* **2015**, *77*, 174–181. [CrossRef]
35. Li, D.; Cheng, T.; Zhou, T.; Li, X. Experimental study of the dynamic strength and fracturing characteristics of marble speci-mens with a single hole under impact loading. *Chin. J. Rock Mech. Eng.* **2015**, *34*, 249–260. (In Chinese)
36. Zhou, L.; Zhu, Z.; Wang, M.; Ying, P.; Dong, Y. Dynamic propaga-tion behavior of cracks emanating from tunnel edges under impact loads. *Soil Dyn. Earthq. Eng.* **2018**, *105*, 119–126. [CrossRef]
37. Zhang, Q.B.; Zhao, J. Determination of mechanical properties and full-field strain measurements of rock material under dynamic loads. *Int. J. Rock Mech. Min. Sci.* **2013**, *60*, 423–439. [CrossRef]
38. Li, X.B. *Rock Dynamics Fundamentals and Applications*; Science Press: Beijing, China, 2014; pp. 18–38.
39. Zhou, Z.; Cai, X.; Li, X.; Cao, W.; Du, X. Dynamic response and energy evolution of sandstone under coupled static–dynamic com-pression: Insights from experimental study into deep rock engineer-ing applications. *Rock Mech. Rock Eng.* **2020**, *53*, 1305–1331. [CrossRef]
40. Gong, F.Q.; Wang, Y.L.; Luo, S. Rockburst proneness criteria for rock materials:review and new insights. *J. Cent. South Univ.* **2020**, *27*, 2793–2821. [CrossRef]
41. Gong, F.Q.; Yan, J.Y.; Li, X.B. A new criterion of rock burst proneness based onthe linear energy storage law and the residual elastic energy index. *Chin. J. Rock Mech. Eng.* **2018**, *37*, 1993–2014. (In Chinese)

materials

MDPI

Article

Using Iron Tailings for Phosphate Removal in Cemented Phosphogypsum (PG) Backfill

Ying Shi [1], Xiaolin Wang [1], Zixuan Qing [1], Yanmei Song [1], Jie Min [1], Yanan Zhou [1], Jing Du [2,3] and Shaofeng Wang [1,*]

1 School of Resources and Safety Engineering, Central South University, Changsha 410083, China
2 School of Materials Engineering, Changshu Institute of Technology, Changshu 215500, China
3 Suzhou Sinoma Design and Research Institute of Non-Metallic Minerals Industry Co., Ltd., Suzhou 215151, China
* Correspondence: sf.wang@csu.edu.cn

Abstract: Compared with the post-treatment of pollutants, such as the removal of phosphate from wastewater, it is more important to develop effective emission control strategies to reduce phosphate pollution. Phosphogypsum (PG) is a typical solid waste byproduct of phosphate production and contains high amounts of residual phosphate. In order to control the phosphate emissions during the recycling of PG aggregates for cemented backfill, another solid waste product—iron tailings (ITs)—was added during the preparation of backfill slurry. The results showed that the ITs effectively accelerated the phosphate removal in cemented PG backfill, enabling the quick reduction in the phosphate concentration to the discharge standard (<0.5 mg/L) within 15 min. This means that the emissions of phosphate to bleeding water were effectively controlled. The adsorption experiment showed that phosphate was adsorbed by the ITs, and the adsorption data fitted well with the Langmuir adsorption model (R^2 = 0.98) and pseudo-second-order kinetic model (R^2 = 0.99), indicating that the phosphate adsorption of ITs was a monolayer chemical adsorption. Furthermore, an unconfined compressive strength (UCS) test was performed on the backfill with the addition of ITs. Compared to the control group (without ITs), the UCS of backfill with 20% ITs increased from 1.08 MPa to 1.33 MPa, indicating that the addition of solid waste could be beneficial to the strength development of the backfill by mitigating the interference of phosphate with the hydration process. The backfill cured for 28 d was selected for the toxic leaching test, and the phosphate concentration in the leachates was always below 0.02 mg/L, indicating that ITs can effectively immobilize phosphate in backfill for a long time.

Keywords: phosphogypsum; cemented backfill; iron tailings; phosphate removal by adsorption

Citation: Shi, Y.; Wang, X.; Qing, Z.; Song, Y.; Min, J.; Zhou, Y.; Du, J.; Wang, S. Using Iron Tailings for Phosphate Removal in Cemented Phosphogypsum (PG) Backfill. *Materials* **2022**, *15*, 8497. https://doi.org/10.3390/ma15238497

Academic Editor: Morena Nocchetti

Received: 25 October 2022
Accepted: 23 November 2022
Published: 29 November 2022

1. Introduction

Secondary pollution should be fully considered in the recycling of solid waste [1]. Phosphogypsum (PG) is a typical solid waste byproduct in the production of phosphoric acid; it is mainly composed of calcium sulfate dihydrate, and also contains high concentrations of soluble phosphate and other impurities (2–4%) [2–4]. Some technologies of PG utilization have been developed. For example, PG is reused as building materials to produce building gypsum powder and the additive to cement [5,6]. PG is also used as fertilizers or soil stabilization amendments in agriculture [7,8]. Furthermore, the PG can be a resource to be considered to recover rare earth elements [9–11]. Cemented backfill technology with PG as an aggregate can massively recycle PG, solving the problem of PG surface storage, however, the high concentration of phosphate will be released again into the bleeding water of the backfill slurry and the underground leachates, causing secondary pollution and threatening the ecological environment [12–14]. Therefore, the efficient removal of phosphate from cemented PG backfill is an issue that cannot be ignored.

At present, there are several methods for removing phosphate, such as chemical precipitation, biological treatment, adsorption, and membrane separation [15–18]. Among them, adsorption has attracted extensive attention due to its simple operation, less residue generation, strong adaptability, and the ability to remove phosphate at low concentrations [19–21]. Zhang et al. [22] prepared a novel type of Fe–Mg–Zr layered bimetallic hydroxide (LDH) via a simple in situ crosslinking method and applied it to wastewater containing phosphate. The maximum phosphate adsorption capacity of the LDH was 21.61 mg/g, and it could still be used after five desorption cycles. Hu et al. [23] prepared a hierarchical CuAl/biomass carbon fiber LDH for phosphate removal, and the removal rate reached up to 99.6% within 50 min. However, most of the materials used for adsorption are expensive, so the development of low-cost materials that can be used for the removal of phosphate via adsorption has attracted intensive interest [24,25]. Li et al. [26] found that the maximum capacity of the drinking-water treatment residuals (DWTRs) to adsorb phosphate reached 19.6 mg/g, which is mainly due to the great amounts of amorphous iron or aluminum in DWTRs. Letshwenyo et al. [27] studied the ability of copper smelting slag to remove phosphate from wastewater through batch experiments; the theoretical and experimental phosphate adsorption capacity reached 0.16 mg/g and 0.26 mg/g, respectively. Obviously, the use of waste and industrial byproducts for phosphate removal can greatly reduce the cost.

Iron tailings (ITs) are a kind of solid waste with a low content of valuable elements produced during the exploration of iron resources. It can be used for the production of road materials, building brick, trace-element fertilizers and concrete [28–31]. Furthermore, ITs have become a research hotspot among many scholars due to their superior phosphate adsorption performance and low cost. Zeng et al. [32] studied the adsorption isotherm, adsorption kinetics, desorption, and other relevant properties of ITs with respect to phosphate. It was verified that ITs can act as an adsorbent of phosphate with superior performance (4.6 mg/g), and the adsorption was irreversible. Sima et al. [33] found that ITs can remove phosphate from wastewater at a rate of up to 1.29 mg/g. Furthermore, ITs have great regeneration potential, as their phosphate removal rate is still 56% after three cycles, and they can be used as a good phosphate adsorbent.

However, most of the previous studies used ITs for the removal of phosphate from sewage, and there has been no research on the use of ITs for controlling phosphate pollution in cemented PG backfills. In fact, compared to the treatment of phosphate after the drainage of bleeding water, it is more meaningful to control phosphate emissions from backfill slurry. Therefore, this study aimed to study the feasibility of using ITs for the immobilization of phosphate in cemented PG backfills. With the addition of various amounts of ITs during the preparation of the backfill slurry, the adsorption rate and the adsorption mechanism were studied. In addition, the effects of ITs on the performance of backfills were tested by measuring the pH, initial setting time and final setting time, and unconfined compressive strength (UCS) of the backfill. Finally, toxic leaching tests were performed on the backfill after curing for 28 d to evaluate the long-term effects of ITs on the immobilization of phosphate.

2. Materials and Methods

2.1. Materials

The ITs used in the present experiments were collected from a mine in Yunnan, China, and washed thrice with deionized water to remove the adhered dust and soluble substances. After washing, they were oven-dried for 24 h at 105 °C. The PG and binder were collected from a phosphate fertilizer company in Guiyang, China. The PG was a moist gray–black powder with a pH of 1.9. The binder was composed of yellow phosphorous slag, fly ash and cement clinker in 6:1:3 ratio [34]. The particle size distribution of PG, ITs and binder are shown in Figure 1.

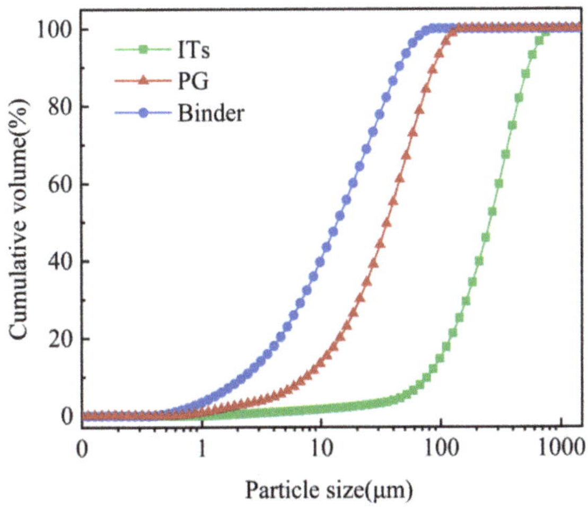

Figure 1. Particle size distribution of PG, binders and ITs.

The diffraction patterns of the ITs and PG were characterized via X-ray diffraction (XRD) using a D8 X-ray automatic diffractometer (Bruker, Billerica, MA, USA) with a resolution of 0.02°/s in the angular domain 5–80°, and the results are shown in Figure 2. The chemical composition of ITs, PG and binder were measured through an X-ray fluorescence analyzer (Bruker, Billerica, MA, USA), as shown in Table 1.

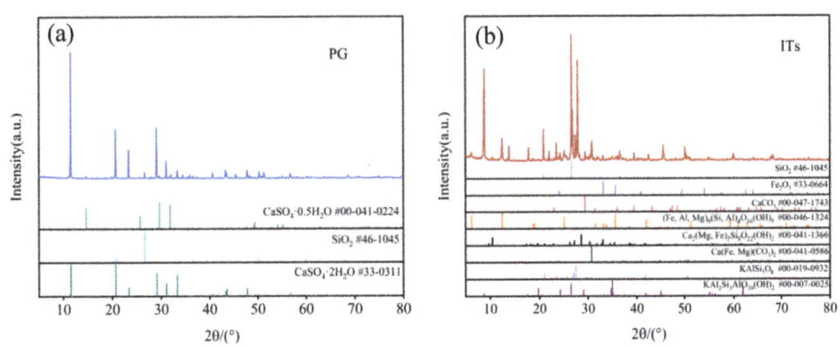

Figure 2. X-ray diffraction patterns of PG (**a**) and ITs (**b**).

Table 1. The main chemical compositions of ITs, PG and binder.

Chemical Composition	ITs (%)	Binder (%)	PG (%)
Na_2O	2.73	0.529	0.12
MgO	2.14	1.5	0.23
Al_2O_3	12.98	5.8	0.485
SiO_2	52.83	18.69	4.393
P_2O_5	1.15	0.184	3.952
SO_3	0.794	4.193	45.68
K_2O	3.763	1.1	0.0966
CaO	4.682	58.69	35.22
TiO_2	1.844	0.562	0.0548
Fe_2O_3	11.47	3.263	0.245
CuO	0.3888	-	0.006
F	-	-	0.42

A scanning electron microscope equipped with an energy-dispersive X-ray spectroscope (SEM-EDS) was used to characterize the ITs and backfill. The microstructure and surface elemental content of the materials were analyzed using a TESCAN MIRA4 scanning electron microscope (TESCAN, Brno, Czechia). Figure 3 shows the SEM image of the surface morphology of the ITs. It can be seen that the surface of the ITs is rough, with irregular distribution of lumps and gaps. This surface morphology means that the ITs have a large specific surface area and high porosity, providing a great amount of active adsorption sites for phosphate removal.

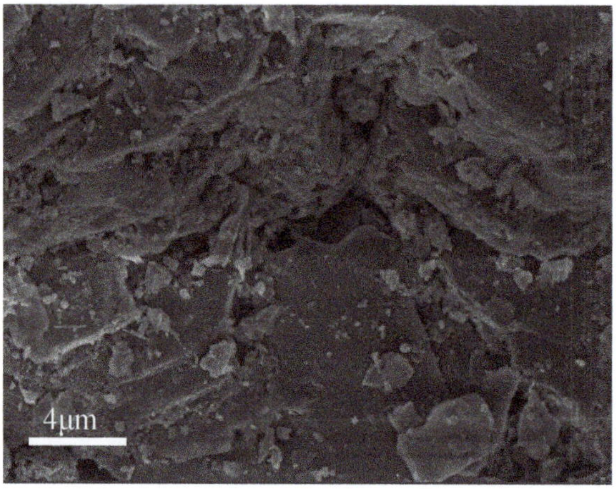

Figure 3. SEM image of ITs.

According to previous studies, the reaction of functional groups on the surface of substances—caused by van der Waals forces—is the reason for the adsorption of the molecules [35]. Therefore, the chemical properties of ITs play an important role in their ability to adsorb phosphate. The XRD result show that ITs (Figure 2b) contained ($KAl_2(AlSi_3O)(OH)_2$), Fe_2O_3 and $Ca(MgFe)(CO_3)_2$. The XRF result shows that ITs (Table 1) contained 11.47% of Fe_2O_3, 12.98% of Al_2O_3, and 4.68% CaO, respectively. Iron (Fe^{3+}), aluminum (Al^{3+}), calcium (Ca^{2+}), and other metal ions contained in ITs have an affinity for phosphate [36]. Therefore, the ITs have the necessary chemical conditions for the adsorption or complexation of phosphate.

2.2. *Preparation of Cemented PG Backfill Slurry*

To investigate the change in phosphate concentration in the cemented backfill, the backfill slurry was prepared with the mixing proportions shown in Table 2. PG aggregate, binder, and water were added to a PP container, where the dry mass ratio of PG to binder was 5:1, and the liquid–solid ratio was 5:1. Three batches of backfill slurry were prepared. After being mixed uniformly, each batch of slurry was immediately divided into two parts: one was combined with various contents of ITs as the experimental group, and the other part was considered as the control group (PBS). Three ITs mixtures (5, 20, and 35 wt.% PG) were tested in this study, denoted as PBSI-5, PBSI-20, and PBSI-35, respectively. Taking the time at which the ITs were added as the zero point, the experimental and control groups were sampled at mixing times of 0, 1, 2, 3, 4, 5, 15, 45, and 105 min. Immediately after sampling, the slurry was filtered through a 0.45 μm microporous membrane to obtain the bleeding water from the backfill slurry for the analysis of pH and phosphate concentration.

Table 2. Mix proportions of the cement PG backfill slurry.

Batch No.	PG (%)	ITs (%)	Binder (%)	Water (%)	ITs Proportion to PG (%)
PBS	14.2	0	2.8	83	0
PBSI-5	14.2	0.71	2.8	83	5
PBSI-20	14.2	2.84	2.8	83	20
PBSI-35	14.2	4.98	2.8	83	35

2.3. *Phosphate Adsorption Experiment*

2.3.1. Adsorption Isotherms of Phosphate on ITs

In order to explore the phosphate adsorption process of ITs, adsorption isotherm experiments were carried out. First, 3 g of ITs was weighed and added to a 100 mL conical flask, and then 50 mL standard solutions of KH_2PO_4 at different concentrations (i.e., 0.5, 2, 5, 10, 15, 25, and 50 mg/L) were added to the flask. At room temperature, the conical flask was placed in an oscillating device and vibrated for 24 h at 150 rpm HCl or NaOH was used to maintain the pH of the solution at 7. Then, the mixture was filtered through a 0.45 µm microporous membrane, and the filtrate was used to measure the phosphate concentration. The data were fitted to the Langmuir and Freundlich isotherm models to analyze the adsorption equilibrium parameters and adsorption behavior.

2.3.2. Adsorption Kinetics of Phosphate on ITs

To study the adsorption kinetics, 3 g of ITs and 150 mL of KH_2PO_4 solution with a phosphate concentration of 20 mg/L were added to a 250 mL conical flask. The pH of the mixed solution was maintained at 7 with HCl or NaOH. The conical flask was placed in an oscillating device at room temperature and shaken at 150 rpm. The supernatant of the sample was filtered after 1, 2, 4, 8, 12, 24, and 30 h. Then, the filtrate's phosphate concentration was measured. The data were fitted to the pseudo-second-order model to determine the phosphate removal kinetics model.

2.4. *Preparation of Cemented PG Backfill*

In order to explore the influence of ITs on the strength development of the cemented PG backfill, backfill with various amounts of ITs was prepared with the mixing proportions shown in Table 3. The mass ratio of aggregate (PG + ITs) to binder was 4:1, and the solid mass concentration was 60%. Four batches of backfill were prepared with different proportions of ITs to PG, i.e., 0, 5, 20, and 35% (PB, PBI-5, PBI-20, and PBI-35, respectively). PG, binder, ITs, and water were first added to a PP container. After mixing for 30 min, the slurry was poured into a plastic mold with a length, width, and height of 40 mm. After the final setting of the backfill, the samples were demolded and placed in a curing box at a constant temperature of 20 ± 2 °C and a humidity of 90 ± 5%.

Table 3. Compositions of the cement PG backfills.

Batch No.	Aggregate		Binders (%)	Water (%)	ITs proportion to PG (%)
	PG (%)	ITs (%)			
PB	48	0	12	40	0
PBI-5	45.6	2.4	12	40	5
PBI-20	38.4	9.6	12	40	20
PBI-35	31.2	16.8	12	40	35

2.5. *Unconfined Compressive Strength Tests*

In order to evaluate the impact of adding ITs on the mechanical properties of the backfill, UCS tests were performed for backfill samples cured for 7 d, 14 d, and 28 d. According to the JGJ/T 70-2009 standard [37], the test was carried out using a WHY-200/10 servo pressure testing machine (Hualong, Shanghai, China) with a loading rate of 0.1 mm/min. Each backfill sample had three replicates, and the average value was used for the calculation of UCS at different curing ages.

2.6. Toxicity Leaching Test

Cemented PG backfill is transported to underground stopes where it is in long-term contact with groundwater [13]. Therefore, it is necessary to determine whether the phosphate can be fixed in backfill for an extended period. Toxicity leaching tests were carried out on the backfill according to the Chinese standard HJ557-2010 [38]. The backfill cured for 28 d was ground and put in an oven for drying at 55 °C. Then, the dried backfill was mixed with deionized water at a liquid–solid ratio of 10:1 in a PP bottle and shaken on a horizontal vibrator at 110 rpm for 8 h. Finally, it was left to stand for 16 h, and the supernatant was filtered to measure the phosphate concentration. Each test was performed in triplicate to ensure the accuracy of the measurement.

2.7. Chemical Measurements

The phosphate concentration measured in this study was the concentration of soluble orthophosphate (PO_4^{3-}-P), which was determined via the ammonium molybdate tetrahydrate spectrophotometry method. First, 100 μL of 10% ascorbic acid solution was added to 5 mL of the sample. After the mixture was shaken well, 200 μL of $H_8MoN_2O_4$ was added in solution. After standing for 15 min, the solution was put into a Shimadzu UV1800 spectrophotometer (Shimadzu, Kyoto, Japan) and measured at the wavelength of 700 nm. The pH of the backfill slurry was measured using a Mettler Toledo FE28 pH meter (Shanghai, China).

3. Results and Discussion

3.1. Phosphate Adsorption of Its in Cemented PG Backfill

3.1.1. Effect of the Addition of ITs on the Removal of Phosphate from the Backfill Slurry

When the ITs were added to the backfill slurry at proportions of 5%, 20%, and 35%, the changes in the phosphate concentration were monitored, as shown in Figure 4. The initial concentration of phosphate in the backfill slurry was 25–28 mg/L. Without the addition of IT (PBS), the phosphate concentration in the slurry decreased gradually. This was because the hydration reaction of the binder produced a large amount of Ca^{2+}, which combined with the phosphate to form precipitates such as calcium phosphate salt [39,40]. In comparison, with the help of ITs, the rate of decrease in the phosphate concentration was more remarkable. For example, as shown in Figure 4a, the phosphate concentration in the control group changed from 24.5 mg/L to 19.5 mg/L in the first minute, decreasing by 20.4%. However, when the amount of ITs added was 5%, the concentration of phosphate changed from 24.5 mg/L to 16.3 mg/L in the first minute, decreasing by 33.5%, indicating that phosphate was adsorbed more quickly due to the addition of ITs. The phosphate removal rates in the other groups (PBSI-20 and PBSI-30) were also faster than the control group. This indicates that the ITs were involved in phosphate adsorption and could remove phosphate from the slurry in a short time.

The amount of ITs added affected the solidification of phosphate in the backfill. During the whole mixing period of 2 h, the phosphate concentrations of the PBS and of the lowest proportion of ITs (i.e., PBSI-5) were 1.63 mg/L and 0.70 mg/L, respectively, which meant that the phosphate concentration in the bleeding water exceeded the discharge standard (<0.50 mg/L) [41]. However, the phosphate concentration of PBSI-20 descended to 0.31 mg/L at 15 min, which was lower than the discharge standard. When the proportion of ITs was increased to 35%, the phosphate removal rate was even higher, and the phosphate concentration was 0.05 mg/L, while the corresponding control group phosphate concentration was 7.10 mg/L after 15 min of mixing. This shows that in the cemented PG backfill, increasing the proportion of ITs can effectively increase the phosphate removal rate and control the phosphate concentration in the bleeding water.

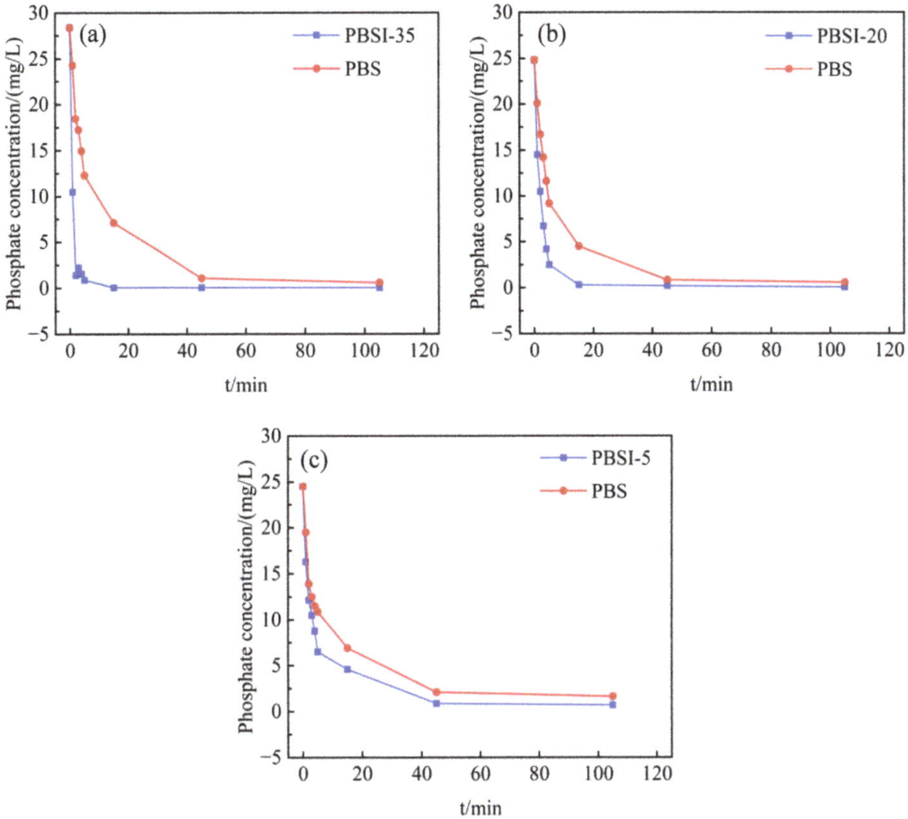

Figure 4. Changes in the phosphate concentration in backfill slurry with different amounts of ITs: (**a**) 35% ITs addition; (**b**) 20% ITs addition; and (**c**) 5% ITs addition.

Based on the above results, during the preparation of the backfill slurry, adding sufficient ITs could quickly and effectively solidify the phosphate, resulting in the phosphate concentration in the bleeding water being below the discharge limit. This means that the post-treatment of wastewater could be avoided, indicating that adsorption by ITs in backfill preparation is an effective emission control strategy for phosphate pollution. This provides a new direction for the removal of phosphate from cemented PG backfill. Furthermore, this method achieves the disposal of waste using other waste, representing a reproducible approach for the large-scale recycling of solid waste.

3.1.2. Effect of ITs on the pH of Backfill Slurry

Figure 5 shows the changes in the pH of the backfill slurry with mixing time. When the low-pH PG aggregate was mixed with the binder, the pH increased gradually from 2.2 to 7–7.3 within 20 min, which was related to the increase in the amount of OH^- due to the hydration reactions of the binder [42]. When the ITs were added to the backfill slurry, the pH increased to even higher levels. After adding 5% ITs, the pH of the slurry increased from 6.35 to 6.67, representing an increase of 5.03%, while the pH of the control group increased from 6.35 to 6.55, representing an increase of 3.14%. The pH of PBSI-35 increased from 7.19 to 8.05, representing an increase of 11.96%, while the corresponding control group only increased from 7.19 to 7.22, representing an increase of only 0.42%. One reason for this is that after adding ITs, the calcium carbonate and metal oxides in the ITs (Figure 2b) can consume some of the hydrogen ions in PG and, thus, increase the pH [43]. Another reason is that ITs can adsorb phosphate. When the pH is within the range of 7–10, the phosphate in the solution mainly exists as $H_2PO_4^-$ and HPO_4^{2-} [44]. When the phosphate was adsorbed by the ITs, according to the principle of ionization balance, the hydrogen ions in the solution decreased and the pH increased in the backfill slurry [45]. As shown in Figures 4 and 5, the more ITs added, the faster the decline in the pollutant concentrations and the greater the pH of the slurry. Therefore, the higher the amount of ITs added, the higher the pH of the backfill slurry.

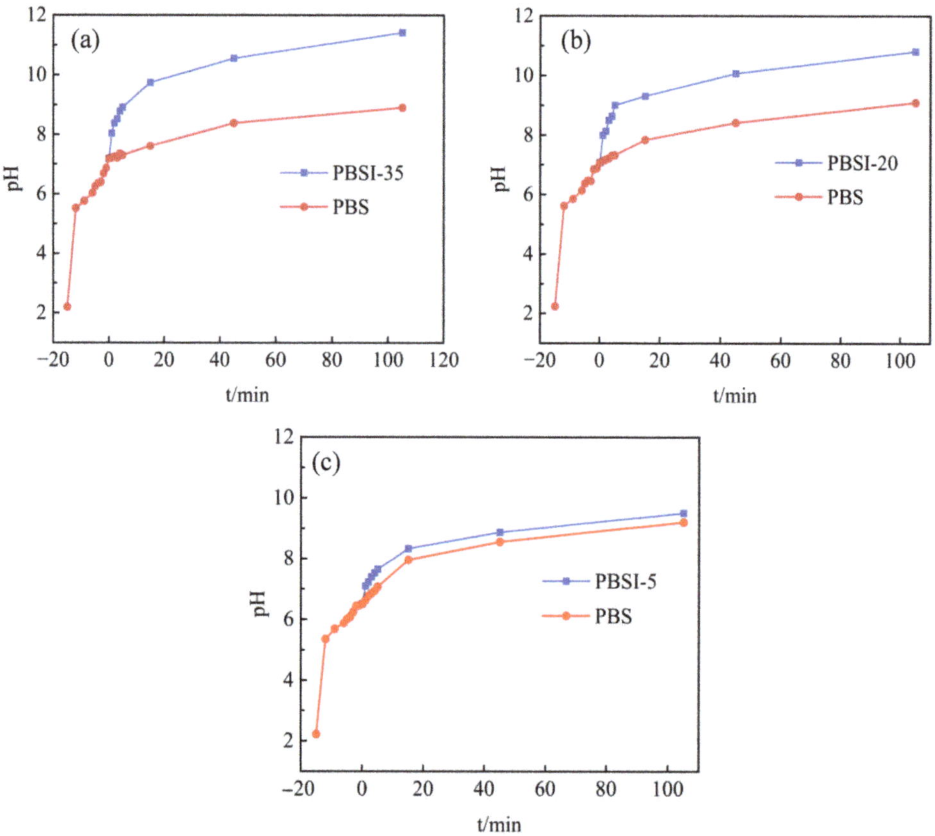

Figure 5. Changes in pH in backfill slurry with different amounts of ITs: (**a**) 35% ITs addition; (**b**) 20% ITs addition; and (**c**) 5% ITs addition.

3.1.3. Adsorption Mechanism of Phosphate by ITs in Backfill

The adsorption kinetic model is usually used to estimate the adsorption rate, which is an important feature to judge the adsorption performance of an adsorbent. Pseudo-first-order and pseudo-second-order kinetic models have been widely used to describe adsorption systems [46,47]. According to the research of Sima et al. [33], ITs are more consistent with the pseudo-second-order kinetic equation. Therefore, the following equation was used to fit the experimental data [48]:

$$\frac{t}{Q_t} = \frac{1}{K_2 {Q_e}^2} + \frac{t}{Q_e} \quad (1)$$

where Q_t is the amount of phosphate adsorbed in time t (mg/g), Q_e is the equilibrium adsorption amount (mg/g), and K_2 is the adsorption equilibrium rate constant of the pseudo-second-order equation (g(mg/h)).

The curve of the adsorption rate of phosphate by ITs was evaluated, and the results are shown in Figure 6. It was observed that the removal of phosphate increased very quickly at first, and then gradually approached constant values. To be specific, the rate of phosphate adsorption was fast in the first 5 h of the adsorption reaction; then, the adsorption rate decreased and reached equilibrium slowly within 10 h. It is possible that with the increase in time, the pores and cation sites of ITs are gradually occupied by phosphate, resulting in no sites being available for phosphate [49].

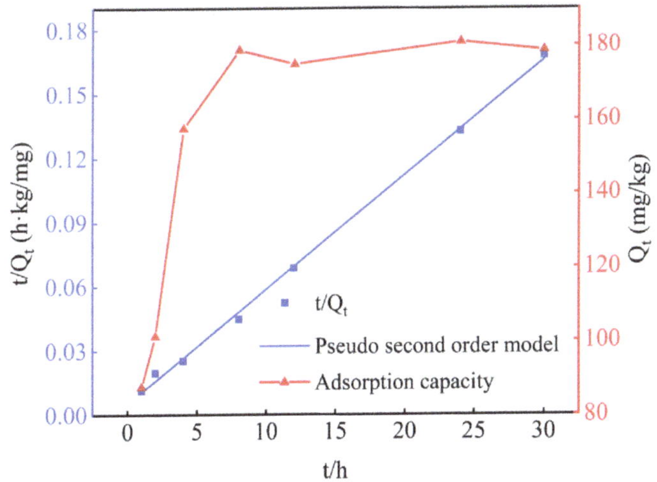

Figure 6. Pseudo-second-order kinetic model and adsorption rate curve of ITs.

The fitting of the pseudo-second-order model of ITs is represented in Figure 6. The correlation coefficient (R^2) of the pseudo-second-order kinetic equation is 0.99, indicating that the phosphate adsorption process of ITs is consistent with the pseudo-second-order kinetic model. This illustrates that the adsorption process is chemical adsorption and that there is a chemical bond between the ITs' surface and the phosphate [50], and the adsorption is relatively stable, with the equilibrium amount of adsorbed phosphate being 185.19 mg/kg. When the amount of ITs added is 70 g and 40 g, the theoretical amount of adsorbed phosphate is 12.9 mg and 7.4 mg, respectively. According to Figure 1, the actual decrease in the amount of phosphate in the backfill slurry with 70 g and 40 g of ITs was about 15.6 mg and 6.7 mg, respectively, which are close to the theoretical values.

Figure 7 shows a comparison of the micromorphology of ITs before and after adsorption in the kinetic experiment. Wang et al. [45] showed that phosphate was adsorbed by iron oxide in ITs through surface complexation and produced iron-based phosphate products. In this study, before adsorption, the surface of the ITs was rough, with a porous structure and many gaps. After adsorption, there were many fine and irregular particles appearing on the surface and pores of the ITs. This might be due to phosphate ions precipitated with ferric iron of ITs forming low-solubility iron–phosphate precipitates [51,52]. Additionally, the protonation of the adsorbent surface in the aqueous solution could make it adsorb phosphorus anions by electrostatic adherence, which was further immobilized by ligand exchange [45]. These indicated that the phosphate in PG was effectively adsorbed by the ITs.

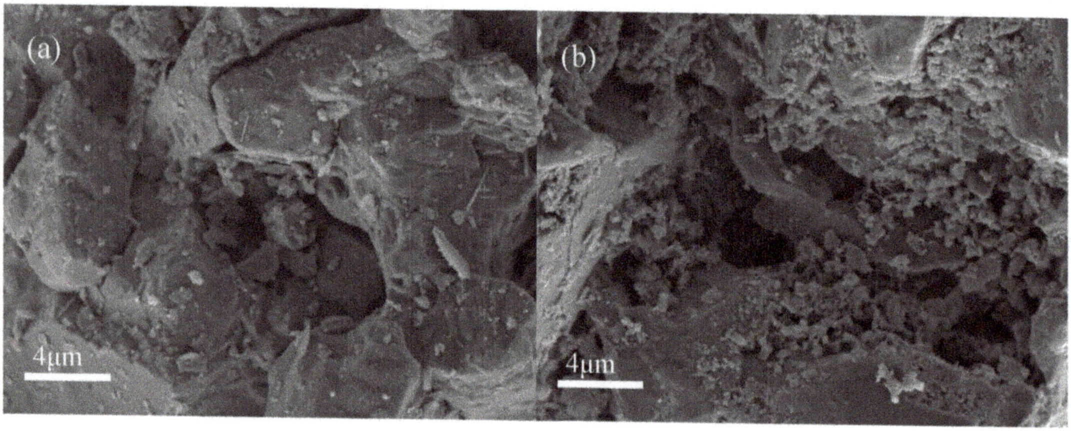

Figure 7. SEM images of ITs before (**a**) and after (**b**) phosphate adsorption.

The adsorption isotherm describes the adsorption processes at the solid–liquid interfaces and can be studied to calculate the adsorption capacity and explain the adsorption mechanism [53,54]. In this study, the adsorption data were fitted according to the Langmuir and Freundlich isotherm models. The linear form of the Langmuir isotherm model can be expressed as follows [55]:

$$\frac{C_e}{Q_e} = \frac{1}{K_L q_m} + \frac{C_e}{q_m} \quad (2)$$

where Q_e is the adsorption capacity of the adsorbent at adsorption equilibrium (mg/kg), q_m is the adsorption capacity when the adsorbent reaches saturation (mg/kg), K_L is the Langmuir equilibrium constant (L/mg), and C_e is the concentration of the adsorbed ion when the adsorption in the solution reaches equilibrium (mg/L). In addition, from the Langmuir equation, a separation factor (R_L) can reflect the feasibility of the adsorption process. This can be expressed as follows:

$$R_L = \frac{1}{1 + K_L C_0} \quad (3)$$

where C_0 is the initial P concentration (mg/L). The correlation coefficient R_L of the Langmuir equation is equal to 1, indicating that the adsorption is linear; when the R_L value is less than 1, it means that it is favorable for adsorption, while $R_L > 1$ indicates that it is not conducive to adsorption [56].

The linear form of the Freundlich isotherm model can be expressed as follows [57]:

$$log q_e = log K_F + \frac{1}{n} log C_e \quad (4)$$

where K_F is the Freundlich constant (mg/kg) and $1/n$ is the adsorption strength parameter.

The fitting results of the experimental data based on the two adsorption models are shown in Figure 8, and the relevant parameters of the Langmuir and Freundlich isotherms are shown in Table 4. The Langmuir and Freundlich correlation coefficients are both larger than 0.89; this shows that both models can explain the adsorption process, indicating that there are both homogeneous and heterogeneous adsorption sites on the ITs. However, the Langmuir correlation coefficient (R^2 = 0.985) is higher than the Freundlich correlation coefficient (R^2 = 0.898), indicating that the Langmuir isotherm model is more suitable for the adsorption process of ITs. The adsorption of phosphate on Its mainly takes place in a single, uniform layer [58]. In addition, in this experiment, R_L was between 0 and 1, indicating that it is conducive to adsorption. According to the curve-fitting results, the equilibrium adsorption capacity of the ITs is 163.93 mg/kg, which is also close to the results of the adsorption kinetics.

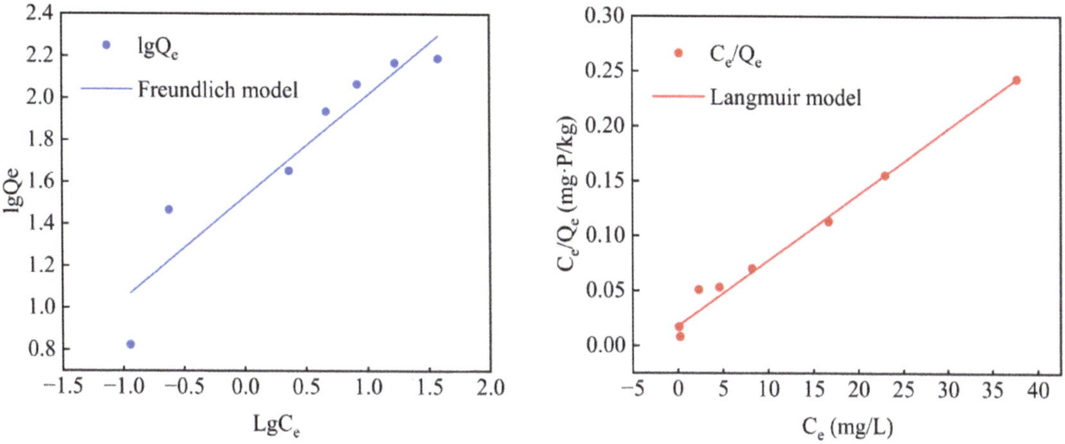

Figure 8. Langmuir and Freundlich models for the adsorption of phosphate onto ITs.

Table 4. Parameters of the Langmuir and Freundlich isotherms for the adsorption of phosphate onto ITs.

Langmuir			Freundlich		
Q_m/(mg/kg)	K_L/(L/mg)	R^2	K_F/(mg/kg)	$1/n$	R^2
163.93	0.43	0.985	34.06	0.49	0.898

The adsorption isotherm and kinetics studies show that the adsorption process of phosphate on ITs is mainly monolayer chemical adsorption. Previous studies have shown that the adsorption process of phosphate on ITs is mainly attributable to the Fe_2O_3 in ITs and includes ligand exchange, precipitation, and electrostatic adherence [45]. In addition to Fe_2O_3, Al^{3+}, Ca^{2+}, and other components in ITs may also promote the removal of phosphate through adsorption or precipitation to some extent [59]. These chemical bonds are formed between the medium surface and the adsorbate, so the adsorption is relatively stable and is not easy to reverse [60]. Combined with the removal of phosphate via precipitation by hydration reactions of the binder during the mixing of the backfill slurry, the cemented PG backfill could effectively and quickly remove phosphate, enabling the phosphate concentration in bleeding water to meet the discharge standard (GB8978-1996).

The changes in phosphate removal efficiency and the contribution of ITs were calculated, as shown in Figure 9. Under the joint action of the binder and ITs, the phosphate removal efficiency increased to 99% within 15 min of mixing (defining the final phosphate removal efficiency as 100% after two hours of mixing). At the first minute, the total removal efficiency was 41%, half of which was attributed to the adsorption by ITs. In the next several minutes, the phosphate removal efficiency attributed to adsorption by ITs remained at 20–35%, while the hydration contributed more and more until the removal efficiency reached 100% at 15 min. Then, with the saturation of ITs' adsorption, hydration gradually began to dominate at 45 min.

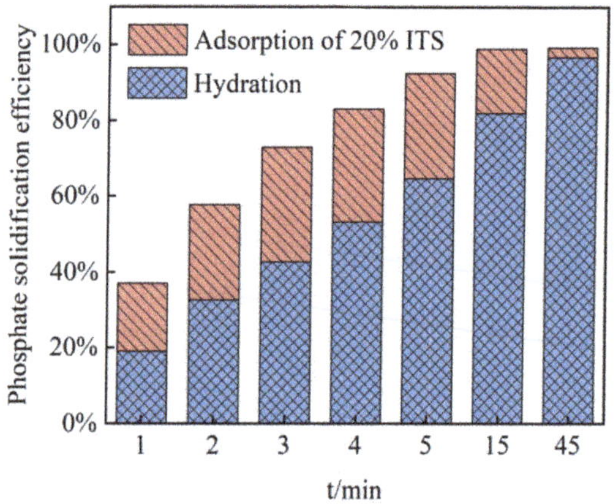

Figure 9. Removal rate of phosphate from slurry by ITs and hydration reaction.

3.2. Strength and Microstructure of Backfill

The development of the early strength has an important relationship with the setting time of the backfill. The initial and final setting times (IST and FST) of the backfill with various proportions of added ITs are shown in Figure 10. The more ITs added, the shorter the IST and FST of the backfill. When the proportion of ITs added was 5%, the IST and FST of the backfill were 34.5 h and 48 h, respectively, while the IST and FST of the control group were 40 h and 56 h, respectively, representing reductions of 13.8% and 16.7%, respectively. When the amount of ITs added increased to 20%, the IFT and FST of the backfill were 26 h and 37 h, respectively, representing reductions of 35% and 33.9% compared with the control group, respectively. This is because phosphate interferes with the hydration process, prolonging the IFT and FST. The introduction of ITs effectively enhanced the phosphate removal efficiency in a short time, thereby improving the hydration process and shortening the setting time of the cemented PG backfill.

The strength of the backfill plays a decisive role in maintaining the stability of mines [61,62]. The UCS test was conducted on the backfill with various proportions of ITs added, as shown in Figure 10. Generally speaking, the strength of all of the backfills developed well, with the lowest UCS at 28 d being above 1 MPa. The addition of ITs effectively promoted the strength development, especially in the early stages. When cured for 7 days, the UCS of the backfill without ITs was 0.26 MPa. The UCS of the backfill with 5% ITs was slightly improved to 0.32 MPa. When the proportions added were increased to 20% and 35%, the UCS of the backfill reached 0.60 MPa and 0.91 MPa, respectively, which are 2.3 and 3.5 times higher than that of the control group, respectively. The SEM of the backfill cured for 7 days is shown in Figure 11. Without the addition of ITs, a relatively small amount of hydration products was sparsely dispersed on the surface of the PG crystal (Figure 11a), corresponding to the low UCS observed in Figure 10. In comparison, with the addition of 20% ITs, the hydration products in the backfill were densely distributed and mutually supported (Figure 11b), helping to improve the strength of the backfill.

The 28 d UCS of cemented PG backfill was also enhanced by ITs. As shown in Figure 10, without ITs, the UCS of the control group was 1.08 MPa at 28 d of curing. However, the UCS of the experimental groups with the addition of 5%, 20%, and 35% ITs was 1.20, 1.33, and 1.38 MPa (representing increases of 11.1%, 23.1%, and 27.8%), respectively. Figure 11 shows SEM images of the backfill cured for 28 days. The hydration products in the backfill without ITs were densely distributed on the surface of the PG crystal (Figure 11c); however, the backfill with 20% ITs had more hydration products, which filled the gaps between aggregates and, thus, formed a whole body (Figure 11d), resulting in higher UCS than the control group.

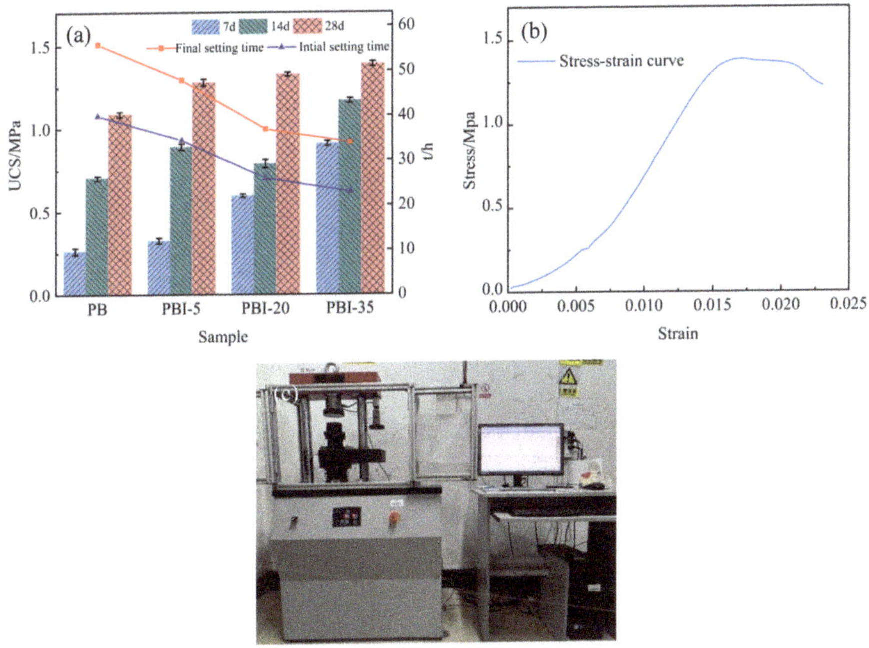

Figure 10. (**a**) UCS and setting time changes of backfill with different amounts of ITs; (**b**) Stress–strain curve of PBI-35 (28 d); (**c**) Servo pressure testing machine for UCS.

Figure 11. SEM images of backfill samples: (**a**) PB cured for 7 d with ITs; (**b**) PBI-20 cured for 7 d; (**c**) PB cured for 28 d; (**d**) PBI-20 cured for 28 d.

In order to determine the distribution of phosphate in the backfill, EDS mapping was performed on the backfill without ITs and with 20% ITs, as shown in Figure 12. The results clearly showed that the content of elemental phosphorus was much less in the backfill with 20% ITs. This is because the phosphorus was firmly adsorbed by the porous structure of the ITs, thereby reducing the phosphorus content on the surface of the backfill. Zhou et al. [63] found that high concentrations of phosphate significantly degraded the UCS of cemented PG backfill, due to the consumption of Ca^{2+} and OH^- derived from the hydration of the binder. When the ITs were introduced into the cemented PG backfill system, the phosphate in the PG aggregate was quickly and firmly adsorbed onto the ITs by ligand exchange, precipitation, and electrostatic adherence. Without the interference of phosphate, more Ca^{2+} and OH^- could participate in the hydration on reaction and produce more hydration products to fill the gaps between aggregates, which benefited the strength development of the backfill. Macroscopically, the slurry was easier to coagulate, and the UCS of the backfill was effectively improved, as discussed above.

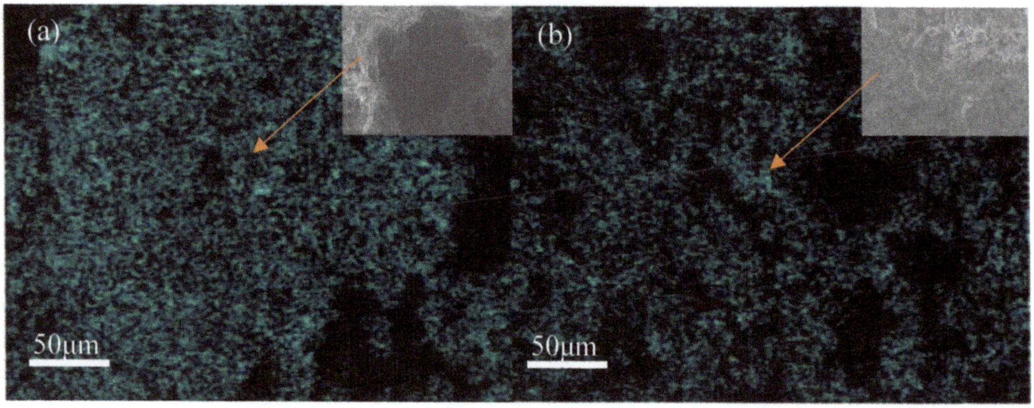

Figure 12. (**a**) The phosphorus mapping diagram of PB cured for 28 d; (**b**) the phosphorus mapping diagram of PBI-20 cured for 28 d.

3.3. Toxic Leaching Test of PG Cemented Backfill

In order to explore the stability of phosphate immobilized by ITs, the backfill cured for 28 d was selected for toxicity leaching tests. With various proportions of ITs in the cemented PG backfill, the phosphate concentration in the leachates was always lower than 0.02 mg/L—far below the discharge standard (0.5 mg/L). This shows that the immobilization of phosphate in the backfill is not disturbed after adding ITs. Therefore, adding ITs to the PG aggregate can solidify the phosphate in the backfill for a long time, which is of great significance for environmental protection.

After the PG aggregate was treated with the ITs as proposed in this research, the phosphate could be effectively immobilized within the ITs, rendering the aggregate harmless and useful. The stable existence of phosphate in the backfill also indicates that the impurities could be well immobilized during the preparation of the backfill, reducing the amount of phosphate released into the bleeding water and the leachates. This greatly reduces the costs of subsequent wastewater treatment. Furthermore, the immobilization of phosphate can enhance the strength performance of the backfill, with substantial significance for the recycling of solid waste.

4. Conclusions

In this study, ITs were used as additives to adsorb phosphate in the process of producing cemented PG backfill. The phosphate immobilization mechanism and the effect of the addition of ITs on the strength of the backfill were investigated. The following conclusions can be drawn:

(1) The addition of ITs to the cemented PG backfill slurry effectively accelerated the removal of phosphate. When the proportion of ITs was 20% or more, the phosphate concentration in the bleeding water could drop to below the discharge standard (0.5 mg/L) within 15 min. The phosphate concentration in the toxic leaching solution of all test blocks was lower than 0.02 mg/L—far lower than the national standard (0.5 mg/L)—indicating that the addition of ITs would not affect the long-term environmental behavior of the backfill.
(2) The linear correlation coefficient in the pseudo-second-order kinetic model was 0.99, indicating that the adsorption of phosphate by ITs takes place through a chemical adsorption process, and there is a strong bond between molecules. ITs are more consistent with Langmuir isotherm model, indicating that the predominant adsorption mode is monolayer adsorption. During the preparation of the backfill, the actual decrease in phosphate attributed to ITs was close to the theoretical values.
(3) The addition of ITs positively affected the UCS of the backfill. With the addition of 20% ITs, the IFT and FST of the backfill were reduced by 35% and 33.9%, respectively, and the 28 d UCS was improved by 23.1%, compared to the control group. Furthermore, it is recommended to study the transportable properties of IT-based cemented PG backfill in the future.
(4) Due to their low cost and good phosphate adsorption performance, ITs have the potential to be used in the preparation of cemented PG backfill to control phosphate pollution, providing new insights into the economical and efficient recycling of solid waste.

Author Contributions: Conceptualization, Y.S. (Ying Shi) and X.W.; Methodology, X.W. and Y.S. (Ying Shi); Software, X.W. and Z.Q.; Validation, Y.S. (Ying Shi) and X.W.; Formal Analysis, Y.S. (Ying Shi) and Y.Z.; Investigation, X.W., Y.S. (Yanmei Song), J.D. and J.M.; Resources, Y.S. (Ying Shi) and S.W.; Data Curation, X.W. and J.D.; Writing—Original Draft Preparation, X.W.; Writing—Review and Editing, Y.S. (Ying Shi) and X.W.; Visualization, X.W. and J.M.; Supervision, Y.S. (Ying Shi); Project Administration, Y.S. (Ying Shi) and S.W.; Funding Acquisition, Y.S. (Ying Shi). All authors have read and agreed to the published version of the manuscript.

Funding: This work was supported by the National Natural Science Foundation of China (NSFC) (Grant No. 42177160 and 51974359).

Institutional Review Board Statement: Not applicable.

Informed Consent Statement: Not applicable.

Data Availability Statement: Not applicable.

Conflicts of Interest: The authors declare no conflict of interest.

References

1. Li, Z.; Hu, Q.; Chen, J.; Liu, X.; Sang, Y.; Wang, J.; Zheng, M.; Ni, H.; Li, Q.; Jiang, Z.; et al. Optimized Strategy for Simultaneous Recovering Bioactive Oligosaccharides and Reusable Perlite From Agar Industrial Waste Residues. *J. Clean Prod.* **2022**, *378*, 134631. [CrossRef]
2. Li, X.; Zhou, Y.; Zhu, Q.; Zhou, S.; Min, C.; Shi, Y. Slurry Preparation Effects On the Cemented Phosphogypsum Backfill through an Orthogonal Experiment. *Minerals* **2019**, *9*, 31. [CrossRef]
3. Fornés, I.V.; Vaičiukynienė, D.; Nizevičienė, D.; Doroševas, V. The Improvement of the Water-Resistance of the Phosphogypsum by Adding Waste Metallurgical Sludge. *J. Build. Eng.* **2021**, *43*, 102861. [CrossRef]
4. Moshynskyi, V.; Malanchuk, Z.; Tsymbaliuk, V.; Malanchuk, L.; Zhomyruk, R.; Vasylchuk, O. Research Into the Process of Storage and Recycling Technogenic Phosphogypsum Placers. *Min. Miner. Deposits.* **2020**, *14*, 95–102. [CrossRef]
5. Zhou, J.; Li, X.; Zhao, Y.; Shu, Z.; Wang, Y.; Zhang, Y.; Shen, X. Preparation of Paper-Free and Fiber-Free Plasterboard with High Strength Using Phosphogypsum. *Constr. Build. Mater.* **2020**, *243*, 118091. [CrossRef]
6. Değirmenci, N. Utilization of Phosphogypsum as Raw and Calcined Material in Manufacturing of Building Products. *Constr. Build. Mater.* **2008**, *22*, 1857–1862. [CrossRef]

7. Papastefanou, C.; Stoulos, S.; Ioannidou, A.; Manolopoulou, M. The Application of Phosphogypsum in Agriculture and the Radiological Impact. *J. Environ. Radioact.* **2006**, *89*, 188–198. [CrossRef]
8. Matveeva, V.A.; Smirnov, Y.D.; Suchkov, D.V. Industrial Processing of Phosphogypsum Into Organomineral Fertilizer. En-viron. *Geochem. Health.* **2022**, *44*, 1605–1618.1. [CrossRef]
9. Lütke, S.F.; Oliveira, M.L.S.; Waechter, S.R.; Silva, L.F.O.; Cadaval, T.R.S.; Duarte, F.A.; Dotto, G.L. Leaching of Rare Earth Elements From Phosphogypsum. *Chemosphere* **2022**, *301*, 134661. [CrossRef]
10. Cánovas, C.R.; Chapron, S.; Arrachart, G.; Pellet-Rostaing, S. Leaching of Rare Earth Elements (Rees) and Impurities From Phosphogypsum: A Preliminary Insight for Further Recovery of Critical Raw Materials. *J. Clean Prod.* **2019**, *219*, 225–235. [CrossRef]
11. Walawalkar, M.; Nichol, C.K.; Azimi, G. Process Investigation of the Acid Leaching of Rare Earth Elements from Phosphogypsum Using HCL, HNO_3, and H_2SO_4. *Hydrometallurgy* **2016**, *166*, 195–204. [CrossRef]
12. Gu, K.; Chen, B.; Pan, Y. Utilization of Untreated-Phosphogypsum as Filling and Binding Material in Preparing Grouting Materials. *Constr. Build. Mater.* **2020**, *265*, 120749. [CrossRef]
13. Shi, Y.; Gan, L.; Li, X.; He, S.; Sun, C.; Gao, L. Dynamics of Metals in Backfill of a Phosphate Mine of Guiyang, China Using a Three-Step Sequential Extraction Technique. *Chemosphere* **2018**, *192*, 354–361. [CrossRef] [PubMed]
14. Min, C.; Shi, Y.; Zhou, Y.; Liu, Z. The Release of Pollutants through the Bleeding of Cemented Phosphogypsum Backfill: Link to Protocols for Slurry Preparation. *Materials* **2022**, *15*, 7126. [CrossRef] [PubMed]
15. Wang, W.; Huang, Y.; Han, G.; Liu, B.; Su, S.; Wang, Y.; Xue, Y. Enhanced Removal of P(V), Mo(Vi) and W(Vi) Generated Oxyanions Using Fe-Mof as Adsorbent From Hydrometallurgical Waste Liquid: Exploring the Influence of Ionic Polymerization. *J. Hazard. Mater.* **2022**, *427*, 128168. [CrossRef]
16. Wagner, E.; Karthikeyan, K.G. Precipitating Phosphorus as Struvite From Anaerobically-Digested Dairy Manure. *J. Clean Prod.* **2022**, *339*, 130675. [CrossRef]
17. Yue, X.; Zhang, T.; Yang, D.; Qiu, F. Direct Separation of Phosphate Under Highly Acidic Conditions Using Mno2@Ceo2 Nanowires Membrane. *Chem. Eng. Process.-Process Intensif.* **2022**, *177*, 108986. [CrossRef]
18. Zhang, L.; Deng, F.; Liu, Z.; Ai, L. Removal of Ammonia Nitrogen and Phosphorus by Biochar Prepared From Sludge Residue After Rusty Scrap Iron and Reduced Iron Powder Enhanced Fermentation. *J. Environ. Manag.* **2021**, *282*, 111970. [CrossRef]
19. Yang, Q.; Wang, X.; Luo, W.; Sun, J.; Xu, Q.; Chen, F.; Zhao, J.; Wang, S.; Yao, F.; Wang, D.; et al. Effectiveness and Mechanisms of Phosphate Adsorption On Iron-Modified Biochars Derived from Waste Activated Sludge. *Bioresour. Technol.* **2018**, *247*, 537–544. [CrossRef]
20. Wu, H.; Wang, J.; Duan, E.; Hu, W.; Dong, Y.; Zhang, G. Phosphorus Removal by Adsorbent Based On Poly-Aluminum Chloride Sludge. *Water Sci. Eng.* **2020**, *13*, 193–201. [CrossRef]
21. Oginni, O.; Yakaboylu, G.A.; Singh, K.; Sabolsky, E.M.; Unal-Tosun, G.; Jaisi, D.; Khanal, S.; Shah, A. Phosphorus Adsorption Behaviors of Mgo Modified Biochars Derived from Waste Woody Biomass Resources. *J. Environ. Chem. Eng.* **2020**, *8*, 103723. [CrossRef]
22. Zhang, Y.; Qin, J.; Wang, X.; Chen, Z.; Zheng, X.; Chen, Y. Advanced Treatment of Phosphorus-Containing Tail Water by Fe–Mg–Zr Layered Double Hydroxide Beads: Performance and Mechanism. *J. Environ. Manag.* **2021**, *296*, 113203. [CrossRef] [PubMed]
23. Hu, F.; Wang, M.; Peng, X.; Qiu, F.; Zhang, T.; Dai, H.; Liu, Z.; Cao, Z. High-Efficient Adsorption of Phosphates from Water by Hierarchical Cual/Biomass Carbon Fiber Layered Double Hydroxide. *Colloids Surf. A Physicochem. Eng. Asp.* **2018**, *555*, 314–323. [CrossRef]
24. Lee, J.; Kim, J.; Yoo, S.; Jho, E.H.; Lee, C.; Park, S. Restoring Phosphorus from Water to Soil: Using Calcined Eggshells for P Adsorption and Subsequent Application of the Adsorbent as a P Fertilizer. *Chemosphere* **2022**, *287*, 132267. [CrossRef]
25. Almeida, P.V.; Santos, A.F.; Lopes, D.V.; Gando-Ferreira, L.M.; Quina, M.J. Novel Adsorbents Based On Eggshell Functionalized with Iron Oxyhydroxide for Phosphorus Removal From Liquid Effluents. *J. Water Process. Eng.* **2020**, *36*, 101248. [CrossRef]
26. Li, X.; Cui, J.; Pei, Y. Granulation of Drinking Water Treatment Residuals as Applicable Media for Phosphorus Removal. *J. Environ. Manag.* **2018**, *213*, 36–46. [CrossRef]
27. Letshwenyo, M.W.; Sima, T.V. Phosphorus Removal From Secondary Wastewater Effluent Using Copper Smelter Slag. *Heliyon* **2020**, *6*, e04134. [CrossRef]
28. Long, J.; Chen, B.; Zhu, Y.; Li, X.; Yue, X.; Zhang, N.; Xia, Y. Mycorrhiza and Iron Tailings Synergistically Enhance Maize Resistance to Arsenic On Medium Arsenic-Polluted Soils through Increasing Phosphorus and Iron Uptake. *Bull. Environ. Contam. Toxicol.* **2021**, *107*, 1155–1160. [CrossRef]
29. Thejas, H.K.; Hossiney, N. Alkali-Activated Bricks Made with Mining Waste Iron Ore Tailings. *Case Stud. Constr. Mater.* **2022**, *16*, e00973.
30. Zhang, N.; Tang, B.; Liu, X. Cementitious Activity of Iron Ore Tailing and its Utilization in Cementitious Materials, Bricks and Concrete. *Constr. Build. Mater.* **2021**, *288*, 123022. [CrossRef]
31. Cao, L.; Zhou, J.; Zhou, T.; Dong, Z.; Tian, Z. Utilization of Iron Tailings as Aggregates in Paving Asphalt Mixture: A Sustainable and Eco-Friendly Solution for Mining Waste. *J. Clean Prod.* **2022**, *375*, 134126. [CrossRef]

32. Zeng, L.; Li, X.; Liu, J. Adsorptive Removal of Phosphate From Aqueous Solutions Using Iron Oxide Tailings. *Water Res.* **2004**, *38*, 1318–1326. [CrossRef] [PubMed]
33. Sima, T.V.; Letshwenyo, M.W.; Lebogang, L. Efficiency of Waste Clinker Ash and Iron Oxide Tailings for Phosphorus Removal From Tertiary Wastewater: Batch Studies. *Environ. Technol. Innov.* **2018**, *11*, 49–63. [CrossRef]
34. Yao, Z. *Technoligical Study and Reliability Analysis of Yellow Phosphorus Slag and Phosphgysum Backfill in Kaiyang Mine*; Central South University: Changsha, China, 2009. (In Chinese)
35. Borah, L.; Goswami, M.; Phukan, P. Adsorption of Methylene Blue and Eosin Yellow Using Porous Carbon Prepared from Tea Waste: Adsorption Equilibrium, Kinetics and Thermodynamics Study. *J. Environ. Chem. Eng.* **2015**, *3*, 1018–1028. [CrossRef]
36. Zhou, R.; Wang, Y.; Zhang, M.; Yu, P.X.; Li, J. Adsorptive Removal of Phosphate From Aqueous Solutions by Thermally Modified Copper Tailings. *Environ. Monit. Assess.* **2019**, *191*, 198. [CrossRef] [PubMed]
37. *JGJ/T70-2009*; Standard for Test Method of Performance on Building Mortar. China Architecture & Building Press: Beijing, China, 2009.
38. *HJ 557-2010*; Solid Waste—Extraction Procedure for Leaching Toxicity—Horizontal Vibration Method. China Environmental Science Press: Beijing, China, 2010.
39. Zhou, Y.; Li, X.; Shi, Y.; Zhu, Q.; Du, J. Reuse of phosphogypsum pretreated with water washing as aggregate for cemented backfill. *Sci. Rep.* **2022**, *12*, 16091. [CrossRef] [PubMed]
40. Holanda, F.D.C.; Schmidt, H.; Quarcioni, V.A. Influence of Phosphorus From Phosphogypsum On the Initial Hydration of Portland Cement in the Presence of Superplasticizers. *Cem. Concr. Compos.* **2017**, *83*, 384–393. [CrossRef]
41. *GB 8978-1996*; Integrated Wastewater Discharge Standard. China Environmental Science Press: Beijing, China, 1996.
42. Shi, Y.; Cheng, L.; Tao, M.; Tong, S.; Yao, X.; Liu, Y. Using Modified Quartz Sand for Phosphate Pollution Control in Cemented Phosphogypsum (Pg) Backfill. *J. Clean Prod.* **2021**, *283*, 124652. [CrossRef]
43. Wang, S.; Yuan, R.; Yu, X.; Mao, C. Adsorptive Removal of Phosphate From Aqueous Solutions Using Lead-Zinc Tailings. *Water Sci. Technol.* **2013**, *67*, 983–988. [CrossRef]
44. Ha, T.; Mahasti, N.N.N.; Lu, M.; Huang, Y. Application of Low-Solubility Dolomite as Seed Material for Phosphorus Recovery From Synthetic Wastewater Using Fluidized-Bed Crystallization (Fbc) Technology. *Sep. Purif. Technol.* **2022**, *303*, 122192. [CrossRef]
45. Wang, Q.; Liao, Z.; Yao, D.; Yang, Z.; Wu, Y.; Tang, C. Phosphorus Immobilization in Water and Sediment Using Iron-Based Materials: A Review. *Sci. Total Environ.* **2021**, *767*, 144246. [CrossRef] [PubMed]
46. Guaya, D.; Cobos, H.; Camacho, J.; López, C.M.; Valderrama, C.; Cortina, J.L. Lta and Fau-X Iron-Enriched Zeolites: Use for Phosphate Removal From Aqueous Medium. *Materials* **2022**, *15*, 5418. [CrossRef]
47. Tao, X.; Huang, T.; Lv, B. Synthesis of Fe/Mg-Biochar Nanocomposites for Phosphate Removal. *Materials* **2020**, *13*, 816. [CrossRef] [PubMed]
48. Du, M.; Zhang, Y.; Wang, Z.; Lv, M.; Xu, Q.; Chen, Z.; Wen, Q.; Li, A. La-Doped Activated Carbon as High-Efficiency Phosphorus Adsorbent: Dft Exploration of the Adsorption Mechanism. *Sep. Purif. Technol.* **2022**, *298*, 121585. [CrossRef]
49. Feng, Y.; Luo, Y.; He, Q.; Zhao, D.; Zhang, K.; Shen, S.; Wang, F. Performance and Mechanism of a Biochar-Based Ca-La Composite for the Adsorption of Phosphate From Water. *J. Environ. Chem. Eng.* **2021**, *9*, 105267. [CrossRef]
50. Mishra, S.R.; Chandra, R.; Kaila, A.J.; Darshi, B.S. Kinetics and Isotherm Studies for the Adsorption of Metal Ions Onto Two Soil Types. *Environ. Technol. Innov.* **2017**, *7*, 87–101. [CrossRef]
51. Kang, S.; Choo, K.; Lim, K. Use of Iron Oxide Particles as Adsorbents to Enhance Phosphorus Removal from Secondary Wastewater Effluent. *Sep. Sci. Technol.* **2003**, *38*, 3853–3874. [CrossRef]
52. Konadu-Amoah, B.; Hu, R.; Ndé-Tchoupé, A.I.; Gwenzi, W.; Noubactep, C. Metallic Iron (Fe0)-Based Materials for Aqueous Phosphate Removal: A Critical Review. *J. Environ. Manag.* **2022**, *315*, 115157. [CrossRef]
53. Huang, W.; Li, D.; Liu, Z.; Tao, Q.; Zhu, Y.; Yang, J.; Zhang, Y. Kinetics, Isotherm, Thermodynamic, and Adsorption Mechanism Studies of La(Oh) 3 -Modified Exfoliated Vermiculites as Highly Efficient Phosphate Adsorbents. *Chem. Eng. J.* **2014**, *236*, 191–201. [CrossRef]
54. Wu, L.; Zhang, S.; Wang, J.; Ding, X. Phosphorus Retention Using Iron (Ii/Iii) Modified Biochar in Saline-Alkaline Soils: Adsorption, Column and Field Tests. *Environ. Pollut.* **2020**, *261*, 114223. [CrossRef]
55. Mekonnen, D.T.; Alemayehu, E.; Lennartz, B. Adsorptive Removal of Phosphate from Aqueous Solutions Using Low-Cost Volcanic Rocks: Kinetics and Equilibrium Approaches. *Materials* **2021**, *14*, 1312. [CrossRef] [PubMed]
56. Rangabhashiyam, S.; Anu, N.; Giri Nandagopal, M.S.; Selvaraju, N. Relevance of Isotherm Models in Biosorption of Pollutants by Agricultural Byproducts. *J. Environ. Chem. Eng.* **2014**, *2*, 398–414. [CrossRef]
57. Jiang, C.; Jia, L.; Zhang, B.; He, Y.; Kirumba, G. Comparison of Quartz Sand, Anthracite, Shale and Biological Ceramsite for Adsorptive Removal of Phosphorus From Aqueous Solution. *J. Environ. Sci.* **2014**, *26*, 466–477. [CrossRef] [PubMed]
58. Wahab, M.A.; Jellali, S.; Jedidi, N. Ammonium Biosorption Onto Sawdust: Ftir Analysis, Kinetics and Adsorption Isotherms Modeling. *Bioresour. Technol.* **2010**, *101*, 5070–5075. [CrossRef] [PubMed]

59. Barca, C.; Gérente, C.; Meyer, D.; Chazarenc, F.; Andrès, Y. Phosphate Removal from Synthetic and Real Wastewater Using Steel Slags Produced in Europe. *Water Res.* **2012**, *46*, 2376–2384. [CrossRef]
60. Jin, H.; Lin, L.; Meng, X.; Wang, L.; Huang, Z.; Liu, M.; Dong, L.; Hu, Y.; Crittenden, J.C. A Novel Lanthanum-Modified Copper Tailings Adsorbent for Phosphate Removal from Water. *Chemosphere* **2021**, *281*, 130779. [CrossRef]
61. Li, X.; Zhou, S.; Zhou, Y.; Min, C.; Cao, Z.; Du, J.; Luo, L.; Shi, Y. Durability Evaluation of Phosphogypsum-Based Cemented Backfill through Drying-Wetting Cycles. *Minerals* **2019**, *9*, 321. [CrossRef]
62. Wang, S.; Li, X.; Yao, J.; Gong, F.; Li, X.; Du, K.; Tao, M.; Huang, L.; Du, S. Experimental Investigation of Rock Breakage by a Conical Pick and its Application to Non-Explosive Mechanized Mining in Deep Hard Rock. Int. *J. Rock Mech. Min. Sci.* **2019**, *122*, 104063. [CrossRef]
63. Zhou, S.; Li, X.; Zhou, Y.; Min, C.; Shi, Y. Effect of Phosphorus On the Properties of Phosphogypsum-Based Cemented Backfill. *J. Hazard. Mater.* **2020**, *399*, 122993. [CrossRef]

materials

MDPI

Article

Research of Dynamic Tensile Properties of Five Rocks under Three Loading Modes Based on SHPB Device

Diyuan Li, Jinyin Ma, Quanqi Zhu * and Bang Li

School of Resources and Safety Engineering, Central South University, Changsha 410083, China
* Correspondence: quanqi_zhu@csu.edu.cn

Abstract: The validity of calculating the dynamic tensile strength of rock materials based on dynamic Brazilian tests is problematic. In order to gain a deeper understanding of the effects of three typical loading methods on the damage mechanism of rock specimens in the dynamic Brazilian tests, five different rocks were selected for the study. In the constant incident energy dynamic Brazilian test, the loading modes had a significant effect on the loading rate and dynamic tensile strength of the specimen, with the highest loading rate and tensile strength of the specimens under mode-III loading, followed by mode-I loading and mode-II loading. A high-speed camera and the digital image correlation (DIC) technique were used to successfully capture the rupture process of the Brazilian disc during impact loading. The evolution of the displacement and strain fields of the specimen was obtained by DIC technique, and four typical failure patterns and two rupture characteristics in the dynamic Brazilian test were summarized. The loading mode determined the crack initiation position of the specimen in the dynamic Brazilian test. The results showed that the mode-III loading is the most consistent with the Brazilian test theory, while the mode-II loading violates the test principle.

Keywords: dynamic Brazilian splitting; digital image correlation; rock materials; SHPB

Citation: Li, D.; Ma, J.; Zhu, Q.; Li, B. Research of Dynamic Tensile Properties of Five Rocks under Three Loading Modes Based on SHPB Device. *Materials* **2022**, *15*, 8473. https://doi.org/10.3390/ma15238473

Academic Editor: Dimitrios Papoulis

Received: 25 October 2022
Accepted: 25 November 2022
Published: 28 November 2022

1. Introduction

Rock masses in deep underground projects are usually subjected to dynamic loads caused by excavation, blasting, and drilling [1–10]. Since local tensile failure often occurs under dynamic loads, many scholars have been attracted to conduct research on the dynamic tensile properties and fracture mechanism of rock materials [11–14]. The test methods for rock tensile strength include direct stretching and indirect stretching, i.e., the Brazilian disc [15,16]. The dynamic Brazilian test based on the split Hopkinson pressure bar (SHPB) is a popular and commonly used method to measure the indirect tensile strength of rock materials [17,18]. Rose et al. [19] were the first to use SHPB system to conduct dynamic Brazilian tests on concrete specimens and obtain the dynamic tensile strength of concrete. In addition, the dynamic tensile strength of ceramics was also measured by Johnstone et al. [20]. Subsequently, scholars have carried out a large number of dynamic Brazilian tests on rock materials. Based on a series of static and dynamic Brazilian tests, Gomez et al. [21] studied the influence of damage degree on tensile strength of concrete and granite. Dai et al. [22] explored several fundamental issues in dynamic Brazilian tests through experimental and numerical studies, such as the dynamic stress balance, the validity of the static standard equation, and the necessity of the arc loading method in dynamic Brazilian tests. Zhang et al. [23] investigated the dynamic mechanical properties and strain field evolution of marble specimens using combination of digital image correlation (DIC) technique and a high-speed camera in the dynamic Brazilian tests. In addition, the reasons for the discrepancy between the results of the dynamic Brazilian test and the direct tensile test were explained based on experimental observation and micromechanical models [24]. Wu et al. [14] investigated the accuracy of the unidirectional strain calculation method in the dynamic Brazilian test using two types of disc specimens based on the theory of stress wave propagation at the interface and verified the reasonableness of the results through tests. Moreover, different factors have been considered in the dynamic Brazilian tests to carry out relevant studies, such as temperature [18], anisotropy [25], and water saturation [26].

In fact, the validity of the dynamic Brazilian test is also problematic. Local crushing failure may occur at the end of the disk under impact loads, which is contrary to the hypothesis of the Brazilian test. Related scholars show different views on this. Wang et al. [27] proposed the flattened Brazilian test, that was, cutting a pair of loading surfaces parallel to each other at both ends of the disc, which could prevent premature failure caused by high stress concentration at the contact end of the specimen and the elastic bar. When the flattened loading angle is greater than 20°, the crack initiation position is in the center of the disk. Dai et al. [22] pointed out through an experimental study that the use of arc loading was not necessary in dynamic Brazilian tests of rocks, since the presence or absence of an arc loading device did not have a large effect on the dynamic tensile strength. Gomez et al. [21] added a pair of liners on the contact surface between the specimen and the elastic bars to prevent the localized failure at the end of the disc caused by point load in the impact test. In addition, some improvements were made to the loading device with the aim of improving the stress condition between the specimen and loading platens, including the addition of curved loading jaws.

Due to the limitations of indoor tests, some test parameters and test conditions were sometimes not satisfied. With the development of computer technology and related theories, many scholars have conducted rich research on dynamic Brazilian tests based on different numerical simulation methods. Ruiz et al. [28] used the 3D (three dimensional) finite element method (FEM) to simulate dynamic Brazilian tests on concrete materials and found that the dynamic strength of concrete was rate-sensitive, with an increasing strain rate. Zhu et al. [29] used the FEM analysis method (RFPA) to study the static and dynamic damage mechanisms of rock disks in Brazilian tests. Mahabadi et al. [30] developed a coupled finite element and discrete element method (FEM/DEM) to research the mechanical behavior of Barre granite based on the Brazilian tests, and the results matched well with the experimental results. In addition, Zhu et al. [29] and Ruiz et al. [28] verified the reliability of quasi-static standard computational equations in dynamic Brazilian tests based on numerical tests.

Throughout the above literature, the current studies on dynamic Brazilian tests are mainly on the dynamic strength and damage modes of specimens under conventional loading methods. Few literatures have systematically investigated the effects of different loading methods on the dynamic mechanical properties and damage behavior of Brazilian specimens. Thus, three different loading methods and five types of rocks were considered, and the deformation and damage patterns of each specimen were studied by using a high-speed camera and the DIC technique in order to obtain more comprehensive results. The key findings of this work are essential for the accurate use of the Brazilian test to obtain tensile strength and other information on rock specimens.

2. Experimental Procedure

2.1. Specimen Preparation

Five typical rock types are selected in this paper, including white sandstone (WS), red sandstone (RS), marble (M), granite (G), and basalt (B), which are consistent with our previous study [31,32]. Three Brazilian disc specimens, with the dimensions of 50 mm × 25 mm (diameter × thickness), were prepared for each group of tests due to three different loading modes. Therefore, nine specimens were prepared for each lithology for a total of 45 specimens, as shown in Figure 1e–g. The dimensions of each specimen were measured, as shown in Table 1, and all specimens met the test requirements. According to the method recommended by the International Society for Rock Mechanics (ISRM) [33], the surfaces of the disc were ground by the diamond grinding wheel so that the parallel error of the two ends was less than 0.25° and the flatness error was less than 0.25 mm. These specimens for each lithology were taken from intact rocks blocks, respectively, without obvious fractures or bedding planes. Before the impact test is carried out, an artificial random speckle should be made on the specimen surface first due to the application of DIC technology and the lack of natural contrast in rocks.

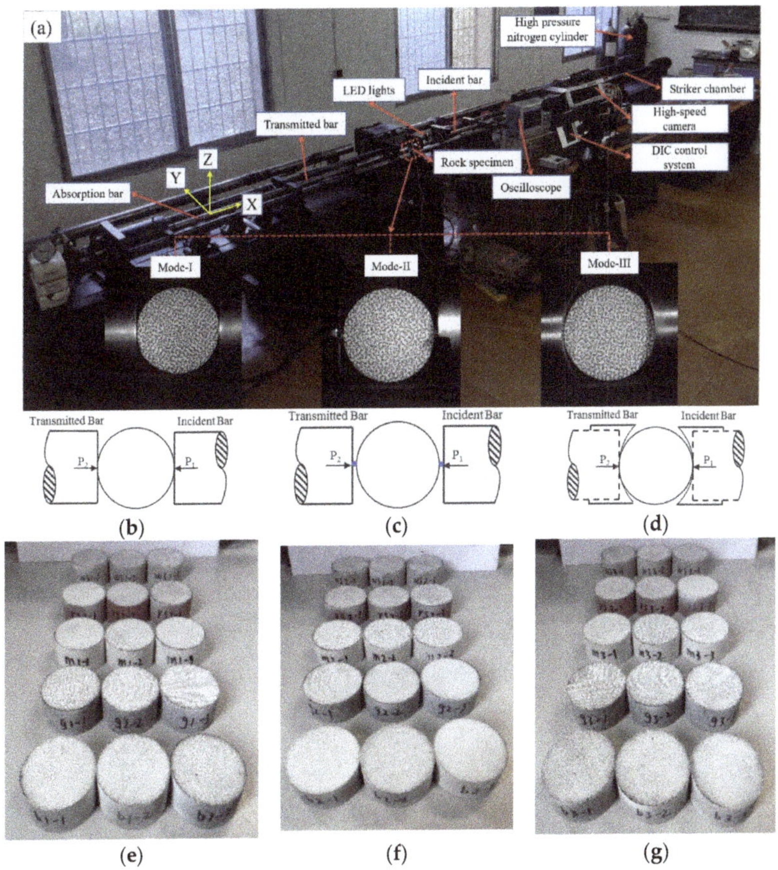

Figure 1. (**a**) Diagram of experimental device for the dynamic Brazilian tests and the high-speed photography system; the diagram for platform loading (**b**), steel bar loading (**c**), and arc loading (**d**); photographic views of the specimen under mode-I loading (**e**), mode-II loading (**f**), and mode-III loading (**g**).

According to the principle of DIC technique, a good speckle pattern usually meets the following conditions: (1) high-contrast; (2) consistent speckle sizes; (3) random distribution; (4) 50% coverage [34–37]. The speckle pattern was made on the specimen surface by means of plate printing and spray painting [38,39]. A printing plate is specially designed to print black spots with diameter of 0.66 mm on the surface of the specimen. First, a brush was used to clean the stains on the specimen surface, and then the white matte paint was used to spray on it to form a thin and uniform primer. After the white paint was solidified, the surface sprayed with white paint was vertically pressed 3–5 times on the printing plate to form random speckles. For each lithology, three standard specimens of 100 mm × 50 mm (height × diameter) were prepared for static uniaxial compression tests. The average uniaxial compressive strength, Young's modulus, and Poisson's ratio of five rock materials were obtained, as shown in Table 2. In Table 2, "σ_c", "E", and "ν" represent the uniaxial compressive strength, Young's modulus, and Poisson's ratio, respectively. The specimen number was recorded as A-B-C, in which A represents the rock type, B represents the loading mode, and C represents the serial number of the specimens.

Table 1. Specimen dimensions and the dynamic Brazilian test results.

Specimens No.	*D*/mm	*t*/mm	P_{fd}/kN	σ_{td}/MPa	Mean/MPa
WS-I-1	49.54	25.03	12.35	6.34	
WS-I-2	49.73	25.05	12.82	6.55	6.35
WS-I-3	49.32	25.04	11.95	6.16	
WS-II-1	49.12	25.03	7.36	3.81	
WS-II-2	49.16	25.08	8.25	4.26	3.91
WS-II-3	49.56	25.04	7.13	3.66	
WS-III-1	49.09	25.10	20.71	10.70	
WS-III-2	49.23	25.09	16.12	8.31	9.52
WS-III-3	49.62	25.06	18.67	9.56	
RS-I-1	49.10	25.02	16.96	8.79	
RS-I-2	49.34	25.11	17.83	9.16	9.88
RS-I-3	49.06	25.04	22.58	11.70	
RS-II-1	49.09	25.03	5.77	2.99	
RS-II-2	49.04	25.02	7.38	3.83	4.24
RS-II-3	49.02	25.03	11.39	5.91	
RS-III-1	49.45	24.98	22.12	11.40	
RS-III-2	49.13	25.01	24.70	12.80	11.53
RS-III-3	49.16	25.08	20.12	10.39	
M-I-1	49.14	25.06	33.46	17.30	
M-I-2	49.23	25.03	30.77	15.90	16.53
M-I-3	49.26	25.06	31.80	16.40	
M-II-1	49.25	25.04	11.35	5.86	
M-II-2	49.10	25.02	12.41	6.43	5.29
M-II-3	49.15	25.11	6.94	3.58	
M-III-1	49.28	25.13	44.16	22.70	
M-III-2	49.32	24.97	28.05	14.50	18.64
M-III-3	49.46	25.02	36.39	18.72	
G-I-1	49.41	25.06	43.18	22.20	
G-I-2	49.26	25.01	35.61	18.40	20.13
G-I-3	49.29	25.07	38.43	19.80	
G-II-1	49.39	25.04	23.19	11.94	
G-II-2	49.43	25.13	17.09	8.76	12.02
G-II-3	49.47	25.09	29.95	15.36	
G-III-1	49.19	25.14	40.40	20.80	
G-III-2	49.22	25.05	45.76	23.63	22.48
G-III-3	49.37	25.13	44.84	23.01	
B-I-1	49.45	25.11	40.57	20.80	
B-I-2	49.43	24.97	45.75	23.60	22.70
B-I-3	49.39	25.11	46.17	23.70	
B-II-1	49.51	25.01	18.61	9.57	
B-II-2	49.44	25.06	17.94	9.22	9.22
B-II-3	49.36	25.11	17.27	8.87	
B-III-1	49.51	25.07	55.21	28.32	
B-III-2	49.32	25.03	44.99	23.20	25.40
B-III-3	49.49	25.05	48.06	24.68	

Table 2. Test results of static uniaxial compression tests.

Rock Types	σ_c (MPa)	E (GPa)	ν
White sandstone	37.7	9.2	0.15
Red sandstone	62.6	14.8	0.18
Marble	79.2	43.7	0.21
Granite	171.7	72.2	0.23
Basalt	158.4	61.9	0.25

2.2. *Experimental Setup and Testing Method*

2.2.1. SHPB System

The dynamic Brazilian test using the split Hopkinson pressure bar (SHPB) system is the ISRM recommended method for determining the dynamic tensile strength of rock materials [40]. The schematic diagram of test principle based on the improved SHPB device and DIC technique is shown in Figure 2. The main components of the SHPB system are three elastic bars with a diameter of 50 mm: the incident bar, the transmitted bar, and the absorption bar. The three elastic bars and the cone-shaped striker were made of 40 Cr alloy with a density of 7817 kg/m^3, an elastic modulus of 233 GPa, and a longitudinal wave velocity of 5458 m/s. In addition, the experimental device also includes a high-pressure nitrogen cylinder, a striker chamber, a dynamic strain meter, an oscilloscope, a high-speed camera, a buffer device, two LED (light-emitting diode) lights, and two control systems. Once the experimental device and specimen were ready, the cone-shaped striker was then ejected from a high-pressure chamber and impacted the free end of the incident bar. A compression wave began to propagate in the incident bar (incident wave, ε_I). When propagating to the interface between the specimen and the incident bar, part of the wave is reflected back to the incident bar (reflected wave, ε_R), and the rest of the wave passes through the specimen to the transmitted bar (transmitted wave, ε_T). The signals of these three stress waves are recorded by the strain gauge bonded on bars, in which the strain gauge on the incident bar records the incident and reflected waves, while the strain gauge on the transmitted bar only records transmitted waves.

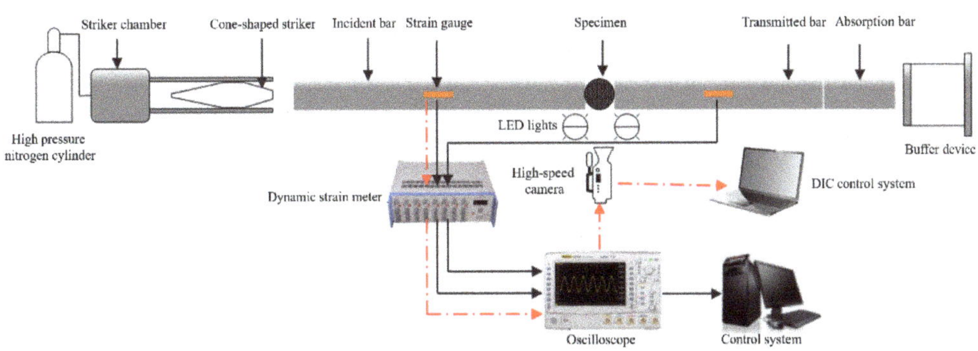

Figure 2. Schematic diagram of experimental device for dynamic Brazilian test and the high-speed photography system.

Based on one-dimensional stress wave theory, the dynamic pressure on the incident end (P_1) and transmitted end (P_2) of the specimen are [41]:

$$P_1 = A_b E_b (\varepsilon_I + \varepsilon_R) \quad (1)$$

$$P_2 = A_b E_b \varepsilon_T \quad (2)$$

The velocities at the contact end face between the incident bar (v_1), transmitted bar (v_2), and the specimen are:

$$v_1 = C_b(\varepsilon_I - \varepsilon_R) \tag{3}$$

$$v_2 = C_b\varepsilon_T \tag{4}$$

Therefore, the displacement of the specimen at the ends of the incident bar (u_1) and transmitted bar (u_2) can be obtained as follows:

$$u_1 = C_b\int_0^t (\varepsilon_I - \varepsilon_R)dt \tag{5}$$

$$u_2 = C_b\int_0^t \varepsilon_T dt \tag{6}$$

In Equations (1)–(6), E_b, A_b, and C_b are the elastic modulus, the cross-sectional area, and the one-dimensional longitudinal stress wave velocity of the elastic bar, respectively; ε_I, ε_R, and ε_T are the incident strain signals, reflected strain signals, and transmitted strain signals, respectively.

Similar to the principle of the static Brazilian test method, the dynamic Brazilian test is based on the same fact that the tensile strength of rock is much lower than the compressive strength. Therefore, when the Brazilian disk is subjected to radial load, the specimen will be failure somewhere in the center due to the tensile stress distributed along the diameter of the loading direction. Assuming that stress balance is satisfied at both ends of the specimen in the dynamic Brazilian test, the dynamic tensile strength (σ_{td}) can be determined by the following formula [24]:

$$\sigma_{td} = \frac{2P_{fd}}{\pi Dt} \tag{7}$$

where P_{fd} is the dynamic peak load of the specimen; D and t are the diameter and thickness of the specimen. According to the Brazilian test theory, the application of this formula is also only applicable to the test of crack initiation in the central region of the disc. The tensile strength values calculated by Equation (7) in this paper are mainly used for comparison under different loading modes.

2.2.2. Loading Mode

As shown in Figure 1, three loading modes are adopted, including platform loading (mode-I), steel bar loading (mode-II), and arc loading (mode-III). Under platform loading, the incident bar and transmitted bar in SHPB system were used to clamp the specimens directly. Two small steel bars made of the same material as the incident bar are fixed at the end of the incident bar and transmitted bar. The dynamic Brazilian test under mode-II were carried out by clamping the specimen with two steel bars. In addition, in order to realize the dynamic Brazilian test of mode-III loading, a pair of arc-shaped loading molds were designed and machined, whose materials were consistent with the incident bar and transmitted bar.

2.2.3. High-Speed Camera and Digital Image Correlation System

A high-speed camera and digital image correlation techniques were used to monitor the deformation and failure characteristics of specimen surface in dynamic Brazilian tests. As shown in Figure 1, the lens of the high-speed camera was perpendicular to the specimen surface, and the resolution of the image was set to 256 × 256 pixels. The image was captured at a frame rate of 79,166 fps (i.e., an inter-frame time of 12.6 μs). In order to achieve the synchronous control of high-speed camera and SHPB system, the external trigger end of the high-speed camera was connected with oscilloscope, and the TTL (transistor transistor logic) signal generated by oscilloscope when recording the stress wave signal of the incident bar would synchronously trigger the high-speed camera. The digital image correlation system is a non-contact technique for identifying the distribution and evolution of strain or displacement. Compared with the traditional strain or displacement measurement method (e.g., strain gauge, extensometer), DIC has many advantages such as non-contact, high reliability, and full-field measurement [42–45]. In this research, the images collected in dynamic Brazilian tests were imported into two-dimensional visual image correlation software (VIC-2D) to obtain the evolution law of the strain and displacement fields on the disk surface.

3. Experimental Results and Discussion

3.1. Dynamic Stress Equilibrium

An effective dynamic Brazilian splitting test must ensure the dynamic stress balance before failure, which can be verified by comparing the dynamic stress history at both ends of the specimen in the impact test. In each loading mode, the dynamic stress history of a typical specimen is shown in Figure 3. The dynamic stress on the incident side is the sum of incident stress and reflected stress, marked as Int+Re in the figure, and the dynamic stress on the transmitted side is transmitted stress. It can be seen from Figure 3a–c that before the peak stress, the dynamic stress had a good consistency at the incident and transmitted sides, which indicated that the specimen has reached the dynamic stress balance under the three loading modes, and tests were effective. The stress at both ends of the specimen became unbalanced due to the complete splitting failure of the specimen after the peak load.

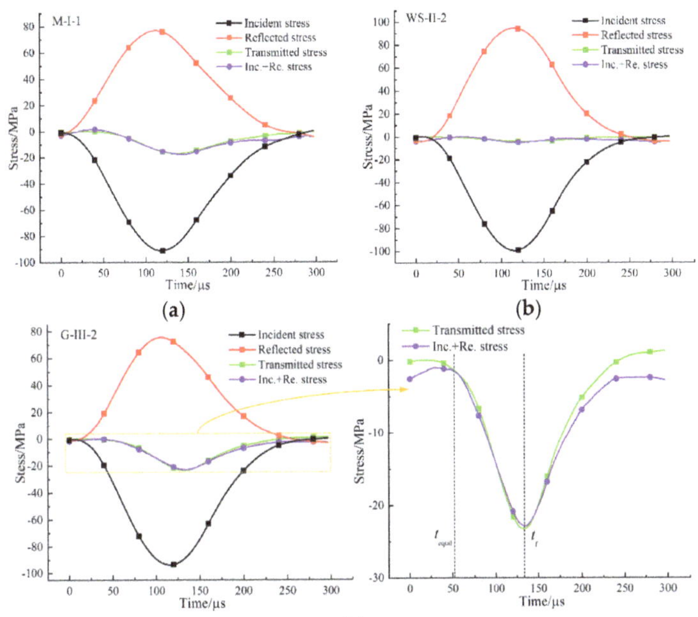

Figure 3. Dynamic stress history of typical specimens under different loading modes: (**a**) M-I-1; (**b**) WS-II-2; (**c**) G-III-2.

3.2. Matching of the High-Speed Camera to Stress Loading Time

Although the high-speed camera and oscilloscope have achieved synchronous control, further processing is needed to make the stress loading time more accurately match with the images taken by the camera. The time when the stress wave propagates from the strain gauge on the incident bar to the contact interface between the specimen and the incident bar is denoted as t_b. There is a time interval when the incident stress wave arrives at the strain gauge and the oscilloscope generates TTL electrical frequency signal and triggers the high-speed camera, and this time interval is denoted as t_{TTL}. Therefore, the number of images collected by the high-speed camera when the specimen begins to be loaded by the incident stress wave is calculated as follows:

$$n = \frac{t_b - t_{TTL}}{t_{\text{frame}}}, t_b = \frac{d}{C_b} \tag{8}$$

where t_{frame} is the time interval for the camera to capture images, namely 12.6 µs; d is the distance between the strain gauge and the incident bar end, which is 1.15 m.

In addition, the TTL electrical frequency signal will be generated synchronously when the voltage signal monitored by the oscilloscope is smaller than −34 mv. Figure 4 shows the original voltage signal variation curve of the specimen G-III-2 during the whole loading process. It can be seen from the figure that t_{TTL} is 40 µs. According to Equation (8), the value of n is 13.7, that is, the specimen has been subjected to about 3.8 µs of the impact load when the 14th image is captured by the high-speed camera. Therefore, the above analysis achieved the match between the stress loading of the specimen and the high-speed camera at the microsecond level, and the time of each image could be accurately determined.

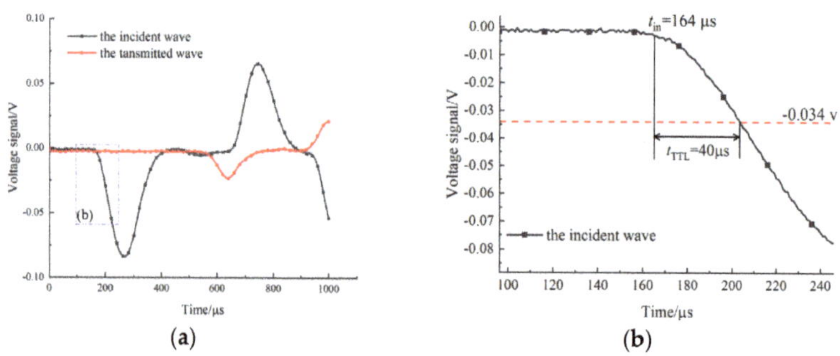

Figure 4. Original voltage signal of the specimen G-III-2 recorded by the oscilloscope (**a**); local magnification of the incident signal (**b**).

3.3. Loading Rate of the Dynamic Brazilian Tests

In conventional dynamic compression tests, the mechanical properties and deformation parameters of materials are generally considered to be dependent on the strain rate. However, in the dynamic Brazilian test, a uniform value of strain rate cannot be obtained due to the uneven distribution of stress and strain in the disc specimen. Since the fixed incident energy is invariant in the dynamic Brazilian test, the loading rate was used to describe the loading condition of the disk in the Brazilian test [24]. The loading rate is determined by the slope of the straight-line segment before the peak value of the tensile stress–time curve in the center of the disc. Taking the specimen G-III-2 as an example, Figure 5a explains the method of determining the loading rate and tensile strength of the specimen. According to the figure, the loading rate and dynamic tensile strength are 389 GPa/s and 23.6 MPa, respectively. The loading rate of other specimens can be obtained according to the same treatment method, and all test data are sorted out and plotted in Figure 5b. It can be found that the loading modes have a significant influence on the loading rate of the specimen. For the same rock type, the loading rate is the highest under mode-I loading, which is the lowest under mode-II loading. In addition, due to the different natural properties of rock specimens, the loading rate of the specimen also shows significant differences.

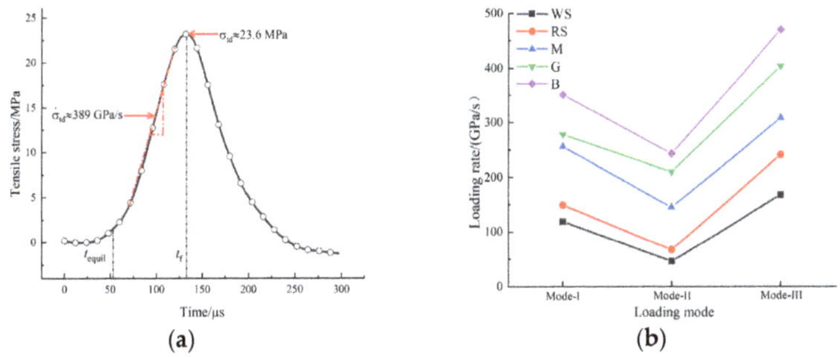

Figure 5. (**a**) History of tensile stress during loading of the specimen G-III-2; (**b**) the loading rate of each rock type under three loading modes.

3.4. Dynamic Tensile Strength and Deformation Characteristics of the Specimen

According to the test data, the dynamic tensile strength of all specimens was calculated by Equation (7). The average dynamic tensile strength of specimens of different lithologies under three loading modes is plotted in Figure 6 and Table 2. It can be seen from the figure that the loading mode has a significant impact on the dynamic tensile strength of the rocks, which is similar to the static test [31]. For all rock types, the dynamic tensile strength obtained by the mode-III loading is the highest, while the mode-II loading measures the lowest value. The dynamic tensile strength obtained by mode-I loading is similar to that of mode-III loading. However, there are also differences between the two tests (dynamic and static Brazilian tests) under different loading modes. First, the tensile strength of the specimen in the dynamic Brazilian test is higher than that in the static Brazilian tests. Second, the dynamic tensile strength obtained by mode-II loading is much lower than that of the other two loading modes. The Brazilian test under impact load seems to amplify the characteristics of these three loading methods, and the steel bar loading may greatly underestimate the dynamic tensile strength of the rock.

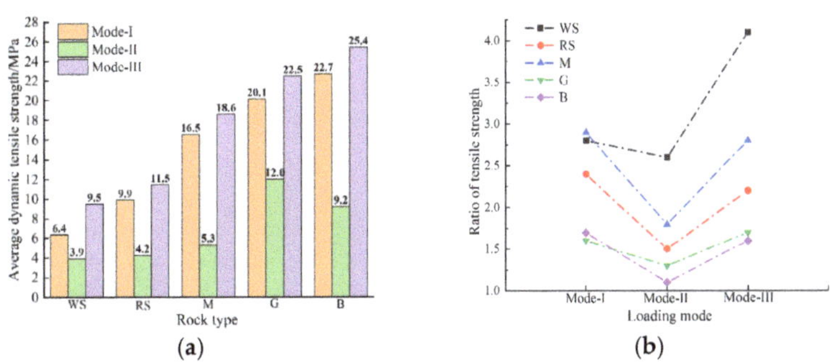

Figure 6. (**a**) Average dynamic tensile strength of different rock types under three loading modes; (**b**) the ratio of tensile strength of different rock types under three loading modes.

The dynamic tensile strength of the specimen was further compared with the static tensile strength, and the ratio of dynamic to static tensile strength was calculated, as shown in Figure 6b. Regardless of the loading modes, dynamic impacts can improve the tensile strength of rocks, and the increased strength of soft rocks (white and red sandstone) is more obvious. In addition, the increase ratio of tensile strength of all specimens under mode-III and mode-I loading are large and similar (except white sandstone), while the increase ratio under mode-II loading is the smallest.

The displacement of the specimen at ends of the incident bar and the transmitted bar can be calculated by Equations (5) and (6). Since there was almost no displacement at the end of the transmitted bar when the specimen was subjected to impact load, only the displacement of the incident bar end was considered. Figure 7a shows the displacement–time curves of white sandstone specimens under three loading modes. The displacements of the specimens under mode-I and mode-III loading are large, while that of the white specimen under mode-II loading is the smallest. This is obviously different from the results of the static test [31], which is mainly because the stress concentration caused by mode-II loading is more intense under dynamic impact, and the steel bar almost has no time to penetrate the end of the specimen. In addition, a linear region was observed at the middle part of three displacement–time curves. In the linear region, the displacement values of the white sandstone specimens under the three loading modes were arranged in descending orders as mode-I, mode-III, and mode-II, respectively. The maximum displacement value of the specimen under mode-I loading was due to the shear breakage phenomenon at the end of incident bar, resulting in an increase in displacement compared with the specimen under mode-III loading. The minimum displacement value of the specimen under mode-II loading was because its loading rate was the lowest, that was, the increase of load per unit time was the least, resulting in the minimum displacement at the same time. A linear segment was observed at the loading–time curve and the displacement–time curve of the specimen, so a linear region should also appear at the load–displacement curve, as shown in Figure 7b. k_1, k_2, and k_3 are the slopes of the straight section of the load–displacement curve of the specimen under the three corresponding loading modes, respectively. The slope can be regarded as the ratio of loading rate to displacement rate in the elastic deformation stage of the specimen under three loading modes, which shows similar characteristics to the load–displacement response of the specimen in the static Brazilian tests [31].

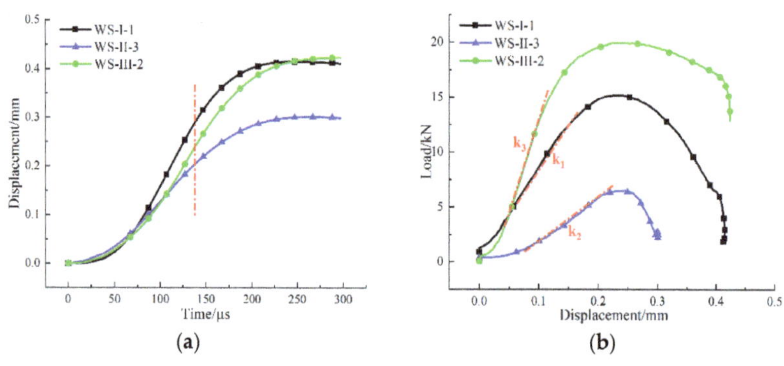

Figure 7. (**a**) Displacement–time curves and (**b**) loading–displacement curves of typical white sandstone specimens under three loading modes.

3.5. Dynamic Failure Analysis of Specimens

An effective dynamic Brazilian test was done to ensure that the specimen cracks at a certain point in the central region of the disc. Therefore, it is necessary to analyze the failure modes of the specimen under three loading modes. Since the high-speed camera was used in the dynamic Brazilian test, the deformation localization and fracture evolution process of the specimen could be captured more accurately than the industrial camera used in the static test [31]. Table 3 shows the full-filed tensile strain contour before the peak load of representative specimens under each loading mode.

Table 3. Contour of tensile strain before peak load of typical specimens under three loading modes.

Rock Type	White Sandstone	Red Sandstone	Marble	Granite	Basalt
Mode-I	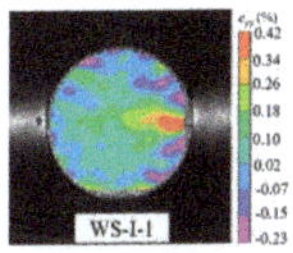	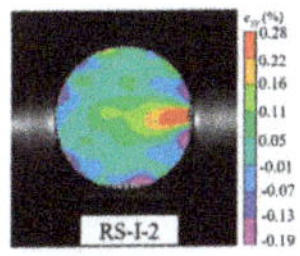	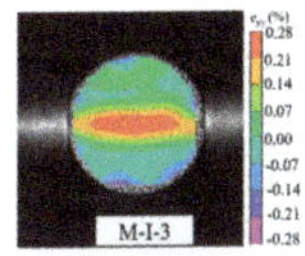	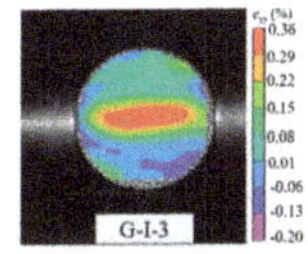	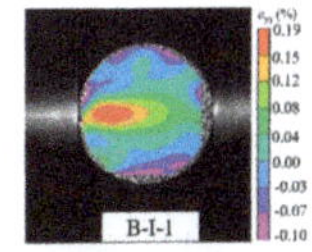
Mode-II	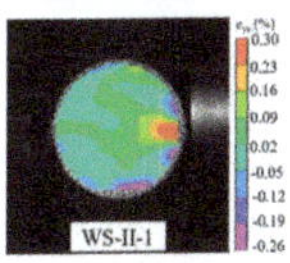	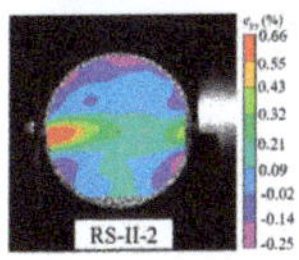	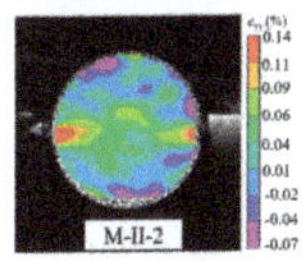	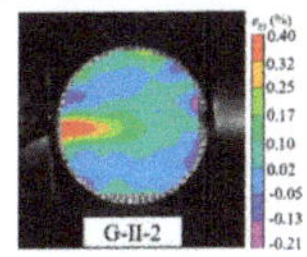	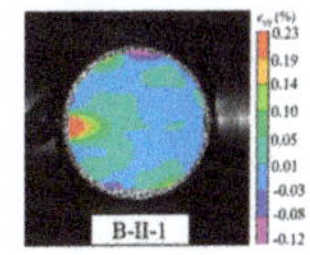
Mode-III	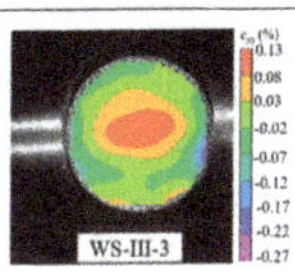	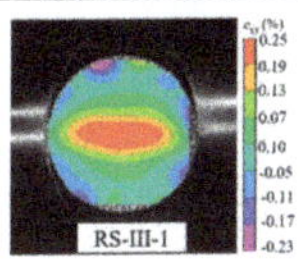	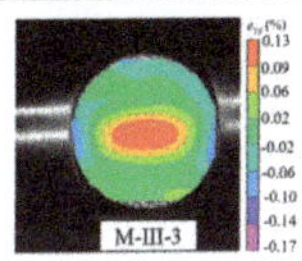	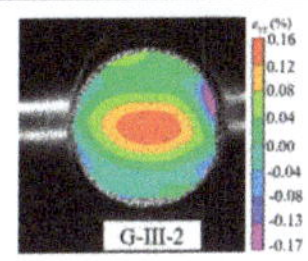	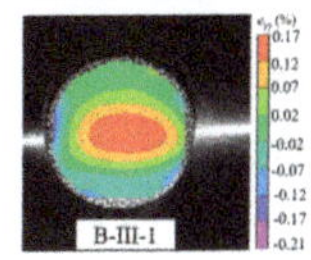

It can be found from the table that the strain localization characteristics of the specimen in the dynamic Brazilian tests are highly similar to those in the static test, showing the same two distribution types. The arc loading mode under dynamic impact was still the most consistent with the Brazilian test theory. However, the strain distribution characteristics of the five rocks (except marble) under mode-I loading differed from the static tests. This may be because when the compressive incident wave propagated to the end of the specimen, relatively weak soft rock was prone to damage at the contact end, and then the damage was aggravated during the loading process. Hard rocks exhibited the opposite condition under transient impact and had sufficient time to reach stress equilibrium. In the following, dynamic failure analysis was carried out on typical specimens that conformed to the above two strain localization characteristics, and the failure process of other specimens was taken as a reference.

3.5.1. Typical Fracture Characteristics I

A typical strain distribution characteristic under arc loading mode was selected for failure analysis. According to relevant theories in the static Brazilian tests, the tensile strain concentration in the center of the specimen can be considered as tensile stress concentration, and the disc specimen will crack from the center. The (The VIC-2D software version number is 6, and VIC-2D software invented by Correlated Solutions Inc. in Irmo, SC, USA) was used to arrange five virtual extensimeters (E1–E5) with equal spacing along the loading axis on the specimen surface (as shown in Figure 8a), and the tensile strain history (taking specimen WS-III-3 as an example) extracted by VIC-2D was plotted in the Figure 9. The tensile strain values of five extensimeters rose slowly, with tensile strain values below 0.25% before 230.6 μs. The increase rate of strain values of extensimeters E2, E3, and E4 started to accelerate, but that of E1 and E5 did not accelerate significantly. This also indicated that the region where the stress concentration occurred in the specimen was first in the center of the disc. After 12.6 μs (image interval), the strain increase rate of the five extensimeters rose sharply. However, the five strain extensimeters in descending order of strain values were E3, E4, E2, E1, and E5. The tensile strain value at the center of the disc was always the maximum. These data suggested that the dynamic crack propagation velocity was extremely rapid. Although the high-speed camera was used, the sequence of strain mutations of five extensimeters could not be accurately identified, and only the sequence of strain mutations of the regional extensimeters could be analyzed. However, the above analysis strongly indicates that the specimen must have started to crack at a point in the center of the disc, and then the fracture spread rapidly to both ends. The data from the virtual extensimeters showed that although the dynamic tensile strength of the specimens under arc loading and plate loading was not significantly different (the former was slightly greater than the latter), it was still necessary to use the arc loading method in the dynamic Brazilian test because it ensured, as much as possible, that the specimen would fracture from the center of the disc.

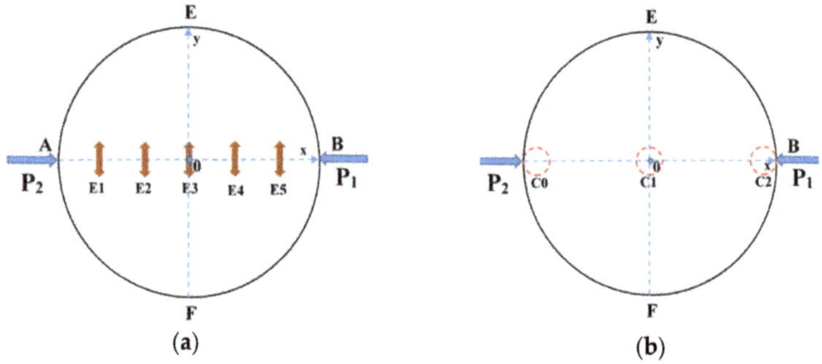

Figure 8. Arrangement of five virtual extensimeters (**a**) and three monitoring circles (**b**) on the specimen surface.

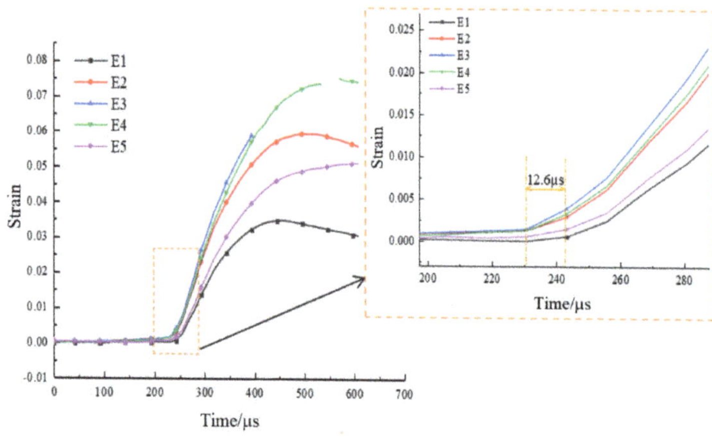

Figure 9. Strain history of five virtual extensimeters for the specimen WS-III-3.

In order to further analyze the deformation and rupture evolution of the specimen, the full-field displacement and strain contours at typical moments were extracted by VIC-2D, which is shown in Figure 10 (the top four images are strain contours and the bottom four images are displacement contours). The leftmost contour was the image of the specimen just before the peak, and the white arrow in the displacement contour represents the displacement vector. Before the peak load, the tensile strain was concentrated in the center of the disc, and the displacement contour showed a good symmetry without y-direction component of the displacement vector (the displacement in the x-direction was due to the compressive stress wave). When the time was 230.6 μs, the strain value increased significantly, while the surrounding strain value became smaller. In addition, the displacement vector in the center of the displacement contour started to have a slight y-direction component, indicating that the specimen was cracked at the center of the disc at this time. The moment of crack initiation point was in full agreement with the analysis by virtual extensimeters. After the crack initiation, the crack started to expand to both ends of the specimen, and the strain contour showed that the strain concentration zone spread to both ends with increasing of strain values. Meanwhile, the process of crack propagation could be well reflected indirectly from the displacement contour: firstly, the displacement concentration areas (red and purple areas) at the upper and lower ends of the specimen gradually spread to both ends with rising displacement values; secondly, the displacement vector in the center region of the disc became longer, and the displacement vector gradually tilted.

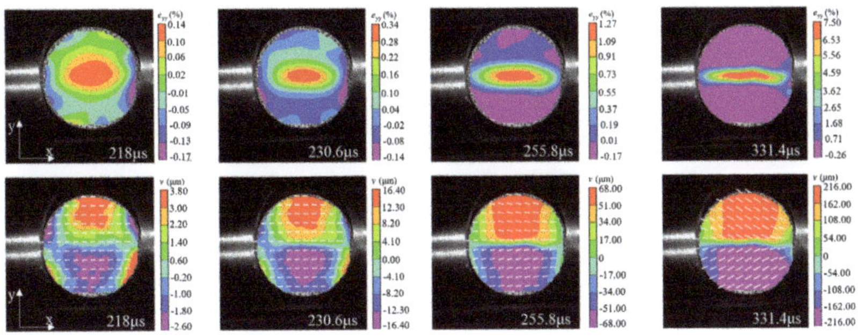

Figure 10. Strain and displacement contours of the specimen WS-III-3 at typical moments (impact load direction from right to left). The positive direction of the x-axis is the impact load direction, and the y-axis is perpendicular to the impact load direction.

3.5.2. Typical Fracture Characteristics II

The strain localization that occurred in the specimen under mode-II loading is another typical fracture characteristic. According to the relevant results in the static test analysis, it was known that for this loading case, the disc tended to have localized premature damage at the end. The following was an example of damage analysis for specimen WS-II-1. Three locations with high probability of crack initiation were marked in the Brazilian disc, as shown by the small red circles in Figure 8b, i.e., C0, C1, and C2. The diameter of these circles is about 5 mm, and their tensile and shear strains with time were extracted using VIC-2D, respectively, as shown in Figure 11. The strain at these three positions was almost unchanged until 66.8 μs. Then, the tensile and shear strains of C2 started to increase significantly, indicating that the shear crushing behavior occurred first due to the high stress concentration. A certain degree of shear stress was also present at the other end of the specimen, C0, but the stress value was small, and no damage occurred. After that, the same strain mutation occurred at C1 and C0, respectively, which indicated that the cracks at the end propagated rapidly to the left along the loading diameter after shear fragmentation occurred at the incident end of the disc. The clear damage sequence shows the superiority of the high-speed camera. In addition, it was worth noting that the moment of damage of the disc under mode-II loading (66.8 μs) was much earlier compared to that under mode-III loading (230.6 μs). The stress concentration at the end due to the steel bar loading method is undoubtedly more prominent under dynamic impact, which leads to lower slope of the load–displacement curve and tensile strength compared to the other two loading modes. Based on the above analysis, the mode-II loading was not suitable for determining the dynamic tensile strength of rock disc specimens.

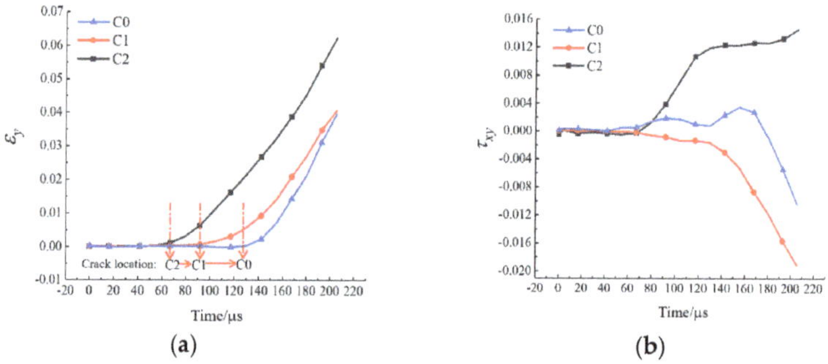

Figure 11. The *y*-direction strain (**a**) and shear strain (**b**) histories of three monitoring circles for the specimen WS-II-1.

The full-field strain and displacement contours at typical moments, as shown in Figure 12 (the top four images are strain contours and the bottom four images are displacement contours), were obtained by using the DIC technique to deeply analyze the deformation and rupture process of the disc under mode-II loading. The leftmost one was the image of the disc just before the peak load, and the white arrows in the displacement contour represent the displacement vectors. As can be seen from Figure 12, minor damage has occurred at the right end of the specimen due to the strain concentration before the peak load. The displacement vector at the right end of the specimen produced a vertical component, and the tensile strain and displacement values are larger here. After the peak load, the specimen underwent more obvious shear damage, and the macroscopic crack was formed at the end. The evolution of the strain contour at the top in Figure 12 reflected dynamic propagation process of the crack, and finally the specimen was completely split and damaged along the loading diameter. Similarly, the process of crack propagation from the incident end to the transmitted end can be well reflected from the displacement contour: firstly, the displacement concentration zone (red and purple areas) near the incident end of the specimen gradually propagated to the left and the displacement value became larger; secondly, the tilted displacement vectors in the right end of the disc gradually became longer and the displacement vectors at the left end gradually began to change from parallel to tilted.

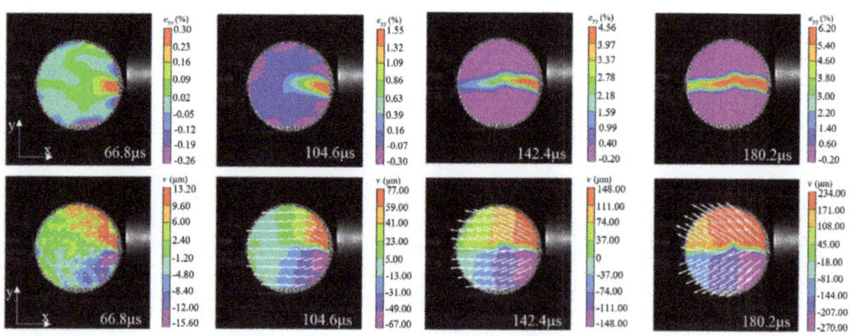

Figure 12. Strain and displacement contours of the specimen WS-II-1 at typical moments (impact load direction from right to left). The positive direction of the x-axis is the impact load direction, and the y-axis is perpendicular to the impact load direction.

3.6. Final Failure Pattern of Specimens

The final failure pattern of typical specimens for each loading method using images captured by the high-speed camera was shown in Table 4, and these specimens were consistent with those in Table 3. The final failure patterns of the researched specimens were summarized into four types by careful observation and combining the results of the analysis related to the DIC technique, as shown in Figure 13.

Table 4. The final failure patterns of typical specimens under three loading modes.

Rock Type	White Sandstone	Red Sandstone	Marble	Granite	Basalt
Mode-I	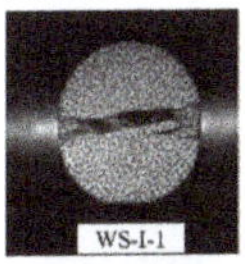	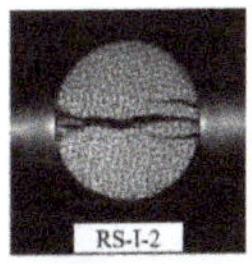	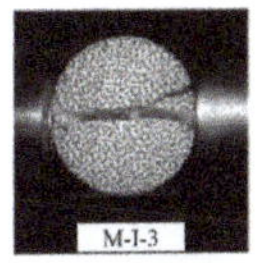	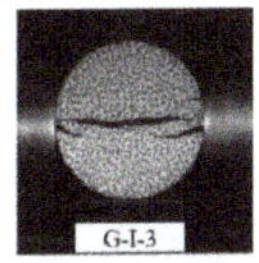	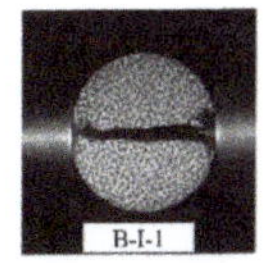
Mode-II	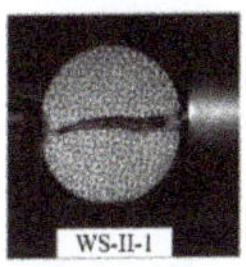	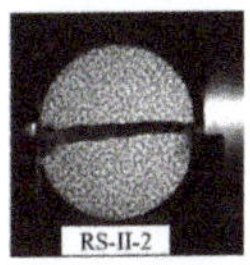	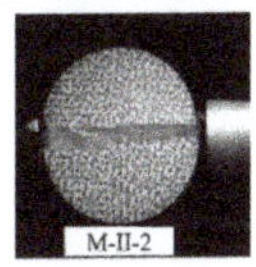	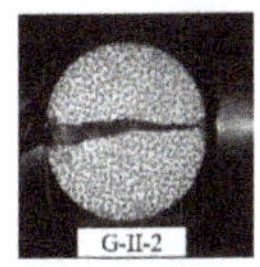	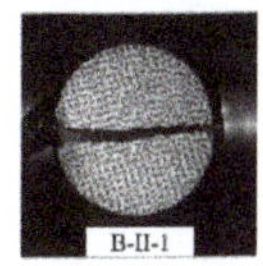
Mode-III	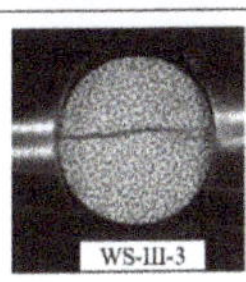	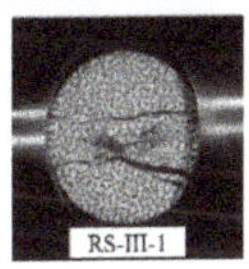	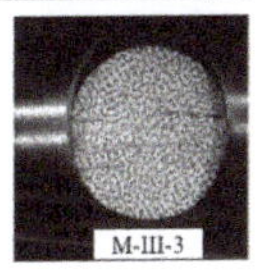	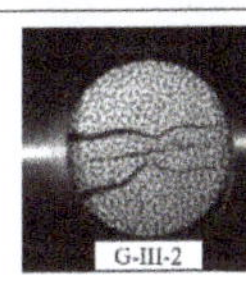	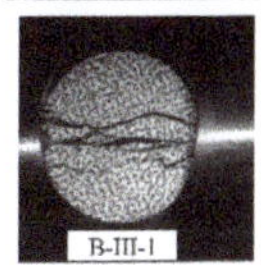

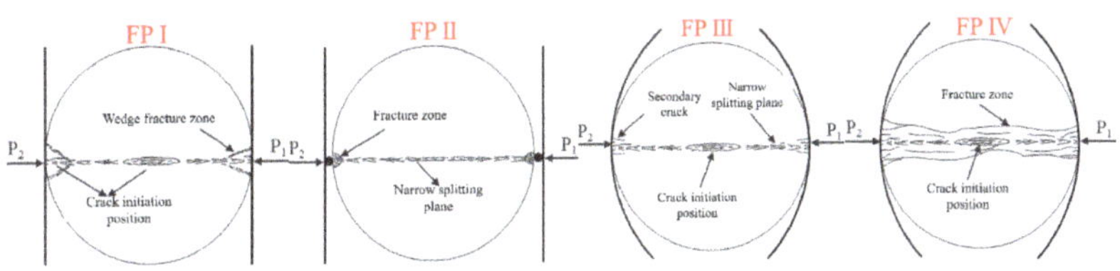

Figure 13. The four failure patterns of specimens in dynamic BD tests.

The failure pattern I (FP I) is a typical damage mode of the specimen under mode-I loading. Regardless of whether the specimen satisfied the central crack initiation, a certain degree of wedge-shaped fragmentation zone appeared at both ends of the specimen, and the central region along the loading diameter showed a narrower cleavage surface. The specimen was cracked at the center and the crack propagated to both ends of the disc causing complete splitting along the loading diameter. However, under the inertial impact of the incident bar, a wedge-shaped fragment would be formed at the end of the specimen due to shearing and twisting effects. The failure pattern II (FP II) is a typical damage mode of the specimen under the mode-II loading mode. Due to the dynamic impact and the strong stress concentration caused by steel bars, tiny crushing zones were generated at both ends of the specimen in contact with the steel bar, and ideal radial splitting was formed in the center of the disc specimen, with a relatively flat rupture surface and no excess secondary cracks generated. The failure pattern III and IV (FP III and IV) are two typical damage modes exhibited by specimens under mode-III loading. However, the FP III was only observed for white sandstone specimens, and the specimens were finally split and damaged radially from the center, accompanied by a very small amount of minor secondary cracks at the specimen ends. For rock specimens other than white sandstone, FP IV was observed. At higher loading rates, in addition to cracking damage along the center of the loading diameter, the specimen developed secondary cracks along the radial penetration at its two ends, thus eventually forming a fragmentation zone along the loading direction. The above analysis of failure modes again strongly suggested that dynamic loading further amplified the uniqueness of each loading method compared to static loading.

4. Conclusions

This paper mainly studied the damage law of dynamic Brazilian test under three loading modes. The relevant principles of the dynamic Brazilian test were introduced first, and the scheme and key operational steps of the test were discussed. The test results were compiled and analyzed in detail, including the tensile strength, deformation characteristics, and the final failure modes of the specimen. The main conclusions are as follows:

(1) The peak load and deformation characteristics of the dynamic Brazilian test were equally strongly influenced by the loading modes. The effects of loading modes on the tensile strength of the specimens were consistent with the static performance, i.e., the dynamic tensile strength determined by mode-III loading was the highest, while it determined that mode-II loading was lowest. In addition, the slope of the load–displacement curve of the specimen under mode-II loading was the smallest, while it was largest in mode-III loading.

(2) There were two typical fracture characteristics in the dynamic Brazilian test. The arc loading method showed outstanding superiority in the dynamic Brazilian test, and the results were more consistent with the Brazilian test theory compared with the other two loading methods. In contrast, the test results under the steel bar loading were unsatisfactory and may greatly underestimate the dynamic tensile strength of the rock.

(3) The specimens exhibited four different damage modes under three loading modes. In addition, it was incomplete or even wrong to collect only the broken pieces of specimens to determine the damage pattern and evaluate the damage mechanism of the specimen. On the contrary, the experiments can help scholars to better understand the evolution of rock fracture and damage mechanism by using the DIC technique.

Author Contributions: Conceptualization, D.L. and J.M.; methodology, Q.Z.; software, Q.Z.; validation, D.L., J.M. and B.L.; formal analysis, B.L.; investigation, J.M.; resources, D.L.; data curation, Q.Z.; writing—original draft preparation, J.M.; writing—review and editing, D.L.; visualization, Q.Z.; supervision, Q.Z.; project administration, D.L.; funding acquisition, D.L. All authors have read and agreed to the published version of the manuscript.

Funding: This research was funded by the National Natural Science Foundation of China, grant number 52074349 and U1934211.

Institutional Review Board Statement: Not applicable.

Informed Consent Statement: Not applicable.

Data Availability Statement: The data presented in this study are available on request from the corresponding author.

Conflicts of Interest: The authors declare no conflict of interest.

References

1. Han, Z.; Li, D.; Li, X. Effects of axial pre-force and loading rate on Mode I fracture behavior of granite. *Int. J. Rock Mech. Min. Sci.* **2022**, *157*, 105172. [CrossRef]
2. Han, Z.; Li, D.; Li, X. Dynamic mechanical properties and wave propagation of composite rock-mortar specimens based on SHPB tests. *Int. J. Min. Sci. Technol.* **2022**, *32*, 793–806. [CrossRef]
3. Han, Z.; Li, D.; Zhou, T.; Zhu, Q.; Ranjith, P. Experimental study of stress wave propagation and energy characteristics across rock specimens containing cemented mortar joint with various thicknesses. *Int. J. Rock Mech. Min. Sci.* **2020**, *131*, 104352. [CrossRef]
4. Xiao, P.; Li, D.; Zhao, G.; Liu, H. New criterion for the spalling failure of deep rock engineering based on energy release. *Int. J. Rock Mech. Min. Sci.* **2021**, *148*, 104943. [CrossRef]
5. Zhu, Q.; Li, X.; Li, D.; Ma, C. Experimental investigations of static mechanical properties and failure characteristics of damaged diorite after dynamic triaxial compression. *Int. J. Rock Mech. Min. Sci.* **2022**, *153*, 105106. [CrossRef]
6. Li, Q.; Ma, D.; Zhang, Y.; Liu, Y.; Ma, Y.; Hu, D. Insights into Controlling Factors of Pore Structure and Hydraulic Properties of Broken Rock Mass in a Geothermal Reservoir. *Lithosphere* **2021**, *2021*, 3887832. [CrossRef]
7. Zhao, R.; Tao, M.; Zhao, H.; Cao, W.; Li, X.; Wang, S. Dynamics fracture characteristics of cylindrically-bored granodiorite rocks under different hole size and initial stress state. *Theor. Appl. Fract. Mech.* **2020**, *109*, 102702. [CrossRef]

8. Wang, S.; Li, X.; Yao, J.; Gong, F.; Li, X.; Du, K.; Tao, M.; Huang, L.; Du, S. Experimental investigation of rock breakage by a conical pick and its application to non-explosive mechanized mining in deep hard rock. *Int. J. Rock Mech. Min. Sci.* **2019**, *122*, 104063. [CrossRef]
9. Wang, S.; Sun, L.; Li, X.; Zhou, J.; Du, K.; Wang, S.; Khandelwal, M. Experimental investigation and theoretical analysis of indentations on cuboid hard rock using a conical pick under uniaxial lateral stress. *Geomech. Geophys. Geo-Energy Geo-Resour.* **2022**, *8*, 34. [CrossRef]
10. Wang, S.; Sun, L.; Tang, Y.; Jing, Y.; Li, X.; Yao, J. Field application of non-blasting mechanized mining using high-frequency impact hammer in deep hard rock mine. *Trans. Nonferrous Met. Soc. China* **2022**, *32*, 3051–3064. [CrossRef]
11. Pei, P.; Dai, F.; Liu, Y.; Wei, M. Dynamic tensile behavior of rocks under static pre-tension using the flattened Brazilian disc method. *Int. J. Rock Mech. Min. Sci.* **2020**, *126*, 104208. [CrossRef]
12. Wu, Y.; Yin, T.; Li, Q.; Zhuang, D.; Chen, Y.; Yang, Z. Analytical investigation on the unstable fracture toughness of fine-grained quartz-diorite rock considering the size effect. *Eng. Fract. Mech.* **2022**, *272*, 108722. [CrossRef]
13. Yang, Z.; Yin, T.; Zhuang, D.; Wu, Y.; Yin, J.; Chen, Y. Effect of temperature on mixed mode I/III fracture behavior of diorite: An experimental investigation. *Theor. Appl. Fract. Mech.* **2022**, *122*, 103571. [CrossRef]
14. Wu, Y.; Yin, T.; Zhuang, D.; Li, Q.; Chen, Y. Research on the effect of thermal treatment on the crack resistance curve of marble using notched semi-circular bend specimen. *Theor. Appl. Fract. Mech.* **2022**, *119*, 103344. [CrossRef]
15. Dorji, J.; Zahra, T.; Thambiratnam, D.; Lee, D. Strength assessment of old masonry arch bridges through moderate destructive testing methods. *Constr. Build. Mater.* **2021**, *278*, 122391. [CrossRef]
16. D'Amato, M.; Luchin, G.; De Matteis, G. A preliminary study on properties of a weak units–strong mortar masonry: The case study of matera tufo masonry (Italy). *Int. J. Archit. Herit.* **2022**, 1–22. [CrossRef]
17. Wu, N.; Fu, J.; Zhu, Z.; Sun, B. Experimental study on the dynamic behavior of the Brazilian disc sample of rock material. *Int. J. Rock Mech. Min. Sci.* **2020**, *130*, 104326. [CrossRef]
18. Yin, T.; Yang, Z.; Wu, Y.; Tan, X.; Li, M. Experimental investigation on the effect of open fire on the tensile properties and damage evolution behavior of granite. *Int. J. Damage Mech.* **2022**, *31*, 1139–1164. [CrossRef]
19. Ross, C.A.; Tedesco, J. Split-Hopkinson pressure-bar tests on concrete and mortar in tension and compression. *Mater. J.* **1989**, *86*, 475–481.
20. Johnstone, C.; Ruiz, C. Dynamic testing of ceramics under tensile stress. *Int. J. Solids Struct.* **1995**, *32*, 2647–2656. [CrossRef]
21. Gomez, J.; Shukla, A.; Sharma, A. Static and dynamic behavior of concrete and granite in tension with damage. *Theor. Appl. Fract. Mech.* **2001**, *36*, 37–49. [CrossRef]
22. Dai, F.; Huang, S.; Xia, K.; Tan, Z. Some fundamental issues in dynamic compression and tension tests of rocks using split Hopkinson pressure bar. *Rock Mech. Rock Eng.* **2010**, *43*, 657–666. [CrossRef]
23. Zhang, Q.; Zhao, J. Determination of mechanical properties and full-field strain measurements of rock material under dynamic loads. *Int. J. Rock Mech. Min. Sci.* **2013**, *60*, 423–439. [CrossRef]
24. Xia, K.; Yao, W. Dynamic rock tests using split Hopkinson (Kolsky) bar system—A review. *J. Rock Mech. Geotech. Eng.* **2015**, *7*, 27–59. [CrossRef]
25. Dai, F.; Xia, K.; Zuo, J.; Zhang, R.; Xu, N. Static and dynamic flexural strength anisotropy of Barre granite. *Rock Mech. Rock Eng.* **2013**, *46*, 1589–1602. [CrossRef]
26. Huang, S.; Xia, K.; Yan, F.; Feng, X. An experimental study of the rate dependence of tensile strength softening of Longyou sandstone. *Rock Mech. Rock Eng.* **2010**, *43*, 677–683. [CrossRef]
27. Wang, Q.; Wu, L. Determination of elastic modulus, tensile strength and fracture toughness of brittle rocks by using flattened Brazilian disk specimen-Part I: Analytical and numerical results. *Chin. J. Rock Mech. Eng.* **2002**, *21*, 199–204.
28. Ruiz, G.; Ortiz, M.; Pandolfi, A. Three-dimensional finite-element simulation of the dynamic Brazilian tests on concrete cylinders. *Int. J. Numer. Methods Eng.* **2000**, *48*, 963–994. [CrossRef]
29. Zhu, W.; Tang, C. Numerical simulation of Brazilian disk rock failure under static and dynamic loading. *Int. J. Rock Mech. Min. Sci.* **2006**, *43*, 236–252. [CrossRef]
30. Mahabadi, O.; Cottrell, B.; Grasselli, G. An example of realistic modelling of rock dynamics problems: FEM/DEM simulation of dynamic Brazilian test on Barre granite. *Rock Mech. Rock Eng.* **2010**, *43*, 707–716. [CrossRef]
31. Li, D.; Li, B.; Han, Z.; Zhu, Q. Evaluation on Rock Tensile Failure of the Brazilian Discs under Different Loading Configurations by Digital Image Correlation. *Appl. Sci.* **2020**, *10*, 5513. [CrossRef]
32. Li, D.; Li, B.; Han, Z.; Zhu, Q.; Liu, M. Evaluation of Bi-modular Behavior of Rocks Subjected to Uniaxial Compression and Brazilian Tensile Testing. *Rock Mech. Rock Eng.* **2021**, *54*, 3961–3975. [CrossRef]
33. Bieniawski, Z.; Hawkes, I. Suggested methods for determining tensile strength of rock materials. *Int. J. Rock Mech. Min. Sci.* **1978**, *15*, 99–103.
34. Ma, J.; Li, D.; Zhu, Q.; Liu, M.; Wan, Q. The mode I fatigue fracture of fine-grained quartz-diorite under coupled static loading and dynamic disturbance. *Theor. Appl. Fract. Mech.* **2021**, *117*, 103140. [CrossRef]
35. Zhu, Q.; Ma, C.; Li, X.; Li, D. Effect of Filling on Failure Characteristics of Diorite with Double Rectangular Holes Under Coupled Static–Dynamic Loads. *Rock Mech. Rock Eng.* **2021**, *54*, 2741–2761. [CrossRef]
36. Han, Z.; Li, D.; Li, X. Experimental study on the dynamic behavior of sandstone with coplanar elliptical flaws from macro, meso, and micro viewpoints. *Theor. Appl. Fract. Mech.* **2022**, *120*, 103400. [CrossRef]

37. Zhu, Q.; Li, C.; Li, X.; Li, D.; Wang, W.; Chen, J. Fracture mechanism and energy evolution of sandstone with a circular inclusion. *Int. J. Rock Mech. Min. Sci.* **2022**, *155*, 105139. [CrossRef]
38. Zhang, C.; Li, D.; Wang, C.; Ma, J.; Zhou, A.; Xiao, P. Effect of confining pressure on shear fracture behavior and surface morphology of granite by the short core in compression test. *Theor. Appl. Fract. Mech.* **2022**, *121*, 103506. [CrossRef]
39. Luo, P.; Wang, L.; Li, D.; Yang, J.; Lv, X. Deformation and failure mechanism of horizontal soft and hard interlayered rock under uniaxial compression based on digital image correlation method. *Eng. Fail. Anal.* **2022**, *142*, 106823. [CrossRef]
40. Zhou, Y.; Xia, K.; Li, X.; Li, H.; Ma, G.; Zhao, J. Suggested Methods for Determining the Dynamic Strength Parameters and Mode-I Fracture Toughness of Rock Materials. In *The ISRM Suggested Methods for Rock Characterization, Testing and Monitoring: 2007–2014*; Springer: Berlin/Heidelberg, Germany, 2011; pp. 35–44. [CrossRef]
41. Kolsky, H. An investigation of the mechanical properties of materials at very high rates of loading. *Proc. Phys. Soc. Sect. B* **1949**, *62*, 676. [CrossRef]
42. Li, D.; Ma, J.; Wan, Q.; Zhu, Q.; Han, Z. Effect of thermal treatment on the fracture toughness and subcritical crack growth of granite in double-torsion test. *Eng. Fract. Mech.* **2021**, *253*, 107903. [CrossRef]
43. Sutton, M.A.; Orteu, J.J.; Schreier, H. *Image Correlation for Shape, Motion and Deformation Measurements: Basic Concepts, Theory and Applications*; Springer Science & Business Media: Berlin/Heidelberg, Germany, 2009. [CrossRef]
44. Li, D.; Zhang, C.; Zhu, Q.; Ma, J.; Gao, F. Deformation and fracture behavior of granite by the short core in compression method with 3D digital image correlation. *Fatigue Fract. Eng. Mater. Struct.* **2022**, *45*, 425–440. [CrossRef]
45. Ma, J.; Li, D.; Luo, P.; Zhu, Q.; Zhang, C.; Su, X. Subcritical crack growth and fracture behavior of rocks and long-term strength estimation. *Theor. Appl. Fract. Mech.* **2022**, *122*, 103664. [CrossRef]

materials

MDPI

Article

An Evaluation Index of Fracability for Reservoir Rocks Based on Fracture Process Zone

Hongran Chen [1,2], Jingrui Niu [3,*] and Mengyang Zhai [1,2]

1 Key Laboratory of Shale Gas and Geoengineering, Institute of Geology and Geophysics, Chinese Academy of Sciences, Beijing 100029, China
2 Innovation Academy for Earth Science, CAS, Beijing 100029, China
3 Chinese Society for Rock Mechanics and Engineering, Beijing 100029, China
* Correspondence: niujingrui@mail.iggcas.ac.cn

Abstract: A reliable evaluation method for the fracability (i.e., ability to generate abundant cracks) of reservoir rocks is a critical issue for maximum hydraulic fracturing efficiency. Most previous fracability indices lacked enough rationality and practicability, and thus could not consistently provide a reliable evaluation. We suggest a new fracability index called crack tolerance, which is represented by the maximum radius of the fracture process zone at the crack tip of a cracked chevron notched Brazilian disk specimen, which corresponds to the critical state for unstable propagation of the notched crack. In experiments and simulations based on the discrete element method, we showed quantitative methods to conveniently determine the value of the crack tolerance and showed that specimens with a greater crack tolerance generated more cracks before rupture and had complex morphologies, which would indicate stronger fracability. The crack tolerance can well characterize the effects of structural and loading conditions, including the grain size heterogeneity, bedding orientation, and environmental temperature, on fracability, and the inherent heterogeneity of rock is the physical basis for it as a fracability evaluation index. Our studies showed the rationality and practicability of this index and provide hints for how to produce abundant complex cracks in reservoirs.

Keywords: fracability; fracture process zone; crack tolerance; chevron notched disk; discrete element method

Citation: Chen, H.; Niu, J.; Zhai, M. An Evaluation Index of Fracability for Reservoir Rocks Based on Fracture Process Zone. *Materials* **2022**, *15*, 8485. https://doi.org/10.3390/ma15238485

Academic Editor: Christophe Pinna

Received: 25 October 2022
Accepted: 24 November 2022
Published: 28 November 2022

1. Introduction

Human beings' mining engineering and energy resource exploitation extensively involve the generation and propagation of cracks within rock materials. Hydraulic fracturing is widely used to enhance the fluid conductivity of reservoirs of oil, gas, and geothermal resources. A reliable evaluation of the rock fracability (i.e., ability to generate abundant cracks) is important for hydraulic fracturing [1].

Brittleness, which is generally viewed as a property (or ability) of solid material that ruptures with little appreciable permanent deformation, has long been considered approximately equivalent to fracability, because it shows empirical relevance to the possibility of crack propagation: reservoir comprising brittle rocks usually responds well to stimulation, whereas preexisting and hydraulic fractures tend to heal rather than to propagate in a less brittle reservoir. This is probably attributed to less energy consumed by the ductile deformation of brittle rock materials [2].

In the past decades, a variety of brittleness indices have been developed to evaluate its effect [3,4], which can be classified into several broad categories:

(1) Based on mineral composition (e.g., [5]), especially the weight or volume proportion of hard minerals such as quartz: a positive correlation seems to exist between the brittleness and mineral contents of rocks. However, such indices do not consider

many other factors that also contribute to brittleness, such as grain size and loading conditions.

(2) Based on elastic parameters (e.g., [6]): for example, rocks with a large Young's modulus and small Poisson's ratio are assumed more brittle. However, such indices can be controversial because many laboratories and in situ observations [7,8] contradict this assumption.
(3) Based on strength: for example, one such index is the ratio of tensile and compressive strengths [9]. Such indices are easily measured, but they lack a physical correlation to brittleness and cracking propagation mechanisms. Thus, these indices may return similar values for various types of rocks with different levels of brittleness.
(4) Based on characteristics of the stress–strain curve such as the relative stress drop, post-peak modulus, and various combinations [10,11]: these indices characterize rock brittleness well and are widely used in predicting the rockburst proneness. However, high brittleness does not consistently represent strong fracability because brittle rock can also act as a barrier to hydraulic fracturing [12].

In summary, many brittleness indices currently popular in fracability evaluation for reservoir lacks mechanical relevance to the rock cracking process. On the other hand, the evaluation indices used in other areas (e.g., those used to estimate rock cuttability [13]) are usually inapplicable for reservoir fracability evaluation owing to the essential differences of physical meaning between brittleness and fracability. Thus far, few evaluation indices of rock fracability meet the following requirements [3]:

(1) Has a firm physical basis;
(2) Consider the heterogeneity of rock material;
(3) Be convenient to measure;
(4) Characterize the effects of loading conditions.

To address this issue, we propose a new evaluation index for rock fracability that we call the *crack tolerance*. See Section 2 for its definition. Sections 3 and 4 show the experimental measurement of this new index and the corresponding numerical simulation results, respectively, to demonstrate the rationality of the index. Based on these analyses, the effects of several characteristics of the rock materials on the crack tolerance are discussed in Section 5.

This study demonstrated the physical rationality of the crack tolerance as an evaluation index and analyzed the effects of the rock structure and loading conditions on the crack tolerance in an effort to extend our understanding of rock fracability and provide hints for how to produce more cracks in the reservoir.

2. Fracture Process Zone and Crack Tolerance

Numerous researchers have revealed that the propagation of macroscopic cracks within rock under tension is attributed to progressive generation, interaction, and nucleation of micro-cracks from the macroscopic crack tips as follows [14]. When the imposed tensile load is small, only a few independent micro-cracks can arise around each crack tip (Figure 1a). As the tensile load increases, the distribution range of the micro-cracks expands, and their density increases. They interact with each other and coalesce (Figure 1b) to cause a gradual macroscopic propagation of the preexisting crack (Figure 1c,d). These micro-cracks indicate nonlinear deformation in the region around a crack tip preceding crack unstable propagation, which is referred to as a fracture process zone (FPZ) [14].

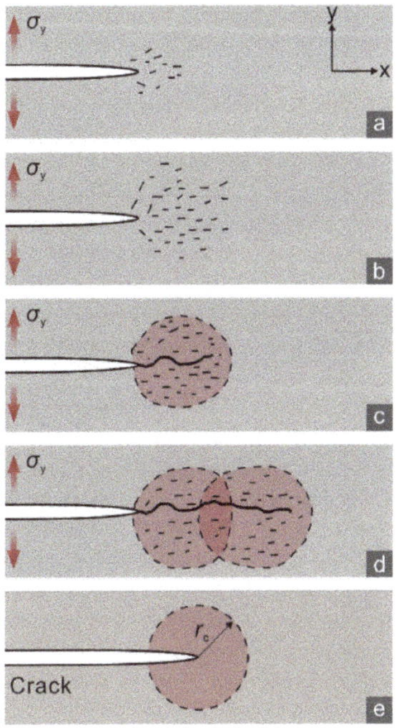

Figure 1. (**a–d**) Formation of a fracture process zone (FPZ), shaded in red, and the propagation of macro fractures under the tensile stress σ_y (denoted by red arrows). (**e**) Schematic of an FPZ with the size r_c.

Crack propagation in tensile mode is most common in hydraulic fracturing because the effect of hydraulic pressure imposed on the crack surface approximates remote tensile stress in nature; additionally, rocks have a much lower tensile strength than compressive and shear strengths. Thus, cracks easily propagate driven by an injected fluid. The principal stresses at a tensile crack tip can be described as [15]

$$\begin{bmatrix}\sigma_1\\ \sigma_2\end{bmatrix} = \frac{K_I}{\sqrt{2\pi r}}\begin{bmatrix}\cos\frac{\theta}{2}\left(1+\sin\frac{\theta}{2}\right)\\ \cos\frac{\theta}{2}\left(1-\sin\frac{\theta}{2}\right)\end{bmatrix} \tag{1}$$

where σ_1 and σ_2 are maximum and intermediate principal stresses, K_I is tensile stress intensity factor, r and θ are polar radius and polar angle for polar coordinate system from the tip. Note that the minimum principal stress not listed here equals to zero. The range of FPZ (i.e., its size) is calculated based on the hypothesis that nonlinear deformation occurs within a region around crack tip when the local stress state satisfies a certain criterion (e.g., tensile strength criterion for rock materials, von Mises criterion for metal materials). The tensile cracks are assumed to propagate parallel to their own plane (i.e., $\theta = 0$) when the σ_1 reaches the tensile strength of the rock (σ_t), because the critical state of crack propagation is attained, which corresponds to the maximum size of the FPZ:

$$\sigma_t = \frac{K_{IC}}{\sqrt{2\pi r_c}}\cos 0(1+\sin 0) \tag{2}$$

which leads to

$$r_c = (K_{IC}/\sigma_t)^2/(2\pi), \tag{3}$$

where K_{IC} is the tensile fracture toughness, and r_c is the maximum FPZ size. In this context, the FPZ is represented by a circle centered on a fracture tip [14] (Figure 1e), and r_c is the radius of a circular FPZ.

A large r_c would indicate that micro-cracks are distributed within a large FPZ in front of a preexisting crack tip. It would also suggest a considerable number of micro-cracks within the FPZ because a preexisting crack will not propagate until the micro-crack density is high enough to reach a critical level [16]. Therefore, r_c may characterize the maximum number of micro-cracks generated in the preparation stage for macroscopic crack propagation. In other words, r_c can be used to indicate the ability of a rock to tolerate micro-cracks before crack unstable propagation. For this reason, we refer to r_c as the *crack tolerance*. The crack morphology may also depend on the crack tolerance because a large r_c would indicate an extensive distribution of micro-cracks, which would likely result in irregular and branch cracks.

The concepts of the FPZ and r_c derive from the propagation process of a single crack with specific boundary conditions. Nevertheless, this process represents the inherent mechanical rule of crack generation within rocks because each crack started as an FPZ. Based on this understanding, the crack tolerance may reflect the potential of a given rock stratum to generate abundant cracks. Recent studies [17,18] have shown that rock specimens with a larger FPZ produce more fragments, which suggests greater fracability and provides evidence supporting our hypothesis. The maximum FPZ radius has exhibited dependence on the structure [19,20] and loading conditions [21] of rock. Thus, we conducted experiments and numerical simulations to analyze their effects on crack tolerance and demonstrate its rationality as an evaluation index.

3. Experiments

3.1. Specimens

We used marble, shale, and sandstone collected from Xishan, Beijing for experiments because marble was observed in some geothermal reservoirs, and shale and sandstone are representative lithologies comprising oil and gas reservoirs. The marble was divided into types A and J (Table 1): marble A totally constituted by calcite had a greater average grain size and was more heterogeneous as defined by Han et al. [22], and marble J mainly consisting of dolomite had an equigranular texture. The microscopy observation and X-ray diffraction (XRD) analysis showed that the shale with fine grains consisted of quartz (55.4%), plagioclase (6.2%), and clay minerals (38.4%, brown grains in Table 1). The quartz and clay minerals were alternatively layered. The sandstone consisted of quartz (69.5%), plagioclase (22.1%), and potassium feldspar (8.4%), and these xenomorphic grains have similar sizes (~2 mm). Most plagioclase grains experienced sericitization.

Table 1. Microstructures of the rock specimens.

Microscopy Observation	Qualitative Description	Average Size (mm)	Grain Size Distribution
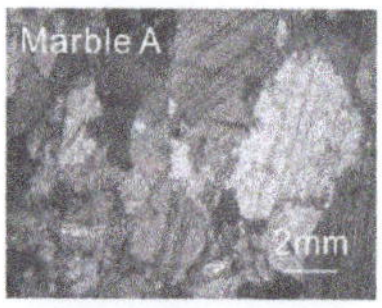	Medium-coarse grains	3.4	3–5 mm: 70% 1–3 mm: 30%
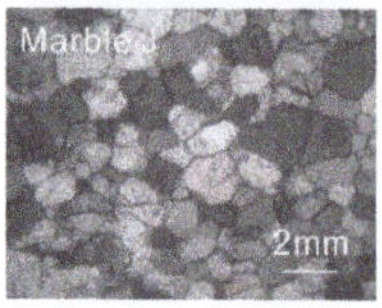	Medium grains	1.5	1–2 mm

Table 1. *Cont.*

Microscopy Observation	Qualitative Description	Average Size (mm)	Grain Size Distribution
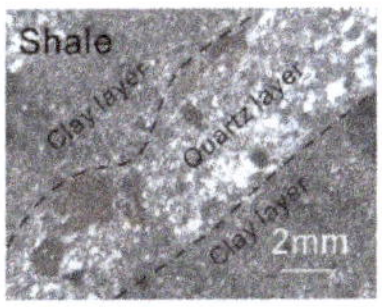	Fine grains	<0.1	/
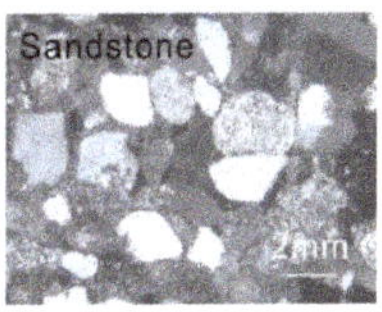	Coarse grains	2	1–3 mm

3.2. Experimental Methodology

The cracked chevron notched Brazilian disk (CCNBD) test involves the formation of FPZs at the two tips of a prefabricated notched crack, which is analogous to a natural crack. Therefore, the CCNBD test is applicable to evaluating crack tolerance. According to Equation (3), quantifying the crack tolerance requires determining the tensile fracture toughness K_{IC} and tensile strength σ_t, which are measured by the CCNBD and Brazilian disk (BD) tests, respectively, as recommended by the International Society for Rock Mechanics (ISRM) and American Society of Testing Materials (ASTM).

The notched crack of each CCNBD specimen was created by a 1 mm thick circular diamond saw. To ensure cutting accuracy, the expected locations of the circular center and the initial and final chevron notched cracks were marked on each disk. We measured the actual values of the parameters shown in Figure 2a,b and confirmed that the dimensionless parameters α_1 and α_B of all CCNBD specimens were within the valid range (Figure 2c). The method reported by Fowell et al. [23] was used to calculate the K_{IC}:

$$K_{IC} = P_{max} Y^{*}{}_{min} / BR^{1/2}, \quad (4)$$

where P_{max} is the peak applied axial load in the CCNBD test and $Y^{*}{}_{min}$ is the critical dimensionless stress intensity value. This is determined by

$$Y^{*} = u \cdot \exp(v \cdot \alpha_1), \quad (5)$$

where u and v are geometric constants that are determined by α_0 and α_B as reported by Fowell et al. [23].

The thickness (B') and diameter (D') of the BD specimens were set identical to those of the CCNBD specimens to eliminate the size effect on the calculated crack tolerance. The B'-to-D' ratio was within the range recommended by the ASTM of 0.2–0.75 [24]. The σ_t was calculated as follows:

$$\sigma_t = 2P'_{max} / \pi B'D', \quad (6)$$

where P'_{max} is the peak applied axial load in the BD test.

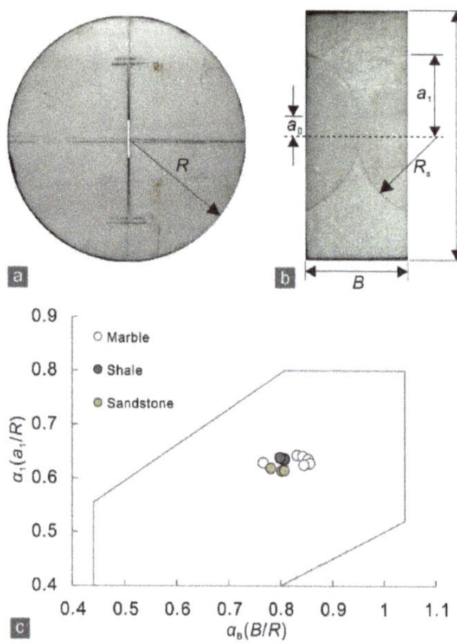

Figure 2. (**a**) Orthographic and (**b**) side profiles of a marble CCNBD specimen; (**c**) valid range for dimensionless parameters α_1 and α_B (outlined in gray) [23] and the distribution of parameter values for all of the prepared CCNBD specimens. Geometric parameters: diameter D = 75 mm, radius R = 37.5 mm, thickness B = 30 mm, saw radius R_s = 25 mm, initial chevron notched crack length a_0 = 8.45 mm, and final chevron notched crack length a_1 = 23.5 mm.

Each CCNBD or BD test (Figure 3a,b) was performed at a constant displacement rate of 0.06 mm/min by an MTS servo-control testing machine (series CMT) with a maximum loading force of 100 kN. This machine is equipped with an SNAS GDS-300 environmental chamber controlled by a WK650 controller (Figure 3c,d). These apparatuses permit environmental temperatures within the chamber up to 200 °C by electrical heaters (Figure 3b). To investigate the effect of temperature, several sandstone specimens were placed in the chamber at 75 or 125 °C for 1 h before the tests began, so that the notched crack propagated within rocks under higher temperatures. Other tests were performed at room temperature (~25 °C). The bedding planes of the shale specimens were set perpendicular (horizontal) or parallel (vertical) to the notched cracks to analyze the effect of the bedding orientation.

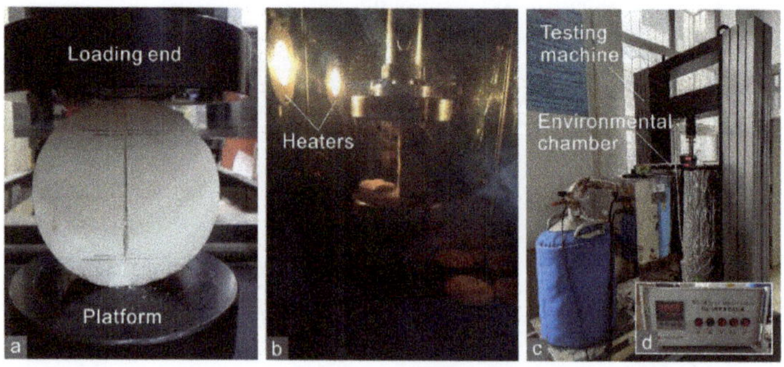

Figure 3. (**a**) CCNBD test on a marble A specimen at room temperature; (**b**) BD test on a sandstone specimen in the chamber at 125 °C; (**c**) MTS servo-control testing machine, SNAS GDS-300 environmental chamber, and (**d**) WK650 temperature controller.

BD tests were conducted on at least three specimens in parallel with the same lithology, bedding orientation, and temperature, and the average strength was taken as the tensile strength for the corresponding set of conditions. The K_{IC} of each CCNBD specimen and the above average σ_t were used in Equation (3) to calculate the crack tolerance.

3.3. Experimental Results

In the CCNBD test, the marble A specimens with stronger heterogeneity had greater tensile strength, fracture toughness, and crack tolerance than the marble J specimens, with relatively homogeneous small grains (Figure 4a). White patches indicating FPZs [25] appeared in front of the notched crack tips (Figure 5a) as the peak loads of the marble A specimens were approached. The patches corresponded to the sparkling areas on the rupture surface (Figure 5b), which may imply breaking cleavages of grains. However, such patches were not observed for the marble J specimens (Figure 5c,d), and neither were the discernible sparkling areas. Furthermore, the main cracks in the marble A specimens had branches causing more fragments (Figure 5b) while the crack in the marble J specimens propagated along a straight path (Figure 5c). These phenomena suggest that a more heterogeneous grain size corresponds to a larger crack tolerance and thus a stronger ability for crack generation.

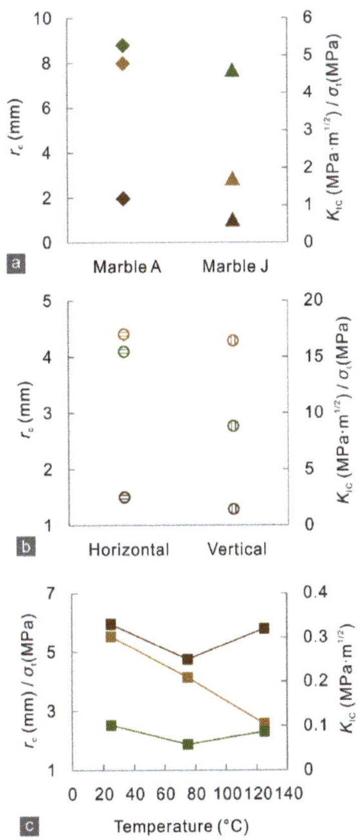

Figure 4. Mean tensile strength (brown), tensile fracture toughness (green), and crack tolerance (sienna) of the (**a**) marble A (diamond) and J (triangle), (**b**) shale (circle) with horizontal and vertical orientations and (**c**) sandstone (square).

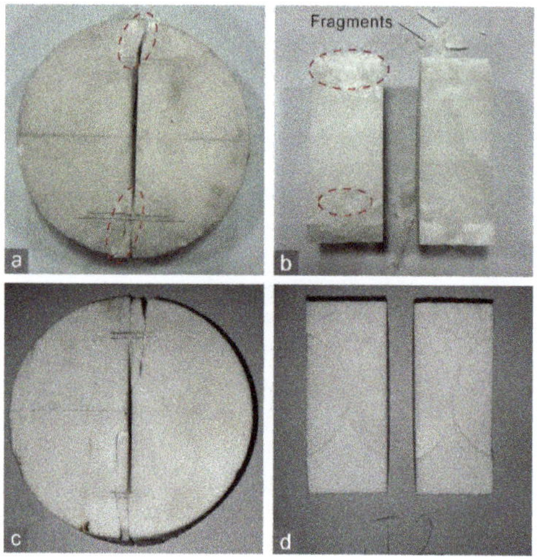

Figure 5. Crack morphologies of (**a**,**b**) marble A and (**c**,**d**) J. The red dashed ellipses in (**a**) denote white patches around the notched crack tips.

The mean crack tolerance of the shale specimens was less with a vertical bedding orientation than with a horizontal orientation (Figure 4b). The tensile strength and fracture toughness displayed similar variation trends with bedding orientation. Similar results can be acquired based on the data from Wang [26]. With a vertical orientation, the main crack of the specimen propagated along the bedding planes (Figure 6a), which generated a smooth rupture surface (Figure 6b). In contrast, with a horizontal orientation, the main crack spanned across bedding planes, and the path with steps was more irregular (Figure 6c,d). This is because the main crack was offset or even bifurcated when it encountered a bedding plane. The branch cracks were captured by bedding planes and then propagated along them, thereby their morphologies were smooth.

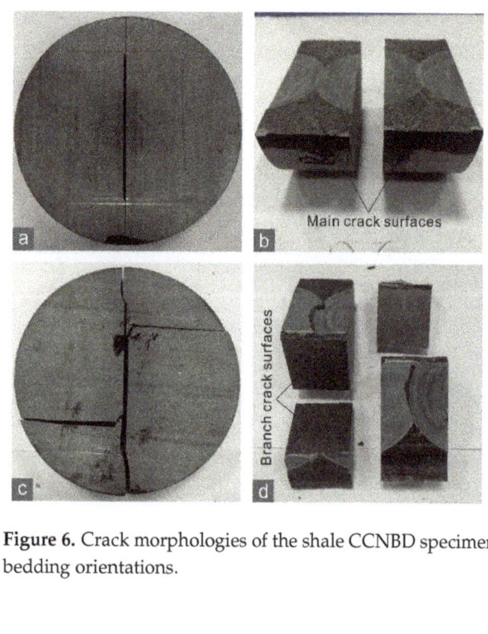

Figure 6. Crack morphologies of the shale CCNBD specimens with (**a**,**b**) horizontal and (**c**,**d**) vertical bedding orientations.

The crack tolerance value of the sandstone specimens consistently declined as the environmental temperature rises from 25 °C to 125 °C, while the tensile strength and fracture toughness exhibited V-shaped trends within this temperature range (Figure 4c). It is difficult to identify changes in crack morphology with the rising temperature with the naked eye (Figure 7a,c,e). According to the edge of their rupture surface, we speculated that the main cracks in the specimens at 125 °C may propagate along less curved paths than the specimens at lower temperatures did (Figure 7b,d,f). The variations of the crack tolerance value and crack morphologies imply that high temperatures possibly reduce rock fracability.

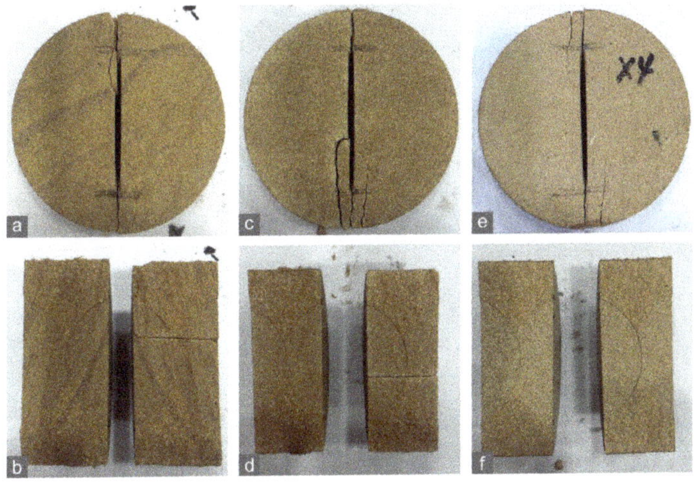

Figure 7. Crack morphologies of the sandstone specimens in the CCNBD tests at (**a**,**b**) 25 °C, (**c**,**d**) 75 °C, and (**e**,**f**) 125 °C.

The crack tolerance preliminarily showed an ability to address the aforementioned problems of previous fracability indices. Firstly, this index has a firm physical basis derived from the FPZ size, representing the nonlinearity of deformation due to micro-crack generation before the macroscopic propagation of the crack. Additionally, the formation of the FPZ is the inherent mechanical behavior of heterogeneous rock materials, and the FPZ size highly depends on the degree of heterogeneity as the previous [27] and present experiment results revealed. From the aspect of practicability, the crack tolerance value can be determined conveniently in the laboratory because BD and CCNBD tests are very common rock mechanical tests, and the small-size specimens they use can be easily obtained from cores. Finally, this index may characterize the effects of structure and loading conditions on fracability to an extent, as the tests on the shale and sandstone showed.

4. Numerical Analysis with the Discrete Element Method

4.1. Particle Flow Code

To test the rationality of the crack tolerance as an evaluation index for rock fracability, we adopt particle flow code in two dimensions (PFC^{2D}), which is widely used for discrete element method (DEM). Rock was modeled as a dense packing of non-uniform-sized and inter-bonded circular particles using this method, and its mechanical behavior relied on the microscale properties and constitutive relations of the bonded contacts between the particles. Following Newton's laws of motion, the force acting at each contact were updated with the particle movements during the simulation process, and the breakage of bonds representing crack generation [28] occurred when a component of the contact force satisfied a certain criterion.

We used the experimental results for marble A and J as examples for the DEM simulation because the marble contained polygonal minerals >1 mm in size, which allowed us to implement a grain-based model (GBM) with a polygon-tessellation grain boundaries [29]. Such a model takes the mineral grain texture into account (Figure 8a), making a simulation more vivid. The modeling method for GBM of marble refers to [27]. Soft-bonded [30] and smooth-joint models [31] were employed to express bonded and unbonded behaviors characteristic of intra-grain and inter-grain contacts (Figure 8b–d), respectively.

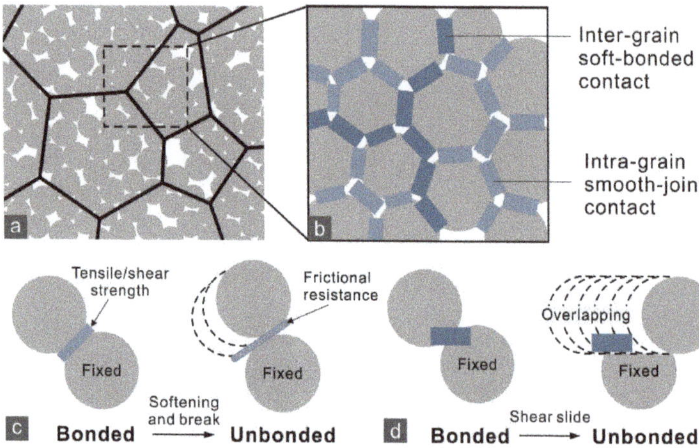

Figure 8. (**a**) Grain-based model for marble and (**b**) the implement of inter-grain soft-bonded and intra-grain smooth-joint contacts. Schematic of bonded and unbonded state for (**c**) soft-bonded and (**d**) smooth-joint contacts. (**b**) is the zoomed-in view of the dashed box in (**a**).

4.2. Model Setup and Parameter Calibration

Based on the grain size distributions in Table 1, four circular 75 mm GBMs were created representing the marble A and J specimens in the BD and CCNBD tests. Each model comprised ~20,000 circular basic particles with a 0.2–0.3 mm radius. Since even a single mineral crystal is anisotropic along different atomic lattices [32], we set the strength and deformation parameters of the soft-bonded contacts to follow the Weibull distribution, and the shape parameter representing heterogeneity was set to 3 and 5 for marble A and J, respectively. A small value for the shape parameter indicates strong heterogeneity.

Before conducting the simulation of the CCNBD tests, the microscale parameters of the particles and contacts required iteratively calibrating through trial and error referring to the BD test results and the previous work [27,33,34]. The GBMs of the BD specimens were positioned between two stiff walls representing the loading end and platform of a compression machine, and the walls moved toward each other at the same constant velocity to result in a quasi-static loading rate. The calibration completes until the simulated load–displacement curves and crack morphology fit well with the observations in the BD tests (Figure 9). The calibrated microscale parameters (Table 2) were used to simulate the marble specimens in the CCNBD test.

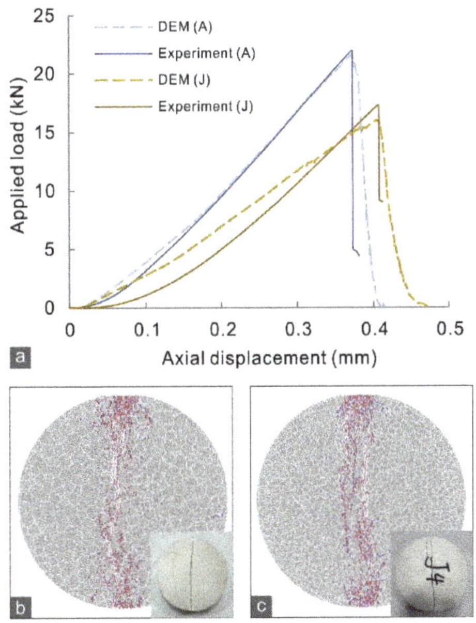

Figure 9. (**a**) Load–displacement relationship of experimental and numerical results for marble A and J in the BD test. (**b**,**c**) Crack morphologies of the marble A and J specimens in the numerical simulation of BD test and the corresponding experiment results.

Table 2. Calibrated microscale parameters of marble A and J specimens.

Elements	Microscale Parameters	Marble A	Marble J
Particles	Density (kg/m^3)	2690	2690
	Effective modulus (GPa)	70	70
	Normal to shear stiffness ratio	1.5	1.5
	Friction coefficient	0.6	0.6
Soft-bonded intra-grain contact	Effective modulus (GPa)	15.0	10.0
	Normal to shear stiffness ratio	2.0	2.0
	Friction coefficient	0.6	0.6
	Tensile strength (MPa)	28.0	18.0
	Cohesion (MPa)	112.0	72.0
	Friction angle (°)	45	45
	Softening factor	0.1	0.1
	Softening tensile strength factor	0.7	0.7
Smooth-joint inter-grain contact	Tensile strength coefficient	0.3	0.3
	Cohesion coefficient	0.8	0.8
	Friction angle coefficient	0.6	0.6
	Friction adjustment coefficient	0.5	0.5
	Normal stiffness coefficient	0.8	0.8
	Shear stiffness coefficient	0.15	0.15

4.3. Simulation Results

In the numerical simulations of the CCNBD test, when the applied load reached a certain level, micro-cracking was initiated near the notched crack tips of the specimens (Figures 10a and 11a). At the peak loads (P_{max}) of the marble A and J specimens, the micro-cracks around the crack tips tended to coalesce to form new macroscopic cracks (Figures 10b and 11b). After that, the notched crack propagated dramatically, which caused a rapid post-peak drop in the applied load and the specimen to rupture (Figures 10c and 11c). Therefore, the preparation stage for dramatic propagation of a notched crack can be defined as from the initiation of micro-cracking to the reaching of the peak load, during which micro-cracks generate to develop the FPZ. The FPZ is the area near the crack tip with a dense micro-crack distribution when the peak load is reached that stays in the critical state of macroscopic rupture. As mentioned in Section 2, the crack tolerance is characterized by the size of the FPZ.

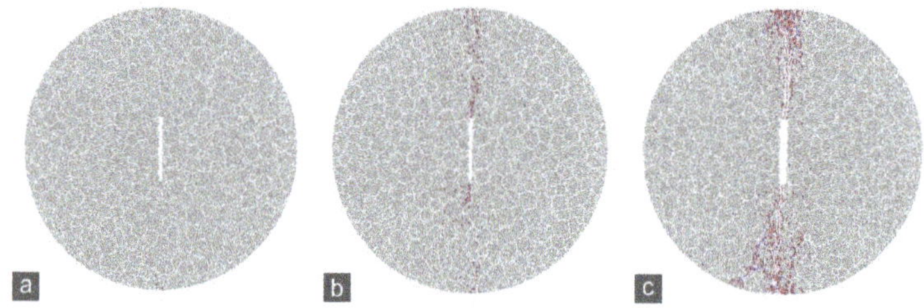

Figure 10. Simulated cracking process of marble A CCNBD specimen. (**a**) Cracking initiation at notched crack tips (~66% P_{max}); (**b**) reaching P_{max}; (**c**) specimen rupture.

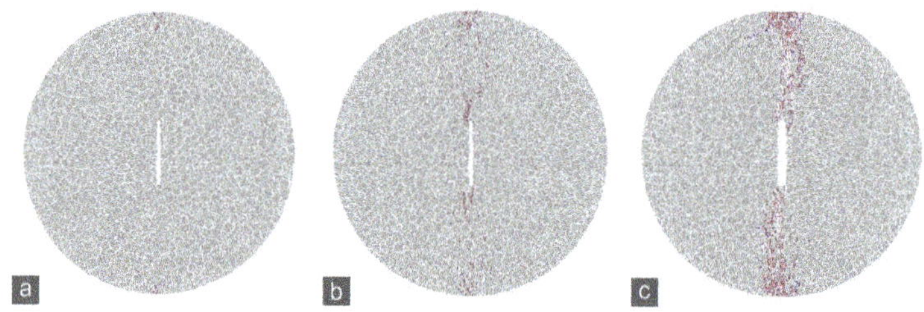

Figure 11. Simulated cracking process of marble J CCNBD specimen. (**a**) Cracking initiation at notched crack tips (~70% P_{max}); (**b**) reaching P_{max}; (**c**) specimen rupture.

As the simulation results (Figures 10b and 11b) showed, the micro-crack density at the vicinity of notched crack tip was especially high in the whole specimen, owing to the nonlinear deformation brought by stress concentration. The FPZ was assumed as a tip-centered circle with radius of r_c [14] that covered the area with high micro-crack density. Therefore, with increasing distance (radius) R_f from the tip and the diminishing intensity of the stress concentration, the deformation transitions from nonlinear inside the FPZ to quasilinear outside the FPZ, and thus the micro-crack density outside the FPZ declined to the background density of the rock [35].

To simplify the analysis, we assumed that the distributions of micro-crack inside and outside the FPZ are uniform but have different density. On the basis of this, the total micro-crack number N within a certain circular statistical range with R_f radius can be formulated as

$$N(R) = \begin{cases} \pi, \mid R_f < r_c \\ \pi {r_c}^2(d - d_0) + \pi R_f^2 d_0, \mid R_f \geq r_c \end{cases} \quad (7)$$

where d and d_0 are the average micro-crack density inside and outside the FPZ, respectively. N displayed a positively correlation with R_f; however, the curves of N deflected when R_f increased to r_c that defined the boundary of the FPZ (Figure 12). This is because the micro-crack density d inside the FPZ can be up to ~15 times as great as the background density d_0 [36]; the increasing rate of N will decelerate once the statistical range extend outside the FPZ. Such a deflection became more identifiable with the increasing ratio of d/d_0. Thus, the crack tolerance of the specimens can be determined by the radius corresponding to the deflection point.

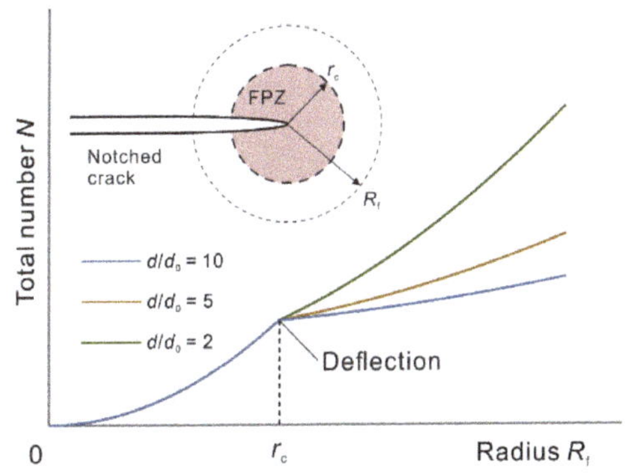

Figure 12. Variation trend of total micro-crack number inside tip-centered circles with various radii R_f. The three curves correspond to cases under different ratios of d/d_0.

As the above calculation predicts, the micro-crack number N in Figures 10b and 11b for the marble A and J specimens increased with R_f, and the N–R_f curves deflected at radii of 7 and 4 mm, respectively (Figure 13), which were closed to the mean values of the crack tolerance of A (~8 mm) and J (~3 mm) measured by experiments. These results demonstrate that calculating the crack tolerance using the tensile fracture toughness and average tensile strength in Equation (3) leads to reliable results. The simulations also showed that marble A had a greater crack tolerance than marble J, and the FPZ of the former contained more micro-cracks than that of the latter preceding specimen rupture (Figure 13c,d). Correspondingly, the GBM of marble A generates 1473 micro-cracks in the whole loading process, more than that of marble J (1404). These results showed that a greater crack tolerance can represent a stronger ability to generate micro-cracks.

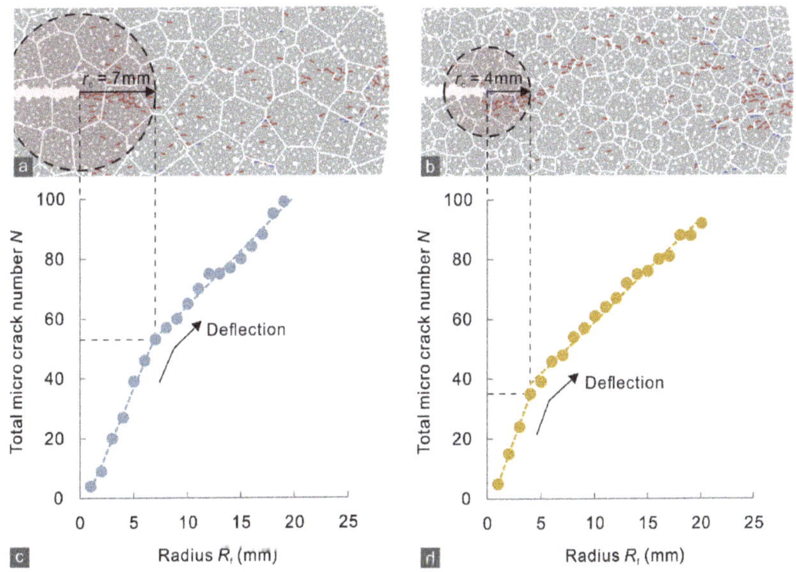

Figure 13. Micro-crack distributions near the notched crack tips in the (**a**) marble A and (**b**) J specimens at the peak applied load in the CCNBD simulation. (**c**,**d**) Total number of micro-cracks in tip-centered circles with different radii. The red and blue curves in (**a**,**b**) indicate intra- and inter-grain micro-cracks. The dashed lines in (**c**,**d**) are fitting lines for the distribution of the solid circles.

The strike angle of micro-cracks (Figure 14) for the marble J mainly distributed in the range of 70–100°, which was narrower than that, 60–110°, for the marble A. This result suggests that more micro-cracks deviated from the loading direction (90°) in the marble A. The coalescence of such micro-cracks with various strike angles resulted in macroscopic cracks that propagated along irregular even branched paths and radiated from the notched crack tips (Figures 5 and 10). Otherwise, the macro-cracks will develop primarily parallel to the loading direction, and thus their morphologies were less complex, as the marble J specimen showed. Therefore, these observations confirmed the assumption that fracability can be characterized by the crack tolerance.

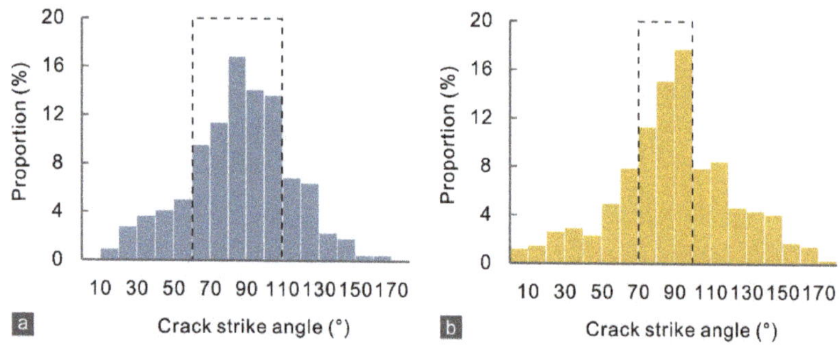

Figure 14. Strike angle distribution of micro-cracks in the CCNBD simulations for (**a**) marble A and (**b**) J, corresponding to Figures 10b and 11b. The dashed boxes define the main distribution ranges.

5. Discussion

5.1. Effect of Grain Size on Crack Tolerance

The grain size greatly influences the cracking behavior of rocks [37]. Regarding rock consisting of grains with various size, the grain size greatly differs across parts of the rock specimen, so does the microscopic strength, which enhances the rock heterogeneity. This is why the heterogeneity index (shape parameter for the Weibull distribution) of marble A was set as smaller than that for marble J in the GBMs.

The microscopic strength in different parts of a strongly heterogeneous rock specimen can distribute in a wide range, so a small increment in the stress near a crack tip can easily cause micro-cracking within such rocks. Therefore, the initiation of micro-cracking was earlier, i.e., corresponding to a smaller ratio between the applied load and the peak load, in marble A than in marble J (Figures 10 and 11). However, cracking can also be arrested easily because it probably encounters stronger local parts soon. Therefore, the rupture of strongly heterogeneous specimens will not occur until there are adequate micro-cracks to create FPZs and macroscopic cracks. In summary, strong heterogeneity strengthens the crack tolerance of rocks.

For rock specimens with a homogeneous grain size distribution, its microscopic strength in different parts can be generally closed to a certain level. Thus, only a few micro-cracks arise before the stress near the notched crack tips reaches that strength level. Once the strength is reached, the cracks propagate dramatically, which causes a rapid rupture. Macroscopic cracks spanning across the specimen form nearly instantaneously following the generation of a small FPZ. For these reasons, relatively homogeneous marble J had a smaller crack tolerance than marble A.

5.2. Effects of the Bedding Orientation and Environmental Temperature

With a vertical bedding orientation, micro-cracking naturally initiates within bedding planes in front of crack tips and propagates along them because the tensile strength of shale bedding planes is usually much smaller than that of layers between planes. Therefore, the tensile strength and fracture toughness were relatively low. Since the micro-cracks are limited to thin bedding planes, the corresponding crack tolerance is also small. With a horizontal bedding orientation, layers comprising various mineral grains make the distribution of the microscopic tensile strength near each tip more heterogeneous, so micro-cracking may be scattered among the layers. The behavior of micro-cracks can be complex when they cross bedding planes: cracks may branch along bedding planes and result in curved macroscopic cracks. Although hydraulic fracturing involves with many factors affecting the interaction between cracks and bedding planes [38], it is generally recognized that a crack propagating along a bedding plane is the most unfavorable situation for generating complex fracture networks [39,40].

Thermal treatment of sandstone leads to dehydration and the thermal expansion of minerals, which promotes the brittle–ductile transition of minerals [41]. Dehydration occurs at ~100 °C, which is when absorbed water escapes from the mineral surface [42], and enhances the friction and bonding strength between minerals. Thermal expansion takes effect when the temperature exceeds 100 °C and closes preexisting micro-cracks in rocks [43], which enhances the tensile strength of mineral grains and boundaries within a certain temperature range. Owing to these effects arising exceeding 100 °C, the sandstone at 125 °C had higher tensile strength and fracture toughness than that at 75 °C (Figure 4c). However, the dehydration and thermal expansion of minerals reduce the structural heterogeneity and rock fracability. In addition, fracability also weakens when sandstone transitions from brittle to ductile [44]. These observations suggest that increasing the temperature from 25 °C to 125 °C should be unfavorable for crack generation in sandstone. Such a negative effect of temperature was also observed in Longmaxi shale [45], a commercial shale gas reservoir in Chongqing, China. Considering that the downhole temperature, especially in deep and geothermal wells, can be much higher than the surface temperature, more attention should be given to the effect of temperature over a broader range on rock fracability.

5.3. *Implications in Hydraulic Fracturing*

The FPZ indicates nonlinear deformation (i.e., micro-crack generation) within a rock, which originates from rock heterogeneity. Nonlinear deformation diminishes and transitions into linear deformation with decreasing rock heterogeneity, which would reduce the FPZ size (i.e., crack tolerance). Complex cracks barely form in the absence of micro-cracks and their interactions. Because the formation of the FPZ is intrinsic to heterogeneous rock, the association between the FPZ size and heterogeneity is the physical basis for the crack tolerance as an evaluation index of fracability. The crack tolerance can reflect the effects of structural and environmental factors because they influence rock heterogeneity [46,47].

A large crack tolerance indicates many micro-cracks within the FPZ, which would tend to cause an irregular morphology of macroscopic cracks and wider zones of micro-cracks along both sides of cracks. These characteristics allow more cracks and pores in the rocks to connect with the main cracks during crack propagation driven by fluid. This increases the volume of cracks, which enhances the fluid conductivity of rock to exploit oil, gas, and geothermal resources.

The present study mainly provided laboratory observations on the effects of three factors to support the rationality of crack tolerance in reservoir fracability evaluation. However, the complexity of crack networks is dependent on various factors. Further investigations that consider other environmental and structural effects (e.g., magnitude and direction of crustal stress) are required to test this fracability index. On the other hand, the reliability and practical applicability of this index need to be further tested using rock specimens collected from reservoirs being exploited.

6. Conclusions

Aiming to the challenge of lacking a reliable index of fracability evaluation for hydraulic fracturing, we suggest that the crack tolerance, i.e., the maximum radius of the FPZ, may be used to characterize the fracability.

Crack tolerance originates from the unique heterogeneity of rock and inherent rules of crack generation, and thus it has a clear physical meaning and firm mechanical basis.

This index can be conveniently quantified in the laboratory using BD and CCNBD tests. We showed that the crack tolerance is positively correlated with the grain size heterogeneity and is negatively correlated with the environmental temperature (25–125 °C). The crack tolerance of shale specimens was greater with a horizontal bedding orientation than with a vertical orientation. Thus, crack tolerance can well characterize the effects of certain rock properties and environmental factors.

In summary, crack tolerance is promising in serving as a reliable evaluation index of rock fracability in terms of rationality and practicability, which has significant engineering implications for efficient hydraulic fracturing.

Author Contributions: Conceptualization, J.N. and H.C.; methodology, J.N., M.Z. and H.C.; validation, J.N.; formal analysis, H.C.; investigation, H.C., M.Z. and J.N.; resources, J.N.; writing—original draft preparation, H.C.; writing—review and editing, J.N., M.Z. and H.C.; visualization, H.C. and M.Z.; funding acquisition, J.N. and H.C. All authors have read and agreed to the published version of the manuscript.

Funding: This research was funded by the National Natural Science Foundation of China, grant numbers 42107184 and 41602330.

Institutional Review Board Statement: Not applicable.

Informed Consent Statement: Not applicable.

Data Availability Statement: All the data required to evaluate the conclusions of this study are present in the paper. The authors will provide additional data related to this paper upon request.

Conflicts of Interest: The authors declare no conflict of interest.

Nomenclature

List of symbols

a_0	Initial chevron notched crack length
a_1	Final chevron notched crack length
ASTM	American Society of Testing Materials
B	Thickness of CCNBD specimen
B'	Thickness of BD specimen
BD	Brazilian disk
CCNBD	Cracked chevron notched Brazilian disk
d	Average micro-crack density inside FPZ
D	Diameter of CCNBD specimen
D'	Diameter of BD specimen
d_0	Average micro-crack density outside FPZ
DEM	Discrete element method
FPZ	Fracture process zone
GBM	Grain-based model
ISRM	International Society for Rock Mechanics
$K_{\rm I}$	Tensile stress intensity factor
$K_{\rm IC}$	Tensile fracture toughness
N	Total micro-crack number
PFC	Particle flow code
$P_{\rm max}$	Peak applied axial load in CCNBD test
$P'_{\rm max}$	Peak applied axial load in BD test
R	Radius of CCNBD specimen
r	Polar radius
$r_{\rm c}$	Radius of circular FPZ

R_f	Radius of circular statistical range
R_s	Saw radius
u	Geometric constants
v	Geometric constants
XRD	X-ray diffraction
Y^*_{min}	Critical dimensionless stress intensity
Greek symbols	
α_0	Dimensionless parameter
α_1	Dimensionless parameter
α_B	Dimensionless parameter
θ	Polar angle
σ_1	Maximum principal stress
σ_2	Intermediate principal stress
σ_t	Tensile strength

References

1. Mullen, M.J.; Enderlin, M.B. Fracability index—More than rock properties. In Proceedings of the SPE Annual Technical Conference and Exhibition, San Antonio, TX, USA, 8–10 October 2012; pp. 1–10.
2. Wang, S.; Li, X.; Yao, J.; Gong, F.; Li, X.; Du, K.; Tao, M.; Huang, L.; Du, S. Experimental investigation of rock breakage by a conical pick and its application to non-explosive mechanized mining in deep hard rock. *Int. J. Rock Mech. Min. Sci.* **2019**, *122*, 104063. [CrossRef]
3. Meng, F.; Wong, L.N.Y.; Zhou, H. Rock brittleness indices and their applications to different fields of rock engineering: A review. *J. Rock Mech. Geotech. Eng.* **2021**, *13*, 221–247. [CrossRef]
4. Zhang, D.; Ranjith, P.G.; Perera, M.S.A. The brittleness indices used in rock mechanics and their application in shale hydraulic fracturing: A review. *J. Pet. Sci. Eng.* **2016**, *143*, 158–170. [CrossRef]
5. Jarvie, D.M.; Hill, R.J.; Ruble, T.E.; Pollastro, R.M. Unconventional shale-gas systems: The Mississippian Barnett Shale of north-central Texas as one model for thermogenic shale-gas assessment. *AAPG Bull.* **2007**, *91*, 475–499. [CrossRef]
6. Rickman, R.; Mullen, M.J.; Petre, J.E.; Grieser, W.V.; Kundert, D. A practical use of shale petrophysics for stimulation design optimization: All shale plays are not clones of the Barnett shale. In Proceedings of the SPE Annual Technical Conference and Exhibition, Denver, CO, USA, 21–24 September 2008; pp. 1–9.
7. Bai, M. Why are brittleness and fracability not equivalent in designing hydraulic fracturing in tight shale gas reservoirs. *Petroleum* **2016**, *2*, 1–19. [CrossRef]
8. Nasehi, M.J.; Mortazavi, A. Effects of in-situ stress regime and intact rock strength parameters on the hydraulic fracturing. *J. Pet. Sci. Eng.* **2013**, *108*, 211–221. [CrossRef]
9. Hucka, V.; Das, B. Brittleness determination of rocks by different methods. *Int. J. Rock Mech. Min. Sci. Geomech. Abstr.* **1974**, *11*, 389–392. [CrossRef]
10. Meng, F.; Zhou, H.; Zhang, C.; Xu, R.; Lu, J. Evaluation methodology of brittleness of rock based on post-peak stress–strain curves. *Rock Mech. Rock Eng.* **2015**, *48*, 1787–1805. [CrossRef]
11. Xia, Y.; Zhou, H.; Zhang, C.; He, S.; Gao, Y.; Wang, P. The evaluation of rock brittleness and its application: A review study. *Eur. J. Environ. Civ. Eng.* **2022**, *26*, 239–279. [CrossRef]
12. Jin, X.; Shah, S.N.; Roegiers, J.-C.; Zhang, B. Fracability evaluation in shale reservoirs—An integrated petrophysics and geomechanics approach. In Proceedings of the SPE Hydraulic Fracturing Technology Conference, The Woodlands, TX, USA, 4–6 February 2014; pp. 1–14.
13. Wang, S.; Sun, L.; Li, X.; Zhou, J.; Du, K.; Wang, S.; Khandelwal, M. Experimental investigation and theoretical analysis of indentations on cuboid hard rock using a conical pick under uniaxial lateral stress. *Geomech. Geophys. Geo Energ. Geo Resour.* **2022**, *8*, 34. [CrossRef]
14. Atkinson, B.K. *Fracture Mechanics of Rock*; Academic Press: London, UK, 1987.
15. Lawn, B.R.; Wilshaw, T.R. *Fracture of Brittle Solids*; Cambridge University Press: New York, NY, USA, 1975.
16. Saleur, H.; Sammis, C.G.; Sornette, D. Discrete scale invariance, complex fractal dimensions, and log-periodic fluctuations in seismicity. *J. Geophys. Res. Solid Earth* **1996**, *101*, 17661–17677. [CrossRef]
17. Ghamgosar, M.; Bahaaddini, M.; Erarslan, N.; Williams, D.J. A new experimental approach to quantify microfractures in the Fracture Process Zone (FPZ) under various loading conditions. *Eng. Geol.* **2021**, *283*, 106024. [CrossRef]
18. Ghamgosar, M.; Erarslan, N.; Williams, D.J. Experimental investigation of fracture process zone in rocks damaged under cyclic loadings. *Exp. Mech* **2017**, *57*, 97–113. [CrossRef]
19. Guha Roy, D.; Singh, T.N.; Kodikara, J. Influence of joint anisotropy on the fracturing behavior of a sedimentary rock. *Eng. Geol.* **2017**, *228*, 224–237. [CrossRef]
20. Zietlow, W.K.; Labuz, J.F. Measurement of the intrinsic process zone in rock using acoustic emission. *Int. J. Rock Mech. Min. Sci.* **1998**, *35*, 291–299. [CrossRef]

21. Erarslan, N. Microstructural investigation of subcritical crack propagation and Fracture Process Zone (FPZ) by the reduction of rock fracture toughness under cyclic loading. *Eng. Geol.* **2016**, *208*, 181–190. [CrossRef]
22. Han, Z.; Zhang, L.; Azzam, R.; Zhou, J.; Wang, S. A statistical index indicating the degree and mechanical effects of grain size heterogeneity in rocks. *Eng. Geol.* **2021**, *293*, 106292. [CrossRef]
23. Fowell, R.J.; Xu, C.; Dowd, P.A. An update on the fracture toughness testing methods related to the cracked chevron-notched Brazilian disk (CCNBD) specimen. *Pure Appl. Geophys.* **2006**, *163*, 1047–1057. [CrossRef]
24. American Society of Testing Materials. *Standard Test Method for Splitting Tensile Strength of Intact Rock Core Specimens*; American Society of Testing Materials: West Conshohocken, PA, USA, 2016; pp. 1–5. [CrossRef]
25. Wong, L.N.Y.; Einstein, H.H. Crack coalescence in molded gypsum and Carrara marble: Part 2—Microscopic observations and interpretation. *Rock Mech. Rock Eng.* **2009**, *42*, 513–545. [CrossRef]
26. Wang, H. Experimental Study of Shale Rock Mechanical Properties and Volume Fracturing. Ph.D. Thesis, University of Chinese Academy of Sciences, Beijing, China, 2015.
27. Chen, H.; Niu, J.; Zhai, M. Characteristics of the fracture process zone for reservoir rock with various heterogeneity. *Energies* **2022**, *15*, 8332. [CrossRef]
28. Potyondy, D.O.; Cundall, P.A. A bonded-particle model for rock. *Int. J. Rock Mech. Min. Sci.* **2004**, *41*, 1329–1364. [CrossRef]
29. Potyondy, D.O. A grain-based model for rock: Approaching the true microstructure. In Proceedings of the Rock Mechanics in the Nordic Countries 2010, Kongsberg, Norway, 9–12 June 2010; pp. 225–234.
30. Itasca Consulting Group Inc. *PFC2D—Particle Flow Code in 2 Dimensions*; Itasca Consulting Group Inc: Minneapolis, MN, USA, 2019.
31. Pierce, M.; Cundall, P.; Potyondy, D.; Ivars, D.M. A synthetic rock mass model for jointed rock. In Proceedings of the 1st Canada—U.S. Rock Mechanics Symposium, Vancouver, BC, Canada, 27–31 May 2007.
32. Whitney, D.L.; Broz, M.; Cook, R.F. Hardness, toughness, and modulus of some common metamorphic minerals. *Am. Miner.* **2007**, *92*, 281–288. [CrossRef]
33. Wang, S.; Zhou, J.; Zhang, L.; Han, Z.; Zhang, F. Parameter studies on the mineral boundary strength influencing the fracturing of the crystalline rock based on a novel Grain-Based Model. *Eng. Fract. Mech.* **2021**, *241*, 107388. [CrossRef]
34. Zhou, J.; Lan, H.; Zhang, L.; Yang, D.; Song, J.; Wang, S. Novel grain-based model for simulation of brittle failure of Alxa porphyritic granite. *Eng. Geol.* **2019**, *251*, 100–114. [CrossRef]
35. Moore, D.E.; Lockner, D.A. The role of microcracking in shear-fracture propagation in granite. *J. Struct. Geol.* **1995**, *17*, 95–114. [CrossRef]
36. Janssen, C.; Wagner, F.; Zang, A.; Dresen, G. Fracture process zone in granite: A microstructural analysis. *Int. J. Earth. Sci.* **2001**, *90*, 46–59. [CrossRef]
37. Eberhardt, E.; Stimpson, B.; Stead, D. Effects of grain size on the initiation and propagation thresholds of stress-induced brittle fractures. *Rock Mech. Rock Eng.* **1999**, *32*, 81–99. [CrossRef]
38. Zhang, Q.; Zhang, X.; Sun, W. A review of laboratory studies and theoretical analysis for the interaction mode between induced hydraulic fractures and pre-existing fractures. *J. Nat. Gas. Sci. Eng.* **2021**, *86*, 103719. [CrossRef]
39. Lin, C.; He, J.; Li, X.; Wan, X.; Zheng, B. An experimental investigation into the effects of the anisotropy of shale on hydraulic fracture propagation. *Rock Mech. Rock Eng.* **2017**, *50*, 543–554. [CrossRef]
40. Zhao, Y.; Zhang, Y.; Wang, C.; Liu, Q. Hydraulic fracturing characteristics and evaluation of fracturing effectiveness under different anisotropic angles and injection rates: An experimental investigation in absence of confining pressure. *J. Nat. Gas. Sci. Eng.* **2022**, *97*, 104343. [CrossRef]
41. Sha, S.; Rong, G.; Tan, J.; He, R.; Li, B. Tensile strength and brittleness of sandstone and granite after high-temperature treatment: A review. *Arab. J. Geosci.* **2020**, *13*, 598. [CrossRef]
42. Zhang, W.; Sun, Q.; Hao, S.; Geng, J.; Lv, C. Experimental study on the variation of physical and mechanical properties of rock after high temperature treatment. *Appl. Eng.* **2016**, *98*, 1297–1304. [CrossRef]
43. Sirdesai, N.N.; Mahanta, B.; Ranjith, P.G.; Singh, T.N. Effects of thermal treatment on physico-morphological properties of Indian fine-grained sandstone. *Bull. Eng. Geol. Env.* **2019**, *78*, 883–897. [CrossRef]
44. Zuo, J.; Zhou, H.; Xie, H. Fracture characteristics of sandstone under thermal effects. *Eng. Mech.* **2008**, *25*, 124–130.
45. Niu, J. Research on the Fracture Mechanics Index of Rock Fracability Evaluation. Ph.D. Thesis, University of Chinese Academy of Sciences, Beijing, China, 2015.
46. Chen, B.; Barboza, B.R.; Sun, Y.; Bai, J.; Thomas, H.R.; Dutko, M.; Cottrell, M.; Li, C. A review of hydraulic fracturing simulation. *Arch. Comput. Method E.* **2022**, *29*, 1–58. [CrossRef]
47. Liu, P.; Ju, Y.; Gao, F.; Ranjith, P.G.; Zhang, Q. CT identification and fractal characterization of 3-D propagation and distribution of hydrofracturing cracks in low-permeability heterogeneous rocks. *J. Geophys. Res. Sol. Ea.* **2018**, *123*, 2156–2173. [CrossRef]

materials

Article

The Release of Pollutants through the Bleeding of Cemented Phosphogypsum Backfill: Link to Protocols for Slurry Preparation

Chendi Min, Ying Shi *, Yanan Zhou and Zhixiang Liu

School of Resources and Safety Engineering, Central South University, Changsha 410083, China
* Correspondence: shiyingfriend@csu.edu.cn

Abstract: The present study investigated the effects of protocols for slurry preparation on the release of pollutants into bleeding water from cemented phosphogypsum (PG) backfill. Backfill slurry was prepared using four different protocols in which different parameters varied, including binder/PG ratio, solid concentration, binder type and mixing procedure. The concentrations of phosphate, fluoride and sulfate and the pH values of the obtained bleeding water were measured. The results demonstrated that the slurry preparation protocols affected the quantities of pollutants through the concentrations of pollutants in bleeding water and the bleeding rate. On the one hand, the binder/PG ratio was the key factor influencing the concentrations of all pollutants in bleeding water. Comparatively speaking, the binder type and mixing procedure had an obvious influence on the fluoride concentration but had little influence on the phosphate and sulfate concentrations in the bleeding water. On the other hand, the protocols for slurry preparation affected the bleeding rate by determining the water retention and water content of the backfill slurry. The most effective protocol for slurry preparation for cemented PG backfill could reduce the bleeding rate and enhance the immobilization of pollutants, minimizing the phosphate concentration in bleeding water to below 0.2 mg/L. However, it appeared that the fluoride concentration was still tens of milligrams per liter (over the limit of 10 ten milligrams per liter), to which attention should be paid.

Keywords: bleeding water; phosphogypsum; cemented backfill; pollutant; backfill slurry

Citation: Min, C.; Shi, Y.; Zhou, Y.; Liu, Z. The Release of Pollutants through the Bleeding of Cemented Phosphogypsum Backfill: Link to Protocols for Slurry Preparation. *Materials* **2022**, *15*, 7126. https://doi.org/10.3390/ma15207126

Academic Editor: Geo Paul

Received: 30 August 2022
Accepted: 7 October 2022
Published: 13 October 2022

1. Introduction

Cemented backfill is the most efficient approach for recycling the solid waste produced in mining and mineral processes and for reducing rock instabilities caused by high excavation-induced stress [1–4]. In cemented backfill, backfill slurry is prepared by mixing the aggregate (solid wastes), binder and water in certain proportions. After being mixed homogeneously, the slurry is transported to the stopes through the pipeline. To promote fluidity, excessive water is usually added when preparing backfill slurry [5]. As a result, the water in the backfill slurry can be characterized as free water and water on the particle surface layer [6]. When the slurry is placed in the stopes, the free water gradually separates from the backfill slurry, accumulating as bleeding water [7]. Previous studies showed that the bleeding property is widespread in backfill slurry [8,9]. However, due to the use of solid waste as raw material, the bleeding water derived from the backfill slurry might contain large quantities of pollutants, thus threatening the surrounding environment. Clearly, the bleeding of backfill is an important pathway for the release of pollutants, which is often overlooked.

As an efficient method for phosphogypsum (PG) management, cemented PG backfill is applied in several mines in China, including the Kaiyang phosphate mine, the Jinchuan nickel mine and the Huangmailing phosphate mine [2,10–12]. PG is a type of solid waste produced via phosphoric acid production, which can be simplified with the following equation [13,14]:

$$Ca_5(PO_4)_3F + 5H_2SO_4 + 10H_2O \rightarrow 3H_3PO_4 + 5CaSO_4(H_2O)_2 + HF + heat \quad (1)$$

According to Equation (1), calcium sulfate dihydrate ($CaSO_4{\cdot}2H_2O$) accounts for more than 90% of PG and the main impurities in PG include H_3PO_4, HF and unreacted H_2SO_4 [15]. Due to the strong demand for phosphoric acid in the fertilizer industry, the global production of PG is huge, especially in the main phosphate rock-producing countries, including China, the USA and Morocco [16]. For instance, the annual PG production in China, the USA and Morocco is approximately 75 million tons, 40 million tons and 15 million tons, respectively [17–19]. Due to the lack of a large-scale utilization method, only 15% of the produced PG is reused to produce agricultural fertilizers [20], set retarders for cement [21,22], soil amendments [23], plaster [24,25] and cementitious materials [26]. The remaining 85% of PG is directly stockpiled, resulting in serious environmental damage to soil, water and the atmosphere [27]. On the other hand, increasingly stringent environmental regulations in China result in the high maintenance costs of stockpiled land, reducing the profitability of the phosphate fertilizer industry. It was reported that cemented PG backfill could utilize 60% of the produced PG [2], which could greatly reduce the costs of PG management. As a waste-based material, the pollutant solidification/stabilization (S/S) efficiency of cemented PG backfill raises concerns. An SPLP test on cemented PG backfill showed that the heavy metals in leachate were within the limit values of Chinese standard DZ/T 0290-2015 [2]. A three-step sequential extraction test indicated that cemented PG backfill effectively solidified/stabilized the metals in PG [28]. The phosphate in the leachate of cemented PG backfill was also efficiently immobilized [29]. However, previous studies mainly focused on leachate of hardened cemented PG backfill rather than on bleeding water.

The release of pollutants through bleeding water could be very different from that of hardened backfill. The bleeding phenomenon occurs immediately after the backfill slurry is placed into the stopes before the S/S process is fully effective. Thus, compared with leachate of hardened backfill, bleeding water could pose a high risk for its release of pollutants. The contents of phosphate, fluoride and sulfate in PG are rather high; active PO_4^{3-}, F^- and SO_4^{2-} could be the main pollutants in bleeding water. Moreover, due to the specific physiochemical properties of the aggregate PG, the water content in cemented PG backfill is rather high to obtain the desired slurry fluidity. Backfill slurry is usually prepared with a solid concentration of 40–65% [2,30,31]. Previous studies showed that the slurry of cemented PG backfill has a relatively high bleeding rate [29,31,32]. For instance, Zhou et al. showed that the bleeding rate was 25–51%, depending on the content of soluble phosphate in PG [29]. The high bleeding rate could further increase the damage of bleeding water to the surrounding environment. Figure 1 illustrates how the pollutants are released into the surrounding environment via bleeding water. The backfill slurry is prepared and firstly transported to the underground stopes. Then, the backfill slurry bleeds and solidifies, providing strength for mining. For cemented backfill using active waste as aggregate such as PG, part of the pollutants in PG are immobilized in the inner backfill matrix via the hydration of binder, and the rest migrate from the backfill slurry to the bleeding water. The massive bleeding water can be drained and enter the groundwater environment. As a result, the pollution caused by cemented PG backfill via bleeding water is worthy of attention.

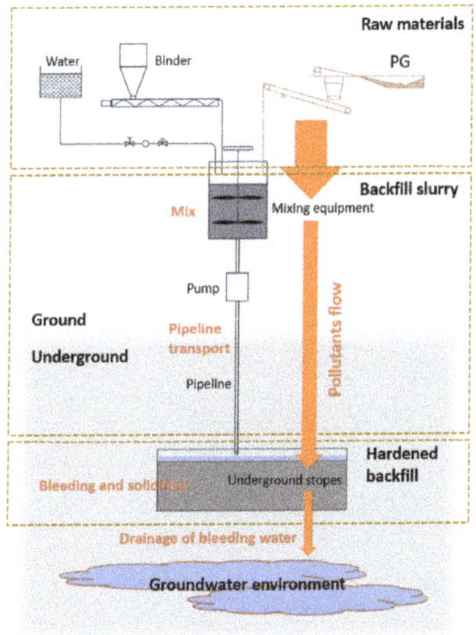

Figure 1. How pollutants are released into the surrounding environment via bleeding water.

It is widely known that the protocols for slurry preparation (binder/aggregate ratio, solid concentration, etc.) have an effect on the strength of hardened backfill. Moreover, previous studies indicated that the bleeding property also varies with the protocols for slurry preparation, which could affect the release of pollutants. Wang et al. found that the bleeding rate of backfill slurry with a low solid concentration of 74% was 20.44% higher than that obtained with a solid concentration of 82% [33]. Furthermore, the binder type could be a factor affecting the bleeding rate. Yao et al. showed that when the fly ash content in the binder was increased from 0 to 20%, the bleeding rate of the backfill slurry increased from 2.45 to 6.07% [34]. In addition, it was also observed that water-reducing agents could lead to a change in the bleeding rate [35]. However, previous studies only used the bleeding rate as an index to evaluate the water retention and stability of the slurry. There are few studies considering the environmental pollution caused by bleeding water derived from slurry prepared using different protocols.

The present study aimed to investigate the effect of different protocols for slurry preparation on the release of pollutants from cemented PG backfill slurry into the bleeding water. Backfill slurry was prepared using different binder/PG ratios, solid concentrations, binder types and mixing procedures. The bleeding rate was tested and the bleeding water was sampled. The release of pollutants was characterized based on the concentrations of PO_4^{3-}, F^- and SO_4^{2-} in bleeding water and the quantities of pollutants released per ton of solid in backfill. This study enhanced the understanding of the environmental damage caused by the bleeding of cemented PG backfill.

2. Materials and Methods

2.1. Materials

PG was collected from a backfill plant in Guizhou, China. Considering that the pH value of PG is in the range of 1–7, three batches of PG with pH values of 1.68, 4.15 and 6.75 (measured by mixing PG and deionized water in a 1:1 ratio) were collected in this study. PG was used as an aggregate for cemented backfill. A composite cementitious agent (CCA; prepared with phosphorous slag, fly ash cement clinker and slaked lime [36]), 32.5R composite Portland cement (CPC) and S95 ground granulated blast furnace slag (GGBFS) were used as binders in this study. Figure 2 and Table 1 show the particle size distribution and physical properties of the raw materials, respectively, and indicate that the three batches of PG were coarser than the binders. In particular, GGBFS had the finest particle size. Table 2 and Figure 3 show the main chemical components (measured using X-ray fluorescence; Bruker, Billerica, MA, USA) and crystal phases (measured using advance D8 X-ray automatic diffractometer; Bruker, Billerica, MA, USA) of the PG samples, respectively. The results indicated that the main component of PG was gypsum. In addition, it was observed that the PG sample with a low pH value contained more P_2O_5 and F.

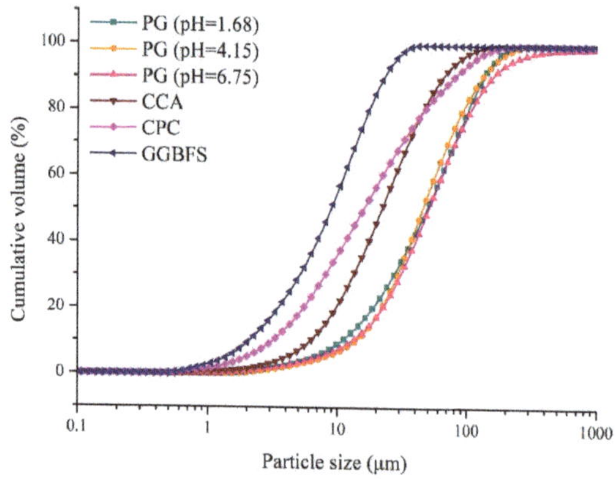

Figure 2. Particle size distribution of PG, CCA, PC and GGBFS [30,37].

Table 1. Physical properties of PG, CCA, PC and GGBFS.

Sample	D10 (μm)	D30 (μm)	D60 (μm)	Cu	Cc
PG (pH = 1.68)	10.53	27.40	62.91	6.08	1.13
PG (pH = 4.15)	13.65	29.02	56.38	4.13	1.09
PG (pH = 6.75)	13.56	33.15	62.76	4.63	1.29
CCA	6.12	13.26	27.74	4.53	1.04
CPC	3.33	8.14	22.60	6.79	0.88
GGBFS	2.00	5.55	11.94	5.97	1.29

Table 2. Main chemical components of the three batches of PG.

Chemical Component	PG (pH = 1.68) (%)	PG (pH = 4.15) (%)	PG (pH = 6.75) (%)
SO_3	44.76	42.19	42.05
CaO	34.93	34.04	34.01
SiO_2	3.23	2.73	2.31
P_2O_5	2.61	0.80	0.65
F	0.81	0.43	0.35
Fe_2O_3	0.28	0.25	0.27
Ba	0.17	0.17	0.09
MgO	0.14	0.04	0.05
Na_2O	0.13	0.03	-
K_2O	0.10	0.08	0.05

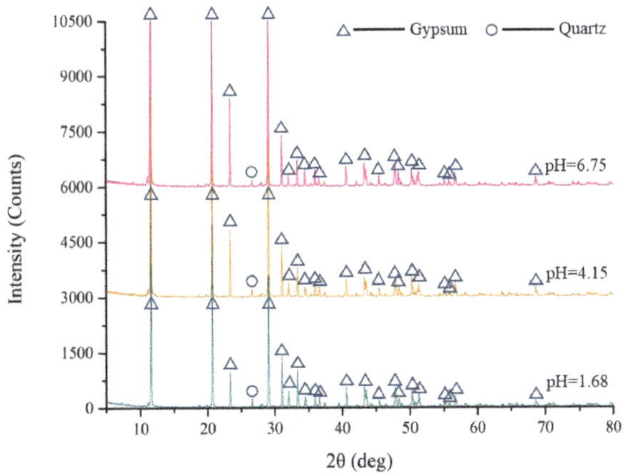

Figure 3. X-ray diffraction analysis of PG.

2.2. *Methods*

2.2.1. Preparation of Backfill Slurry

Batch A: To investigate bleeding water from backfill slurry with different binder/PG ratios, the binder/PG ratios were set to 1:2, 1:3, 1:4; 1:5, and 1:6. PG, binder and tap water were mixed for 30 min at a constant solid concentration of 50%. The PG with a pH value of 1.68 and CCA were used as aggregate and binder, respectively.

Batch B: To obtain bleeding water from the backfill slurry with different solid concentrations, the backfill slurry was prepared at solid concentrations of 40, 45, 48, 50, 52, 55, 58 and 60% with a mixing time of 30 min. The binder/PG ratio was 1:4. The PG with a pH value of 4.15 and CCA were used as aggregate and binder, respectively.

Batch C: This batch was used to obtain bleeding water from backfill slurry with different binder types. GGBFS was used to partially substitute PC as the binder. The GGBFS replacement proportions were 0, 20, 40, 60 and 90%, and the mixing time was 30 min. The binder/PG ratio was 15:85, and the solid concentration of backfill slurry was 60%. The PG with a pH of 6.75 was used as the aggregate.

Batch D: To obtain bleeding water from backfill slurry using different mixing times, the slurry was sampled after mixing times of 5, 15, 30, 45, 60, 120, and 240 min. The binder/PG ratio was 1:4, and the solid concentration was set to 65%. The PG with a pH value of 4.15 and CCA were used as backfill materials, respectively.

2.2.2. Toxic Leaching Test

The toxic leaching test was conducted to evaluate the quantities of impurities in the PG used in this study according to Chinese standard HJ 557-2010 [38]. The three batches of PG (dry mass) and deionized water were mixed in a ratio of 1:10. The mixture was put on a horizontal shaker to shake at 110 rpm/min for 8 h and then was moved to a table for 16 h. The supernatant was collected and filtered through a 0.45 mm filter for further analysis. The concentrations of PO_4^{3-} (as mg/L P) and SO_4^{2-} in bleeding water were measured with a spectrophotometer (Shimadzu, Japan). The concentration of F^- was measured using a fluorine ion-selective electrode (Leici, Shanghai, China). The pH and total dissolved solids (TDS) were tested using a pH meter (Ohaus, Parsippany, NJ, USA) and a TDS meter (Ohaus, Parsippany, NJ, USA), respectively.

2.2.3. Bleeding Rate

The bleeding rate test refers to Chinese standard GB/T 50080-2016 [39]. The backfill slurry was poured into a container with lid after mixing for the designed time. The supernatant water was aspirated using an injector at 10 min intervals in the first hour and at 30 min intervals thereafter until no bleeding occurred. The bleeding rate was calculated using Equation (2), where B is the bleeding rate, V_b is the volume of bleeding water, W is the mass of water used to prepare the backfill slurry, G_1 is the mass of backfill slurry and G_2 is the mass of backfill slurry in the container.

$$B = \frac{V_b}{(W/G_1)G_2} \times 100 \tag{2}$$

2.2.4. Sampling and Chemical Analyses of Bleeding Water

Backfill slurry was collected after the designed mixing times. Homogeneous backfill slurry was injected into the tubes of a centrifuge. The configuration of the centrifuge was set to a rotating speed of 4000× g r/min and a rotating time of 2 min. Supernatant water was collected and passed through a 0.45 μm filter for chemical analyses. The pH and the concentrations of PO_4^{3-}, F^- and SO_4^{2-} in bleeding water were measured according to the methods given in Section 2.2.2. The pH of bleeding water reflected the pH of backfill slurry.

2.2.5. Quantities of Pollutants Released Per Ton of Solid in Backfill

The quantities of pollutants released in bleeding water per ton of solid were used to evaluate the migration of pollutants from backfill slurry into the bleeding water. The quantities of pollutants released per ton of solid were calculated using Equation (3), where Q is the quantity of a pollutant released per ton of solid, C_p is the concentration of a pollutant in bleeding water, B is the bleeding rate of backfill slurry, W is the mass of water used to prepare backfill slurry and M_s is the mass of the solid in backfill slurry.

$$Q = \frac{C_p(B \times W)}{M_s} \tag{3}$$

3. Results

3.1. Chemical Properties of PG

Figure 4 shows the concentrations of pollutants in the leachate of PG with different pH values. Overall, the pH appeared to be inversely correlated with the TDS value and the concentrations of PO_4^{3-}, F^- and SO_4^{2-} in the leachate of PG, indicating that the PG with a low pH value contained more soluble ions. As the pH of PG increased from 1.68 to 6.75, the concentrations of PO_4^{3-}, F^- and SO_4^{2-} in leachate decreased by 99.5, 98.5 and 79.2%, respectively. Nevertheless, in the case of the PG with a pH of 6.75, the concentration of PO_4^{3-} was still 29.6 times higher than the limit value (0.5 mg/L) according to the Chinese standard GB 8978-1996 [40]. The results indicated that low pH was closely related to high quantities of soluble pollutants, while high pH values meant fewer pollutants in PG.

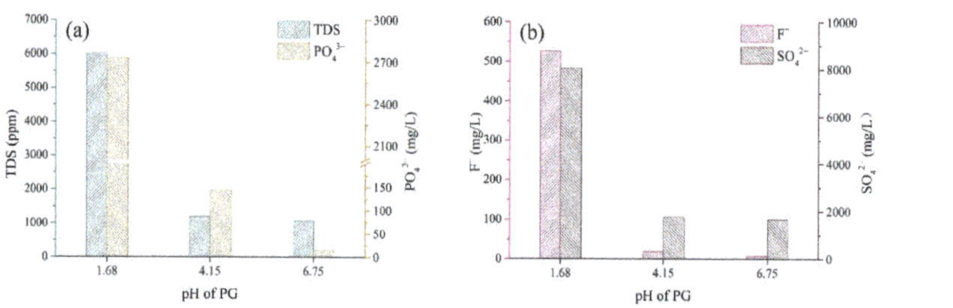

Figure 4. Concentrations of pollutants in leachate of PG with different pH values. (**a**) TDS and PO_4^{3-}; (**b**) F^- and SO_4^{2-}.

3.2. Binder/PG Ratio

(1) pH and bleeding rate of backfill slurry

The binder/PG ratio is a factor determining the strength of hardened backfill. In cemented PG backfill, the binder/PG ratio could also affect the pH and bleeding properties of the backfill slurry. Figure 5a shows the variation in the pH of backfill slurry according to different binder/PG ratios. When the binder/PG ratio varied from 1:2 to 1:4, the pH exhibited a drastic decrease from 11.89 to 7.28. Then, the pH further decreased to 6.86 with the reduction in the binder/PG ratio to 1:6. It can be seen that the presence of more alkali provided by the high content of binder could effectively neutralize the residual acids of PG, while a low content of binder led to low-pH backfill slurry. Because the hydration of the binder usually requires a high pH value, slurry with a binder/PG ratio above 1:3 could smoothly carry out hydration [41].

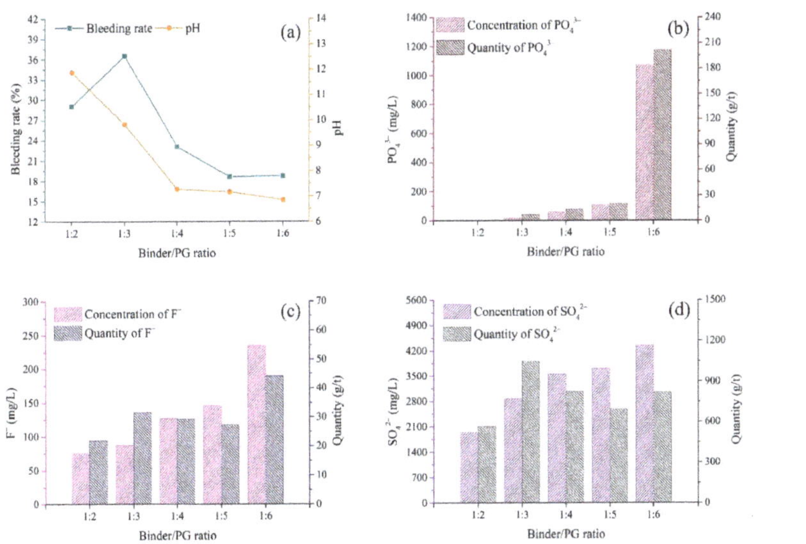

Figure 5. Variations in bleeding rate, pH and pollutants in bleeding water according to different binder/PG ratios of backfill slurry. (**a**) Bleeding rate and pH; (**b**) PO_4^{3-}; (**c**) F^-; (**d**) SO_4^{2-}.

The variation in the bleeding rate according to the binder/PG ratios is shown in Figure 5a. When the binder/PG ratio was 1:2, the bleeding rate was 29.0%, which was lower than that when the binder/PG ratio was 1:3. This could be explained by the water retention of the binder being better than that of PG, as this ratio (1:2) contained more binder with fine particles, as shown in Figure 2 [42,43]. When the binder/aggregate ratio varied from 1:3 to 1:6, the bleeding rate showed a gradual decrease from 36.5 to 18.76%, which could be associated with the pH of the backfill slurry. Previous studies showed that low pH means low zeta potential and repulsion force between the particles [44,45]. The poor fluidity induced by a low zeta potential may be related to the low bleeding rate [44,46]. These results indicated that the high amount of acids in the aggregate PG increased the water retention capability, thus hindering the bleeding of the backfill slurry.

(2) Pollutants in bleeding water

The variations in the concentrations of pollutants in bleeding water according to different binder/PG ratios are shown in Figure 5b–d. Overall, the concentrations of three typical pollutants (PO_4^{3-}, F^- and SO_4^{2-}) significantly increased with the decrease in binder content. In backfill slurry with a binder/PG ratio of 1:2, the concentration of PO_4^{3-} was only 0.01 mg/L. When the binder/PG ratio decreased to 1:5, the concentration of PO_4^{3-}

increased to 106.4 mg/L, followed by a drastic increase to 1070.0 mg/L in bleeding water when the binder/PG ratio decreased to 1:6. This increasing rule was also applicable to F^- and SO_4^{2-}. The concentrations of F^- and SO_4^{2-} increased by 209.2 and 122.9% with the decrease in the binder/PG ratio from 1:2 to 1:6. The results indicated that a high binder content had a positive effect on the immobilization of pollutants in the backfill slurry. It is worth noting that the SO_4^{2-} in backfill slurry with pH values higher than 11 (binder/PG ratio of 1:2) could still be released from the dissolution of the aggregate PG because its concentration in bleeding water (about 1500 mg/L) was close to the solubility of calcium sulfate.

The quantities of pollutants released per ton of solid in backfill for different binder/PG ratios are shown in Figure 5b–d. The quantity of PO_4^{3-} in bleeding water increased from 0.003 g/t to 200.7 g/t in the selected binder/PG ratio range, indicating the acceleration of the release of PO_4^{3-}. However, compared with PO_4^{3-}, the released quantities of F^- and SO_4^{2-} showed different patterns with respect to the binder/PG ratio. Figure 5c,d show that when the binder/PG ratio was 1:3, the quantities of F^- and SO_4^{2-} exhibited rather high values, 31.8 g/t and 1046.5 g/t, respectively, which might be attributed to the high bleeding rate causing the high quantities of pollutants released into the bleeding water. High quantities of F^- and SO_4^{2-} were also observed when the binder/PG ratio was 1:6, which was likely related to high concentrations of F^- and SO_4^{2-} in bleeding water.

3.3. Solid Concentration

(1) pH and bleeding rate of backfill slurry

As shown in Figure 6a, the pH of the slurry was independent of the solid concentration, while the bleeding rate significantly varied with the solid concentrations. As the solid concentration increased from 40 to 52%, the bleeding rate rapidly decreased from 60.3 to 31.0%. The greater amount of free water introduced via the low solid concentration could be the reason for the high bleeding rate, which was also observed in the study by Yao et al. [42]. When the solid concentration was further increased from 52 to 60%, the bleeding rate only slightly declined from 31.0 to 30.5%. These results indicated that the free water of slurry with a solid concentration of less than 52% more easily accumulated and separated from the slurry and then transformed into the bleeding water.

(2) Pollutants in bleeding water

The concentrations of PO_4^{3-}, F^- and SO_4^{2-} in bleeding water are shown in Figure 6b–d. The concentration of PO_4^{3-} was measured to be in the range of 0.03–0.06 mg/L, which was much lower than the limit value of 0.5 mg/L [40]. The results indicated that the binder could effectively solidify/stabilize PO_4^{3-}. For F^- and SO_4^{2-}, it is clearly shown in Figure 6c,d that the lower the solid concentration was, the lower concentrations of pollutants in bleeding water were. This could be the result of the dilution effect of the greater amount of abundant bleeding water in backfill slurry with a low solid concentration. When the solid concentration increased from 40 to 60%, the concentrations of F^- and SO_4^{2-} gradually increased from 60 to 77 mg/L and from 1575 to 1970 mg/L, respectively.

Figure 6b–d show that the quantities of pollutants released (per ton of solid in backfill) were negatively correlated with the solid concentrations, and their trend was opposite to that of the pollutant concentrations. When the solid concentration increased from 40% to 60%, the quantities of F^- and SO_4^{2-} released per ton of backfill rapidly decreased rapidly from 54.3 to 15.6 g/t and from 1424.6 to 400.2 g/t, respectively. The results indicated that although a greater amount of bleeding water decreased the pollutant concentrations in bleeding water, the high bleeding rate contributed to the release of more pollutants from the backfill slurry into the bleeding water.

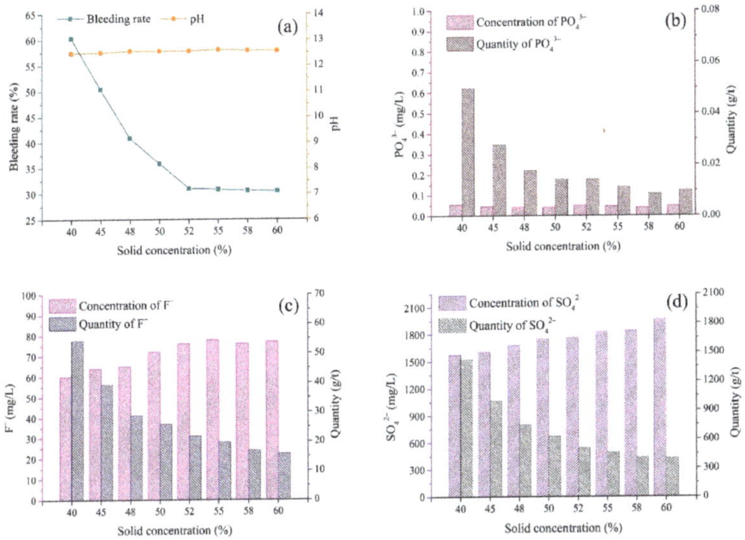

Figure 6. Variations in bleeding rate, pH and pollutants in bleeding water according to different solid concentrations of backfill slurry. (**a**) Bleeding rate and pH; (**b**) PO_4^{3-}; (**c**) F^-; (**d**) SO_4^{2-}.

3.4. Binder Type

(1) pH and bleeding rate of backfill slurry

Different types of binder showed different immobilization effects on pollutants. The variations in the bleeding rate and pH according to the GGBFS replacement proportion are shown in Figure 7a. When the GGBFS replacement proportion was increased from 0% to 20%, the bleeding rate rapidly dropped from 47.1 to 39.1%. As the GGBFS replacement proportion further increased to 90%, the bleeding water slightly decreased from 38.9 to 37.1%. The results indicated that GGBFS had better water retention capability than PC, as GGBFS was finer than PC (Figure 2) [42,43]. Figure 7a shows that the pH slightly decreased from 11.95 to 11.60 when the GGBFS replacement proportion increased from 0 to 90%, which could be attributed to the neutral characteristic of GGBFS. Nevertheless, the pH values of all slurry samples were always above 11, indicating that the incorporation of alkali was sufficient to activate GGBFS [41].

(2) Pollutants in bleeding water

The concentration of PO_4^{3-} was in the range of 0.06–0.08 mg/L (Figure 7b), indicating that both PC and PC-GGBFS immobilized PO_4^{3-} at a low level. Figure 7c shows that the concentration of F^- exhibited an increase from 18 mg/L to 26 mg/L in the range of GGBFS replacement proportion from 0 to 90%. The increase in the concentration of F^- might be related to the decrease in cement proportion. The results indicated that cement had a better immobilization effect on pollutants than GGBFS. The concentration of SO_4^{2-}, which was mainly derived from the dissolution of $CaSO_4{\cdot}2H_2O$, exhibited similar values, around 1400 mg/L, for five levels of GGBFS content, as shown in Figure 7d.

As shown in Figure 7b, the quantity of PO_4^{3-} released was only around 0.02 g/t, which could be the result of the low concentration of PO_4^{3-}. As for F^-, as shown in Figure 7c, compared with the significant increase in the concentration of F^-, the release of F^- per ton of solid showed only a slight increase for various binder types, which was likely due to the offset effect of the decrease in the bleeding rate. Figure 7d shows that the quantity of SO_4^{2-} exhibited a decrease in the selected range of GGBFS replacement proportion. This phenomenon could be explained by the combined effect of the stable concentration of SO_4^{2-} and the decrease in the bleeding rate for various binders.

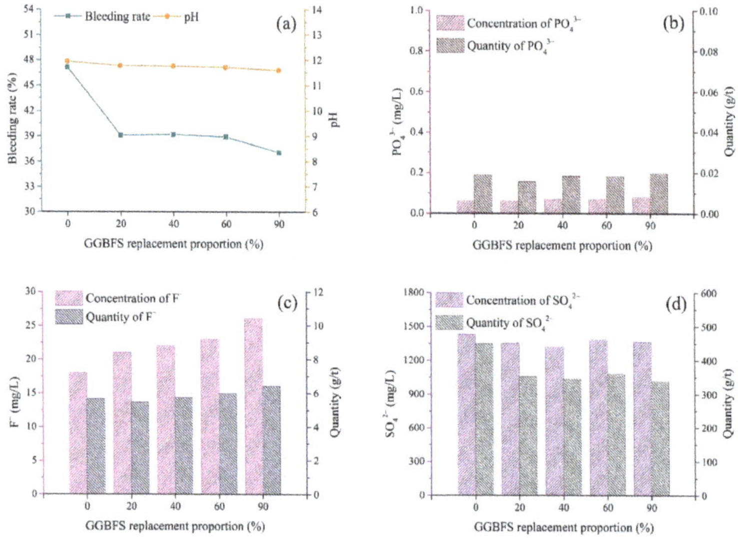

Figure 7. Variations in bleeding rate, pH and pollutants in bleeding water according to GGBFS replacement proportion. (**a**) Bleeding rate and pH; (**b**) PO_4^{3-}; (**c**) F^-; (**d**) SO_4^{2-}.

3.5. Mixing Procedure

(1) pH and bleeding rate of backfill slurry

The slurry of cemented PG backfill requires a rather long mixing time to obtain the proper fluidity for pipeline transport [37]. The properties of the backfill slurry were affected by the mixing time. Figure 8a shows the variations in the pH and bleeding rate of backfill slurry according to different mixing times. The bleeding rate increased over the 5–60 min interval, followed by a decrease when the mixing time was extended to 240 min [37]. The increase in the bleeding rate before 60 min was the result of the instability of the slurry due to mixing, which can reduce the water retention of the slurry [47]. On the other hand, a lower amount of free water induced via hydration with the extension of the mixing time could be the explanation for the decrease in the bleeding rate after 60 min [37]. As for the pH of the backfill slurry, it varied in the small range of 12.68–12.79, indicating that the binder improved the pH of the whole system in the initial stage of mixing.

(2) Pollutants in bleeding water

Figure 8b,c show the changes in the concentrations of PO_4^{3-} and F^- according to the mixing time. It is worth noting that in the first 5 min, the concentrations of PO_4^{3-} and F^- in bleeding water were far below the leachate of PG, indicating that the immobilization of PO_4^{3-} and F^- was accomplished in the first several minutes of mixing. After the first immobilization, the concentrations of both PO_4^{3-} and F^- showed several changes according to the mixing time. For PO_4^{3-}, the concentration increased a little in the first 30 min of mixing and then remained at low levels (about 0.18 mg/L) in the following experimental time intervals. The concentration of F^- showed a positive correlation with the extension of the mixing time, as shown in Figure 8c, indicating that the mixing procedure promoted the release of F^-. The increase in the concentrations of PO_4^{3-} and F^- with the extension of the mixing time could be attributed to the gradual dissolution of $Ca_3(PO_4)_2$ and CaF_2, which was related to the consumption of $Ca(OH)_2$ via the hydration of the binder. As shown in Figure 8d, the concentration of SO_4^{2-} showed a relatively high value, 2269 mg/L, in the first 5 min and decreased to 1806 mg/L in 10 min, followed by a relatively stable value after a mixing time of 15 min. The results indicated that the immobilization of SO_4^{2-} occurred in the first 15 min.

The quantities of pollutants released per ton of solid in backfill are shown in Figure 8b–d. As expected, 1 ton of raw solid in backfill only released 0.005–0.018 g/t of PO_4^{3-}, while the quantity of F^- exhibited a gradual increase with the extension of mixing time, as shown in Figure 8c. The quantity of F^- was in the range of 4–16 g/t. It is worth noting that the quantities of F^- and SO_4^{2-} after 60 min of mixing time showed relatively high values, which could be explained by the maximum value of the bleeding rate after the mixing time of 60 min.

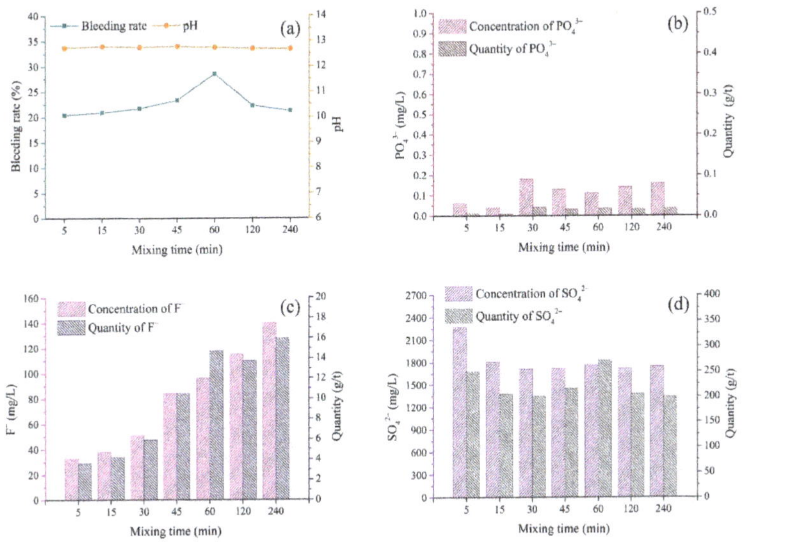

Figure 8. Variations in bleeding rate, pH and pollutants in bleeding water according to mixing time. (**a**) Bleeding rate [37] and pH; (**b**) PO_4^{3-}; (**c**) F^-; (**d**) SO_4^{2-}.

4. Discussion

4.1. Release of Pollutants from Backfill Slurry into the Bleeding Water

The quantities of pollutants released were determined by the pollutant concentrations in bleeding water and the bleeding rate.

The pollutant concentrations in the bleeding water were directly determined via the chemical properties of the slurry, such as the pH value. Figure 5 shows that the backfill slurry with low pH released more pollutants into the bleeding water, indicating the presence of high quantities of pollutants in raw materials. More specifically, because of the high content of pollutants in the aggregate PG with low pH (Figure 4), the backfill slurry with a low pH value usually contained more pollutants, which meant that more pollutants were released into the bleeding water, whereas, slurry pH above 11 meant that the pollutants were well immobilized, thus allowing the pollutant concentrations in the bleeding water to remain at relatively low levels. This phenomenon could be explained by the fact that the sufficient Ca^{2+} introduced by the binder transformed active ions such as PO_4^{3-}, F^- and SO_4^{2-} into undissolved forms, reducing the quantities of pollutants in the backfill slurry [29]. In cemented PG backfill, PO_4^{3-} could be better solidified/stabilized than F^- and SO_4^{2-}. To be more specific, PO_4^{3-} in the bleeding water could be minimized to below 0.2 mg/L when the pH value of backfill slurry was increased over 11 (Figures 5–8), which was less than the limit value of 0.5 mg/L [40]. As for F^-, the concentration was tens of milligrams per liter despite the backfill slurry exhibiting a high pH value, which was higher than the limit value of ten milligrams per liter [40]. It was also observed that the concentration of SO_4^{2-} in the bleeding water was always higher than 1300 mg/L, even though the pH value of the backfill slurry was higher than 11; this was due to the dissolution of the aggregate PG.

The release of pollutants was also affected by the physical properties of the backfill slurry, such as the bleeding rate. For instance, it was observed (Figure 6) that the backfill slurry with a high bleeding rate, 60.3%, released 54 g of F^- into the bleeding water per ton of solid in the backfill, which was 3.5 times the quantity released by the backfill slurry with the bleeding rate of 30.5%. It was clear that a high bleeding rate meant more pollutants were released into the environment. It is worth noting that the high bleeding rate meant that more bleeding water was drained, so the pollutants were diluted. For example, as shown in Figure 6, the backfill slurry with a bleeding rate of 30.5% exhibited 77 mg/L F^- in the bleeding water, which was 17 mg/L higher than the concentration with a bleeding rate of 60.3%. Nevertheless, considering the combined effect of the pollutant concentration and the bleeding rate, the total quantity of pollutants released was much higher when severe bleeding occurred.

4.2. Effect of Preparation Protocols on Chemical Properties of Bleeding Water

The chemical properties of bleeding water, including pH and concentrations of pollutants, varied with the protocols for slurry preparation.

The pH of bleeding water was mainly affected by the pH of the aggregate PG. When low-pH PG (1.68) was used as the aggregate, the binder/PG ratio showed a significant effect on the pH of the bleeding water. As shown in Figure 5, the pH of the bleeding water was in the range of 6.86–12.70. A high binder/PG ratio introduced more alkali, resulting in the improvement of the pH of the bleeding water. However, the backfill slurry prepared with high-pH PG always had high pH, which was due to the fact that less binder was needed to neutralize the residual acid in PG, therefore sufficient OH^- was present in the backfill slurry. For instance, when the high-pH PG (4.15 and 6.18) was used as the aggregate in the studied batches (different solid concentrations, binder types and mixing procedures, as shown in Figures 6–8), the bleeding water always showed rather high pH values, above 11.

The preparation protocols demonstrated a relationship with the concentrations of pollutants in the bleeding water. As for the binder/PG ratio, high binder dosage could more efficiently solidify/stabilize pollutants, and fewer pollutants were released. As shown in Figure 5, when the binder/PG ratio increased from 1:6 to 1:2, the S/S efficiency of PO_4^{3-}, F^- and SO_4^{2-} was improved by 100%, 67.7% and 30.7%, respectively. For solid concentration, a greater amount of bleeding water, due to the low concentration, could dilute the pollutants, causing low pollutant concentrations. For binder type and mixing time, it seemed that the concentration of F^- positively related to GGBFS content and mixing time, while the concentrations of PO_4^{3-} and SO_4^{2-} varied little with these two protocols, as shown in Figures 7 and 8.

4.3. Effect of Preparation Protocols on Bleeding Rate of Backfill Slurry

The bleeding rate was affected by the preparation protocols in two aspects: (i) the initial water content and (ii) the water retention. On the one hand, it was clear that the initial water content was the critical factor in terms of solid concentration. As shown in Figure 6, when the initial water content increased from 48 to 60%, the bleeding rate increased by about 50%. On the other hand, the water retention was affected by the protocols for slurry preparation. As for the binder/PG ratio, the water retention of the backfill slurry was intensified by the low pH related to the low binder/PG ratio, as shown in Figure 5. As for the binder type, the GGBFS replacement proportion could increase the water retention, which was related to the large specific surface area of the binder. The fine GGBFS particles increased the requirement for water on the surface layer and decreased the proportion of free water, increasing the water retention and reducing the bleeding rate [42,43].

In the cemented PG backfill, the backfill slurry exhibited a rather high bleeding rate ranging from 18.6 to 60.3%, which was much higher than that of the backfill slurry obtained using tailings as aggregates (lower than 15% to improve the roof-contacted filling ratio) [48]. The high bleeding rate could be attributed to the high initial water content that was required to obtain the desired fluidity. Different from the tailings, the specific gravity of PG is much lower [49–51]. With the same solid concentration of backfill slurry, the particle number of aggregate PG can be much larger than that of aggregate tailings. The larger number of particles in the cemented PG backfill slurry results in a larger total specific surface area and requires greater water content to ensure fluidity [42]. In practice, in mines, the solid concentration of cemented PG backfill is generally 40–65%, which is much lower than that of cemented tailings backfill (which typically ranges from 70 to 85%) [2,31,48,52,53].

A schematic illustration of how the protocols for slurry preparation affected the release of pollutants into the bleeding water is shown in Figure 9. The protocols for slurry preparation affected the concentrations of pollutants in bleeding water and the bleeding rate of backfill slurry. Increasing the pH of the backfill slurry by adding more binder reduced the concentrations of pollutants in bleeding water. On the other hand, increasing the solid concentration and the content of fine raw materials reduced the bleeding water produced. Furthermore, the results of this study indicated that the high pollutant concentrations and the high bleeding rate of the backfill slurry contributed to the release of pollutants into the bleeding water.

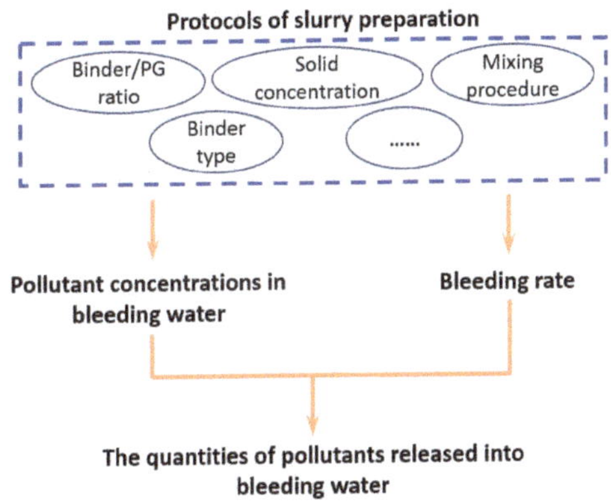

Figure 9. Schematic illustration of how the protocols for slurry preparation affected the release of pollutants from backfill slurry into bleeding water.

5. Conclusions

In this study, four different protocols for slurry preparation were considered to investigate the release of pollutants from cemented PG backfill slurry into the bleeding water. Based on the above results, conclusions could be drawn as shown below:

(1) The bleeding water from cemented PG backfill contained PO_4^{3-}, F^- and SO_4^{2-}, which pose a risk of damage to the surrounding environment. The high pollutant concentrations and the high bleeding rate resulted in higher quantities of pollutants being released from the backfill slurry;

(2) The pollutant concentrations in the bleeding water could be minimized via proper protocols for slurry preparation, such as increasing the binder/PG ratio. A greater amount of more binder efficiently transformed PO_4^{3-}, F^- and SO_4^{2-} into undissolved forms. The dilution effect induced by the low solid concentration also helped to reduce the pollutant concentrations;

(3) The poor water retention and high initial water content of the backfill slurry increased the bleeding rate. Introducing more fine raw materials into the backfill slurry, such as binder and GGBFS, could bond more surface layer water and increase the water retention. The pH of the backfill slurry is also negatively related to water retention;

(4) Cemented PG backfill could solidify/stabilize PO_4^{3-} in the bleeding water well when the pH value was higher than 11. However, the concentration of F^- in the bleeding water always exceeded the limit value of the national standard (10 mg/L); therefore, further studies are needed for the S/S of F^-.

Author Contributions: Conceptualization, C.M. and Y.S.; Methodology, C.M.; Software, C.M. and Y.Z.; Validation, C.M. and Y.S.; Formal Analysis, Z.L. and Y.Z.; Investigation, C.M.; Resources, Y.S. and Z.L.; Data Curation, C.M.; Writing—Original Draft Preparation, C.M.; Writing—Review and Editing, Y.S. and C.M.; Visualization, C.M. and Y.Z.; Supervision, Y.S.; Project Administration, Y.S. and Z.L.; Funding Acquisition, Y.S. All authors have read and agreed to the published version of the manuscript.

Funding: This work was supported by the National Natural Science Foundation of China (NSFC) (Grant No. 42177160 and 51974359), the State Key Research Development Program of China (Grant No. 2018YFC1800400), the Postgraduate Scientific Research Innovation Project of Hunan Province (Grant No. CX20210294) and the Fundamental Research Funds for the Central Universities of Central South University (Grant No. 2021zzts0273).

Institutional Review Board Statement: Not applicable.

Informed Consent Statement: Not applicable.

Data Availability Statement: Not applicable.

Conflicts of Interest: The authors declare no conflict of interest.

References

1. Wang, S.; Li, X.; Yao, J.; Gong, F.; Li, X.; Du, K.; Tao, M.; Huang, L.; Du, S. Experimental investigation of rock breakage by a conical pick and its application to non-explosive mechanized mining in deep hard rock. *Int. J. Rock Mech. Min. Sci.* **2019**, *122*, 104063. [CrossRef]
2. Li, X.; Du, J.; Gao, L.; He, S.; Gan, L.; Sun, C.; Shi, Y. Immobilization of phosphogypsum for cemented paste backfill and its environmental effect. *J. Clean. Prod.* **2017**, *156*, 137–146. [CrossRef]
3. Fall, M.; Benzaazoua, M.; Saa, E.G. Mix proportioning of underground cemented tailings backfill. *Tunn. Undergr. Space Tech.* **2008**, *23*, 80–90. [CrossRef]
4. Huynh, L.; Beattie, D.A.; Fornasiero, D.; Ralston, J. Effect of polyphosphate and naphthalene sulfonate formaldehyde condensate on the rheological properties of dewatered tailings and cemented paste backfill. *Miner. Eng.* **2006**, *19*, 28–36. [CrossRef]
5. Jiang, G.; Wu, A.; Wang, Y.; Li, J. The rheological behavior of paste prepared from hemihydrate phosphogypsum and tailing. *Constr. Build. Mater.* **2019**, *229*, 116870. [CrossRef]
6. Zhang, C.; Wang, A.; Tang, M.; Liu, X. The filling role of pozzolanic material. *Cem. Concr. Res.* **1996**, *26*, 943–947. [CrossRef]
7. Koohestani, B. Effect of saline admixtures on mechanical and microstructural properties of cementitious matrices containing tailings. *Constr. Build. Mater.* **2017**, *156*, 1019–1027. [CrossRef]
8. Wei, H.; Xiao, B.; Gao, Q. Flow properties analysis and identification of a fly ash-waste rock mixed backfilling slurry. *Minerals* **2021**, *11*, 576. [CrossRef]
9. Xiao, B.; Wen, Z.; Wu, F.; Li, L.; Yang, Z.; Gao, Q. A simple L-shape pipe flow test for practical rheological properties of backfill slurry: A case study. *Powder Technol.* **2019**, *356*, 1008–1015. [CrossRef]
10. Wang, S.; Sun, L.; Huang, L.; Li, X.; Shi, Y.; Yao, J.; Du, S. Non-explosive mining and waste utilization for achieving green mining in underground hard rock mine in China. *Trans. Nonferr. Metal. Soc. China* **2019**, *29*, 1914–1928. [CrossRef]
11. Yang, X. *Development of Phosphogypsum—Based Early Strength Cementitious Materials and Resource Utilization of Waste*; University of Science and Technology Beijing: Beijing, China, 2017. (In Chinese)
12. Jiang, J. *The Application Research of Phosphogypsum-Based Cementing Filling Marerials in Huang Mailing Mine*; Wuhan University of Technology: Wuhan, China, 2016. (In Chinese)
13. Bisone, S.; Gautier, M.; Chatain, V.; Blanc, D. Spatial distribution and leaching behavior of pollutants from phosphogypsum stocked in a gypstack: Geochemical characterization and modeling. *J. Environ. Manag.* **2017**, *193*, 567–575. [CrossRef]
14. Pérez-López, R.; Macías, F.; Cánovas, C.R.; Sarmiento, A.M.; Pérez-Moreno, S.M. Pollutant flows from a phosphogypsum disposal area to an estuarine environment: An insight from geochemical signatures. *Sci. Total Environ.* **2016**, *553*, 42–51. [CrossRef] [PubMed]
15. Rashad, A.M. Potential use of phosphogypsum in alkali-activated fly ash under the effects of elevated temperatures and thermal shock cycles. *J. Clean. Prod.* **2015**, *87*, 717–725. [CrossRef]
16. U.S. Geological Survey. Mineral Commodity Summaries 2022. 2022; p. 125. Available online: https://pubs.usgs.gov/periodicals/mcs2022/mcs2022.pdf (accessed on 29 August 2022).
17. Ye, X. The using current situation, form analysis and measures of phosphogypsum utilization in China in 2017. *Sulphuric Acid Ind.* **2018**, *8*, 1–4. (In Chinese)
18. Rusch, K.A.; Malone, R.F. Final Report: Searching for Optimum Composition of Phosphogypsum: Fly ash: Cement Composites for Oyster Culch Materials. 2004. Available online: https://cfpub.epa.gov/ncer_abstracts/index.cfm/fuseaction/display.abstractDetail/abstract_id/5886/report/F (accessed on 29 August 2022).

19. Essabir, H.; Nekhlaoui, S.; Bensalah, M.O.; Rodrigue, D.; Bouhfid, R.; Qaiss, A.E.K. Phosphogypsum Waste Used as Reinforcing Fillers in Polypropylene Based Composites: Structural, Mechanical and Thermal Properties. *J. Polym. Environ.* **2017**, *25*, 658–666. [CrossRef]
20. Hentati, O.; Abrantes, N.; Caetano, A.L.; Bouguerra, S.; Goncalves, F.; Roembke, J.; Pereira, R. Phosphogypsum as a soil fertilizer: Ecotoxicity of amended soil and elutriates to bacteria, invertebrates, algae and plants. *J. Hazard. Mater.* **2015**, *294*, 80–89. [CrossRef] [PubMed]
21. Rosales, J.; Perez, S.M.; Cabrera, M.; Gazquez, M.J.; Bolivar, J.P.; de Brito, J.; Agrela, F. Treated phosphogypsum as an alternative set regulator and mineral addition in cement production. *J. Clean. Prod.* **2020**, *244*, 118752. [CrossRef]
22. Singh, M. Treating waste phosphogypsum for cement and plaster manufacture. *Cem. Concr. Res.* **2002**, *32*, 1033–1038. [CrossRef]
23. Degirmenci, N.; Okucu, A.; Turabi, A. Application of phosphogypsum in soil stabilization. *Build. Environ.* **2007**, *42*, 3393–3398. [CrossRef]
24. Ennaciri, Y.; Zdah, I.; El Alaoui-Belghiti, H.; Bettach, M. Characterization and purification of waste phosphogypsum to make it suitable for use in the plaster and the cement industry. *Chem. Eng. Commun.* **2020**, *207*, 382–392. [CrossRef]
25. Diwa, R.R.; Tabora, E.U.; Palattao, B.L.; Haneklaus, N.H.; Vargas, E.P.; Reyes, R.Y.; Ramirez, J.D. Evaluating radiation risks and resource opportunities associated with phosphogypsum in the Philippines. *J. Radioanal. Nucl. Chem.* **2022**, *331*, 967–974. [CrossRef]
26. Gijbels, K.; Pontikes, Y.; Samyn, P.; Schreurs, S.; Schroeyers, W. Effect of NaOH content on hydration, mineralogy, porosity and strength in alkali/sulfate-activated binders from ground granulated blast furnace slag and phosphogypsum. *Cem. Concr. Res.* **2020**, *132*, 106054. [CrossRef]
27. Tayibi, H.; Choura, M.; Lopez, F.A.; Alguacil, F.J.; Lopez-Delgado, A. Environmental impact and management of phosphogypsum. *J. Environ. Manag.* **2009**, *90*, 2377–2386. [CrossRef]
28. Shi, Y.; Gan, L.; Li, X.; He, S.; Sun, C.; Gao, L. Dynamics of metals in backfill of a phosphate mine of Guiyang, China using a three-step sequential extraction technique. *Chemosphere* **2018**, *192*, 354–361. [CrossRef]
29. Zhou, S.; Li, X.; Zhou, Y.; Min, C.; Shi, Y. Effect of phosphorus on the properties of phosphogypsum-based cemented backfill. *J. Hazard. Mater.* **2020**, *399*, 122993. [CrossRef] [PubMed]
30. Min, C.; Shi, Y.; Liu, Z. Properties of cemented phosphogypsum (PG) backfill in case of partially substitution of composite Portland cement by ground granulated blast furnace slag. *Constr. Build. Mater.* **2021**, *305*, 124786. [CrossRef]
31. Cao, Z.; Liu, B.; Li, X.; Li, D.; Dong, L. Experimental study on backfilling mine goafs with chemical waste phosphogypsum. *Geofluids* **2019**, *2019*, 9218916. [CrossRef]
32. Li, X.; Zhou, Y.; Zhu, Q.; Zhou, S.; Min, C.; Shi, Y. Slurry preparation effects on the cemented phosphogypsum backfill through an orthogonal experiment. *Minerals* **2019**, *9*, 31. [CrossRef]
33. Wang, Z.; Feng, G.; Qi, T.; Guo, Y.; Du, X. Evaluation of static segregation of cemented gangue-fly ash backfill material using electrical resistivity method. *Measurement* **2020**, *154*, 107483. [CrossRef]
34. Yao, Y.; Sun, H. A novel silica alumina-based backfill material composed of coal refuse and fly ash. *J. Hazard. Mater.* **2012**, *213*, 71–82. [CrossRef] [PubMed]
35. Guo, Y.; Wang, P.; Feng, G.; Qi, T.; Liu, G.; Ren, A. Performance of coal gangue-based cemented backfill material modified by water-reducing agents. *Adv. Mater. Sci. Eng.* **2020**, *2020*, 2302895. [CrossRef]
36. Yao, Z. *Technoligical Study and Reliability Analysis of Yellow Phosphorus Slag and Phosphgysum Backfill in Kaiyang Mine*; Central South University: Changsha, China, 2009. (In Chinese)
37. Min, C.; Li, X.; He, S.; Zhou, S.; Zhou, Y.; Yang, S.; Shi, Y. Effect of mixing time on the properties of phosphogypsum-based cemented backfill. *Constr. Build. Mater.* **2019**, *210*, 564–573. [CrossRef]
38. *HJ 557-2010*; Solid Waste—Extraction Procedure for Leaching Toxicity—Horizontal Vibration Method. China Environmental Science Press: Beijing, China, 2010.
39. *GB/T 50080-2016*; Standard for Test Method of Performance on Ordinary Fresh Concrete. China Architecture & Building Press: Beijing, China, 2016.
40. *GB 8978-1996*; Integrated Wastewater Discharge Standard. China Environmental Science Press: Beijing, China, 1996.
41. Song, S.; Sohn, D.; Jennings, H.M.; Mason, T.O. Hydration of alkali-activated ground granulated blast furnace slag. *J. Mater. Sci.* **2000**, *35*, 249–257. [CrossRef]
42. Yao, H.; Liu, D. Utilization of low activity CFBC ash in cemented paste backfill containing phosphate tailings. *Min. Metall. Explor.* **2021**, *38*, 2485–2492. [CrossRef]
43. Zhang, X.; Lin, J.; Liu, J.; Li, F.; Pang, Z. Investigation of hydraulic-mechanical properties of paste backfill containing coal gangue-fly ash and its application in an underground coal mine. *Energies* **2017**, *10*, 1309. [CrossRef]
44. Simon, D.; Grabinsky, M. Apparent yield stress measurement in cemented paste backfill. *Int. J. Min. Reclam. Environ.* **2013**, *27*, 231–256. [CrossRef]
45. Elakneswaran, Y.; Nawa, T.; Kurumisawa, K. Zeta potential study of paste blends with slag. *Cem. Concr. Res.* **2009**, *31*, 72–76. [CrossRef]
46. Mikanovic, N.; Jolicoeur, C. Influence of superplasticizers on the rheology and stability of limestone and cement pastes. *Cem. Concr. Res.* **2008**, *38*, 907–919. [CrossRef]

47. Mehdipour, I.; Razzaghi, M.S.; Amini, K.; Shekarchi, M. Effect of mineral admixtures on fluidity and stability of self-consolidating mortar subjected to prolonged mixing time. *Constr. Build. Mater.* **2013**, *40*, 1029–1037. [CrossRef]
48. Yang, L.; Qiu, J.; Jiang, H.; Hu, S.; Li, H.; Li, S. Use of Cemented Super-Fine Unclassified Tailings Backfill for Control of Subsidence. *Minerals* **2017**, *7*, 11. [CrossRef]
49. Chen, Q.; Zhang, Q.; Qi, C.; Fourie, A.; Xiao, C. Recycling phosphogypsum and construction demolition waste for cemented paste backfill and its environmental impact. *J. Clean. Prod.* **2018**, *186*, 418–429. [CrossRef]
50. Niroshan, N.; Sivakugan, N.; Veenstra, R.L. Flow Characteristics of Cemented Paste Backfill. *Geotech. Geol. Eng.* **2018**, *36*, 2261–2272. [CrossRef]
51. Kesimal, A.; Yilmaz, E.; Ercikdi, B. Evaluation of paste backfill mixtures consisting of sulphide-rich mill tailings and varying cement contents. *Cem. Concr. Res.* **2004**, *34*, 1817–1822. [CrossRef]
52. Cao, S.; Song, W.; Yilmaz, E. Influence of structural factors on uniaxial compressive strength of cemented tailings backfill. *Constr. Build. Mater.* **2018**, *174*, 190–201. [CrossRef]
53. Fall, M.; Benzaazoua, M.; Ouellet, S. Experimental characterization of the influence of tailings fineness and density on the quality of cemented paste backfill. *Miner. Eng.* **2005**, *18*, 41–44. [CrossRef]

materials

MDPI

Article

Research on the Attenuation Characteristics of High-Frequency Elastic Waves in Rock-Like Material

Xiling Liu [1,2], Feng Xiong [1], Qin Xie [1], Xiukun Yang [1], Daolong Chen [1] and Shaofeng Wang [1,*]

1 School of Resources and Safety Engineering, Central South University, Changsha 410083, China
2 State Key Laboratory of Coal Resources and Safe Mining, China University of Mining and Technology, Xuzhou 221116, China
* Correspondence: sf.wang@csu.edu.cn

Abstract: In order to study the frequency-dependent attenuation characteristics of high-frequency elastic waves in rock-like materials, we conducted high-frequency elastic wave attenuation experiments on marble, granite, and red sandstone rods, and investigated the frequency dependence of the attenuation coefficient of high-frequency elastic waves and the frequency dependence of the attenuation of specific frequency components in elastic waves. The results show that, for the whole waveform packet of the elastic wave signal, the attenuation coefficient and the elastic wave frequency have an approximate power relationship, with the exponents of this power function being 0.408, 0.420, and 0.384 for marble, granite, and red sandstone, respectively, which are close to 1/2 the exponent value obtained theoretically by the Kelvin–Voigt viscoelastic model. However, when the specific frequency components are tracked during the elastic wave propagation, the exponents of the power relationship between the attenuation coefficient and frequency are 0.982, 1.523, and 0.860 for marble, granite, and red sandstone, respectively, which indicate that the relationship between the attenuation coefficient and frequency is rock-type dependent. Through the analysis of rock microstructure, we demonstrate that this rock-type-dependent relationship is mainly caused by the scattering attenuation component due to the small wavelength of the high-frequency elastic wave. Therefore, the scattering attenuation component may need to be considered when the Kelvin–Voigt model is used to describe high-frequency elastic wave attenuation in rock-like materials. The results of this research are of good help for further understanding the attenuation characteristics of high-frequency elastic waves in rock-like materials.

Keywords: elastic wave; frequency dependent attenuation; attenuation coefficient; rock microstructures

Citation: Liu, X.; Xiong, F.; Xie, Q.; Yang, X.; Chen, D.; Wang, S. Research on the Attenuation Characteristics of High-Frequency Elastic Waves in Rock-Like Material. *Materials* **2022**, *15*, 6604. https://doi.org/10.3390/ma15196604

Academic Editor: Carmelo Majorana

Received: 25 July 2022
Accepted: 20 September 2022
Published: 23 September 2022

1. Introduction

When an elastic wave propagates in a medium, attenuation will occur, which is accompanied by a decrease in energy, amplitude, and frequency with increasing distance. The causes of attenuation can be attributed to geometrical spreading, internal friction, mode conversion, and scattering during propagation [1]. Thereinto, the attenuation caused by internal friction is strongly correlated with the properties of the medium, which is the main reason for the attenuation of elastic waves.

The understanding of wave attenuation in rock media originates from the study of seismic waves. As early as 1941, Born estimated the attenuation coefficients of the earth based on seismic reflection records [2], and the quality factor Q was introduced to describe the absorption of elastic waves by the material and the inelasticity of the material, and to characterize the attenuation. Q is defined as the ratio of the total energy E_0 to the energy ΔE consumed by the vibration in a cycle [3]. Futterman [4] first discussed the basic characteristics of seismic wave attenuation in a rock mass and pointed out that its amplitude attenuation was related to the quality factor, the velocity, frequency of seismic waves, etc. Since then, many scholars have conducted research on the attenuation characteristics in various rock masses as well as on the influences of wave velocity, frequency, temperature, pressure, fluid viscosity and permeability, and propagation type (P wave or S wave) [5–9]. An in-depth understanding of elastic wave attenuation will facilitate the acquisition of the source characteristics according to the data collected by the sensor; additionally, the internal properties of the rock mass may be inverted based on the understanding of attenuation [10]. Similarly, the study of the attenuation characteristics of elastic waves in small-scale rock samples may elucidate the physical properties of rocks [11–16].

In laboratory rock deformation tests, the acoustic emission (AE) technique is often used to study the failure characteristics of rocks by recording the elastic waves released during crack initiation, propagation, and penetration [17–23]. It is also necessary to deeply understand the attenuation characteristics of the AE signal in the rock sample [24–27] and then obtain the characteristics of the fracture source in the rock sample. As the rock AE signal is generally higher than 100 kHz [28], the attenuation of elastic waves with a high frequency is also a subject of concern. In research on the attenuation of high-frequency AE signals in rock-like materials, Dobrin et al. [29] summarized that the amplitude of the AE signal obeys the exponential attenuation law, based on a large number of laboratory experiments. Furthermore, as the medium through which elastic waves propagate in a rock-like material is viscoelastic, the viscoelasticity of the medium is often studied, and many viscoelastic medium theories have been proposed [30,31].

As the attenuation is frequency dependent, many frequency-dependent attenuation theories have also been proposed. Müller believed that in rocks containing pore fluids, the squirting flow mechanism and wave-induced flow also affect the attenuation of elastic waves, which is frequency dependent [32]. The viscosity effect is also frequency dependent, and considering the viscoelasticity of the rock mass, the attenuation caused by internal friction is also frequency dependent [33,34]. The continuous development and progress of measurement technology have resulted in more accurate attenuation characteristic parameters being obtained to explore the frequency dependence of attenuation [35]. However, a high-frequency elastic wave signal attenuates rapidly, which makes the attenuation measurement of such elastic wave more difficult. Therefore, there have been many studies focusing on the attenuation of elastic waves at the seismic exploration frequency, but the attenuation of high-frequency elastic waves is not fully understood.

Figure 1 shows the distribution of various rock attenuation coefficients at different frequencies, which was obtained by summarizing a large amount of experimental data by Schön et al. [36]. In the low-frequency range, the attenuation coefficient increases approximately linearly with the increasing of frequency, while in the high-frequency range (higher than 10^5 Hz), the attenuation coefficients of the various rock types are not well correlated with the frequency. Furthermore, in research on the attenuation characteristics in rocks, small samples are analyzed according to the characteristics of the signal collected by the sensor after the elastic wave is generated by the seismic source and passes through the sample; however, such an experimental design cannot fully describe the detailed attenuation process in the propagation path of elastic waves, and the acquisition of the attenuation process in the propagation path would greatly facilitate the understanding of the propagation characteristics of high-frequency signals in rock samples.

In this study, we researched the frequency dependence of attenuation in the frequency range of 100 kHz–1 MHz. By conducting attenuation experiments on elastic waves with different frequencies and using rock rods of a different lithology, we determined the attenuation coefficients (α) of elastic waves in rocks. We then analyzed the relationship between α and f based on the Kelvin–Voigt viscoelasticity theory. Moreover, we explained the differences in attenuation velocity and frequency dependence in different lithological rocks by examining the microstructure of the rocks. Furthermore, we explored the relationship between the attenuation coefficient and frequency of dry rocks in the high-frequency range, and discussed the attenuation process during the propagation of elastic waves in rocks. The results of this study will facilitate an understanding of the frequency-dependent characteristics of attenuation and also provide a theoretical basis for inverting the source characteristics with the signal collected by the sensor in rock AE tests.

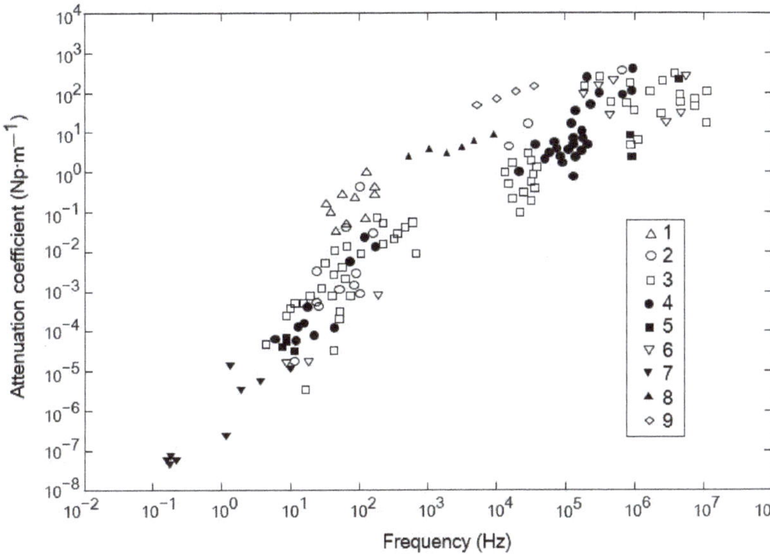

Figure 1. Relationship between the attenuation coefficients of various rocks and frequency: 1—uncemented sedimentary rock; 2—semi-cemented sedimentary rock; 3—solidified sedimentary rock; 4,5—magmatic rock; 6: metamorphic rock; 7—results of deep seismic reflections; 8—limestone; 9—sandstone (dry).

2. Frequency-Dependent Theories of α

In wave science, there are many theories characterizing wave propagation [6,37–39]. During wave propagation, the medium will cause a loss in wave energy, and the attenuation caused by the absorption and scattering accounts for a large part of the total attenuation. Scattering is significant when the wavelength is close to the particle size of the medium. The absorption attenuation of dry rock is mainly the friction dissipation caused by the relative motion of the grain boundary and the crack surface and the energy consumption caused by the residual strain. The absorption attenuation in a rock medium is often related to the viscoelasticity of the rock and is frequency dependent. Absorption attenuation in rock media can be described using a viscoelastic medium model. In viscoelastic media, the stress–strain relationship is no longer a simple linear relationship and many assumptions have been put forward. Different incomplete elastic medium models have been established. The Maxwell and Kelvin–Voigt models are classic.

The Maxwell model comprises an elastic element in series with a viscous element. When stress is applied to the Maxwell model, it is applied equally on the two components, and the strain includes two parts: an elastic body and a viscous body. Owing to the series connection of the elements, the strain in the model is transient. For a stress σ_0, the strain ε of the model is expressed as follows:

$$\varepsilon = \frac{\sigma_0}{E} + \frac{\sigma_0}{\eta}t \tag{1}$$

where E is the elastic modulus, η is the viscosity coefficient of the viscous element, and t is time. According to Equation (1), as long as a tiny force acts on the Maxwell body, the strain will increase infinitely, and when a tiny stress is unloaded, the model will exhibit irreversible viscous deformation, which is inconsistent with the actual characteristics of the rock mass disturbed by elastic waves.

The elements contained in the Kelvin–Voigt model are the same as those in the Maxwell model (Figure 2), except that the elastic and viscous elements are connected parallelly. When stress is applied, the strains of the two elements are the same, and the total stress is the sum of the two stresses. When the Kelvin–Voigt model is subjected to a stress of σ_0,

$$\sigma_0 = E\varepsilon + \eta\frac{d\varepsilon}{dt} \tag{2}$$

this ordinary differential equation should be solved:

$$\varepsilon = \frac{\sigma_0}{E}\left[1 - exp\left(-\frac{E}{\eta}t\right)\right] \tag{3}$$

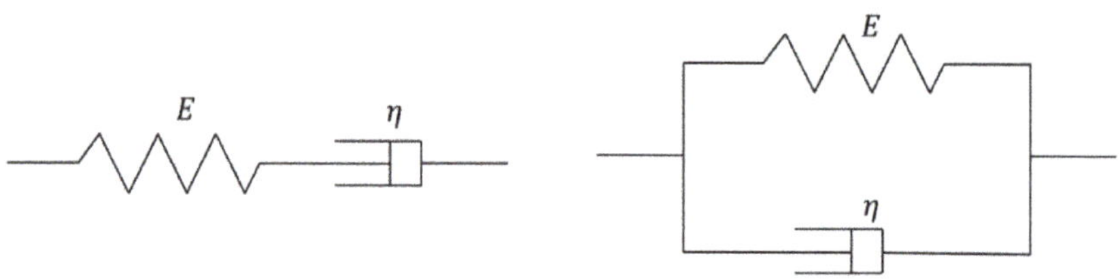

(a) (b)

Figure 2. Classic viscoelastic medium models: (**a**) Maxwell model; (**b**) Kelvin–Voigt model.

The strain of the Kelvin–Voigt body is not instantaneous after being stressed, and it also shows the hysteresis characteristics of strain after stress disappears, which can better reflect the stress–strain relationship of the internal particles, owing to the viscoelastic characteristics of the rock material under the action of elastic waves. Kelvin–Voigt bodies are commonly used in seismic exploration. Based on this model, some scholars have used numerical calculations to simulate the propagation law of seismic wave fields in viscoelastic media [40,41].

The Kelvin–Voigt model is brought into the elastic equation of space:

$$\left.\begin{array}{l} \sigma_x = \lambda\theta + 2\mu\frac{\partial u}{\partial x} + \lambda'\frac{\partial\theta}{\partial t} + 2\mu'\frac{\partial}{\partial t}\frac{\partial u}{\partial x} \\ \sigma_y = \lambda\theta + 2\mu\frac{\partial v}{\partial y} + \lambda'\frac{\partial\theta}{\partial t} + 2\mu'\frac{\partial}{\partial t}\frac{\partial v}{\partial y} \\ \sigma_z = \lambda\theta + 2\mu\frac{\partial w}{\partial z} + \lambda'\frac{\partial\theta}{\partial t} + 2\mu'\frac{\partial}{\partial t}\frac{\partial w}{\partial z} \\ \tau_{xy} = \mu\left(\frac{\partial v}{\partial x} + \frac{\partial u}{\partial y}\right) + \mu'\frac{\partial}{\partial t}\left(\frac{\partial v}{\partial x} + \frac{\partial u}{\partial y}\right) \\ \tau_{yz} = \mu\left(\frac{\partial w}{\partial y} + \frac{\partial v}{\partial z}\right) + \mu'\frac{\partial}{\partial t}\left(\frac{\partial w}{\partial y} + \frac{\partial v}{\partial z}\right) \\ \tau_{zx} = \mu\left(\frac{\partial u}{\partial z} + \frac{\partial w}{\partial x}\right) + \mu'\frac{\partial}{\partial t}\left(\frac{\partial u}{\partial z} + \frac{\partial w}{\partial x}\right) \end{array}\right\} \tag{4}$$

where $\sigma_x, \sigma_y,$ and σ_z are the normal stress components in the x-, y-, and z-axes, respectively; τ_{xy}, τ_{yz}, and τ_{zx} are the shear stress components in the x-, y-, and z-axes, respectively; u, v, and w are the displacement components in the x-, y-, and z-axes, respectively; λ and μ are the Lamé coefficients; and λ' and μ' are the viscosity coefficients that characterize the viscous properties of the medium.

The equilibrium equation can be obtained from Equation (4), regardless of the body force:

$$\left.\begin{array}{l} \rho\frac{\partial^2 u}{\partial t^2} = (\lambda+\mu)\frac{\partial\theta}{\partial x} + \mu\nabla^2 u + (\lambda'+\mu')\frac{\partial}{\partial t}\frac{\partial\theta}{\partial x} + \mu'\frac{\partial}{\partial t}\nabla^2 u \\ \rho\frac{\partial^2 v}{\partial t^2} = (\lambda+\mu)\frac{\partial\theta}{\partial y} + \mu\nabla^2 v + (\lambda'+\mu')\frac{\partial}{\partial t}\frac{\partial\theta}{\partial y} + \mu'\frac{\partial}{\partial t}\nabla^2 v \\ \rho\frac{\partial^2 w}{\partial t^2} = (\lambda+\mu)\frac{\partial\theta}{\partial z} + \mu\nabla^2 w + (\lambda'+\mu')\frac{\partial}{\partial t}\frac{\partial\theta}{\partial z} + \mu'\frac{\partial}{\partial t}\nabla^2 w \end{array}\right\} \tag{5}$$

where ρ is the density of the medium, θ is the volume strain, and ∇^2 means $\frac{\partial^2}{\partial x^2} + \frac{\partial^2}{\partial y^2} + \frac{\partial^2}{\partial z^2}$.

This is the differential equation of motion in a viscoelastic body based on the Kelvin–Voigt model, regardless of the body force. It describes the absorption of waves by the medium through the first-order partial derivatives of the two displacements with respect to time.

Consider a plane harmonic P-wave propagating along the positive x-direction, which only contains the u component in the x-direction and is independent of the y- and z-directions of the components. Owing to the irrotational characteristics of the P-wave, we can simplify Equation (5) and obtain the wave equation of the P-wave in a viscoelastic medium:

$$\rho \frac{\partial^2 u}{\partial t^2} = (\lambda + 2\mu)\frac{\partial^2 u}{\partial x^2} + (\lambda' + 2\mu')\frac{\partial^2}{\partial x^2}\frac{\partial u}{\partial t} \tag{6}$$

Then we can write the displacement equation of the P wave:

$$u(x,t) = Ae^{-\alpha x}e^{-i(\omega t - kx)} \tag{7}$$

where ω is the angular frequency, k is the wave number, and i is the imaginary number. By substituting Equation (7) into Equation (6) and making the real part equal to the imaginary part, the expression of the attenuation coefficient α can be obtained:

$$\alpha = \left[\frac{\rho\beta\omega^2}{2\left(\beta^2 + \beta'^2\omega^2\right)}\left(\sqrt{1 + \frac{\beta'^2\omega^2}{\beta^2}} - 1\right)\right]^{1/2} \tag{8}$$

where $\beta = \lambda + 2\mu$, $\beta\prime = \lambda' + 2\mu'$, and β and β' are constant. Equation (7) shows that the attenuation coefficient α is a function of frequency; that is, the attenuation of elastic waves at different frequencies in a viscoelastic medium is different. When the frequency of the wave is high,

$$\frac{\beta'^2\omega^2}{\beta^2} \gg 1$$

Additionally, $\omega = 2\pi f$, so Equation (8) can be approximated as

$$\alpha \approx \left(\frac{\rho\pi f}{\beta'}\right)^{1/2} \tag{9}$$

According to Equation (9), the attenuation coefficient α is proportional to the square root of the frequency f. When the frequency increases, the attenuation coefficient increases as well. With the propagation of the elastic wave, high-frequency components are quickly absorbed.

3. Experimental Setup

The samples used in the experiment were long rectangular rods consisting of three different types of rocks (marble, granite, and red sandstone). Relevant information on the samples is presented in Table 1. Ultrasonic transducers (ϕ = 25 mm) were used to generate elastic waves at six different frequencies (50 kHz, 100 kHz, 250 kHz, 500 kHz, 750 kHz, and 1 MHz, and the generated elastic waves are of the sine type, as shown in Figure 3, taking 50 kHz and 100 kHz excitations as examples). Vallen AMSY-6 AE equipment with 32 channels was used as the acquisition device. The signals were collected by a VS45-H-type wide-frequency response piezoelectric sensor with a resonant frequency of 20 kHz to 450 kHz. The preamplifier gain, threshold value, and sampling rate of the signal acquisition system were set to 34 dB, 35 dB, and 10 MHz, respectively.

Table 1. Densities and dimensions of the rock samples.

Sample	Length/cm	Width/cm	Height/cm	Density/g·cm^{-3}
Granite	200	4	4	2.62
Marble	196	4	4	2.60
Red sandstone	181	4	4	2.35

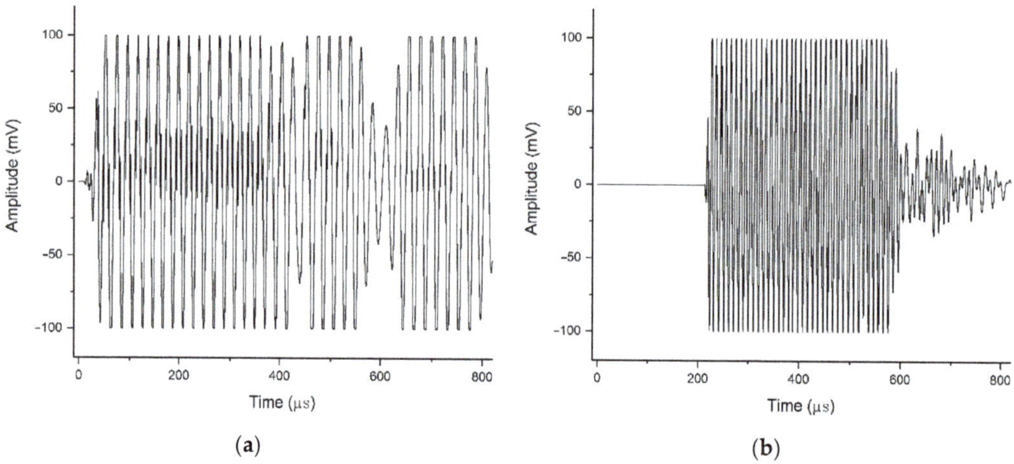

Figure 3. The waveform of the generated signal in (**a**) 50 kHz and (**b**) 100 kHz.

In the test, the transducer was placed at the end of the rock rod. To maintain the position of the transducer in the center of the rock rod end face, suitable cushion blocks were selected to hold the rock rod during the test. To explore the attenuation changes of elastic waves with different frequencies over distance, signal collection sensors were arranged at different locations on one side of the rock rod, and another signal collection sensor was placed at the other rock rod end face opposite to the one where the excitation transducer was fixed. The deployment of the elastic wave excitation transducer and the signal collection sensor is shown in Figure 4 and Table 2. The coupling agent was applied between the sensor, transducer, and contact surface of the rock, and the sensor was fixed with a magnetic suction fixture to prevent it from falling off during the test. After all sensors were pasted, their sensitivity was calibrated individually using lead break tests.

Table 2. Signal collection sensor layout scheme.

Sample	Distance (cm)											
	1	2	...	7	8	9	10	11	12	...	15	16
Granite	5	15	...	65	80	95	110	125	140	...	185	200
Marble	5	15	...	65	80	95	110	125	140	...	185	196
Red sandstone	5	15	...	65	75	85	95	105	120	...	165	181

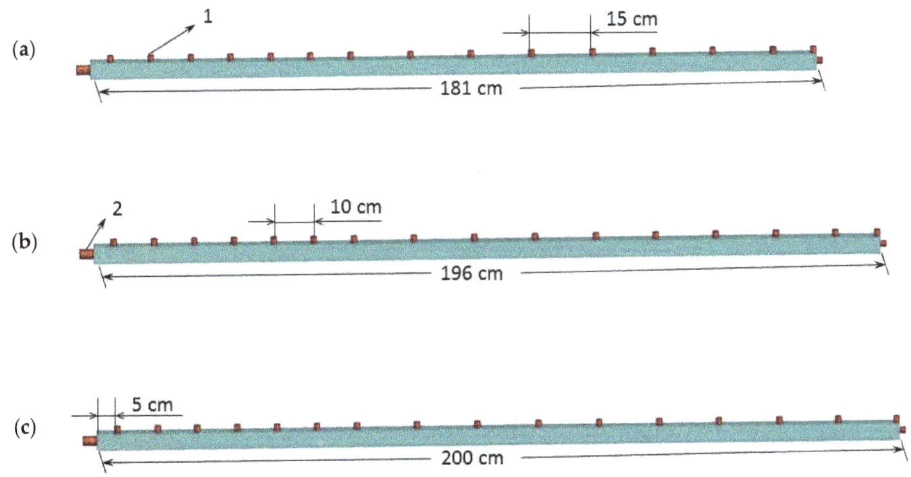

Figure 4. Sensor arrangement for (**a**) red sandstone, (**b**) marble, and (**c**) granite: 1—sensors for elastic wave collection; 2—transducer for elastic wave excitation with various frequencies.

4. Results

4.1. Attenuation Characteristics of the Excitation Sources with Different Frequencies

Figure 5 shows the attenuation of the amplitude with distance for marble, granite, and red sandstone at different frequencies. Owing to the upper limit of the amplitude collected by the sensor, the amplitudes at the first several AE sensors on the marble and granite samples were all 100 dB. The amplitude of the wave in the red sandstone was significantly attenuated, and the sensors located at the back could not receive valid waveform information. As the frequency increased, the amplitude collected by the sensor at the same distance decreased. The decreasing speed of the curve was also accelerated with an increase in frequency, and the attenuation was accelerated. Additionally, the attenuation coefficient of the rock sample increased with an increase in frequency. There were also differences in attenuation between the different rock samples. Based on the inclination of the curve, the attenuation in red sandstone was the highest and the attenuation in marble was the lowest.

The attenuation coefficient in the rock sample can be calculated according to the amplitude distribution obtained in the test, and the value of the attenuation coefficient can be obtained using Equation (10).

$$A(x) = A_0 exp(-\alpha x) \tag{10}$$

where A_0 is the amplitude at the source, α is the attenuation coefficient, x is the distance, and $A(x)$ is the amplitude of the wave propagating to x from the source. The unit of amplitude should be the voltage amplitude (mV), which can be obtained from the waveform data collected by the sensor. The calculated attenuation coefficients of different rocks at different frequencies are listed in Table 3.

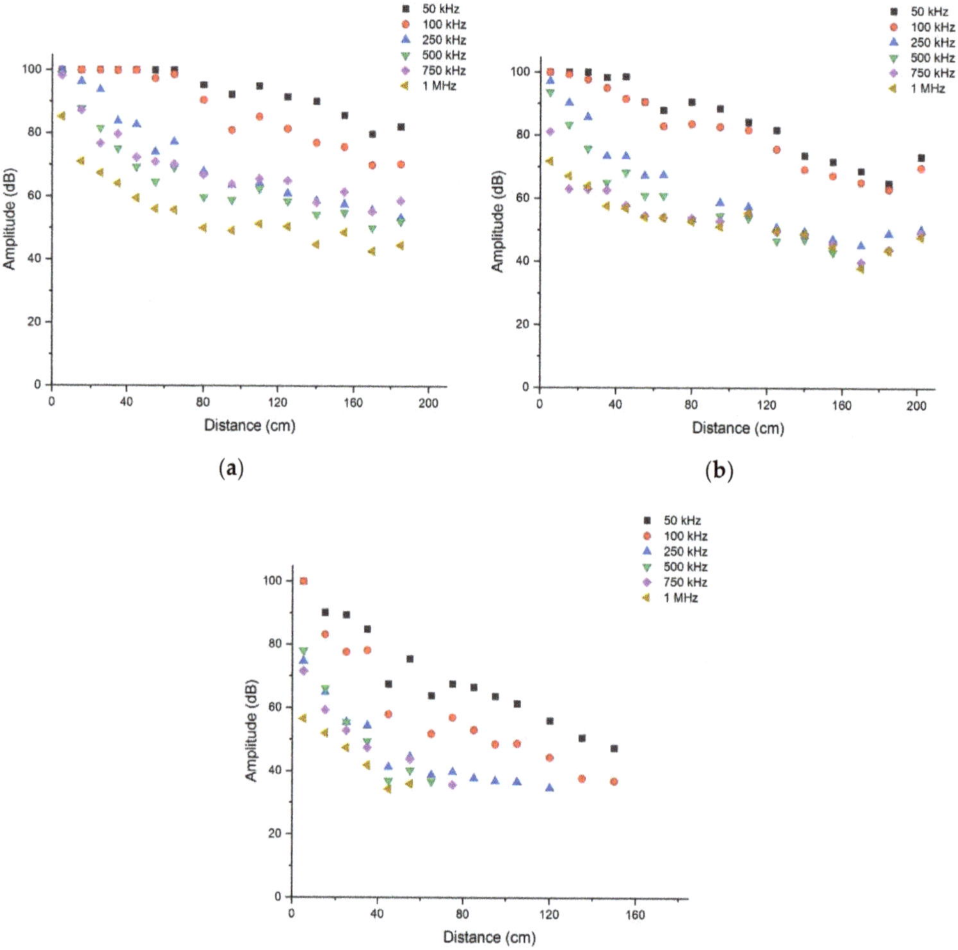

Figure 5. Attenuation of amplitudes in (**a**) marble, (**b**) granite, and (**c**) red sandstone at different frequencies.

Table 3. Attenuation coefficients of the samples (Np·cm^{-1}).

Frequency	Marble	Granite	Red Sandstone
50 kHz	0.0098	0.0197	0.0367
100 kHz	0.0250	0.0278	0.0440
250 kHz	0.0351	0.0398	0.0531
500 kHz	0.0378	0.0533	0.0870
750 kHz	0.0342	0.0421	0.0925
1 MHz	0.0344	0.0295	0.0564

As is shown in Table 3, the attenuation coefficients of the three types of rocks increased significantly with an increase in frequency. At the initial stage of the frequency increase, the increase in attenuation coefficients was roughly proportional. However, when the frequency exceeded 500 kHz, the attenuation coefficient did not increase significantly at 750 kHz and 1 MHz. This may be related to the response frequency range of the broadband sensor for waveform collection. The resonant frequency of the signal collection sensor used in the test was 20–450 kHz; thus, the response effect of the elastic wave signal with a higher frequency may be poor. Based on the distribution characteristics of the attenuation coefficient at different frequencies (hereinafter, we use f to represent the frequency of the generated elastic wave), we fitted the correlation between α and f. The attenuation data obtained when the response was poor at frequencies of 750 kHz and 1 MHz were omitted during the fitting. The fitting results are presented in Figure 6 and Table 4.

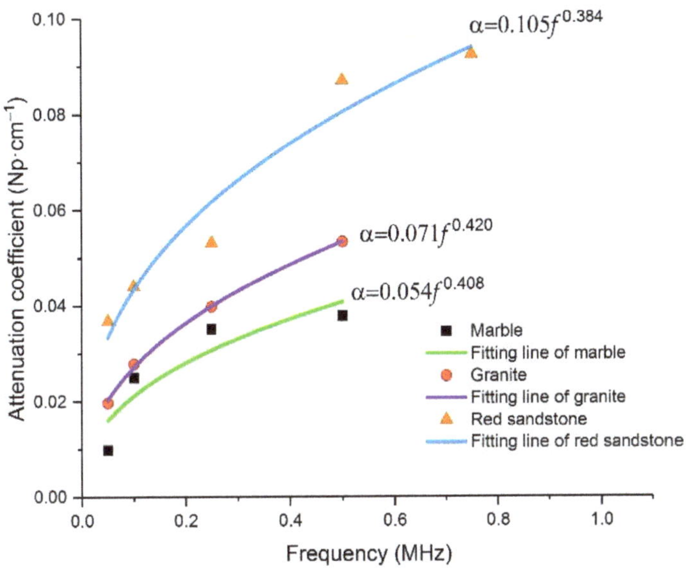

Figure 6. Measured attenuation coefficients of red sandstone, granite, and marble at different frequencies, and the fitted curves. It is obvious that the relationship between α and f has a similar form.

Table 4. Fitting relationship between α and f.

	Marble	Granite	Red Sandstone
Relationship	$\alpha = 0.054f^{0.408}$	$\alpha = 0.071f^{0.420}$	$\alpha = 0.105f^{0.384}$
R^2	0.833	0.998	0.949

By fitting the correlation between α and f, the attenuation coefficients in marble, granite, and red sandstone showed a power function relationship with frequency, with the exponent of the power function being very similar (0.408, 0.420, and 0.384, respectively). The value of R^2 is greater than 0.8, and the fitting degree of the granite and red sandstone fitting curves is greater than 0.9. However, the coefficients before f are different (0.105, 0.071, and 0.054 for red sandstone, granite, and marble, respectively), which lead to highest amplitude attenuation for red sandstone, followed by granite and marble. For the whole elastic wave energy packet, the attenuation mechanism of the medium due to internal friction is approximately the same. The relationship between α and f has the same formalization, but the extent of the attenuation is different due to the different composition of the medium. As a matter of fact, the effects as hysteresis and viscoelastic damping persist throughout the whole propagation process, with energy dissipation accumulating and intensifying, resulting in an exponential increase in attenuation over distance.

4.2. Attenuation Characteristics of Elastic Waves with Specific Frequency Components

When an elastic wave propagates in the rock, a dispersion phenomenon occurs. The different frequency components of the wave were separated owing to the different phase velocities, and the different components of the wave had different attenuation extents. In the analysis in Section 4.1, the change in amplitude with distance reflects the attenuation of elastic waves excited with a specific frequency as a whole in the time domain, and the amplitudes used to calculate the attenuation coefficient are the maximum voltage of the whole waveform packet. In fact, as the elastic wave propagates in rock rods, the high-frequency components of the signal are gradually filtered out, but we just treat the signal of the different frequency components as a whole energy packet to investigate its attenuation characteristics, and the change in specific frequency components of the collected signal in different distances cannot be known. In this section, we will focus on the attenuation of specific frequency components in the waveform during the propagation process. In attenuation tests at various excitation frequencies, we can track the excited frequency component by performing fast Fourier transform (FFT) on the collected time-domain waveforms at different locations. Taking the 500 kHz excitation frequency attenuation test on marble as an example, as shown in Figure 7, the peak frequency (frequency with maximum amplitude in amplitude spectrum) moved from approximately 500 kHz to approximately 100 kHz, and the amplitude of the 500 kHz frequency component gradually decreases with the propagation of the elastic wave. By tracking the specific frequency component in attenuation tests at various excitation frequencies, the amplitude change of the same frequency component with distance can be obtained, thereby allowing us to analyze the attenuation characteristics of the specific frequency component.

As shown in Table 5, the attenuation coefficient (here, we denote this attenuation coefficient as α_s, which is calculated by tracking a specific frequency component at various distances) increases gradually with the increase in frequency, showing a similar trend in three kinds of rock rods, and the attenuation coefficient α_s and f also follow a power function relationship (as shown in Table 6 and Figure 8). However, the exponents of the power functions between α_s and f of the three kinds of rock rods are 0.982, 1.523, and 0.860 for marble, granite, and red sandstone, respectively, which vary a lot and is quite different with the results in Section 4.1, where the exponents of the power functions between α and f are almost the same. This indicates that the exponent of the power function between the attenuation coefficient and frequency is rock-type dependent; we will further discuss this phenomenon in the following section.

Table 5. Attenuation coefficients of the specific frequency components (Np·cm^{-1}).

Frequency	Attenuation Coefficients α_s (Np·cm^{-1})		
	Marble	Granite	Red Sandstone
50 kHz	0.0171	0.0321	0.0368
100 kHz	0.0334	0.0397	0.0742
250 kHz	0.0560	0.0550	0.2008
500 kHz	0.1420	0.1301	0.3267
750 kHz	0.1940	0.2612	—
1 MHz	0.2655	0.4151	—

Notes: α_s is also calculated through Equation (10), with the amplitudes obtained by tracking a specific frequency component at various distances.

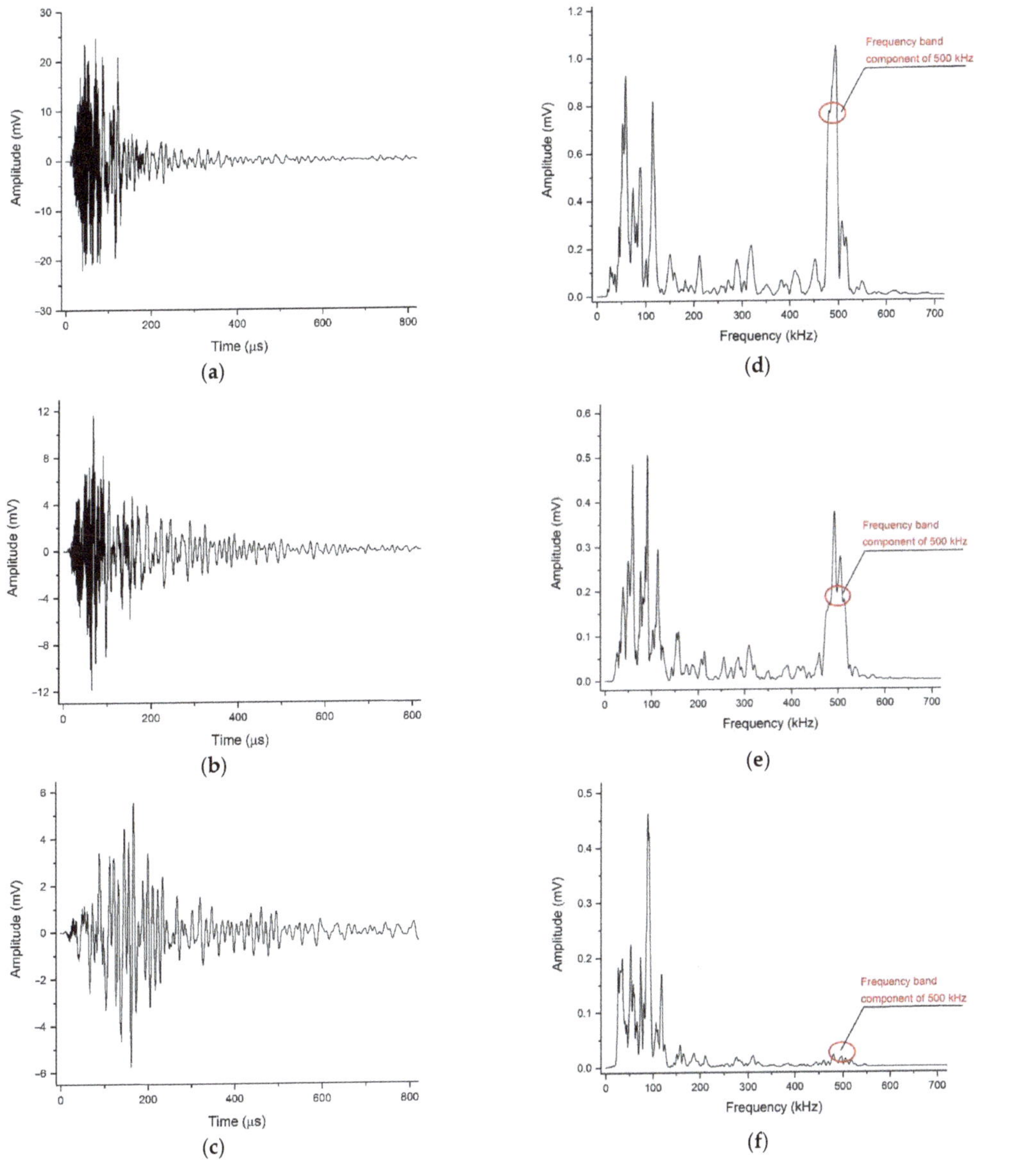

Figure 7. Waveforms (**a**–**c**) and amplitude spectra (**d**–**f**) at 15 cm, 25 cm, and 35 cm in marble at a frequency of 500 kHz. The red circles represent the 500 kHz components.

Table 6. Fitting relationship between α_s and f of the frequency components in waves.

	Marble	**Granite**	**Red Sandstone**
Relationship	$\alpha_s = 0.264f^{0.982}$	$\alpha_s = 0.410f^{1.523}$	$\alpha_s = 0.603f^{0.860}$
R^2	0.994	0.985	0.990

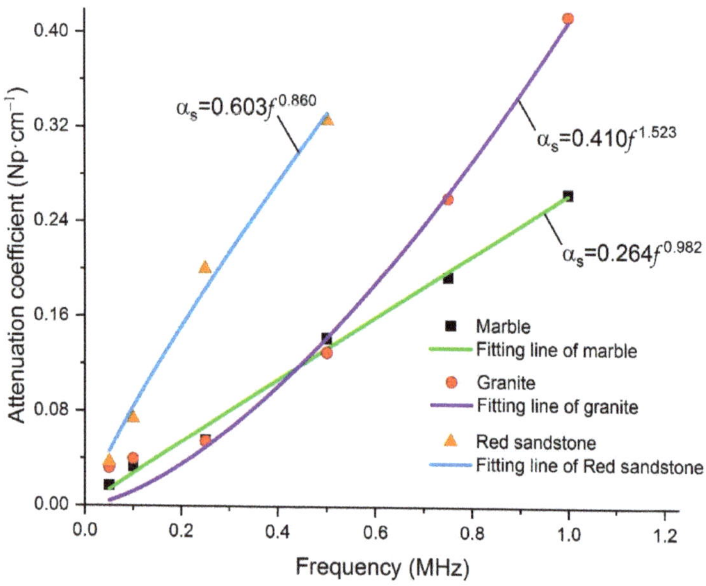

Figure 8. Relationship between α_s of the specific frequency components and f.

5. Discussion

Through the analysis of the experimental data, we obtained the frequency dependence of the attenuation coefficient, reflecting the amplitude of elastic waves in different rocks. In the analysis of the Kelvin–Voigt model, as shown in Equation (9), the exponent of the power function between α and f approximately equals 1/2, and the coefficient before f is determined by the density of the medium and its viscous characteristics. In Section 4.1, the exponents of the power function between α and f of the attenuation tests on marble, granite, and red sandstone are shown to be 0.408, 0.420, and 0.384, respectively, which are very close to that 1/2 theoretical value, while the coefficient before f has a lot to do with the rock type. The influence of the differences between different lithological rocks seems only to involve the coefficient of f and do not affect the exponent of power function between α and f; this is unified in the theoretical analyzing model and the experiment. This demonstrates that the Kelvin–Voigt model can describe well the correlation between the attenuation coefficient and frequency in high-frequency waves and provides a theoretical explanation for the obtained experimental results.

However, it should be noted that the amplitudes used to calculate the attenuation coefficient in Section 4.1 are the maximum voltage of the whole waveform packet, and the change in amplitude with distance reflects the attenuation of elastic waves as a whole energy packet. In Section 4.2, we also tracked the specific frequency component of the waveform in the attenuation tests at various excitation frequencies; the results show that the attenuation coefficient α_s of a specific frequency and f also follow a power function relationship, but the exponents of the power functions between α_s and f of the three kinds of rock rods were 0.982, 1.523, and 0.860 for marble, granite, and red sandstone, respectively, which vary a lot and is quite different to the results in Section 4.1. This indicates that the exponent of the power function between the attenuation coefficient and frequency is rock-type dependent. To further investigate this phenomenon, we obtained the microstructures of several samples through refractive index experiments of transparent sheets (Figure 9). The density of the red sandstone is the lowest; the mineral particles were small, and the gap between the particles was large. It is easy to produce relative sliding when disturbed, such that the elastic energy is converted into thermal energy dissipation, resulting in the absorption of elastic waves. The mineral particles in marble are also small, but closely combined; therefore, the elastic wave is less absorbed when propagating in the medium. This is also beneficial for reducing the attenuation caused by the reflection and refraction of waves at the particle boundary. The mineral particles in the granite sample are larger and the integrity of the particles is good; however, the internal structure was relatively broken, and there were many long-shaped gaps between the particles. This microstructure provides favorable conditions for frictional sliding between particles and increases the attenuation of elastic waves in the medium. This may also be the reason the attenuation in granite was greater than that in marble of a similar density.

(**a**) (**b**) (**c**)

Figure 9. Microstructure of (**a**) red sandstone, (**b**) granite, and (**c**) marble obtained from the transparent refractive index experiments. The particle size and distribution of red sandstone and marble are relatively uniform but vary greatly in granite.

Joints and fissures often exist in rock materials, and voids exist between mineral particles. These discontinuities provide favorable conditions for the sliding of adjacent fissures or particle surfaces when elastic waves pass through the medium [42]. When the elastic wave propagates in the rock material, it causes a change in the particle stress and strain, and the normal stress between the arriving interfaces increases, resulting in an increase in friction and energy loss caused by relative motion. This slippage continues until the wave passes through the interface. At this point, the direction of the frictional shear stress changes, the cracks or pores return to the original equilibrium position, and friction work is generated again. As the frequency of the wave increases, the reciprocating cycle of this friction force accelerates, resulting in a faster attenuation velocity. Therefore, the main factor causing elastic wave attenuation is the internal friction. Internal friction and scattering are medium-related factors affecting elastic wave attenuation. Separation between frictional attenuation and scattering is challenging. Matsushima et al. [43] proposed a technique to separate the two attenuations under the assumption of linear dependence of frictional attenuation on frequency, and only vertical seismic profiling (VSP) data. However, attenuation due to scattering becomes significant when the wavelength of the elastic wave is comparable to the diameter of the mineral particle. It can be seen from Figure 9 that the microstructure of red sandstone and marble exhibit similar characteristics—a uniform small particle size and distribution—while the microstructure of granite, which has a large particle size, has obvious differences with the above two kinds of rocks. When the frequency of the excited elastic wave is larger than 500 kHz, the wavelength of these high-frequency elastic waves is equivalent to the size of the mineral particles in granite, and the scattering attenuation will be significant, resulting in a higher attenuation coefficient compared to marble (the α_s of granite becomes larger than that of marble when the excited elastic wave frequency exceeds 500 kHz, as seen in Figure 8). As the mineral particles in red sandstone and marble are small, the wavelength of the excited frequency elastic waves in this test is far less comparable with the diameter of the mineral particles in red sandstone and marble; correspondingly, the scattering attenuation is not significant in these two kinds of rocks.

This is also the reason why the exponent of the power functions between the α_s and f of granite is different from marble and red sandstone.

6. Conclusions

Elastic waves attenuate during propagation inside a rock and their amplitude and energy decrease with increasing distance. The attenuation velocities of the elastic waves with different frequencies were different in this study. Based on the experiments conducted, if the elastic waves are treated as a whole energy packet to conduct the attenuation test in rock materials, the experimental results are in good agreement with the viscoelasticity assumption based on the Kelvin–Voigt model, which deduced that the exponent of the power function relationship between the attenuation coefficient and frequency is about 1/2 and independent of rock type. However, if the specific frequency components of the elastic wave are tracked in the attenuation tests, the exponent of the power function relationship between the attenuation coefficient and frequency is rock-type dependent, which does not match the result based on the Kelvin–Voigt model. As a matter of fact, the component of scattering attenuation is not reflected in the Kelvin–Voigt model, and it cannot be reflected in the attenuation analysis of elastic waves as a whole energy packet. The results of this paper show that the attenuation analysis by tracking specific frequency components can well demonstrate the detailed causes of attenuation, and the attenuation characteristics can also be well related with the internal structure of rock material. Therefore, attention should be paid to the scattering attenuation component in high-frequency elastic wave attenuation tests in rock-like materials.

Furthermore, high-frequency signal attenuation testing requires targeted collection sensors; that is, the resonant frequency of the signal collection sensors affects the results of this type of research. Among the attenuation coefficients of the different rock samples at different frequencies, as shown in Table 3, α did not increase significantly when the frequency was higher than 500 kHz, which may be related to the poor response of the selected sensor to frequencies above 500 kHz. On the one hand, if a broadband sensor is selected, the response to the different frequency signals is not good. When selecting a narrowband sensor, the response to a certain frequency signal is good, but the response to other frequency components is poor. Based on the experimental results, it is worthwhile to consider sensors with different resonant frequencies at different locations.

Author Contributions: Conceptualization, X.L. and S.W.; methodology, X.L. and F.X.; software, F.X., Q.X., X.Y.; validation, X.L.; formal analysis, F.X. and D.C.; writing—original draft preparation, X.L. and F.X.; writing—review and editing, X.L. and S.W; funding acquisition, X.L. All authors have read and agreed to the published version of the manuscript.

Funding: This research was funded by the National Natural Science Foundation of China (Grant No. 42172316), Natural Science Foundation of Hunan Province (Grant No. 2021JJ30810) and the Research Fund of The State Key Laboratory of Coal Resources and Safe Mining, CUMT (SKL-CRSM21KF005).

Institutional Review Board Statement: Not applicable.

Informed Consent Statement: Not applicable.

Data Availability Statement: The origin date of attenuation test of elastic wave in rock rods has been uploaded to a public repository called figshare (https://figshare.com/articles/book/amplitude_and_attenuation_coefficient_xlsx/19213035) on 22 February 2022, and it is also available by contacting the corresponding author at sf.wang@csu.edu.cn.

Acknowledgments: This research was funded by the National Natural Science Foundation of China (Grant No. 42172316), Natural Science Foundation of Hunan Province (Grant No. 2021JJ30810) and the Research Fund of The State Key Laboratory of Coal Resources and Safe Mining, CUMT (SKLCRSM21KF005), for which the authors are very thankful.

Conflicts of Interest: The authors declare no conflict of interest.

References

1. Hardy, H.R., Jr. *Acoustic Emission/Microseismic Activity*; A. A. Balkema Publishers: Lisse, PA, USA, 2003.
2. Born, W.T. The attenuation constant of earth materials. *Geophysics* **1941**, *6*, 132–148. [CrossRef]
3. Knopoff, L.; MacDonald, G.J.F. Attenuation of small amplitude stress waves in solids. *Rev. Mod. Phys.* **1958**, *30*, 1178–1192. [CrossRef]
4. Futterman, W.I. Dispersive Body Waves. *J. Geophys. Res. Atmos.* **1962**, *67*, 5279–5291. [CrossRef]
5. Jackson, D.D. Elastic relaxation model for seismic wave attenuation in the earth. *Phys. Earth Planet. Inter.* **1969**, *2*, 30–34. [CrossRef]
6. Liu, J.; Ba, J.; Ma, J.; Yang, H. An analysis of seismic attenuation in random porous media. *Sci. China Phys. Mech. Astron.* **2010**, *53*, 628–637. [CrossRef]
7. Johnston, D.H. *The Attenuation of Seismic Waves in Dry and Saturated Rocks*; Massachusetts Institute of Technology: Cambridge, MA, USA, 1973.
8. Biot, M.A. Theory of Propagation of Elastic Waves in a Fluid-Saturated Porous Solid. I. Low-Frequency Range. *J. Acoust. Soc. Am.* **1956**, *28*, 168–178. [CrossRef]
9. Bouchaala, F.; Guennou, C. Estimation of viscoelastic attenuation of real seismic data by use of ray tracing software: Application to the detection of gas hydrates and free gas. *C. R. Geosci.* **2012**, *344*, 57–66. [CrossRef]
10. Jacobs, L.J.; Owino, J.O. Effect of aggregate size on attenuation of Rayleigh surface waves in cement-based materials. *J. Eng. Mech.* **2000**, *126*, 1124–1130. [CrossRef]
11. Johnston, D.H.; Toksöz, M.N.; Timur, A. Attenuation of seismic waves in dry and saturated rocks: II. Mechanisms. *GEOPHYSICS* **1979**, *44*, 691–711. [CrossRef]
12. Ma, R.; Ba, J. Coda and intrinsic attenuation from ultrasonic measurements in tight siltstones. *J. Geophys. Res. Solid Earth* **2020**, *125*, e2019JB018825. [CrossRef]
13. Winkler, K.; Nur, A. Pore fluids and seismic attenuation in rocks. *Geophys. Res. Lett.* **1979**, *6*, 1–4. [CrossRef]
14. Dong, L.; Tong, X.; Ma, J. Quantitative Investigation of Tomographic Effects in Abnormal Regions of Complex Structures. *Engineering* **2021**, *7*, 1011–1022. [CrossRef]
15. Tisato, N.; Madonna, C. Attenuation at low seismic frequencies in partially saturated rocks: Measurements and description of a new apparatus. *J. Appl. Geophys.* **2012**, *86*, 44–53. [CrossRef]
16. Adam, L.; Batzle, M.; Lewallen, K.T.; van Wijk, K. Seismic wave attenuation in carbonates. *J. Geophys. Res.* **2009**, *144*, 1–14. [CrossRef]
17. Lei, X.-L.; Kusunose, K.; Nishizawa, O.; Cho, A.; Satoh, T. On the Spatio-Temporal Distribution of Acoustic Emissions in Two Granitic Rocks Under Triaxial Compression: The Role of Pre-Existing Cracks. *Geophys. Res. Lett.* **2000**, *27*, 1997–2000. [CrossRef]
18. Thompson, B.D.; Young, R.P.; Lockner, D.A. Fracture in Westerly Granite under AE Feedback and Constant Strain Rate Loading: Nucleation, Quasi-static Propagation, and the Transition to Unstable Fracture Propagation. *Pure Appl. Geophys.* **2006**, *163*, 995–1019. [CrossRef]
19. Dong, L.; Tong, X.; Hu, Q.; Tao, Q. Empty region identification method and experimental verification for the two-dimensional complex structure. *Int. J. Rock Mech. Min. Sci.* **2021**, *147*, 104885. [CrossRef]
20. Liu, X.-l.; Li, X.-b.; Hong, L.; Yin, T.b.; Rao, M. Acoustic emission characteristics of rock under impact loading. *J. Cent. South Univ.* **2015**, *22*, 3571–3577. [CrossRef]
21. Dong, L.; Luo, Q. Investigations and new insights on earthquake mechanics from fault slip experiments. *Earth-Sci. Rev.* **2022**, *228*, 104019. [CrossRef]
22. Liu, X.; Liu, Z.; Li, X.; Gong, F.; Du, K. Experimental study on the effect of strain rate on rock acoustic emission characteristics. *Int. J. Rock Mech. Min. Sci.* **2020**, *133*, 104420. [CrossRef]
23. Xie, Q.; Li, S.-X.; Liu, X.-L.; Gong, F.-Q.; Li, X.-B. Effect of loading rate on fracture behaviors of shale under mode I loading. *J. Cent. South Univ.* **2020**, *27*, 3118–3132. [CrossRef]
24. Liu, X.; Han, M.; He, W.; Li, X.; Chen, D. A New b Value Estimation Method in Rock Acoustic Emission Testing. *J. Geophys. Res. Solid Earth* **2020**, *125*, e2020JB019658. [CrossRef]
25. Dong, L.; Hu, Q.; Tong, X.; Liu, Y. Velocity-Free MS/AE Source Location Method for Three-Dimensional Hole-Containing Structures. *Engineering* **2020**, *6*, 827–834. [CrossRef]
26. Dong, L.; Chen, Y.; Sun, D.; Zhang, Y. Implications for rock instability precursors and principal stress direction from rock acoustic experiments. *Int. J. Min. Sci. Technol.* **2021**, *31*, 789–798. [CrossRef]
27. Chen, D.; Liu, X.; He, W.; Xia, C.; Gong, F.; Li, X.; Cao, X. Effect of attenuation on amplitude distribution and b value in rock acoustic emission tests. *Geophys. J. Int.* **2021**, *229*, 933–947. [CrossRef]
28. Scholz, C.H. *Microfracturing of Rock in Compression*; Massachusetts Institute of Technology: Cambridge, MA, USA, 1967.
29. Dobrin, M.B. *Introduction to Geophysical Prospecting*; McGraw-Hill Book Company: New York, NY, USA, 1960.
30. Emmerich, H.; Korn, M. Incorporation of attenuation into time-domain computations of seismic wave fields. *Geophysics* **1987**, *52*, 1252–1264. [CrossRef]
31. Carcione, J.M.; Kosloff, D.; Kosloff, R. Wave propagation simulation in a linear viscoacoustic medium. *Geophys. J. Int.* **1988**, *93*, 393–401. [CrossRef]
32. Müller, T.M.; Gurevich, B.; Lebedev, M. Seismic wave attenuation and dispersion resulting from wave-induced flow in porous rocks—A review. *Geophysics* **2010**, *75*, 75A147–75A175. [CrossRef]

33. Nur, A.; Simmons, G. The effect of viscosity of a fluid phase on velocity in low porosity rocks. *Earth Planet. Sci. Lett.* **1969**, *7*, 99–108. [CrossRef]
34. Attewell, P.B.; Ramana, Y.V. Wave attenuation and internal friction as functions of frequency in rocks. *Geophysics* **1966**, *31*, 1049–1056. [CrossRef]
35. Zhang, G.; Li, X.; Zhang, S.; Kundu, T. Investigation of frequency-dependent attenuation coefficients for multiple solids using a reliable pulse-echo ultrasonic measurement technique. *Measurement* **2021**, *177*, 109270. [CrossRef]
36. Schön, J.H. *Physical Properties of Rocks: Fundamentals and Principles of Petrophysics*; Pergamon Press: Oxford, UK, 1996.
37. Tiwari, R.; Mukhopadhyay, S. On electromagneto-thermoelastic plane waves under Green–Naghdi theory of thermoelasticity-II. *J. Therm. Stresses* **2017**, *40*, 1040–1062. [CrossRef]
38. Tiwari, R.; Misra, J.C.; Prasad, R. Magneto-thermoelastic wave propagation in a finitely conducting medium: A comparative study for three types of thermoelasticity I, II, and III. *J. Therm. Stresses* **2021**, *44*, 785–806. [CrossRef]
39. Yamamoto, M. Phenomenological Theory of Visco-elasticity of Three Dimensional Bodies. *J. Phys. Soc. Jpn.* **1959**, *14*, 313–330. [CrossRef]
40. Červený, V.; Pšenčík, I. Time-averaged and time-dependent energy-related quantities of harmonic waves in inhomogeneous viscoelastic anisotropic media. *Geophys. J. Int.* **2007**, *170*, 1253–1261. [CrossRef]
41. Carcione, J.M.; Poletto, F.; Gei, D. 3-D wave simulation in anelastic media using the Kelvin–Voigt constitutive equation. *J. Comput. Phys.* **2004**, *196*, 282–297. [CrossRef]
42. Bouchaala, F.; Ali, M.Y.; Matsushima, J. Attenuation study of a clay-rich dense zone in fractured carbonate reservoirs. *Geophysics* **2019**, *84*, B205–B216. [CrossRef]
43. Matsushima, J.; Ali, M.Y.; Bouchaala, F. A novel method for separating intrinsic and scattering attenuation for zero-offset vertical seismic profiling data. *Geophys. J. Int.* **2017**, *211*, 1655–1668. [CrossRef]

materials

Article

Experimental Study on Mechanical Properties and Failure Laws of Granite with Artificial Flaws under Coupled Static and Dynamic Loads

Guang Li [1,2], Shuaiqi Liu [3,*], Rong Lu [4], Fengshan Ma [1,2] and Jie Guo [1,2]

1 Key Laboratory of Shale Gas and Geoengineering, Institute of Geology and Geophysics, Chinese Academy of Sciences, Beijing 100029, China
2 Institutions of Earth Science, Chinese Academy of Sciences, Beijing 100029, China
3 Economics & Technology Research Institute, China National Petroleum Corporation, Beijing 100011, China
4 Beijing Urban Construction Design & Development Group Co., Ltd., Beijing 100029, China
* Correspondence: lifangrui21@mails.ucas.ac.cn; Tel.: +86-153-1164-1640

Abstract: Rock is the main construction material of rock engineering, such as the engineering of mines and tunnels; in addition, its mechanical properties and failure laws are of great significance to the stability evaluation of rock engineering, especially under the conditions of coupled static–static stresses. In this study, granite specimens were manufactured with artificial flaws. Coupled static and dynamic loads tests were carried out with a modified split Hopkinson pressure bar (SHPB) apparatus; and six typical levels of axial pre-stresses and three crack inclination angles were designed. Three-dimensional digital image correlation (3D-DIC) was also applied to record and analyze the fracturing process and damage evolution of the specimens. The test results show that there was no compaction stage in the stress–strain curve under combined dynamic and static loading. The dynamic strength of the specimens increased first and then decreased with the increase in the static pressure; moreover, the specimens reached the maximum dynamic strength when the static pressure was 10% UCS. The dynamic strength decreased first and then increased with the increase in the crack inclination angle; and the lowest strength appeared when the inclination angle was 45°. The change in axial compression had a significant influence on the failure mode, and the failure mode gradually transformed from shear–tensile failure to shear failure with the increase in the pre-stress. The tensile strain was usually generated at the end of the fractures or near the rock bridge. When the axial pressure was small, the tensile strain zone parallel to the loading direction was easily generated; and when the axial pressure was large, a shear strain zone developed, extending along the diagonal direction. The research results can provide a theoretical reference for the correct understanding of the failure mechanisms of granite and its engineering stability under actual conditions.

Keywords: coupled static and dynamic loads; artificial flaw; SHPB; 3D-DIC; failure laws

Citation: Li, G.; Liu, S.; Lu, R.; Ma, F.; Guo, J. Experimental Study on Mechanical Properties and Failure Laws of Granite with Artificial Flaws under Coupled Static and Dynamic Loads. *Materials* **2022**, *15*, 6105. https://doi.org/10.3390/ma15176105

Academic Editor: Saeed Chehreh Chelgani

Received: 4 August 2022
Accepted: 1 September 2022
Published: 2 September 2022

Publisher's Note: MDPI stays neutral with regard to jurisdictional claims in published maps and institutional affiliations.

1. Introduction

In the fields of resource exploitation, transportation, and underground engineering, the construction of deep rock engineering is more and more frequent; in addition, the research of deep rock mass mechanics has received extensive attention. Due to the influence of in situ stress, geological structure, weathering, human activities, and other factors, there are many defects in rock that weaken its ultimate bearing capacity; this includes cracks, holes, joints, foliation, faults, and so on [1,2]. Rock fractures usually start at the tip of the original defect, extend through each other, and finally form a whole failure [3,4]. In addition to the high ground stress, in the process of deep engineering construction, rocks experience different disturbances and are often subjected to strong dynamic loads; such as blasting vibration, excavation, drilling, etc., which is a combination of dynamic and static loads [5,6]. Both rock fractures and engineering disturbances are important factors that affect the mechanical properties of a deep high-stress rock mass. Moreover, deep hard rock has significant brittleness; this is more likely to cause engineering disasters, such as rock burst and rock engineering instability. Therefore, it is necessary to study the mechanical properties and failure behavior of defective rocks under combined dynamic and static loading.

Laboratory testing of rock fracture and damage under combined dynamic and static loading is an important method to understand this problem. Based on the improved split Hopkinson pressure bar (SHPB), Li et al. conducted impact experiments on marble with a single fracture; the results showed that the dynamic failure mode of a fractured rock mass was shear failure [7]. Wang et al. carried out one-dimensional dynamic and static combined loading tests on coal samples with different moisture content [8]. Li et al. studied the mechanical and failure characteristics of sandstone samples with prefabricated holes under dynamic–static loading [9]. Weng et al. reproduced the crack propagation and failure process of granite with round or square holes [10]. Li et al. conducted one-dimensional dynamic and static combined loading tests on sandstone samples with prefabricated holes to study the cracking behavior and its mechanism [11]. Yin et al. explored the influence of heat treatment temperature and loading rate on the dynamic fracture behavior of Brazilian disc granite containing fractures under a pre-static load [12]. In addition, some scholars have used the acoustic emission, speckle, and mathematical programming methods to explore the strength, fracture mechanism, and failure evolution law of defective rocks [13–16]. However, the above researchers have mostly selected relatively low-strength rock samples for the tests; moreover, the studies on the mechanical properties of deep hard rock are still insufficient. Meanwhile, there have been few studies on the role of static confining pressure and the inclination angle of fracture in the process of rock failure under coupled static and dynamic loads.

Therefore, a series of laboratory tests were designed and conducted to study the mechanical and failure characteristics of fractured rock under the combined dynamic and static loading; based on an improved SHPB system and digital image correlation (DIC) technology. These provide a theoretical reference for understanding the failure mechanism of granite and evaluating the stability of engineering surrounding rock.

2. Coupled Static and Dynamic Loading Tests

2.1. Creating the Specimens

Rock samples in the following tests were all taken from the Sanshandao gold mine, China. Granite with good integrity and homogeneity was used; with a density of 2580 kg/m^3; an elastic modulus of 9.86 GPa; and a uniaxial compressive strength (UCS) of 139.66 MPa. A series of core slices were identified using optical microscopy. The rock sample composition is approximately 34% plagioclase, 27% potassium feldspar, 30% granular quartz, and 6% flaky biotite; and accessory minerals, such as calcite, white mica, and brown epidote, and so on, as shown in Figure 1 [17,18]. Rock cores were processed into cuboid samples, with a size of 45 mm × 45 mm × 20 mm (length × height × thickness), were divided into complete samples and crack samples with double parallel joints. A cutting machine was used to precast cracks in the granite specimens. The length of the fractures was 10 mm and the width was 1 mm, as shown in Figure 2. The specimens(University of South China, Hengyang, China) with cracks included three different inclination angles, which were 0°, 45°, and 90°, respectively.

(a) (b) (c)

Figure 1. Microscopic scanning of the rock sample: (**a**) a picture of the scanner; (**b**) the microscopic scanning of the rock sample; (**c**) the element content characteristics of the rock sample.

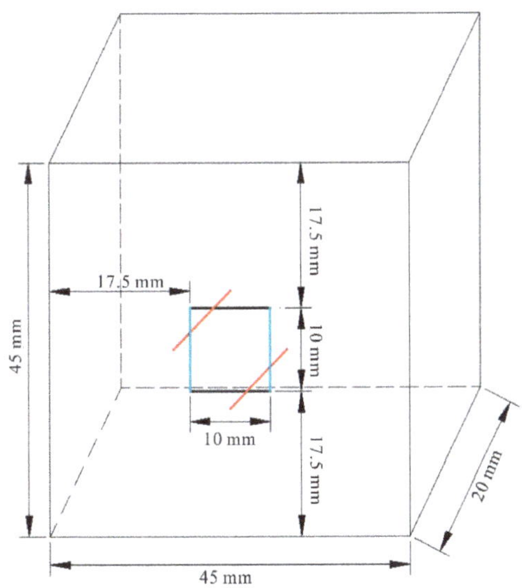

Figure 2. Specimen size.

2.2. Testing Equipment

In this experiment, the SHPB device(Central South University, Changsha, China) was used to simulate the mechanical behavior of the specimens; with parallel double cracks under simultaneous dynamic and static loading. The SHPB device is mainly composed of three parts, namely the power drive system; the pressure rod system; and the data collection system, as shown in Figure 3 [19]. The pressure rod system includes an incident rod, a transmission rod, and an absorption rod. The specimen was placed between the incident rod and the transmitted rod; and the impact process was realized using air pressure to accelerate the warhead. Two sets of strain gauges were attached to the incident and transmitted rods to record the strain. In order to achieve the purpose of the constant strain rate loading, the spindle punch was used to realize the half sine wave loading. Meanwhile, the system was equipped with an ultra-dynamic strain gauge and oscilloscope; this can realize the acquisition and record the stress wave signal and data processing. Based on the one-dimensional stress wave theory, the average dynamic stress, strain, and strain rate of the sample can be calculated by the following formula [20]:

$$\sigma(t) = \frac{A_e}{2A_s}[\sigma_I(t) - \sigma_R(t) + \sigma_T(t)]$$

$$\varepsilon(t) = \frac{1}{\rho_e C_e L_s}\int_0^t [\sigma_I(t) + \sigma_R(t) - \sigma_T(t)]d_t$$

$$\dot{\varepsilon}(t) = \frac{1}{\rho_e C_e L_s}[\sigma_I(t) + \sigma_R(t) - \sigma_T(t)]$$

where $\sigma_I(t)$, $\sigma_R(t)$, and $\sigma_T(t)$ are the incident stress, reflected stress, and transmitted stress of the rod at time t; A_e, ρ_e, and C_e are the cross-sectional area, density, and longitudinal wave velocity of the rod; and A_s and L_s are the cross-sectional area and length of the sample, respectively.

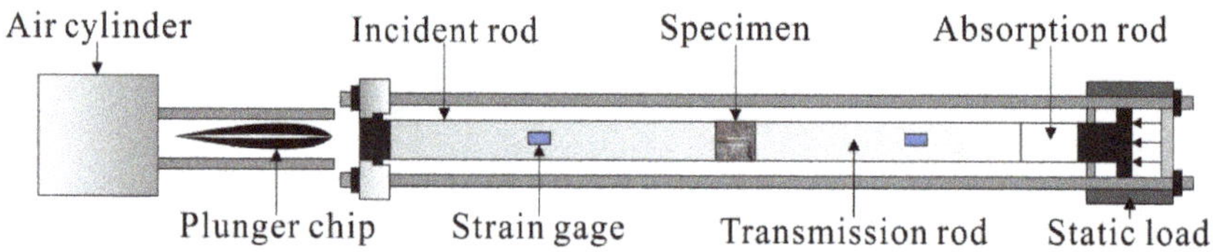

Figure 3. Testing equipment.

DIC was used to observe the failure process of the rock under combined dynamic and static loading; which can measure the deformation process of the samples without direct contact, compared to other deformation measurement methods [21]. During the test, two high-intensity light-emitting diode (LED) lights were used to fill the light, two high-speed cameras were used to shoot, and DIC dynamic high-speed cameras(CSI Technology Group, West Hartford, CT, USA) were used to record the progressive failure process. After the artificial speckles were sprayed on the sample surface, the three-dimensional coordinates of the measured points on the object surface before and after the deformation could be obtained based on the principle of binocular stereo vision. Based on a series of continuous speckle images collected from different perspectives, the correlation matching operation was performed on all the markers of the two images before and after the deformation; in addition, the 3D displacement field and strain field on the surface of the sample during loading were finally obtained.

2.3. *Experimental Project*

The pre-loaded axial static pressure rates of the sample were set as 14.0, 27.9, 41.9, 69.8, and 83.8 MPa; these corresponded to 0%, 10%, 20%, 30%, 50%, and 60% of the UCS, and are represented by S_0, S_A, S_B, S_C, S_D, and S_E, respectively. The sample containing prefabricated cracks was referred to as 'Flaw'. In other words, sample S_B-flaw 45° represents a rock specimen containing 45° cracks with a pre-loaded axial static pressure of 10% UCS.

In order to ensure the correctness of the SHPB test results, dynamic stress balance had to be reached at both sides of the sample before rock failure under the dynamic load. The stress equilibrium curves of the S_B-flaw specimens were taken as examples. As shown in Figure 4, the superposition wave of the incident wave and reflected wave was basically consistent with the transmitted wave, especially the section before the peak; this indicates that the rock sample reached the stress equilibrium state during the loading period. In the deformation process of the rock samples, the assumption of the stress balance condition was satisfied; moreover, the improved SHPB device was consistent with the one-dimensional stress wave transfer characteristics, which can effectively eliminate the wave dispersion and inertia effects. Thus, this proves the validity of the test results.

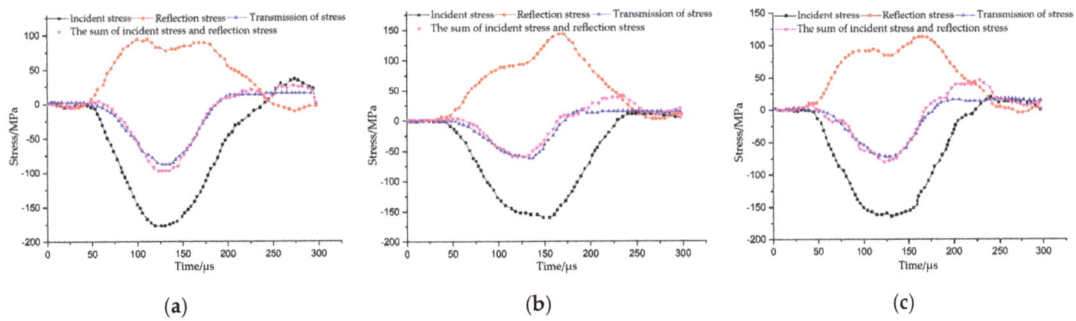

Figure 4. Stress equilibrium curves of the S_B-flaw specimens: (**a**) S_B-flaw 0°; (**b**) S_B-flaw 45°; and (**c**) S_B-flaw 90°.

3. Results and Discussion

3.1. Test Results

The test results of the dynamic and static combined loading are shown in Table 1; where the dynamic strength is the peak stress of the dynamic stress–strain curve, reflecting the impact resistance of the sample. The combined strength was the sum of the axial pre-static load and dynamic strength, reflecting the actual peak strength when the sample was damaged [22,23].

Table 1. Dynamic–static loading test results.

Sample	Static Pressure/MPa	Dynamic Strength/MPa	Combined Strength/MP	Peak Strain/10^{-3}	Strain Rate/s^{-1}
S_0-flaw 0°	0	155.60	155.60	8.83	109.16
S_0-flaw 45°		132.36	132.36	5.02	105.55
S_0-flaw 90°		141.32	141.32	6.07	136.89
S_A-flaw 0°	10% UCS	196.50	210.50	6.62	98.37
S_A-flaw 45°		152.72	166.72	5.70	135.10
S_A-flaw 90°		177.50	191.50	7.01	96.38
S_B-flaw 0°	20% UCS	194.15	222.05	5.28	113.72
S_B-flaw 45°		132.27	160.17	5.46	144.60
S_B-flaw 90°		159.76	187.66	5.11	119.33
S_C-flaw 0°	30% UCS	189.45	231.35	5.53	114.61
S_C-flaw 45°		132.68	174.58	5.37	154.26
S_C-flaw 90°		152.70	194.60	5.01	143.12
S_D-flaw 0°	50% UCS	166.90	236.70	4.28	140.65
S_D-flaw 45°		90.96	160.76	3.21	141.67
S_D-flaw 90°		133.18	202.98	4.18	103.60
S_E-flaw 0°	60% UCS	149.32	233.12	4.04	131.82
S_E-flaw 45°		106.25	190.05	3.57	162.34
S_E-flaw 90°		122.97	206.77	3.41	144.48

3.2. Deformation Characteristics

The stress–strain relationship of the sample was obtained from the signal transformation collected by the strain gauge, as shown in Figure 5a–f. Overall, different from general static load stress–strain curves, the stress–strain curves under dynamic and static loading did not have a compaction phase, and began directly at the elastic stage. This is because the pre-loaded static pressure had a compaction effect on the sample before the dynamic load loading, and the micro defects inside the sample were closed. The stress–strain curve of the combined dynamic and static loading can be divided into the pre-failure zone and post-failure zone; the stress peak point being the boundary. The pre-failure zone consisted of two stages, namely the elastic stage and the yield stage. Due to the fast loading rate, the rock samples quickly reached the peak strength and then completely failed; in addition, there was nearly no strain-softening stage or residual strength stage.

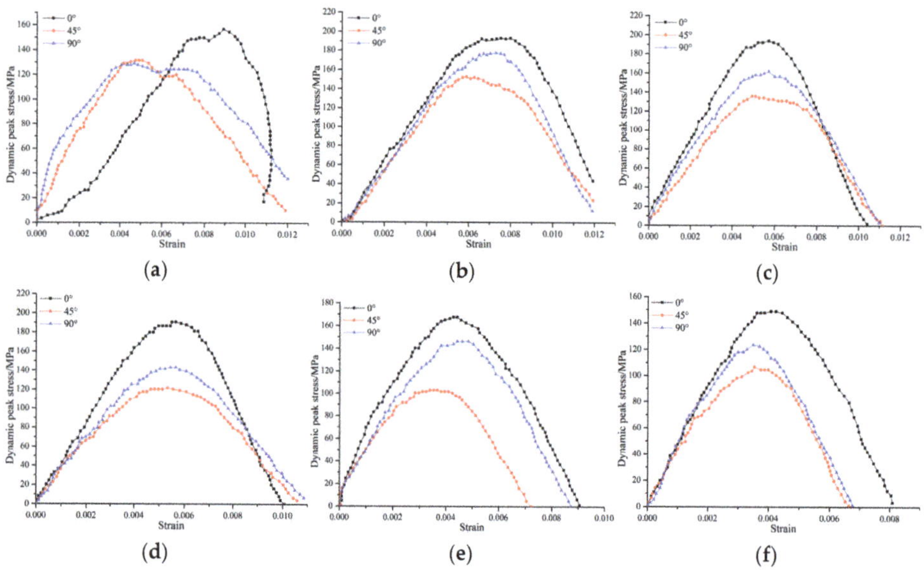

Figure 5. Dynamic stress–strain curves of different inclination angles under static axial compression: (**a**) 0 MPa; (**b**) 10% UCS; (**c**) 20% UCS; (**d**) 30% UCS; (**e**) 50% UCS; and (**f**) 60% UCS.

The results are shown in Table 1 and Figure 5. When the crack angle was constant and the axial static pressure increased gradually in the range of 0~83.8 MPa (0%~60% UCS), the dynamic strength of the specimen rose first and then declined. When the static pressure was 10% UCS, the dynamic strength of the specimens reached the maximum. The pre-stress played the role of compaction on the specimen and increased its ability to resist the dynamic force. However, with the increase in the pressure value, the static pressure itself caused damage to the specimen; thus, reducing the strength of the specimen. Under the same static pressure, the dynamic strength of the sample first decreased; it then increased with the increase in the crack inclination angle. The strength was weakest as the inclination angle of the fissure was 45°. Therefore, in actual engineering, more attention should be paid to the change trend of the fissure dip angle in the surrounding rock; this is conducive to our judgment on engineering stability.

3.3. Failure Modes

Figure 6 shows the maximum principal strain cloud diagram of the fractured granite specimens with a 0° dip angle under different axial pressures. It can be seen from the figure that the axial compression significantly affected the failure mode of the sample. When the static load was 0, three tensile strain zones parallel to the loading direction were generated in the specimen, two of which ran through the cracks; the final result was obvious tensile failure (Figure 6a). When the static load was 10% UCS, two inclined tensile strain bands led to the sample failure; in addition, a large tensile strain zone was generated between the parallel double cracks (Figure 6b). When the static load was 20% UCS, an inclined shear strain zone through the sample appeared in the upper part of the specimen; and a compound strain zone with shear strain and tensile strain developed around the lower crack (Figure 6c). When the static load was 30% UCS, a set of "X"-shaped shear strain zones centered on the fracture developed in the specimen; moreover, two parallel fractures were connected due to the shear strain (Figure 6d). When the static load was 50% UCS, the upper part of the specimen generated a penetrated tensile strain zone parallel to the loading direction; the lower part formed a composite zone of tensile and shear strain through the cracks; and a large local tensile strain failure occurred on the lower right corner (Figure 6e). When the static load was 60% UCS, the failure of the specimen was caused by a primary shear strain zone connecting the lower crack; and a secondary shear strain zone connecting the upper crack of the specimen. In addition, a large tensile strain zone was generated between the parallel cracks and the upper left part of the specimen (Figure 6f).

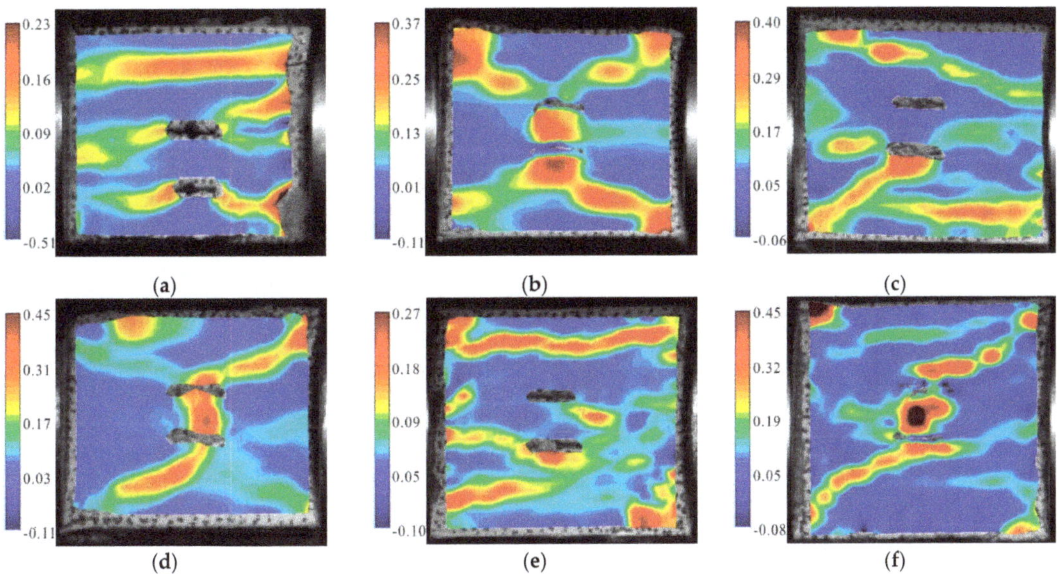

Figure 6. Maximum principal strain cloud diagram of the granite specimens with 0° cracks: (**a**) S_0-flaw 0°; (**b**) S_A-flaw 0°; (**c**) S_B-flaw 0°; (**d**) S_C-flaw 0°; (**e**) S_D-flaw 0°; and (**f**) S_E-flaw 0°.

Figure 7 shows the maximum principal strain cloud diagram of the fractured granite specimens with a 45° dip angle under different axial pressures. It can be seen from the figure that anti-wing shear strain zones always developed in the sample; and tensile strain zones formed at the tip of the parallel double fracture and were parallel to the loading direction. When the static load was 0, two tensile strain concentration zones developed on the upper left corner and the lower right corner of the specimen, respectively; they connected with the counter-wing shear strain zone, resulting in the failure (Figure 7a). When the static load was 10% UCS, an inclined shear strain zone from the upper left to the lower right and a horizontal tensile strain zone formed in the specimen; in addition, a violent deformation failure occurred between the two strain concentration zones (Figure 7b). When the static load was 20% UCS, a horizontal tensile strain zone connected to the fracture tip formed in the upper and lower parts of the specimen, respectively; together, these cracks led to a shear–tensile failure (Figure 7c). When the static load was 30% UCS, except for an anti-wing shear strain zone along the diagonal through the specimen, a small shear strain zone also appeared along the fracture. The two strain concentration zones connected with each other and extended to the boundary of the specimen to cause damage (Figure 7d). When the static load was 50% UCS, there was a small shear strain zone at the upper crack tip and a tensile strain zone at the lower crack tip; together, they caused the sample failure (Figure 7e). When the static load was 60% UCS, a typical anti-wing shear strain zone along the diagonal direction formed in the specimen; while a horizontal tensile strain region also appeared in the upper part (Figure 7f).

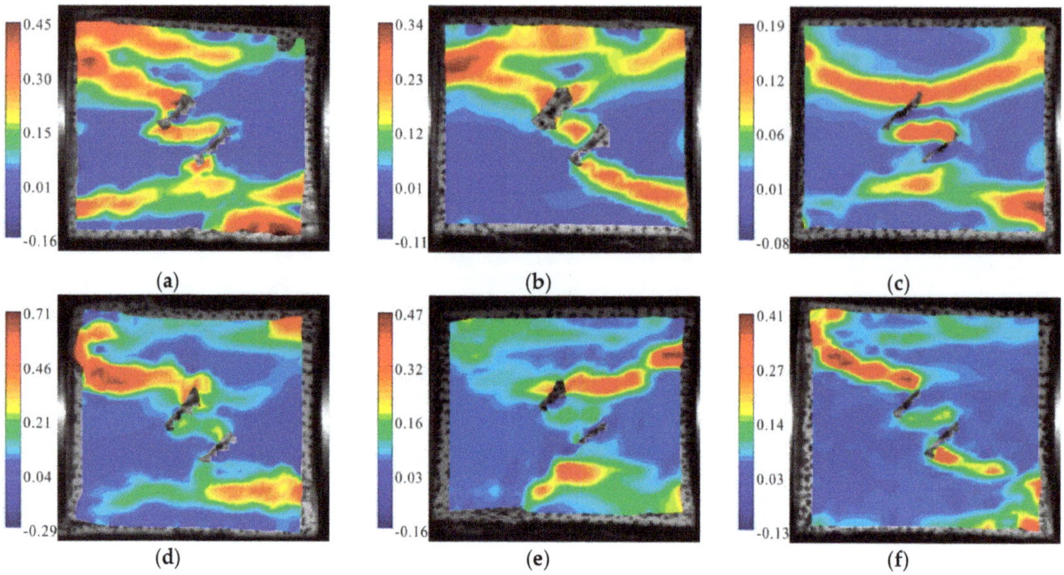

Figure 7. Maximum principal strain cloud diagram of the granite specimens with 45° cracks: (**a**) S_0-flaw 45°; (**b**) S_A-flaw 45°; (**c**) S_B-flaw 45°; (**d**) S_C-flaw 45°; (**e**) S_D-flaw 45°; and (**f**) S_E-flaw 45°.

Figure 8 shows the maximum principal strain cloud diagram of the fractured granite specimens with a 90° dip angle under different axial pressures. It can be seen from the Figure that when the static load was 0, there were two main tensile strain zones and one secondary tensile strain zone in the specimen. One of the main tensile strain zones ran through the tip of the double fractures, and the expansion of the three strain concentration zones jointly led to the final failure (Figure 8a). When the static load was 10% UCS, a tensile crack through the cracks was generated in the upper part of the specimen and connected with the upper tensile strain zone at the right boundary of the specimen; while a shear strain zone was generated near the lower part of the cracks (Figure 8b). When the static load was 20% UCS, there were two tensile strain zones in the lower part of the specimen: one of which was approximately parallel to the loading direction and went through the specimen; while the other one developed through the lower tip of the cracks. Meanwhile, a shear strain zone developed in the upper part and along the inclined direction (Figure 8c). When the static load was 30% UCS, two shear strain zones were generated in the upper left and lower left of the specimen, respectively; both of which tended to develop toward the fracture tip. In addition, a principal tensile strain zone was formed between the upper tip of the left fracture and the left boundary (Figure 8d). When the static load was 50% UCS, the sample failure was induced by two shear strain zones connecting the upper and lower tips of the double fracture, respectively (Figure 8e). When the static load was 60% UCS, a tensile strain zone formed in the lower part of the specimen; and a shear strain zone developed from two sides of the specimen to the cracks. These two strain concentration zones connected with each other and eventually lead to the failure (Figure 8f).

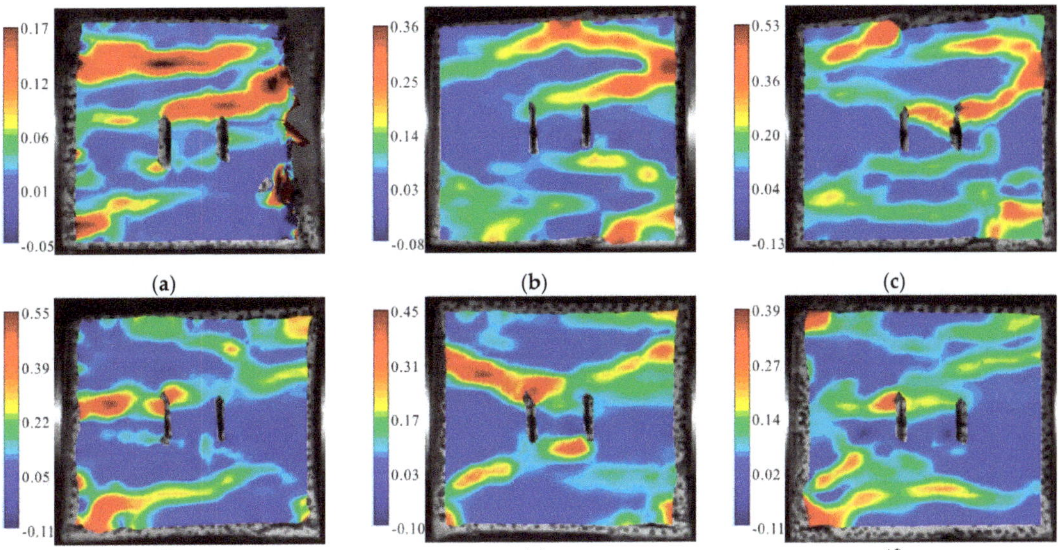

Figure 8. Maximum principal strain cloud diagram of the granite specimens with 90° cracks: (**a**) S_0-flaw 90°; (**b**) S_A-flaw 90°; (**c**) S_B-flaw 90°; (**d**) S_C-flaw 90°; (**e**) S_D-flaw 90°; and (**f**) S_E-flaw 90°.

4. Conclusions

1. The dynamic strength of the specimens increased first and then decreased with an increase in static pressure; in addition, the specimen reached the maximum dynamic strength when the static pressure was 10% UCS. The pre-stress played the role of compaction on the specimen and increased its ability to resist the dynamic force. However, with the increase in the pressure value, the static pressure itself caused damage to the specimen; thus, reducing the strength of the specimen. The dynamic strength decreased first and then increased with the growth of the crack inclination angle; moreover, the lowest strength appeared when the inclination angle was 45°. The angle of 45° was conducive to the generation of shear strain zones along the diagonal direction, which often led to the final failure of the specimen.
2. The change in the axial compression had a significant influence on the failure mode. In terms of the specimen with 0° cracks, the specimen exhibited tensile failure in the absence of axial compression. Under uniaxial compression of 20% and 50%, the typical tensile shear failure occurred; and when the uniaxial compression was 10%, 30%, or 60%, the shear failure was predominant. The failure mode of the specimens with a 45° fracture was mainly a counter-wing shear strain developing around the end of the fractures; in addition, this shear mode became more and more obvious with the increase in the axial compression. The failure mode gradually transformed from shear–tensile failure to shear failure. The failure mode of the specimens with a 90° crack was mainly characterized by the tensile strain at the tips of the fracture. With the increase in the axial compression, the length and number of shear cracks increased; and the failure mode transformed from tensile failure to shear–tensile failure.
3. The inclination angle of the double parallel cracks played a key role in the formation and development of the tensile strain zone. The tensile strain was usually generated at the end of the fractures or near the rock bridge. When the axial pressure was small, the tensile strain zone parallel to the loading direction was easily generated; moreover, when the axial pressure was large, the shear strain zone extending along the diagonal direction developed.

Author Contributions: Data curation, G.L.; formal analysis, S.L. and J.G.; methodology, G.L. and R.L.; software, G.L. and S.L.; writing—original draft, G.L.; writing—review and editing, G.L. and F.M.; experiments, G.L., R.L., and S.L. All authors have read and agreed to the published version of the manuscript.

Funding: This research was supported by the National Science Foundation of China (grant nos. 41831293 and 42072305).

Institutional Review Board Statement: Not applicable.

Informed Consent Statement: Not applicable.

Data Availability Statement: Not applicable.

Acknowledgments: The authors are grateful to the assigned editor and anonymous reviewers for their enthusiastic help and valuable comments, which have greatly improved this paper.

Conflicts of Interest: The authors declared that they have no conflict of interest to this work. We declare that we do not have any commercial or associative interest that represents a conflict of interest in connection with the work submitted.

References

1. Liu, G.; Chen, Y.; Du, X.; Wang, S.; Fernández-Steeger, T.M. Evolutionary Analysis of Heterogeneous Granite Microcracks Based on Digital Image Processing in Grain-Block Model. *Materials* **2022**, *15*, 1941. [CrossRef]
2. Yang, S.Q.; Huang, Y.H.; Tian, W.L.; Zhu, J.B. An experimental investigation on strength, deformation and crack evolution behavior of sandstone containing two oval flaws under uniaxial compression. *Eng. Geol.* **2017**, *217*, 35–48. [CrossRef]
3. Zou, C.; Wong, L.N.; Loo, J.J.; Gan, B.S. Different mechanical and cracking behaviors of single-flawed brittle gypsum specimens under dynamic and quasi-static loadings. *Eng. Geol.* **2016**, *201*, 71–84. [CrossRef]
4. Li, G.; Ma, F.; Guo, J.; Zhao, H. Experimental research on deformation failure process of roadway tunnel in fractured rock mass induced by mining excavation. *Env. Earth Sci.* **2022**, *82*, 243. [CrossRef]
5. Wang, W.; Wang, H.; Li, D.; Li, H.; Liu, Z. Strength and failure characteristics of natural and water-saturated coal specimens under static and dynamic loads. *Shock. Vib.* **2018**, *2018*, 1–15. [CrossRef]
6. Li, G.; Ma, F.; Guo, J.; Zhao, H.; Liu, G. Study on deformation failure mechanism and support technology of deep soft rock roadway. *Eng. Geol.* **2020**, *264*, 105262. [CrossRef]
7. Chen, H.; Shao, Z.; Fujii, Y. An Experimental Investigation on the Creep Behavior of Deep Brittle Rock Materials. *Materials* **2022**, *15*, 1877. [CrossRef]
8. Li, X.; Zhou, T.; Li, D. Dynamic strength and fracturing behavior of single-flawed prismatic marble specimens under impact loading with a split-hopkinson pressure bar. *Rock Mech. Rock Eng.* **2016**, *50*, 1–16. [CrossRef]
9. Tang, L.Z.; Cheng, L.P.; Wang, C.; Shu, J.B.; Li, W.J.; Chen, Y. Dynamic characteristics of serpentinite under condition of high static load and frequent dynamic disturbance. *Rock Soil Mech.* **2016**, *37*, 2737–2745.
10. Li, D.; Xiao, P.; Han, Z.; Zhu, Q. Mechanical and failure properties of rocks with a cavity under coupled static and dynamic loads. *Eng. Fract. Mech.* **2018**, *225*, 106195. [CrossRef]
11. Weng, L.; Li, X.; Taheri, A.; Wu, Q.; Xie, X. Fracture evolution around a cavity in brittle rock under uniaxial compression and coupled static-dynamic loads. *Rock Mech. Rock Eng.* **2018**, *51*, 531–545. [CrossRef]
12. Li, Y.; Peng, J.; Zhang, F.; Qiu, Z. Cracking behavior and mechanism of sandstone containing a pre-cut hole under combined static and dynamic loading. *Eng. Geol.* **2016**, *213*, 64–73. [CrossRef]
13. Yin, T.; Bai, L.; Li, X.; Li, X.; Zhang, S. Effect of thermal treatment on the mode I fracture toughness of granite under dynamic and static coupling load. *Eng. Fract. Mech.* **2018**, *199*, 143–158. [CrossRef]
14. Minliang, C.; Hongwen, J.; Xiujun, M.; Haijian, S.; Mingrui, D.; Tantan, Z. Fracture evolution characteristics of sandstone containing double fissures and a single circular hole under uniaxial compression. *Int. J. Min. Sci. Technol.* **2017**, *27*, 499–505.
15. Khatri, K.; Lal, A. Stochastic XFEM based fracture behavior and crack growth analysis of a plate with a hole emanating cracks under biaxial loading. *Theor. Appl. Fract. Mech.* **2018**, *96*, 1–22. [CrossRef]
16. Feng, P.; Dai, F.; Liu, Y.; Xu, N.; Fan, P. Effects of coupled static and dynamic strain rates on mechanical behaviors of rock-like specimens containing pre-existing fissures under uniaxial compression. *Can. Geotech. J.* **2018**, *55*, 640–652. [CrossRef]
17. Li, G.; Wang, Z.; Ma, F.; Guo, J.; Liu, J.; Song, Y. A Case Study on Deformation Failure Characteristics of Overlying Strata and Critical Mining Upper Limit in Submarine Mining. *Water* **2022**, *14*, 2465. [CrossRef]
18. Liu, S.; Ma, F.; Zhao, H.; Guo, J.; Duan, X.; Sun, Q. Numerical Investigation of a Hydrosplitting Fracture and Weak Plane Interaction Using Discrete Element Modeling. *Water* **2020**, *12*, 535. [CrossRef]
19. Liu, X.; Yang, S.; Huang, Y.; Cheng, J. Experimental study on the strength and fracture mechanism of sandstone containing elliptical holes and fissures under uniaxial compression. *Eng. Fract. Mech.* **2019**, *205*, 205–217. [CrossRef]
20. Li, X.; Zhou, Z.; Lok, T.S.; Hong, L.; Yin, T. Innovative testing technique of rock subjected to coupled static and dynamic loads. *Int. J. Rock Mech. Min. Sci.* **2008**, *45*, 739–748. [CrossRef]

21. Zhao, Y.; Zhang, L.; Wang, W.; Pu, C.; Wan, W.; Tang, J. Cracking and stress–strain behavior of rock-like material containing two flaws under uniaxial compression. *Rock Mech. Rock Eng.* **2016**, *49*, 2665–2687. [CrossRef]
22. Song, H.; Zhang, H.; Fu, D.; Kang, Y.; Huang, G.; Qu, C.; Cai, Z. Experimental study on damage evolution of rock under uniform and concentrated loading conditions using digital image correlation. *Fatigue Fract. Eng. Mater. Struct.* **2013**, *36*, 760–768. [CrossRef]
23. Zou, C.; Wong, L.N. Experimental studies on cracking processes and failure in marble under dynamic loading. *Eng. Geol.* **2014**, *173*, 19–31. [CrossRef]

materials

MDPI

Article

The Confinement-Affected Strength Variety of Anisotropic Rock Mass

Songfeng Guo [1,2,3,*], Shengwen Qi [1,2,3], Bowen Zheng [1,2,3], Lei Xue [1,2,3], Xueliang Wang [1,2,3], Ning Liang [1,2,3], Yu Zou [1,2,3], Fengjiao Tang [1,2,3], Waqar Muhammad Faisal [1,2,3], Weiluan Wen [1,2,3], Yongchao Li [1,2,3] and Xin Yu [1,2,3]

1 Key Laboratory of Shale Gas and Geoengineering, Institute of Geology and Geophysics, Chinese Academy of Sciences, Beijing 100029, China
2 Innovation Academy for Earth Science, Chinese Academy of Sciences, Beijing 100029, China
3 University of Chinese Academy of Sciences, Beijing 100049, China
* Correspondence: guosongfeng@mail.iggcas.ac.cn; Tel.: +86-010-8299-8022

Abstract: It has been recognized that the anisotropic structures dominate the deformation and strength properties of laminated rock masses. The resultant strength anisotropy is strongly affected by confining pressures beyond anisotropic structures. Nevertheless, the effects of confinement are inconsistent among existing experiments and not fully understood. This study focuses on the effects of confining pressure on strength anisotropy through theoretical derivation together with experimental results analysis. The variations in the possibility of anisotropic structural plane dominant failure and strength anisotropy degree under different confining pressures are discussed. The different types of anisotropic structural planes, i.e., the fresh contact discontinuity or soft, thick layer, are found as the key factor resulting in different confinement effects. The strength anisotropy weakens gradually and vanishes eventually as confining stress increases for the anisotropic rock mass with the structural plane of fresh contact discontinuity. On the other hand, the strength does not vanish at very high confining stress and the anisotropic strength difference even rises as confining stress increases for the anisotropic rock mass with the anisotropic structural plane of the soft layer. This study improves the understanding of anisotropic rock mass mechanical behavior, especially at high confining stress, and may promote the development of excavation and supporting techniques for underground projects.

Keywords: rock mass; anisotropy; confining pressure; strength; failure mechanism; discontinuity

Citation: Guo, S.; Qi, S.; Zheng, B.; Xue, L.; Wang, X.; Liang, N.; Zou, Y.; Tang, F.; Faisal, W.M.; Wen, W.; et al. The Confinement-Affected Strength Variety of Anisotropic Rock Mass. *Materials* **2022**, *15*, 8444. https://doi.org/10.3390/ma15238444

Academic Editors: Angelo Marcello Tarantino, Shaofeng Wang, Tianshou Ma, Linqi Huang, Jie Zhou and Changjie Zheng

Received: 22 October 2022
Accepted: 25 November 2022
Published: 27 November 2022

1. Introduction

It is commonly believed that rock mass is heterogeneous and discontinuous, containing fissures, joints, faults, cleavage planes, and bedding planes and these structural planes dominate the mechanical behaviors of rock mass [1–7]. Different loading directions concerning structural planes orientations always make rock mass anisotropic and more problematic during engineering construction [8–17].

Numerous types of rock mass have inherent or structural anisotropy, such as parametamorphic and sedimentary rocks, and other discontinuity-induced stratiform-like rock masses. Extensive compression tests have been put forward to explore the strength and failure features of various anisotropic rock masses at uniaxial and triaxial compression conditions in the laboratory, e.g., slate [18–20], schist [13,20–24], phyllite [13,25], marble [13,23], sandstone [3], shale [26–28], limestone [28–30], mudrock [31], columnar basalt [32], and artificially anisotropic rock specimens [3,33]. Numerical studies have also been conducted to investigate the mechanical behavior of anisotropic rock mass intrinsically [12,34]. These effective studies reflect that the deformation and strength properties are largely dominated by the angle between the normal anisotropic structural plane and the direction of minimum principal stress (α). As shown in Figure 1, most of the anisotropic rock mass had maximum strength at a critical angle around $\alpha = 0°$ or $90°$ and failed as the rock block fractured, while they had a minimum around $\alpha = 45° + \varphi_{an}/2$ and failed while sliding along the anisotropic structural plane, where φ_{an} is the friction angle of anisotropic structural plane.

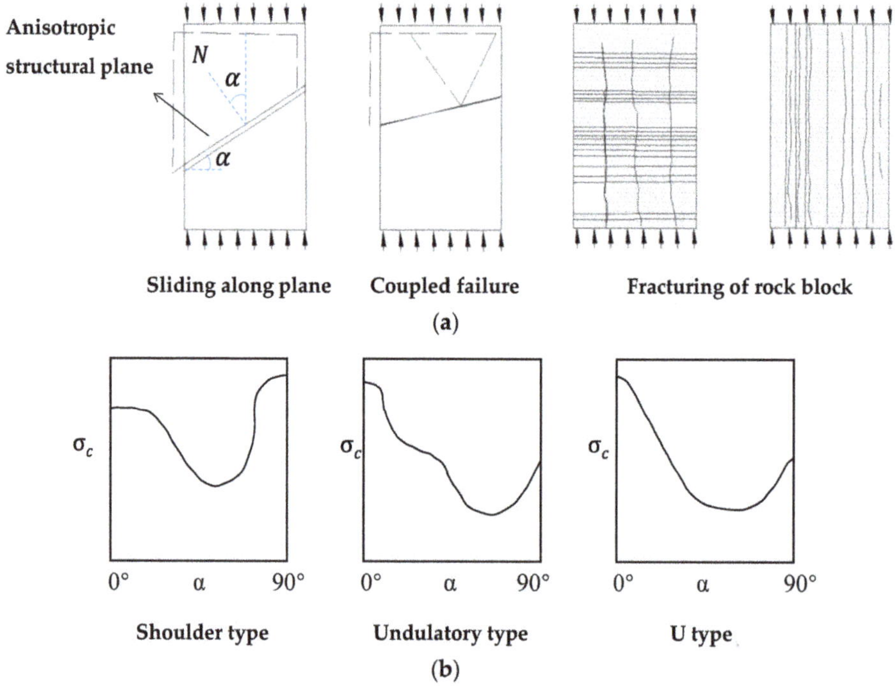

Figure 1. The direction dominant mechanical behavior of anisotropic rock mass. (**a**) The failure types anisotropy; (**b**) the strength anisotropy (modified based on [35]).

As the depth of engineering construction tends to be greater and greater in recent years, the environmental geostress of rock mass reaches as high as tens of MPa. The mechanical properties of anisotropic rock mass under high confining stress are more concerning among rock mass geomechanical researchers. The existing research has indicated that the confining stress greatly affects the strength anisotropy by contributing to the normal stress on the weak plane [24,36–38]. Sun (1988) noted that although the rock mass structure dominates the mechanical behavior of the rock mass at low confining stress state, the domination of the rock mass structure is restricted as confining stress increases [2]. His viewpoint was supported by some triaxial compression tests, which showed that the strength anisotropy decreased gradually with confining pressure increases and finally disappeared at high enough confining pressure [38,39]. The transformation critical confining stress of rock mass strength from anisotropy to isotropy was proposed and comparable with the experiments [40]. The recently developed classification methods for anisotropic rock mass consider the weaken effects of confining pressure on the anisotropy degree [41,42]. However, some experiments also indicate that strength anisotropy exists even at very high confining pressure, e.g., dolomitic limestone has strength anisotropy at a confining stress of 80 MPa [29]. Thus, the strength anisotropy properties of rock mass under various confining pressure are not fully understood and need further discussions.

This study focuses on the effects of confining pressure on strength anisotropy through theoretical derivation along with experimental results analysis. The variations in the possibility of anisotropic structural plane dominant failure and strength anisotropy degree under different confining pressures are discussed. The different types of anisotropic structural plane are the key factors resulting in different confinement effects.

2. Theoretical Analysis on Strength of Anisotropic Rock Mass

Several researchers have developed strength prediction methods for anisotropic rock mass in order to depict the strength variations with the orientation of anisotropic structural plane under different confining stresses. These methods are mostly modified based on Griffith theory [43], Mohr–Coulomb strength criterion [1,28,36,44–52], Hoek-Brown strength criterion [23,41,53,54], and other empirical formulas [19,22,25,55] or the fuzzy method [21,56]. Among these methods, the Mohr–Coulomb strength criterion based on the single plane of weakness theory [1] is one of the most widely used. The anisotropic structural plane is a well-defined parallel discontinuity, and the rock block is regarded as isotropic, both are depicted by Mohr–Coulomb criterion in the classic Jaeger's criterion. In this study, the widely used Mohr–Coulomb strength threshold is adopted to depict the strength of rock block and anisotropic structural plane. The strength and failure mode of an anisotropic rock specimen under confining pressure (σ_3) are exhibited with Mohr cycles in Figure 2. The strength of anisotropic rock specimen changes with the inclined angle α of anisotropic structural plane. The potential maximum and minimum strengths can be signified as Equations (1) and (2).

$$\sigma_{1max} = \frac{1+\sin\varphi_{rb}}{1-\sin\varphi_{rb}}\sigma_3 + \frac{2c_{rb}\cos\varphi_{rb}}{1-\sin\varphi_{rb}} \quad \text{when } \alpha = 45^\circ + \frac{1}{2}\varphi_{rb} \tag{1}$$

$$\sigma_{1min} = \frac{1+\sin\varphi_{as}}{1-\sin\varphi_{as}}\sigma_3 + \frac{2c_{as}\cos\varphi_{as}}{1-\sin\varphi_{as}} \quad \text{when } \alpha = 45^\circ + \frac{1}{2}\varphi_{as} \tag{2}$$

where σ_{1max} and σ_{1min} denote the potential maximum and minimum strength of anisotropic rock specimen strength, respectively. φ_{rb} and c_{rb} denote the internal friction angle and cohesion of the rock block, respectively; while φ_{as} and c_{as} denote the friction angle and cohesion of the anisotropic structural plane, respectively.

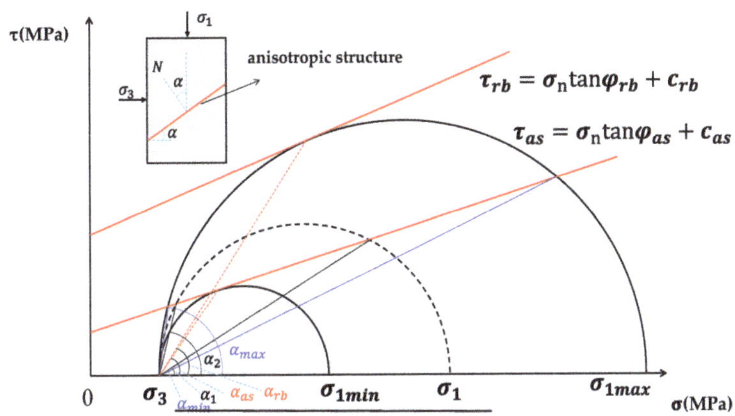

Figure 2. The Mohr circle analysis on limit failure condition of anisotropic rock mass. and τ_{as} denote the shear strength of rock block and anisotropic structural plane, respectively, while σ_n denotes the normal stress on shear failure plane.

As shown in Figure 2, the failure of anisotropic rock mass should be sheared through the rock block when $\alpha < \alpha_{min}$ or $\alpha > \alpha_{max}$ and its strength can be presented by Equation (1). On the other hand, the failure of anisotropic rock mass is most likely to slide or shear along the anisotropic structural plane when $\alpha_{min} < \alpha < \alpha_{max}$, and its strength is between the results reached by Equations (1) and (2), which is shown in Equation (3).

$$\sigma_{1as} = \sigma_3 \tan\alpha \cot(\alpha - \varphi_{as}) + \frac{c_{as}}{\cos\alpha(\sin\alpha - \cos\alpha\tan\varphi_{as})} \quad \text{when } \alpha_{\min} < \alpha < \alpha_{\max} \tag{3}$$

where σ_{1as} denotes the strength of anisotropic rock mass, α_{max} and α_{min} denote the maximum and minimum inclined angle α between which anisotropic rock mass strength is dominant by anisotropic structural plane.

This indicates that the strength relates closely to the applied stress direction α in Equation (3). The minimum strength expressed in Equation (2) is a special case of Equation (3) at $\alpha = 45° + \frac{1}{2}\varphi_{as}$. In addition, the strength reached in Equation (3) is equivalent to that reached in Equation (1) when $\alpha = \alpha_{min}$ or $\alpha = \alpha_{max}$, and the anisotropic structural plane and rock block dominate the failure mechanism cooperatively. The two-angle thresholds α_{min} and α_{max} can be obtained based on the equivalence of Equations (1) and (3), and the Mohr circle analysis, see Equations (4) and (5).

$$\alpha_{min} = \frac{1}{2}\sin^{-1}\left\{\left[1 + \frac{(c_{as}\cot\varphi_{as} + \sigma_3)(1 - \sin\varphi_{rb})}{\sigma_3\sin\varphi_{rb} + c_{rb}\cos\varphi_{rb}}\right]\sin\varphi_{as}\right\} + \frac{1}{2}\varphi_{as} \quad (4)$$

$$\alpha_{max} = \frac{\pi}{2} + \varphi_{as} - \alpha_{min} \quad (5)$$

This indicates from the equations that the critical angles α_{min} and α_{max} are not only the functions of strength parameters but also for confining stress.

3. Verification Study of Existing Experimental Data

The structural planes that influence the anisotropic property can be generally grouped into two types according to mechanical properties, i.e., weak and hard discontinuities (Figure 3). The former includes bedding planes in sedimentary rocks, weak intercalated layers in rock matrix. The latter mainly refers to fresh and clean fractures (or relatively stiff interlayers sometimes) within the rock matrix. In this section, the experimental results involving different types of discontinuities in literatures are presented to assess the availability of above analysis. The dolomitic limestone in Section 3.1 and sandstone in Section 3.2 are sedimentary rocks and thus their anisotropic structural planes are mainly bedding planes regarded as weak planes. Comparatively, the anisotropic structural planes of plaster of Paris in Section 3.3 are an artificial contact joint, regarded as hard discontinuities.

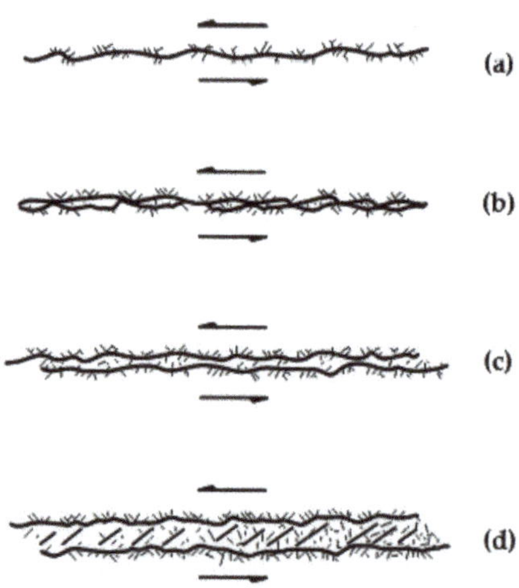

Figure 3. The types of discontinuity. The fresh rough contact joint is well-exhibited in (**a**), while the filled layer becomes thicker and thicker (**b**–**d**) denoting the soft discontinuous layer.

3.1. Dolomitic Limestone

A large number of experimental studies on the strength of layered rock mass were conducted under varied confining pressures [29]. The dolomitic limestone of Manlius formation was prepared as a cylindrical rock specimen with a length of 10 cm and diameter of 1.2 cm. The tests were carried out under four levels of confining pressures of 20 MPa, 40 MPa, 60 MPa, and 80 MPa, respectively, and the inclined angles of layers (α) ranged from 0° to 90°. The internal friction angle (φ_{rb}) and cohesion (c_{rb}) of the matrix (rock block) were 34.5° and 78.23 MPa, respectively, while the internal friction angle (φ_{as}) and cohesion (c_{as}) of limestone layer (anisotropic structural plane) were 27° and 63.56 MPa, respectively (Table 1). It should be noted that the specimen size is much smaller than the standard specimen with diameter of 5 cm and length of 10 cm at least, and as the size effect the strengths are much higher than those reported in other literatures, e.g., 15.765–124.74 MPa in [57]. The experimental results and the estimated values based on Equations (1)–(3) are presented in Figure 4 as hollow squares and lines, respectively. It can be seen that the estimated strengths are generally comparable with the experimental results.

Table 1. Strength parameters of anisotropic rock mass.

Rock Type	c_{rb} (MPa)	φ_{rb} (°)	c_{as} (MPa)	φ_{as} (°)	$c_{rb}\cot\varphi_{rb}$ (Mpa)	$c_{as}\cot\varphi_{as}$ (Mpa)	Variation with Increasing Confinement: Possibility of Anisotropic Structural Plane-Controlled Strength	Anisotropic Strength Ratio	Anisotropic Strength Difference
Dolomitic limestone	78.23	34.5	63.56	27	113.8	124.8	Increase	Decrease	Increase
Sandstone	26.92	47.92	16.45	33.5	25.57	24.93	Decrease	Decrease	Increase
Jointed plaster of Paris	3.67	21.26	0.86	32.56	9.30	1.35	Decrease	Decrease	Decrease

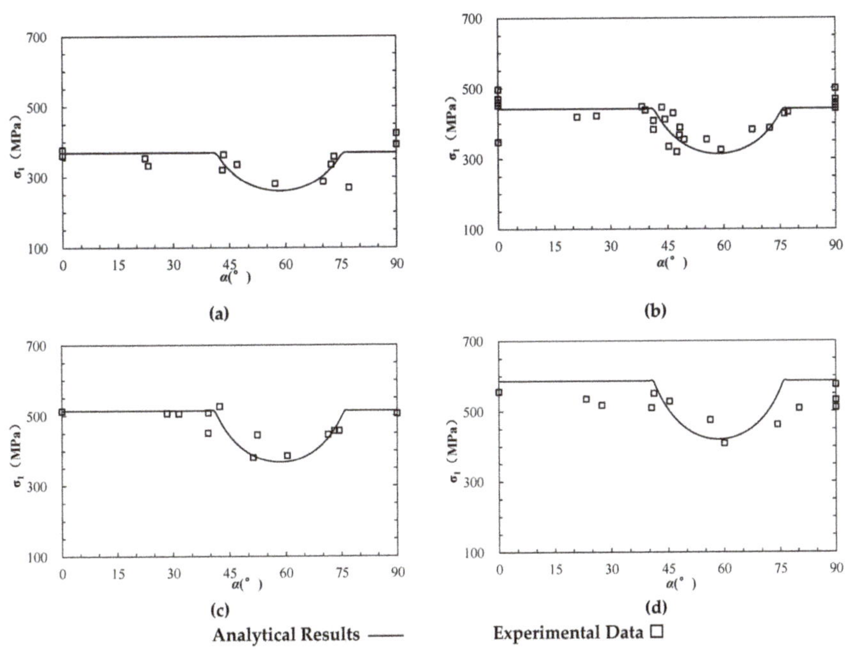

Figure 4. The analytical and experimental strength of anisotropic dolomitic limestone (modified based on [34]). (**a**–**d**) represent the results under triaxial compression at different confining pressure (σ_3) of 20, 40, 60, 80 MPa, respectively.

3.2. Lyonian Sandstone

Sandstone is a typical anisotropic rock stem from a bedding plane formed during the depositional process. A series of triaxial compression tests were performed on Lyonian sandstone in different directions [58]. The specimens were prepared as cylindrical shapes with height of 10.8 cm and a diameter of 5.4 cm. The inclined angle of the bedding plane ranged from 0° to 90°, and confining pressures incorporated four levels of 0, 10.5 MPa, 21 MPa, and 31.5 MPa. The internal friction angle (φ_{rb}) and cohesion (c_{rb}) of the rock block were 47.92° and 26.92 MPa, respectively; while the internal friction angle (φ_{as}) and cohesion (c_{as}) of the anisotropic structural plane were 33.5° and 16.45 MPa, respectively (Table 1). The experimental results and the estimated values based on Equations (1)–(3) are presented in Figure 5 as hollow squares and lines, respectively, which are compared with each other.

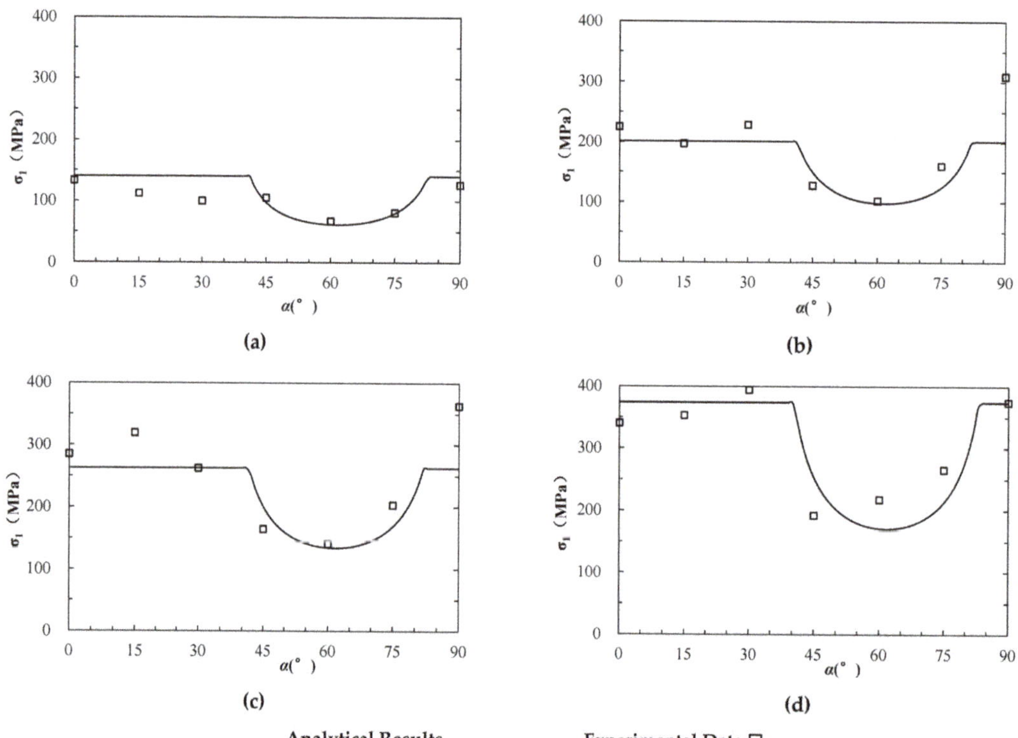

Figure 5. The analytical and experimental strength of anisotropic sandstone (modified based on [58]). (**a–d**) represent the results under uniaxial compression and triaxial compression at different confining pressure (σ_3) of 10.5, 21 and 31.5 MPa, respectively.

3.3. Plaster of Paris with Artificial Contact Joint

Ramamurthy and Arora (1994) conducted confined compression tests on the plaster of Paris with artificial contact discontinuities [3]. The specimens were prepared with a diameter of 38 mm and height of 76 mm. A number of cleanly and roughly broken joints at various inclinations (α = 0°, 10°, 20°, 30°, 40°, 50°, 60°, and 90°) were developed by breaking the specimens in the direction of the prenotch. Different confining pressures of 0.3 MPa, 0.5 MPa, 1 MPa, 1.5 MPa, 2 MPa, 5 MPa, and 7 MPa were applied. The cohesion and internal friction angle of plaster of Paris are 3.67 MPa and 21.26°, respectively. Comparatively, the cohesion and frictional angle of discontinuities are about 0.86 MPa and 32.56°, respectively (Table 1). The estimated strength based on Equations (1)–(5) was presented at different confining pressures as well as the corresponding experimental data

(Figure 6). The results show an acceptable fit between the estimated and real data at different confining levels.

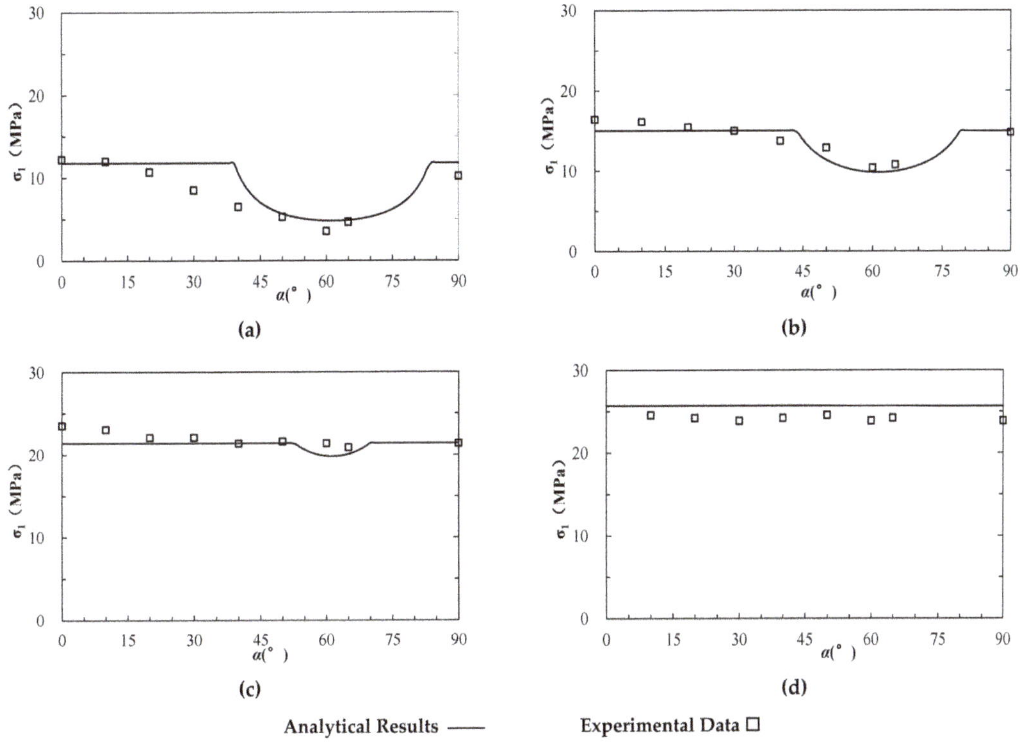

Figure 6. The analytical and experimental strength of anisotropic jointed plaster of Paris (modified based on [3]). (**a**–**d**) represent the results under triaxial compression at different confining pressure (σ_3) of 0.5, 2, 5, and 7 MPa, respectively.

4. Discussion

4.1. The Effects of Confinement on the Possibility of Anisotropic Structural Plane-Controlled Strength

The scope of incline angle α falling between α_{min} and α_{max} can indicate the possibility of anisotropic structural plane-controlled failure, which shows different trends as confining stress increases in various cases, e.g., the data in Figures 4–6. The causing factors needs a thorough discussion. If using $A(\sigma_3)$ substitutes for the formula for confining stress (σ_3) in Equation (4), we can reach the derivative of $A(\sigma_3)$ as follows.

$$\begin{aligned}\frac{\partial A(\sigma_3)}{\partial \sigma_3} &= \frac{\partial\left[1+\frac{(c_{as}\cot\varphi_{as}+\sigma_3)(1-\sin\varphi_{rb})}{\sigma_3\sin\varphi_{rb}+c_{rb}\cos\varphi_{rb}}\right]}{\partial\sigma_3} \\ &= \frac{(c_{rb}\cot\varphi_{rb}-c_{as}\cot\varphi_{as})(1-\sin\varphi_{rb})\sin\varphi_{rb}}{(\sigma_3\sin\varphi_{rb}+c_{rb}\cos\varphi_{rb})^2}\end{aligned} \quad (6)$$

The equation indicates that the positive or negative is dominated by the relationship between $c_{rb}\cot\varphi_{rb}$ and $c_{as}\cot\varphi_{as}$ that signifies the values of the intersection of the shear strength and lateral axis for rock block and anisotropic structural plane, respectively, shown in Figure 2.

① When $c_{rb}\cot\varphi_{rb} > c_{as}\cot\varphi_{as}$, $\frac{\partial A(\sigma_3)}{\partial\sigma_3}$ is positive, which means that $A(\sigma_3)$ increases with σ_3. This signifies that $A(\sigma_3)$ (or α_{min}) obtains its minimum value at $\sigma_3 = 0$. The scope of $\alpha_{min} < \alpha < \alpha_{max}$ narrow down as confining stress (σ_3) increases, which means the confinement reduces the anisotropic structural plane failure possibilities under this condition.

② When $c_{rb} \cot \varphi_{rb} < c_{as} \cot \varphi_{as}$, $\frac{\partial A(\sigma_3)}{\partial \sigma_3}$ is negative, which means that $A(\sigma_3)$ decreases with σ_3. This signifies $A(\sigma_3)$ (or α_{min}) obtains its maximum value at $\sigma_3 = 0$. The scope of $\alpha_{min} < \alpha < \alpha_{max}$ expands as confining stress (σ_3) increases, which means that the confinement increases the anisotropic structural plane failure possibilities under this condition.

For the three experimental cases presented above, the corresponding $c_{rb} \cot \varphi_{rb}$ and $c_{as} \cot \varphi_{as}$ are shown in Table 1. The results indicate that $c_{rb} \cot \varphi_{rb}$ is the relatively small one for dolomitic limestone, while the relatively big one is for sandstone and jointed plaster of Paris. Thus, the possibility of an anisotropic structural plane-controlled failure increases for dolomitic limestone and decreases for sandstone and jointed plaster of Paris as confinement increases.

We use the ratio of confining pressure σ_3 to maximum potential strength σ_{1max} to non-dimensionalize the confinement. The estimated thresholds of anisotropic structural plane inclination (α_{min} and α_{max}) with dimensionless confinement are presented in Figure 7. The tendency of α_{min} decreases while α_{max} increases for dolomitic limestone, which means the scope of $\alpha_{min} \sim \alpha_{max}$ expands up as confinement increases. Conversely, the tendency of α_{min} increases while α_{max} decreases for dolomitic limestone and jointed plaster of Paris, which means the scope of $\alpha_{min} \sim \alpha_{max}$ narrows down as confinement increases. The results agree well with the estimation in Table 1. Moreover, the inclination thresholds of dolomitic limestone and sandstone change slightly with confinement; conversely, the jointed plaster of Paris has a much more obvious change tendency. The α_{min} reaches α_{max} for jointed plaster of Paris when σ_3/σ_{1max} exceeds 0.61, which signifies that the possibility of anisotropic structure-controlled strength is zero at high confining stress. This phenomenon may have a relation with the different degree between $c_{rb} \cot \varphi_{rb}$ and $c_{as} \cot \varphi_{as}$ of the three rock types (Table 1).

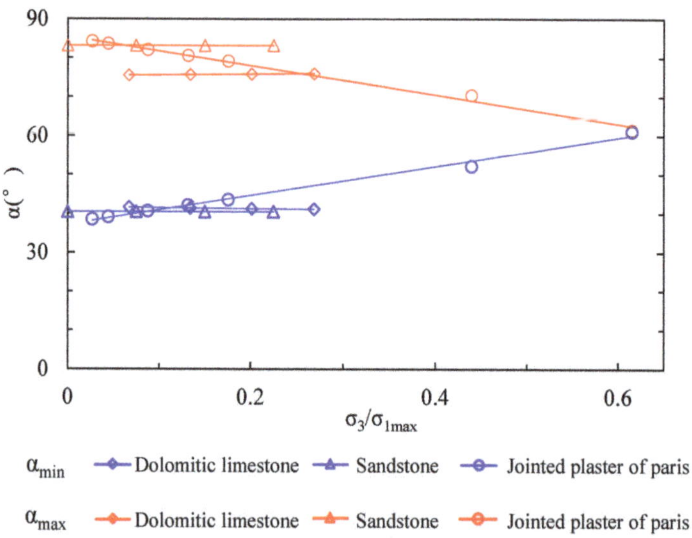

Figure 7. The relation between possibility of anisotropic structural plane-controlled failure and confinement.

4.2. The Effects of Confinement on Strength Anisotropy Degree

Researchers have proposed several indexes to determine the anisotropy degree of anisotropic rock mass, e.g., the uniaxial compressive strength anisotropy index [35], and the point load strength anisotropy index [59,60]. In this study, we adopt anisotropic strength ratio (ASR), i.e., the ratio of potential maximum to minimum compressive strength, to discuss the effects of confinement on anisotropy degree (Equation (7)).

$$\text{ASR} = \frac{\sigma_{1max}}{\sigma_{1min}} \tag{7}$$

We calculated ASRs of the three rock types under different levels of confinement, and exhibit them as blue lines in Figure 8. The results indicate that ASRs decrease for all three rock types as confinement increases, which means the confinement can weaken the strength anisotropy of rock mass. Among them, the ASR of jointed plaster of Paris tends toward 1 at high confining stresses, possessing strength isotropy. This is in accordance with many existing researches [39,40].

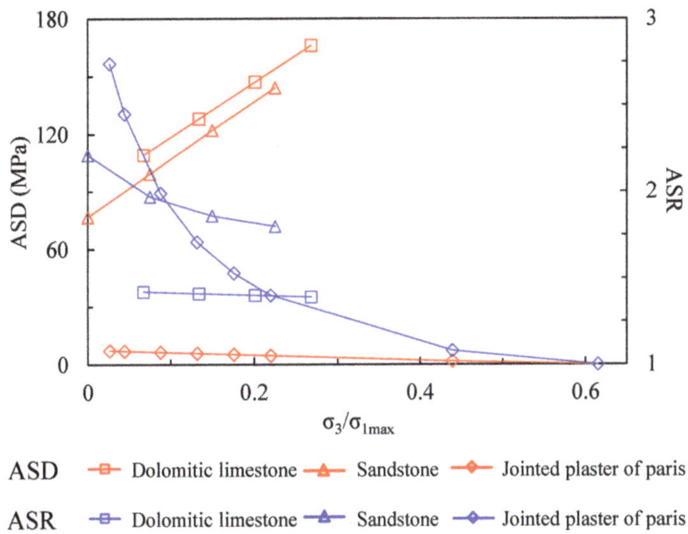

Figure 8. The relation between anisotropy degree of anisotropic rock mass and confinement.

Despite isotropization under high confinement using the index of anisotropic strength ratio, i.e., the absolute value between potential maximum and minimum compressive strength, i.e., the anisotropic strength difference (ASD) (Equation (8)) can reflect whether the rock mass can possess a total strength isotropy when the confinement is high enough.

$$\text{ASD} = \sigma_{1max} - \sigma_{1min} \tag{8}$$

The calculated ASDs are shown as red lines in Figure 8. This indicates that ASD increases for dolomitic limestone and sandstone as confinement increases, and decreases for jointed plaster of Paris. This phenomenon signifies that although the tendency of ASD predicts an anisotropy weakening for dolomitic limestone and sandstone, the potential minimum strength never reaches the maximum strength, and the absolute difference becomes even larger as confinement increases.

The effects of confinement on strength anisotropy are mainly dependent on the formula regarding to the friction strength of rock block and anisotropic structural plane (Equations (1) and (2)). The tendency of ASD with variable confinement can be directly determined by a comparison between $\frac{1+\sin\varphi_{rb}}{1-\sin\varphi_{rb}}$ and $\frac{1+\sin\varphi_{as}}{1-\sin\varphi_{as}}$. ASD increases under the condition of $\varphi_{rb} > \varphi_{as}$, and decreases under the condition of $\varphi_{rb} < \varphi_{as}$. For the exceptional cases of $\varphi_{rb} = \varphi_{as}$, ASD is constant. On the other hand, the factors influencing the ASR tendency are more complicated.

4.3. Further Discussions

The above discussion reflects that the strength anisotropy of an anisotropic rock mass under varied confining pressures is closely related to the strength properties of rock block and anisotropic structural plane. The commonly used Mohr–Coulomb criterion is adopted to depict the strength of rock block and anisotropic structural plane in this study. A number of other strength criteria can also be applied following a similar process, such as the Hoek–Brown strength formula [61].

The effects of the anisotropic structural plane on rock mass strength are not always weakened gradually as confining pressure increases, but variable according to the strength properties. This study indicates that the effect of the anisotropic structural plane on rock mass strength does not vanish even under very high confining pressure for the rock mass with relatively soft anisotropic structural planes, e.g., sedimentary rock mass or weak intercalated rock mass (see Figure 3b–d). The strengths of such discontinuities are highly dependent on the thickness and strength of interlayers. On the other hand, the fresh contact discontinuities may lose their effects gradually on the rock mass strength as confinement increases (see Figure 3a). The strength of such discontinuities is dominated mainly by the joint roughness and wall strength [36,62–65]. Byerlee (1978) performed statistical research on the friction properties of such contact discontinuities. Based on a thorough analysis of rock friction experiments, he concluded that the shear strength of sliding one rock over another varies widely dependent on surface roughness at low normal stress up to 5 MPa [66], and is nearly independent of rock type and surface roughness at high normal stress. The shear strength equations at different normal stresses are presented as Equations (9)–(11).

$$\mathrm{T} = \sigma_n \tan\left[JRClog_{10}\left(\frac{JCS}{\sigma_n}\right) + \varphi_b\right]\sigma_n \leq 5\ \mathrm{Mpa} \tag{9}$$

$$\tau = 0.85\sigma_n \quad 5\ \mathrm{Mpa} < \sigma_n \leq 200\ \mathrm{Mpa} \tag{10}$$

$$\tau = 0.5 + 0.6\sigma_n \quad \sigma_n > 200\ \mathrm{Mpa} \tag{11}$$

Where JRC, JCS, and φ_b denote joint roughness coefficient, joint wall compressive strength and basic internal friction angle, respectively.

According to Byerlee's formula, we can see that at the depth of traditional civil or mining engineering, the friction angle of fresh discontinuities is around 40° (Equation (10)). This is a referable indicator to estimate whether the anisotropic structural plane holds or loses effects on rock mass strength as confining pressure increases. The anisotropic strength difference (ASD) decreases and anisotropic rock mass tends to show features of isotropy as confining pressure increases when the internal friction angle of rock block is smaller than 40°, and vice versa.

It should be noted that this study focuses on the anisotropic rock mass incorporating one set of joints, and thus the presented formulas have limitations for rock mass anisotropy stemming from two and more sets of joints.

5. Conclusions Remarks

Theoretical analyses were conducted on the strength of anisotropic rock mass based on experimental results under different confining pressures. Some concluding remarks were reached as follows:

(1) The anisotropic structural planes incorporate both weak-filled layers and hard contact discontinuities, which cause strength anisotropy and different failure modes of rock mass at low confining pressures;

(2) The commonly used Mohr–Coulomb strength criterion is adopted to depict the strength of both rock block and anisotropic structural plane, based on which the formulas to estimate anisotropic strength under certain confining pressures are devel-

oped. The formulas compare well with the compression experiments data of various anisotropic rock types under different confining pressures;

(3) The possibility of anisotropic structural plane-controlled rock strength as confining pressure increases is not definite but theoretically related to the comparison between $c_{rb} \cot \varphi_{rb}$ and $c_{as} \cot \varphi_{as}$;

(4) Likewise, the tendency of strength anisotropy degree with increasing confinement is not definite either. As confining pressure increases, the anisotropic strength ratio (ASD) always decreases, while the anisotropic strength difference (ASD) increases or decreases depending on the friction strength of the rock block and anisotropic structural plane;

(5) The different anisotropic structural plane types may lead to distinct behaviors under high confinement, i.e., soft-filled layers or hard contact discontinuities-induced anisotropy.

This research presents an elaborate analysis on confinement-affected strength anisotropy by distinguishing the soft and hard anisotropic structures and extends the knowledge on such a fundamental topic in rock mass geomechanics. The analysis in this study has some tolerable limitations using the linear Mohr–Coulomb strength criterion to depict the strength of rock block and anisotropic structural plane under a wide range of confining pressures. Despite this, the theoretical analysis agrees well with the experiments, and the new understanding in this study can provide guidance for anisotropic rock mass engineering in high geostress environments. The directionally controlled progressive failure and the corresponding engineering measures of anisotropic rock mass are of concern and need further research urgently.

Author Contributions: Conceptualization, S.G. and S.Q.; Methodology, S.G. and B.Z.; Validation, S.G. and Y.L.; Formal analysis, S.G. and Y.Z.; Investigation, S.G. and F.T.; Data curation, S.G. and X.Y.; Writing—original draft, S.G. and W.W.; Writing—review & editing, S.G., L.X., X.W., N.L. and W.M.F.; Funding acquisition, S.G. and S.Q. All authors have read and agreed to the published version of the manuscript.

Funding: This research was financially supported by funds of National Natural Science Foundation of China under grants Nos. 42077266, 41825018, 42141009, 42090051, 41902289 and the Second Tibetan Plateau Scientific Expedition and Research Program (STEP), China under grant No. 2019QZKK0904, and Youth Innovation Promotion Association of the Chinese Academy of Sciences under grant No. 2022062. A special acknowledgement should also be expressed to the China–Pakistan Joint Research Centre on Earth Sciences that supported the implementation of this study.

Institutional Review Board Statement: Not applicable.

Informed Consent Statement: Not applicable.

Conflicts of Interest: The authors declare no conflict of interest.

References

1. Jaeger, J.C. Shear fracture of anisotropic rocks. *Geol. Mag.* **1960**, *97*, 65–72. [CrossRef]
2. Sun, G. *Rock Mass Structure Mechanics*; Science Press: Beijing, China, 1988.
3. Ramamurthy, T.; Arora, V.K. Strength predictions for jointed rocks in confined and unconfined states. *Int. J. Rock Mech. Min. Sci. Geomech. Abstr.* **1994**, *31*, 9–22. [CrossRef]
4. Zangerl, C.; Loew, S.; Eberhardt, E. Structure, geometry and formation of brittle discontinuities in anisotropic crystalline rocks of the Central Gotthard Massif, Switzerland. *Eclogae Geol. Helv.* **2006**, *99*, 271–290. [CrossRef]
5. Bidgoli, M.N.; Jing, L. Anisotropy of strength and deformability of fractured rocks. *J. Rock Mech. Geotech. Eng.* **2014**, *6*, 156–164. [CrossRef]
6. Guo, S.; Qi, S.; Zou, Y.; Zheng, B. Numerical Studies on the Failure Process of Heterogeneous Brittle Rocks or Rock-Like Materials under Uniaxial Compression. *Materials* **2017**, *10*, 378. [CrossRef]
7. Guo, S.; Qi, S.; Zhan, Z.; Ma, L.; Getahun, E.; Zhang, S. Numerical study on the progressive failure of heterogeneous geomaterials under varied confining stresses. *Eng. Geol.* **2022**, *269*, 10556. [CrossRef]
8. Margielewski, W. Structural control and types of movements of rock mass in anisotropic rocks: Case studies in the Polish Flysch Carpathians. *Geomorphology* **2006**, *77*, 47–68. [CrossRef]

9. Wang, S.H.; Lee, C.I.; Ranjith, P.G.; Tang, C.A. Modelling the effects of heterogeneity and anisotropy on the excavation damage/disturbed zone (EDZ). *Rock Mech. Rock Eng.* **2009**, *42*, 229–258. [CrossRef]
10. Zhang, Z.X.; Xu, Y.; Kulatilake, P.H.S.W.; Huang, X. Physical model test and numerical analysis on the behavior of stratified rock masses during underground excavation. *Int. J. Rock Mech. Min. Sci.* **2012**, *49*, 134–147. [CrossRef]
11. Jiang, Q.; Feng, X.; Hatzor, Y.H.; Hao, X.; Li, S. Mechanical anisotropy of columnar jointed basalts: An example from the Baihetan hydropower station, China. *Eng. Geol.* **2014**, *175*, 35–45. [CrossRef]
12. Shen, B.; Siren, T.; Rinne, M. Modelling fracture propagation in anisotropic rock mass. *Rock Mech. Rock Eng.* **2015**, *48*, 1067–1081. [CrossRef]
13. Özbek, A.; Gul, M.; Karacan, E.; Alca, O. Anisotropy effect on strengths of metamorphic rocks. *J. Rock Mech. Geotech. Eng.* **2018**, *10*, 164–175. [CrossRef]
14. Sun, X.; Chen, F.; Miao, C.; Song, P.; Li, G.; Zhao, C.; Xia, X. Physical modeling of deformation failure mechanism of surrounding rocks for the deep buried tunnel in soft rock strata during the excavation. *Tunn. Undergr. Space Technol.* **2018**, *74*, 247–261. [CrossRef]
15. He, J.; Qi, S.; Zhan, Z.; Guo, S.; Li, C.; Zheng, B.; Huang, X.; Zou, Y.; Yang, G.; Liang, N. Seismic response characteristics and deformation evolution of the bedding rock slope using a large-scale shaking table. *Landslides* **2021**, *18*, 2835–2853. [CrossRef]
16. Guo, X.; Deng, P.; Liu, Q.; Xu, X.; Wang, N.; Jiang, Y.; Yu, Y. Progressive fracture and swelling of anisotropic rock masses around deep tunnels: A new floor heave mechanical mechanism. *Arab. J. Geosci.* **2022**, *15*, 1325. [CrossRef]
17. Singh, D.; Singh, P.K.; Kainthola, A.; Pandey, H.K.; Kumar, S.; Singh, T.N. Analysis of failure pattern in cut slopes of bedded sandstone: A case study. *Environ. Earth Sci.* **2022**, *81*, 398. [CrossRef]
18. Donath, F.A. Experimental study of shear failure in anisotropic rocks. *Geol. Soc. Am. Bull.* **1961**, *72*, 985–990. [CrossRef]
19. Saeidi, Q.; Rasouli, V.; Vaneghi, R.G.; Gholami, R.; Torabi, S.R. A modified failure criterion for transversely isotropic rocks. *Geosci. Front.* **2014**, *5*, 215–225. [CrossRef]
20. Acharya, D.; Raina, A.K.; Panthee, S. Relationship between point load index and compressive strength of foliated metamorphic rocks at different loading angels. *Arab. J. Geosci.* **2022**, *15*, 490. [CrossRef]
21. Singh, V.K.; Singh, D.; Singh, T.N. Prediction of strength properties of some schistose rocks from petrographic properties using artificial neural networks. *Int. J. Rock Mech. Min. Sci.* **2001**, *38*, 269–284. [CrossRef]
22. Nasseri, M.H.B.; Rao, K.S.; Ramamurthy, T. Anisotropic strength and deformational behavior of Himalayan schists. *Int. J. Rock Mech. Min. Sci.* **2003**, *40*, 3–23. [CrossRef]
23. Saroglou, H.; Tsiambaos, G. A modified Hoek- Brown failure criterion for anisotropic intact rock. *Int. J. Rock Mech. Min. Sci.* **2008**, *45*, 223–234. [CrossRef]
24. Ghazvinian, A.; Vaneghi, R.G.; Hadei, M.R.; Azinfar, M.J. Shear behavior of inherently anisotropic rocks. *Int. J. Rock Mech. Min. Sci.* **2013**, *61*, 96–110. [CrossRef]
25. Rao, K.S.; Rao, G.V.; Ramamurthy, T. A strength criterion for anisotropic rocks. *Indian. Geotech. J.* **1986**, *16*, 317–333.
26. Niandou, H.; Shao, J.; Henry, J.P.; Fourmaintraux, D. Laboratory investigation of the mechanical behavior of Tournemire shale. *Int. J. Rock Mech. Min. Sci.* **1997**, *34*, 3–16. [CrossRef]
27. Cheng, C.; Li, X.; Qian, H. Anisotropic failure strength of shale with increasing confinement: Behaviors, factors and mechanism. *Materials* **2017**, *10*, 1310. [CrossRef]
28. Tien, Y.M.; Kuo, M.C. A failure criterion for transversely isotropic rocks. *Int. J. Rock Mech. Min. Sci.* **2001**, *38*, 399–412. [CrossRef]
29. McGill, G.E.; Raney, J.A. Experimental study of faulting in an anisotropic, inhomogeneous dolomitic limestone. *Geol. Soc. Am. Bull.* **1970**, *81*, 2949–2958. [CrossRef]
30. Xu, X.; Liu, B.; Li, S.; Yang, L.; Song, J.; Li, M.; Mei, J. Experimental study on conductivity anisotropy of limestone considering the bedding directional effect in the whole process of uniaxial compression. *Materials* **2016**, *9*, 165. [CrossRef]
31. Ajalloeian, R.; Lashkaripour, R.G. Strength anisotropies in mudrocks. *Bull. Eng. Geol. Environ.* **2000**, *59*, 195–199. [CrossRef]
32. Zhang, T.; Xu, W.; Wang, H.; Wang, R.; Yan, L.; Hu, M. Anisotropic mechanical behavior of columnar jointed rock masses subjected to cyclic loading: An experimental investigation. *Int. J. Rock Mech. Min. Sci.* **2021**, *144*, 104954. [CrossRef]
33. Tien, Y.M.; Kuo, M.C.; Juang, C.H. An experimental investigation of failure mechanism of simulated transversely isotropic rocks. *Int. J. Rock Mech. Min. Sci.* **2006**, *43*, 1163–1181. [CrossRef]
34. Wang, P.; Liu, C.; Qi, Z.; Liu, Z.; Cai, M. A rough discrete fracture network model for geometrical modeling of jointed rock mass and the anisotropic behavior. *Appl. Sci.* **2022**, *17*, 1720. [CrossRef]
35. Ramamurthy, T. Strength and modulus responses of anisotropic rocks. In *Comprehensive Rock Engineering, Vol. 1. Fundamentals*; Hudson, J.A., Ed.; Pergamon Press: Oxford, UK, 1993; pp. 313–329.
36. Singh, B.; Geol, R.K.; Mehrotra, V.K.; Gary, S.K.; Allu, M.R. Effect of intermediate principal stress on strength of anisotropic rock mass. *Tunn. Undergr. Space Technol.* **1998**, *13*, 71–79.
37. Guo, S.; Qi, S. Numerical study on progressive failure of hard rock samples with an unfilled undulate joint. *Eng. Geol.* **2015**, *193*, 173–182. [CrossRef]
38. Wu, N.; Liang, Z.; Li, Y.; Li, H.; Li, W.; Zhang, M. Stress-dependent anisotropy index of strength and deformability of jointed rock mass: Insights from a numerical study. *Bull. Eng. Geol. Environ.* **2019**, *78*, 5905–5917. [CrossRef]
39. Zhou, R. *The Effect of Confining Pressure and the Coefficient of Confining Pressure*; Engineering Geomechanics Research; Geological Publishing House: Beijing, China, 1985; pp. 209–214.

40. Guo, S.; Qi, S.; Huang, X. Anisotropy of rock mass strength and its transformation critical confining stress. *Chin. J. Rock Mech. Eng.* **2013**, *32*, 3222–3227. (In Chinese)
41. Saroglou, C.; Qi, S.; Guo, S.; Wu, F. ARMR, a new classification system for the rating of anisotropic rock masses. *Bull. Eng. Geol. Environ.* **2019**, *78*, 3611–3626. [CrossRef]
42. Guo, S.; Qi, S.; Saroglou, C. A-BQ, A classification system for anisotropic rock mass based on China National Standard. *J. Cen. South. Univ.* **2020**, *27*, 3090–3102. [CrossRef]
43. Walsh, J.B.; Brace, W.F. A fracture criterion for brittle anisotropic rocks. *J. Geophys. Res.* **1964**, *69*, 3449. [CrossRef]
44. Nova, R. The failure of transversely isotropic rocks in triaxial compression. *Int. J. Rock Mech. Min. Sci. Geomech. Abstr.* **1980**, *17*, 325–332. [CrossRef]
45. Cazacu, O.; Cristescu, N.D.; Shao, J.F.; Henry, J.P. A new failure criterion for transversely isotropic rocks. *Int. J. Rock Mech. Min. Sci.* **1998**, *35*, 130. [CrossRef]
46. Duveau, G.; Shao, J. A modified single plane of weakness theory for the failure of highly stratified rocks. *Int. J. Rock Mech. Min. Sci.* **1998**, *35*, 807–813. [CrossRef]
47. Pouya, A.; Ghoreychi, M. Determination of rock mass strength properties by homogenization. *Int. J. Numer. Anal. Met.* **2001**, *25*, 1285–1303. [CrossRef]
48. Chen, X.; Yang, Q.; Qiu, K.B.; Feng, J.L. An anisotropic strength criterion for jointed rock masses and its application in wellbore stability analyses. *Int. J. Numer. Anal. Met.* **2008**, *32*, 607–631. [CrossRef]
49. Halakatevakis, N.; Sofianos, A.I. Strength of a blocky rock mass based on an extended plane of weakness theory. *Int. J. Rock Mech. Min. Sci.* **2010**, *47*, 568–582. [CrossRef]
50. Triantafyllidis, T.; Gerolymatou, E. Estimation of the strength of stratified rock mass. *Rock Mech. Rock Eng.* **2014**, *47*, 535–547. [CrossRef]
51. Guo, S.; Qi, S.; Zhan, Z.; Zheng, B. Plastic- strain- dependent strength model to simulate the cracking process of brittle rocks with an existing non-persistent joint. *Eng. Geol.* **2017**, *231*, 114–125. [CrossRef]
52. Shen, J.; Shu, Z.; Cai, M.; Du, S. A shear strength model for anisotropic blocky rock masses with persistent joints. *Int. J. Rock Mech. Min. Sci.* **2020**, *134*, 104430. [CrossRef]
53. Shi, X.; Yang, X.; Meng, Y.; Li, G. Modified Hoek- Brown failure criterion for anisotropic rocks. *Environ. Earth Sci.* **2016**, *75*, 995. [CrossRef]
54. Wang, Z.; Qi, C.; Ban, L.; Yu, H.; Wang, H.; Fu, Z. Modified Hoek- Brown failure criterion for anisotropic intact rock under high confining pressures. *Bull. Eng. Geol. Environ.* **2022**, *81*, 333. [CrossRef]
55. Gao, Z.; Zhao, J.; Yao, Y. A generalized anisotropic failure criterion for geomaterials. *Int. J. Solids. Struct.* **2010**, *47*, 3166–3185. [CrossRef]
56. Asadi, M. Optimized Mamdani fuzzy models for predicting the strength of intact rocks and anisotropic rock masses. *J. Rock Mech. Geotech. Eng.* **2016**, *8*, 218–224. [CrossRef]
57. Wen, L.; Luo, Z.; Yang, S.; Qin, Y.; Wang, W. Correlation of Geo-Mechanics Parameters with Uniaxial Compressive Strength and P-Wave Velocity on Dolomitic Limestone Using a Statistical Method. *Geotech. Geol. Eng.* **2019**, *37*, 1079–1094. [CrossRef]
58. Youash, Y.Y. Experimental deformation of layered rocks. In Proceedings of the 1st ISRM Congress, Lisbon, Portugal, 25 September–1 October 1966.
59. ISRM. *Rock Characterization, Testing and Monitoring, ISRM Suggested Methods*; Pergamon Press: Oxford, UK, 1981.
60. Tsidzi, K. The influence of foliation on point load strength anisotropy of foliated rocks. *Eng. Geol.* **1990**, *29*, 49–58. [CrossRef]
61. Hoek, E.; Brown, E.T. The Hoek-Brown failure criterion and GSI-2018 edition. *J. Rock Mech. Geotech. Eng.* **2019**, *11*, 445–463. [CrossRef]
62. Barton, N.; Choubey, V. The shear strength of rock joints in theory and practice. *Rock Mech.* **1977**, *10*, 1–54. [CrossRef]
63. Byerlee, J. Friction of Rocks. Pure. *Appl. Geophys.* **1978**, *116*, 615–626. [CrossRef]
64. Ge, Y.; Kulatilake, P.H.S.W.; Tang, H.; Xiong, C. Investigation of natural rock joint roughness. *Comput. Geotech.* **2014**, *55*, 290–305. [CrossRef]
65. Zheng, B.; Qi, S.; Luo, G.; Liu, F.; Huang, X.; Guo, S. Characterization of discontinuity morphology based on extensive 3D fractal dimension by integrating laser scanning with ArcGIS. *Bull. Eng. Geol. Environ.* **2021**, *80*, 2261–2281. [CrossRef]
66. Barton, N. Review of a new shear-strength criterion for rock joints. *Eng. Geol.* **1973**, *7*, 287–332. [CrossRef]

MDPI
St. Alban-Anlage 66
4052 Basel
Switzerland
www.mdpi.com

Materials Editorial Office
E-mail: materials@mdpi.com
www.mdpi.com/journal/materials

www.ingramcontent.com/pod-product-compliance
Lightning Source LLC
Chambersburg PA
CBHW061556220926
51850CB00005B/356
9783036591247